Friedrich Wagemann

Crossed Modules

De Gruyter Studies in Mathematics

Volume 82

Friedrich Wagemann

Crossed Modules

DE GRUYTER

Mathematics Subject Classification 2020
Primary: 18G50; Secondary: 16E40, 17B56, 20J06, 17A32, 22E41, 22E60, 55P99, 57R32, 57T10

Author
Dr. Friedrich Wagemann
Université de Nantes
Faculté des Sciences et Techniques
2 rue de la Houssinière
44322 Nantes
France
wagemann@math.univ-nantes.fr

ISBN 978-3-11-075076-8
e-ISBN (PDF) 978-3-11-075095-9
e-ISBN (EPUB) 978-3-11-075099-7
ISSN 0179-0986

Library of Congress Control Number: 2021940888

Bibliographic information published by the Deutsche Nationalbibliothek
The Deutsche Nationalbibliothek lists this publication in the Deutsche Nationalbibliografie; detailed bibliographic data are available on the Internet at http://dnb.dnb.de.

Typesetting: VTeX UAB, Lithuania
Printing and binding: CPI books GmbH, Leck

www.degruyter.com

Preface

The mathematical concept of a crossed module exists for a wide variety of algebraic structures. For the algebraic structure of groups, a *crossed module of groups* is a homomorphism of groups $\mu : M \to N$ together with an action of the group N on the group M by automorphisms, such that on the one hand μ is equivariant with respect to the given action of N on M and the conjugation action of N on itself, and on the other hand, the conjugation action of M on itself is tied to the action of N on M via the prescription that for all $m, m' \in M$, we have $mm'm^{-1} = \mu(m) \cdot m'$. In a similar manner, one may define crossed modules of Lie algebra (replacing, e. g., automorphisms by derivations and the conjugation action by the adjoint action with respect to the bracket) and crossed modules of associative algebras. Full definitions of these are given in Chapters 2 and 3.

Crossed modules of groups were the first crossed modules to be explored and they were invented in the 1940s by J. H. C. Whitehead in order to describe the homotopy 2-type of a space. The general picture of crossed modules of algebraic structures includes an interpretation of 3-cohomology classes in terms of crossed modules and an interpretation of crossed modules in terms of higher structures. Namely, in the case of groups, crossed modules are the simplest examples of *2-groups*.

My personal encounter with crossed modules goes back to a question of J.-L. Loday during one of my seminar talks in Strasbourg in the early 2000s on the subject of the crossed module of Lie algebras:

$$0 \to \mathbb{C} \to \widehat{L\mathfrak{g}} \to \mathfrak{der}(\widehat{L\mathfrak{g}}) \to \mathrm{Vect}(S^1) \to 0.$$

Here, the outer derivations of the universal central extension $\widehat{L\mathfrak{g}}$ of the loop algebra $L\mathfrak{g}$ of a finite dimensional, complex, simple Lie algebra $\mathfrak{g}$ are identified with the Lie algebra of vector fields $\mathrm{Vect}(S^1)$ on the circle S^1, and the above exact sequence stems from the usual crossed module

$$0 \to Z(\mathfrak{h}) \to \mathfrak{h} \xrightarrow{\mu} \mathfrak{der}(\mathfrak{h}) \to \mathfrak{out}(\mathfrak{h}) \to 0,$$

which exists for any Lie algebra $\mathfrak{h}$ and where μ sends an element $h \in \mathfrak{h}$ to the inner derivation $\mathrm{ad}_h \in \mathfrak{der}(\mathfrak{h})$. Now Loday's question was whether this crossed module represents the 3-cohomology class generated by the Godbillon–Vey class $[\theta] \in H^3(\mathrm{Vect}(S^1), \mathbb{C})$. In fact, the answer to this question is "no," and it led me to searching for a crossed module representing this class. The answer will be explained later in this book (in Chapter 2). It generated in return a whole bunch of new questions, which dragged me into the subject of crossed modules. The innovative part of my approach, thanks to Loday, was the search for constructions of crossed modules, while there were few nonelementary explicit crossed modules (other than the examples from homotopy theory, to be explained in Section 1.10) known by then.

Crossed modules of groups and Lie algebras turn up in standard books of homological algebra as interpretations of cohomology classes of cohomological degree 3.

https://doi.org/10.1515/9783110750959-201

Cohomology of groups or Lie algebras provide invariants which can be useful in order to compare these groups or Lie algebras. As the invariants are defined in an abstract way, interpretations of cohomology classes make them more accessible. Such interpretations are usually given in these standard books in detail for 0-, 1- and 2-cohomology classes (in terms of invariants, crossed homomorphism and extensions, respectively), but the interpretation of 3-cohomology classes is left to the reader or done in exercises. One of my goals in this book is to give full proofs and all (sometimes tedious) justifications for the interpretation of 3-cohomology classes in terms of crossed modules. For groups, Lie algebras and associative algebras, the theorem stating an isomorphism between H^3 and the abelian group of equivalence classes of crossed modules is due to Gerstenhaber in 1964, according to Mac Lane. Unfortunately, full proofs of this theorem (including the independence of all auxiliary structure) are hard to find in the literature.

The scope of the present book is twofold: On the one hand, we give an introduction to crossed modules of groups, Lie algebras and associative algebras with fully written out proofs. This is contained in the first three chapters. Therefore, this part is suitable for graduate students or a second year graduate course on homological algebra. We will assume that the reader is familiar with the standard concepts of homological algebra as, for example, the Five Lemma. In the second part, we go on to explore more advanced and less standard subjects as crossed modules of Hopf algebra, Lie groups and racks. This part is designed to take the interested reader to more recent subjects and present day research on crossed modules.

More precisely, in the first three chapters, which constitute the first part of this book, we explore the well-established theory of crossed modules of groups (Chapter 1), crossed modules of Lie algebras (Chapter 2) and crossed modules of associative algebras (Chapter 3). The focus here is on the construction of crossed modules from 3-cohomology classes on the one hand, and on the transformation of crossed modules from one algebraic structure into crossed modules of another algebraic structure under standard functors. Along the way, we explore many, sometimes less well-known properties of crossed modules of groups, Lie algebras and associative algebras. It is easy to come up with new questions about these areas, simply by comparing the properties of crossed modules of groups, Lie algebras and associative algebras. For example, we explain in detail how the construction of crossed modules is related to the Lyndon–Hochschild–Serre and Hochschild–Serre spectral sequences for groups respectively for Lie algebras. But for associative algebras, the corresponding link to a standard spectral sequence is not clear to us.

The three chapters of the second part of this book explore more advanced subjects as crossed modules of Hopf algebras (Chapter 4), crossed modules of algebraic structures, which carry an additional geometric structure (Chapter 5) and crossed modules of racks in Chapter 6. Here, Chapter 5 treats the structure of Lie groups, Lie–Malcev theory and Lie–Rinehart algebras. In Chapter 6, we generalize on the level of crossed modules the classical links between groups, Lie algebras and associative bialgebras to racks, Leibniz algebras and rack bialgebras, leading thus the interested reader to present day research topics. Once again, new research questions may arise very easily just by comparing with the much more complete catalogue of properties of crossed modules of groups, Lie algebras and associative bialgebras.

We do not claim any exhaustivity for these subjects. We had to choose between different subjects in order to keep the volume of this book reasonable. For example, we did not consider crossed modules (of Lie algebras, for example) as an algebraic structure in itself and consider their sums, extensions, cohomology. Another subject which is not treated at all, is to lift the level of abstraction to operads and consider crossed modules of algebras over a certain type of operads. Our reason for not doing so is to keep as a focus the triad groups/Lie-algebras/Hopf-algebras, which is in our belief among the most fundamental algebraic structures in mathematics.

Acknowledgment

First of all, I thank Jean-Louis Loday and Karl-Hermann Neeb. From both, I learned a lot about crossed modules and about mathematics in general. Furthermore, I would like to express my thanks to Hossein Abbaspour, Martin Bordemann, Alissa Crans, Yaël Frégier, Michael Kinyon, Ulrich Krähmer, Manuel Ladra, Joao Faria Martins, Teimuraz Pirashvili, Salim Rivière, Bert Wiest, Christoph Wockel and Chenchang Zhu. Parts of this book were written at Max Planck Institut Bonn whom I thank for excellent working conditions.

Friedrich Wagemann

https://doi.org/10.1515/9783110750959-202

Contents

Preface — V

Acknowledgment — IX

1 Crossed modules of groups — 1
1.1 Definitions — **1**
1.2 Relation to third cohomology I — **5**
1.3 Relation to third cohomology II — **12**
1.4 Relation to general group extensions — **16**
1.4.1 Outer actions and factor systems — **16**
1.4.2 Significance of the zero class in $H^3(G,Z)$ — **22**
1.4.3 The obstruction crossed module — **24**
1.5 Construction of crossed modules — **30**
1.5.1 The principal construction — **30**
1.5.2 End of the proof of Theorem 1.2.1 — **32**
1.5.3 Constructions involving free groups — **34**
1.6 Relation to the Lyndon–Hochschild–Serre spectral sequence — **39**
1.7 Examples — **46**
1.7.1 The string group (after Stolz–Teichner) — **47**
1.7.2 The string group (after Baez–Crans–Schreiber–Stevenson) — **50**
1.7.3 A crossed module of diffeomorphisms — **52**
1.8 Relation to relative cohomology — **54**
1.9 Strict 2-groups — **62**
1.10 Motivation: Classification of homotopy 2-types — **68**
1.10.1 The crossed module $\pi_2 \to \pi_1$ — **68**
1.10.2 Free crossed modules — **71**
1.10.3 Classifying 2-homotopy types — **78**
1.11 Exercises — **81**
1.12 Bibliographical notes — **83**

2 Crossed modules of Lie algebras — 85
2.1 Definitions — **85**
2.2 Relation to third cohomology I — **89**
2.2.1 Gerstenhaber's theorem — **89**
2.3 Relation to third cohomology II — **95**
2.4 Relation to general extensions — **100**
2.4.1 Outer actions and factor systems — **101**
2.4.2 The obstruction crossed module — **106**
2.5 Construction of crossed modules — **112**
2.5.1 The principal construction — **112**

2.5.2 End of the proof of Theorem 2.2.1 — 114
2.6 Relation to Lie algebra kernels — 117
2.6.1 Lie algebra kernels and third cohomology — 117
2.6.2 Finite dimensional kernels — 122
2.7 Relation to the Hochschild–Serre spectral sequence — 128
2.8 Examples — 131
2.8.1 The string Lie algebra — 131
2.8.2 The string Lie algebra revisited — 133
2.8.3 A crossed module representing the Godbillon–Vey cocycle — 145
2.9 Relation to relative cohomology — 151
2.10 Strict Lie 2-algebras — 157
2.10.1 Crossed modules and cat^1-Lie algebras — 158
2.10.2 Strict 2-vector spaces and 2-term complexes — 159
2.10.3 Strict Lie 2-algebras and crossed modules — 159
2.10.4 Semistrict Lie 2-algebras and 2-term L_∞-algebras — 162
2.10.5 Classification of semistrict Lie 2-algebras — 163
2.11 Exercises — 166
2.12 Bibliographical notes — 167

3 Crossed modules of associative algebras — 169
3.1 Definitions — 169
3.1.1 Crossed modules with free algebras — 172
3.2 Relation to third cohomology I — 173
3.2.1 Constructing the map b — 173
3.2.2 Bijectivity of the map b — 178
3.3 Relation to third cohomology II — 185
3.4 Relation to general extensions — 188
3.5 Construction of crossed modules — 192
3.6 Associative algebra kernels — 194
3.6.1 Definition of an associative algebra kernel — 194
3.6.2 Cohomology class associated to a kernel — 195
3.6.3 Relation to general extensions — 198
3.6.4 The different extensions corresponding to a special kernel — 199
3.6.5 Gerstenhaber's theorem for kernels — 202
3.7 Strict associative 2-algebras — 206
3.7.1 cat^1-associative algebras — 206
3.7.2 Crossed modules and category objects in Alg — 208
3.8 Relation between crossed modules of associative and Lie algebras respective groups — 209
3.8.1 Definition of the functors — 209
3.8.2 Crossed modules of groups and of associative algebras — 210
3.8.3 Crossed modules of Lie algebras and of associative algebras — 216

3.9 Examples — 220
3.9.1 The crossed module associated to a cochain algebra — 220
3.9.2 The crossed module associated to a chain complex of vector spaces — 221
3.10 Exercises — 224
3.11 Bibliographical notes — 225

4 Crossed modules of Hopf algebras — 227
4.1 Preliminaries — 227
4.2 Crossed modules of Lie and Hopf algebras — 232
4.3 Crossed modules of groups and Hopf algebras — 237
4.4 Crossed comodules of groups and Hopf algebras — 239
4.5 Integration of crossed modules of Lie algebras — 243
4.5.1 Integration into formal groups — 243
4.5.2 Integration into an algebraic group — 244
4.5.3 Integration of crossed modules — 245
4.6 Crossed modules of Hopf algebras and cat^1-Hopf algebras — 248
4.6.1 Semidirect product of Hopf algebras — 248
4.6.2 cat^1-Hopf algebras — 253
4.6.3 Equivalence between crossed module and cat^1 Hopf algebras — 255
4.7 Strict quantum 2-groups — 259
4.8 Exercises — 271
4.9 Bibliographical notes — 272

5 Crossed modules with geometric structure — 273
5.1 Crossed modules of Lie groups — 273
5.1.1 Smooth outer actions — 273
5.1.2 Extensions of Lie groups — 279
5.1.3 Crossed modules of Lie groups — 288
5.1.4 Significance of the zero class in $H^3(G, Z)$ — 293
5.1.5 The smooth obstruction crossed module — 295
5.1.6 Integration of crossed modules of Lie algebras — 300
5.2 Crossed modules in Malcev theory — 304
5.3 Crossed modules of Lie–Rinehart algebras — 312
5.4 Exercises — 316
5.5 Bibliographical notes — 318

6 Crossed modules of racks — 319
6.1 Racks and related notions — 319
6.1.1 Racks and Leibniz algebras — 319
6.1.2 Racks, Leibniz algebras and rack bialgebras — 327
6.2 Crossed modules of racks — 335

6.3 Crossed modules of rack bialgebras — 341
6.4 Links between the different types of crossed modules — 344
6.4.1 Crossed modules of Lie racks and Leibniz algebras — 345
6.4.2 Crossed modules of rack bialgebras and Leibniz algebras — 347
6.4.3 Crossed modules of racks and rack bialgebras — 350
6.5 Categorical racks — 351
6.5.1 From categories to crossed modules — 351
6.5.2 From crossed modules to categories — 354
6.5.3 Crossed modules of racks and trunks — 356
6.5.4 From crossed modules of racks to trunks — 358
6.6 Topological applications of crossed modules of racks — 359
6.6.1 The rack space of a crossed module of racks — 359
6.6.2 Crossed modules of racks from link coverings — 361
6.7 Exercises — 365
6.8 Bibliographical notes — 366

A Cohomology of groups — 367
A.1 Definitions — 367
A.2 Abelian extensions — 368
A.3 The Lyndon–Hochschild–Serre spectral sequence — 371
A.4 Continuous cohomology — 371

B Cohomology of Lie algebras — 373
B.1 Definitions — 373
B.2 Abelian extensions — 374
B.3 The Hochschild–Serre spectral sequence — 375
B.4 Continuous cohomology — 378

C Cohomology of associative algebras — 379
C.1 Definitions — 379
C.2 Hochschild cohomology — 379
C.3 Square zero extensions — 380

D Lie groups and their cohomology — 383
D.1 Lie groups — 383
D.2 Defining a Lie group structure — 384
D.3 Lie group cohomology — 384

Bibliography — 387

Index — 391

1 Crossed modules of groups

In this first chapter, we discuss crossed modules of groups. The concept of a crossed module of groups originated from homotopy theory. Namely, it was J. H. C. Whitehead who introduced crossed modules in the quest of capturing algebraically homotopy 2-types, while the classification of homotopy 1-types is equivalent to the problem of classifying groups up to isomorphism, using the functors π_1 (the fundamental group) and B (the classifying space) to establish the equivalence. We shift Whitehead's original notation here to the modern indexing, while he spoke of 3-types and 2-types. The classification of homotopy 2-types will be sketched in Section 1.10.

We start this chapter with the relation of crossed modules to third cohomology. Further topics include the relation of crossed modules to the existence of extensions and the link to the Lyndon–Hochschild–Serre spectral sequence (Section 1.6). Going far beyond the elementary examples of crossed modules, we discuss in Section 1.7 two versions of the string group, which are closely related to crossed modules. In Section 1.8, we approach crossed modules of groups from the point of view of relative cohomology. In Section 1.9, we show that crossed modules of groups are strict 2-groups, that is, a categorized version of the notion of a group. The chapter closes with an exercise section.

1.1 Definitions

Definition 1.1.1. A *crossed module of groups* is the data of a homomorphism of groups $\mu : M \to N$ together with an action η of N on M by automorphisms, denoted $\eta : N \to \mathrm{Aut}(M)$ or simply $m \mapsto {}^{n}m$ for all $n \in N$ and all $m \in M$, such that
(a) $\mu({}^{n}m) = n\mu(m)n^{-1}$ for all $n \in N$ and all $m \in M$, and
(b) ${}^{\mu(m)}m' = mm'm^{-1}$ for all $m, m' \in M$.

Remark 1.1.2. Property (a) means that the homomorphism μ is equivariant with respect to the N-action via η on M and the conjugation action on M. Property (b) is called *Peiffer identity*.

Remark 1.1.3. To each crossed module of groups $\mu : M \to N$, one associates a four-term exact sequence

$$0 \to V \stackrel{i}{\to} M \stackrel{\mu}{\to} N \stackrel{\pi}{\to} G \to 1,$$

where $\mathrm{Ker}(\mu) =: V$ and $G := \mathrm{Coker}(\mu)$. This is in fact an exact sequence of groups as explained in the following remark.

Remark 1.1.4.
(a) By property (a), $\mathrm{Im}(\mu)$ is a normal subgroup of N. Therefore, the quotient $G = N/\mathrm{Im}(\mu)$ is a group.

https://doi.org/10.1515/9783110750959-001

(b) By property (b), an element $m \in \mathrm{Ker}(\mu)$ satisfies $m' = mm'm^{-1}$ or $m'm = mm'$ for all $m' \in M$. This means that $\mathrm{Ker}(\mu)$ is contained in the center of N, and in particular $\mathrm{Ker}(\mu)$ is abelian.

(c) The homomorphism π is surjective, so we can lift elements from G to N, and then act with these on M via η. This prescription defines an *outer action* of G on M. More precisely, let $\rho : G \to N$ and $\rho' : G \to N$ be (set-theoretical) sections of π with $\rho(1) = \rho'(1) = 1$ and $x \in G$, then $\eta(\rho(x))$ and $\eta(\rho'(x))$ differ by an inner automorphism conj_m for some $m \in M$. Indeed, for all $m' \in M$,

$$\begin{aligned}\eta(\rho(x)) \circ (\eta(\rho'(x)))^{-1}(m') &= \eta(\rho(x)(\rho'(x))^{-1})(m') \\ &= \eta(\mu(m))(m') \equiv {}^{\mu(m)}m' \\ &= mm'm^{-1} \equiv \mathrm{conj}_m(m').\end{aligned}$$

Here, we used property (b) in the last line, and m exists by exactness of the four-term sequence, because $\rho(x)(\rho'(x))^{-1} \in \mathrm{Ker}(\pi) = \mathrm{Im}(\mu)$.
Moreover, $S := \eta \circ \rho$ satisfies the axioms of a group action also up to an inner automorphism. Indeed, denote for all $x, y \in G$ by

$$\alpha(x,y) := \rho(x)\rho(y)(\rho(xy))^{-1}$$

the failure of ρ to be a group homomorphism. Then $\alpha(x,y) \in \mathrm{Ker}(\pi)$, because π is a group homomorphism and ρ a section of π. Therefore, by exactness of the four-term exact sequence, there exists $\beta(x,y) \in M$ such that

$$\mu(\beta(x,y)) = \alpha(x,y).$$

Then to examine whether $\eta \circ \rho$ is an action, we have to consider $\eta(\rho(x)) \circ \eta(\rho(y)) \circ \eta(\rho(xy))^{-1}$. But as η is a group homomorphism, this gives for all $m \in M$,

$$\eta(\alpha(x,y))(m) = \eta(\mu(\beta(x,y)))(m) = {}^{\beta(x,y)}m = \mathrm{conj}_{\beta(x,y)}(m)$$

again by property (b), and in this sense, the outer action S is an action G on M up to inner automorphisms.
Now property (a) implies that the restriction of this outer action to V induces the structure of a G-module on V.

(d) Note that by this lifting procedure, the action η (resp., the adjoint action) does not in general render M (resp., N) a G-module.

Remark 1.1.5. Let us list some elementary examples of crossed modules:

(a) Each G-module V gives rise to a crossed module $\mu : V \to G$, where μ sends all elements of V to the neutral element $1 \in G$.

(b) Each central extension is a crossed module. Indeed, central extensions correspond exactly to the case where $\mu : M \to N$ is surjective (and then M is a central extension of N by V).

(c) Each (inclusion of a) normal subgroup in some group constitutes a crossed module. Indeed, an inclusion of a normal subgroup corresponds exactly to the case where the map $\mu : M \to N$ is injective.
(d) For each group L, there is a canonical crossed module,

$$\mu : L \to \mathrm{Aut}(L),$$

where μ sends an element $x \in L$ to the inner automorphism conj_x defined by $\mathrm{conj}_x(y) = xyx^{-1}$ for all $y \in L$. The action of $\mathrm{Aut}(L)$ on L is the usual action by automorphisms. The kernel of μ is the center $Z(L)$ of L, and the cokernel of μ is the *outer automorphisms*,

$$\mathrm{Out}(L) := \mathrm{Aut}(L) \,/\, \mathrm{Inn}(L),$$

the quotient of the group of automorphisms $\mathrm{Aut}(L)$ by the group of inner automorphisms $\mathrm{Inn}(L)$ (i. e., those automorphisms of the form conj_x for some $x \in L$).

Definition 1.1.6. Two crossed modules $\mu : M \to N$ (with action η) and $\mu' : M' \to N'$ (with action η') such that $\mathrm{Ker}(\mu) = \mathrm{Ker}(\mu') =: V$ and $\mathrm{Coker}(\mu) = \mathrm{Coker}(\mu') =: G$ are called *elementary equivalent* if there are group homomorphisms $\varphi : M \to M'$ and $\psi : N \to N'$ which are compatible with the actions, that is,

$$\varphi(\eta(n)(m)) = \eta'(\psi(n))(\varphi(m))$$

for all $n \in N$ and all $m \in M$, and such that the following diagram is commutative:

$$\begin{array}{ccccccccccc} 0 & \longrightarrow & V & \xrightarrow{i} & M & \xrightarrow{\mu} & N & \xrightarrow{\pi} & G & \longrightarrow & 0 \\ & & \downarrow{\scriptstyle \mathrm{id}_V} & & \downarrow{\scriptstyle \varphi} & & \downarrow{\scriptstyle \psi} & & \downarrow{\scriptstyle \mathrm{id}_G} & & \\ 0 & \longrightarrow & V & \xrightarrow{i'} & M' & \xrightarrow{\mu'} & N' & \xrightarrow{\pi'} & G & \longrightarrow & 0 \end{array}$$

Note that the relation of elementary equivalence is reflexive and transitive, but not symmetric in general as the homomorphisms φ and ψ need not be invertible. We call *equivalence of crossed modules* the equivalence relation generated by elementary equivalence. One easily sees that two crossed modules are equivalent in case there exists a zig-zag of elementary equivalences going from one to the other (where the arrows do not necessarily all go in the same direction).

Let us denote by $\mathrm{crmod}(G, V)$ the set of equivalence classes of crossed modules with a fixed kernel V and a fixed cokernel G.

Remark 1.1.7. Compare this equivalence relation to the equivalence of abelian extensions; cf. Appendix A. In the framework of extensions, equivalence imposes that the underlying set of the extension is the product set, up to bijection. For crossed modules, the exactness of the sequence does not prescribe the set-theoretical structure of the two middle terms as only the end terms are fixed. This leads to much more different representatives of the same equivalence class of a crossed module.

We now turn to the definition of the *sum of two crossed modules*:

Consider two crossed modules $\mu : M \to N$ and $\mu' : M' \to N'$ with the same kernel V and cokernel G and their corresponding four-term exact sequences

$$0 \to V \xrightarrow{i} M \xrightarrow{\mu} N \xrightarrow{\pi} G \to 1$$

and

$$0 \to V \xrightarrow{i'} M' \xrightarrow{\mu'} N' \xrightarrow{\pi'} G \to 1.$$

Denote by $K := \{(v, -v) \mid v \in V\}$ the kernel of the addition map $V \oplus V \to V$, $(v, v') \mapsto v + v'$. Notice that the diagonal $\Delta_V : V \to V \oplus V$, $v \mapsto (v, v)$, followed by the quotient map $V \oplus V \to (V \oplus V)/K$ identifies V with the quotient $(V \oplus V)/K$. K can be considered as a subspace in $M \times M'$ via $i \times i'$. As V is central in M and M', K is a normal subgroup in $M \times M'$.

On the other hand, denote by $N \times_{\pi \times \pi'} N'$ the pullback group

$$N \times_{\pi \times \pi'} N' := \{(n, n') \in N \times N' \mid \pi(n) = \pi'(n')\}.$$

The restriction to $N \times_{\pi \times \pi'} N'$ of the map $\pi \times \pi'$ has its values in the diagonal $\Delta G := \{(g, g) \in G \times G \mid g \in G\}$, which is isomorphic to G via the (restriction of the) diagonal map $\Delta_G : G \to \Delta G \subset G \times G$.

With these preliminaries, we have the following definition.

Definition 1.1.8. The *sum of two crossed modules* $\mu : M \to N$ and $\mu' : M' \to N'$ such that $\mathrm{Ker}(\mu) = \mathrm{Ker}(\mu') = V$ and $\mathrm{Coker}(\mu) = \mathrm{Coker}(\mu') = G$ is by definition the crossed module:

$$0 \to V \xrightarrow{(i \oplus i') \circ \Delta_V} (M \times M')/K \xrightarrow{\mu \times \mu'} N \times_{\pi \times \pi'} N' \xrightarrow{\Delta_G^{-1} \circ (\pi \times \pi')} G \to 1.$$

The action of $N \times_{\pi \times \pi'} N'$ on $(M \times M')/K$ is induced by the product of the actions of the two factors. The compatibility relations (a) and (b) of Definition 1.1.1 are true in the direct product, and thus in the crossed module sum.

Lemma 1.1.1. *The sum of crossed modules defines an abelian group structure on the set of equivalence classes of crossed modules* $\mathrm{crmod}(G, V)$ *with fixed kernel V and cokernel G.*

Proof. It is clear that the sum of crossed modules is associative and commutative (up to equivalence) as it is induced by the direct product. It is equally clear that the sum is compatible with the equivalence relation as we can take the product of the maps giving the equivalences.

We have to show that there is a zero element and an inverse to every crossed module. We define the *zero crossed module* with given kernel V and cokernel G to be

$$0 \to V \xrightarrow{\mathrm{id}_V} V \xrightarrow{1} G \xrightarrow{\mathrm{id}_G} G \to 1,$$

where the map $1 : V \to G$ maps all elements of V to the unit $1 \in G$. The inverse of a crossed module

$$0 \to V \xrightarrow{i} M \xrightarrow{\mu} N \xrightarrow{\pi} G \to 0$$

is defined to be

$$0 \to V \xrightarrow{-i} M \xrightarrow{\mu} N \xrightarrow{\pi} G \to 0.$$

In order to show that this crossed module is inverse to the given one, notice that $\mathrm{crmod}(G, -)$ is an additive functor. Thus we have for an equivalence class $[\mu : M \to N] \in \mathrm{crmod}(G, V)$

$$(\alpha_1 + \alpha_2)[\mu : M \to N] = \alpha_1[\mu : M \to N] + \alpha_2[\mu : M \to N],$$

where $\alpha_i : V \to V'$ $(i = 1, 2)$ are two G-module morphisms. We refer to [72] Chapter III, Section 5 for more information. This reduces the proof to showing that pushforward by the zero map $0 : V \to V$ gives the zero class. Now, pushforward by the zero map splits up a direct factor V in $(V \times M)/(0 \times (-i(V)))$ and we have then a commutative diagram

$$\begin{array}{ccccccccccc}
0 & \longrightarrow & V & \xrightarrow{i} & M & \xrightarrow{\mu} & N & \xrightarrow{\pi} & G & \longrightarrow & 1\\
 & & \downarrow{\scriptstyle 0} & & \downarrow{\scriptstyle \mathrm{incl}_2} & & \downarrow{\scriptstyle \mathrm{id}_N} & & \downarrow{\scriptstyle \mathrm{id}_G} & & \\
0 & \longrightarrow & V & \longrightarrow & (V\times M)/(0\times(-i(V))) & \xrightarrow{1\times\mu} & N & \xrightarrow{\pi} & G & \longrightarrow & 1\\
 & & \downarrow{\scriptstyle \mathrm{id}_V} & & \downarrow{\scriptstyle \mathrm{proj}_1} & & \downarrow{\scriptstyle \pi} & & \downarrow{\scriptstyle \mathrm{id}_G} & & \\
0 & \longrightarrow & V & \xrightarrow{\mathrm{id}_V} & V & \xrightarrow{1} & G & \xrightarrow{\mathrm{id}_G} & G & \longrightarrow & 1
\end{array}$$

Here, incl_2 and proj_2 are induced by the standard inclusion and projection maps to/from the direct product. This shows that the class $0[\mu : M \to N]$ is the zero class. □

1.2 Relation to third cohomology I

Crossed modules may serve as explicit representatives of third cohomology classes. This is the essence of the following theorem which is due to Gerstenhaber, according to Mac Lane [71].

Theorem 1.2.1. *The map, which associates to a crossed module its 3-cocycle γ (to be defined below), induces an isomorphism of abelian groups*

$$B : \mathrm{crmod}(G, V) \cong H^3(G, V).$$

The proof of Theorem 1.2.1 will be completed in Section 1.5.2. In a first step, we will only discuss in this section how to associate a cohomology class in $H^3(G,V)$ to a given crossed module.

We shall use the notation $B([\mu : M \to N])$ or simply $B([\mu])$ for B applied to a crossed module $\mu : M \to N$ representing the equivalence class $[\mu : M \to N]$.

Let us show how to associate to a given crossed module $\mu : M \to N$ with kernel V and cokernel G a 3-cocycle γ of G with values in V. For this, recall the four-term exact sequence from Remark 1.1.3:

$$0 \to V \xrightarrow{i} M \xrightarrow{\mu} N \xrightarrow{\pi} G \to 1.$$

The first step is to take a section ρ of π and to compute the failure of ρ to be a group homomorphism, that is, denote for all $x_1, x_2, x_3 \in G$ by

$$\alpha(x_1, x_2) := \rho(x_1)\rho(x_2)\rho(x_1x_2)^{-1}$$

this failure. Then $\alpha(x_1, x_2) \in \mathrm{Ker}(\pi)$, because π is a group homomorphism and ρ a section of π. Therefore, by exactness of the four-term exact sequence, there exists $\beta(x_1, x_2) \in M$ such that

$$\mu(\beta(x_1, x_2)) = \alpha(x_1, x_2).$$

The next step is then to form the formal group cohomology coboundary operator $d^M\beta(x_1, x_2, x_3)$ (cf. equation (A.1) and see the proof of Lemma 1.2.2) as if M were a G module (which is not the case in general). Note that M is not an abelian group in general, thus this formal coboundary operator is necessarily a *non-abelian* coboundary operator. Its definition is given in the proof of the following lemma.

Lemma 1.2.2. *$\mu(d^M\beta(x_1, x_2, x_3)) = 1$, where d^M is the formal group cohomology coboundary operator (cf. equation (A.1) and see the proof of this lemma), which we apply to $\beta \in C^2(G,M)$ as if M were a G-module, and where $x_1, x_2, x_3 \in G$.*

Proof. By definition, we have

$$d^M\beta(x_1, x_2, x_3) = \eta(\rho(x_1))(\beta(x_2, x_3))\beta(x_1, x_2x_3)\beta(x_1x_2, x_3)^{-1}\beta(x_1, x_2)^{-1}.$$

This is not precisely the expression of the coboundary given in equation (A.1) in Appendix A, but this is the correct generalization to the non-abelian setting, that is, there has been a choice of the order of the factors suitable for the following computations. Applying μ, using that μ is a group homomorphism and property (a) of a crossed module gives

$$\begin{aligned}&\mu(d^M\beta(x_1, x_2, x_3))\\ &\quad = \mu(\eta(\rho(x_1))(\beta(x_2, x_3))\beta(x_1, x_2x_3)\beta(x_1x_2, x_3)^{-1}\beta(x_1, x_2)^{-1})\end{aligned}$$

$$= \rho(x_1)\alpha(x_2,x_3)\rho(x_1)^{-1}\alpha(x_1,x_2x_3)\alpha(x_1x_2,x_3)^{-1}\alpha(x_1,x_2)^{-1}$$
$$= \rho(x_1)\rho(x_2)\rho(x_3)\rho(x_2x_3)^{-1}\rho(x_1)^{-1}\rho(x_1)\rho(x_2x_3)\rho(x_1x_2x_3)^{-1}$$
$$\rho(x_1x_2x_3)\rho(x_3)^{-1}\rho(x_1x_2)^{-1}\rho(x_1x_2)\rho(x_2)^{-1}\rho(x_1)^{-1}$$
$$= 1.$$ □

This means that $\mu(d^M\beta(x_1,x_2,x_3)) \in \mathrm{Ker}(\mu) = \mathrm{Im}(i)$ and, therefore, there exists an element $\gamma(x_1,x_2,x_3) \in V$ such that $d^M\beta(x_1,x_2,x_3) = i(\gamma(x_1,x_2,x_3))$. The next lemma shows that the cochain $\gamma \in C^3(G,V)$ is a 3-cocycle. As V is an abelian group, this cocycle property refers to the usual (abelian) coboundary operator defined in equation (A.1) in Appendix A.

Lemma 1.2.3. *The cochain $\gamma \in C^3(G,V)$ is a 3-cocycle.*

Proof. It is enough to show that $i(d^V\gamma(x_1,x_2,x_3,x_4)) = 0$ for all $x_1,x_2,x_3,x_4 \in G$ as i is injective.

$$i(d^V\gamma(x_1,x_2,x_3,x_4))$$
$$= i\left(\eta(\rho(x_1))(\gamma(x_2,x_3,x_4)) + \sum_{i=1}^{3}(-1)^i\gamma(\ldots,x_ix_{i+1},\ldots) + \gamma(x_1,x_2,x_3)\right)$$
$$= \eta(\rho(x_1))(d^M\beta(x_2,x_3,x_4))\prod_{i=1}^{3} d^M\beta(\ldots,x_ix_{i+1},\ldots)^{\pm1}d^M\beta(x_1,x_2,x_3)$$
$$= \eta(\rho(x_1))(d^M\beta(x_2,x_3,x_4))d^M\beta(x_1x_2,x_3,x_4)^{-1}$$
$$d^M\beta(x_1,x_2x_3,x_4)d^M\beta(x_1,x_2,x_3)d^M\beta(x_1,x_2,x_3x_4)^{-1},$$

the last equation being true because $i(V)$ is central in M.

Now replace each $d^M\beta(x,y,z)$ by

$$d^M\beta(x,y,z) = \eta(\rho(x))(\beta(y,z))\beta(x,yz)\beta(xy,z)^{-1}\beta(x,y)^{-1}.$$

In order to save some space, let us write in the following more simply

$$\eta(\rho(x))(\beta(y,z)) =: x\cdot\beta(y,z).$$

We obtain with this notation:

$$i(d^V\gamma(x_1,x_2,x_3,x_4))$$
$$= x_1\cdot(x_2\cdot\beta(x_3,x_4)\beta(x_2,x_3x_4)\beta(x_2x_3,x_4)^{-1}\beta(x_2,x_3)^{-1})$$
$$\beta(x_1x_2,x_3)\beta(x_1x_2x_3,x_4)\beta(x_1x_2,x_3x_4)^{-1}[(x_1x_2)\cdot\beta(x_3,x_4)^{-1}]$$
$$x_1\cdot\beta(x_2x_3,x_4)\beta(x_1,x_2x_3x_4)\beta(x_1x_2x_3,x_4)^{-1}\beta(x_1,x_2x_3)^{-1}$$
$$x_1\cdot\beta(x_2,x_3)\beta(x_1,x_2x_3)\beta(x_1x_2,x_3)^{-1}\beta(x_1,x_2)^{-1}$$
$$\beta(x_1,x_2)\beta(x_1x_2,x_3x_4)\beta(x_1,x_2x_3x_4)^{-1}x_1\cdot\beta(x_2,x_3x_4)^{-1}.$$

Now use the action property up to inner automorphisms

$$x_1 \cdot (x_2 \cdot \beta(x_3, x_4)) = \beta(x_1, x_2)[(x_1x_2) \cdot \beta(x_3, x_4)]\beta(x_1, x_2)^{-1}.$$

We then obtain

$$\begin{aligned}
&i(d^V\gamma(x_1, x_2, x_3, x_4)) \\
&= \beta(x_1, x_2)[(x_1x_2) \cdot \beta(x_3, x_4)\beta(x_1, x_2)^{-1} \\
&\quad x_1 \cdot (\beta(x_2, x_3x_4)\beta(x_2x_3, x_4)^{-1}\beta(x_2, x_3)^{-1})] \\
&\quad [\beta(x_1x_2, x_3)\beta(x_1x_2x_3, x_4)\beta(x_1x_2, x_3x_4)^{-1}(x_1x_2) \cdot \beta(x_3, x_4)^{-1}] \\
&\quad [x_1 \cdot \beta(x_2x_3, x_4)\beta(x_1, x_2x_3x_4)\beta(x_1x_2x_3, x_4)^{-1}\beta(x_1, x_2x_3)^{-1}] \\
&\quad [x_1 \cdot \beta(x_2, x_3)\beta(x_1, x_2x_3)\beta(x_1x_2, x_3)^{-1} \\
&\quad \beta(x_1x_2, x_3x_4)\beta(x_1, x_2x_3x_4)^{-1}x_1 \cdot \beta(x_2, x_3x_4)^{-1}].
\end{aligned}$$

Here, we have already placed four terms into brackets: We will now permute the first two of these and separately the last two of them. This is possible, because the middle two terms are cocycle expressions and are thus central in M (cf. Lemma 1.2.2). We obtain

$$\begin{aligned}
&i(d^V\gamma(x_1, x_2, x_3, x_4)) \\
&= \beta(x_1, x_2)[\beta(x_1x_2, x_3)\beta(x_1x_2x_3, x_4)\beta(x_1x_2, x_3x_4)^{-1}(x_1x_2) \cdot \beta(x_3, x_4)^{-1}] \\
&\quad [(x_1x_2) \cdot \beta(x_3, x_4)\beta(x_1, x_2)^{-1}x_1 \cdot (\beta(x_2, x_3x_4)\beta(x_2x_3, x_4)^{-1}\beta(x_2, x_3)^{-1})] \\
&\quad [x_1 \cdot \beta(x_2, x_3)\beta(x_1, x_2x_3)\beta(x_1x_2, x_3)^{-1}\beta(x_1x_2, x_3x_4)\beta(x_1, x_2x_3x_4)^{-1}x_1 \cdot \beta(x_2, x_3x_4)^{-1}] \\
&\quad [x_1 \cdot \beta(x_2x_3, x_4)\beta(x_1, x_2x_3x_4)\beta(x_1x_2x_3, x_4)^{-1}\beta(x_1, x_2x_3)^{-1}] \\
&= \beta(x_1, x_2)[\beta(x_1x_2, x_3)\beta(x_1x_2x_3, x_4)\beta(x_1x_2, x_3x_4)^{-1}] \\
&\quad [\beta(x_1, x_2)^{-1}x_1 \cdot (\beta(x_2, x_3x_4)\beta(x_2x_3, x_4)^{-1})] \\
&\quad [\beta(x_1, x_2x_3)\beta(x_1x_2, x_3)^{-1}\beta(x_1x_2, x_3x_4)\beta(x_1, x_2x_3x_4)^{-1}x_1 \cdot \beta(x_2, x_3x_4)^{-1}] \\
&\quad [x_1 \cdot \beta(x_2x_3, x_4)\beta(x_1, x_2x_3x_4)\beta(x_1x_2x_3, x_4)^{-1}\beta(x_1, x_2x_3)^{-1}],
\end{aligned}$$

where in the second step we have reduced adjacent, mutually inverse terms. We will replace brackets in this expression in order to make clear which terms we exchange next:

$$\begin{aligned}
&i(d^V\gamma(x_1, x_2, x_3, x_4)) \\
&= \beta(x_1, x_2)\beta(x_1x_2, x_3)\beta(x_1x_2x_3, x_4)\beta(x_1x_2, x_3x_4)^{-1} \\
&\quad \beta(x_1, x_2)^{-1}x_1 \cdot (\beta(x_2, x_3x_4)\beta(x_2x_3, x_4)^{-1}) \\
&\quad [\beta(x_1, x_2x_3)\beta(x_1x_2, x_3)^{-1}\beta(x_1x_2, x_3x_4)\beta(x_1, x_2x_3x_4)^{-1}x_1 \cdot \beta(x_2, x_3x_4)^{-1}] \\
&\quad [x_1 \cdot \beta(x_2x_3, x_4)\beta(x_1, x_2x_3x_4)\beta(x_1x_2x_3, x_4)^{-1}\beta(x_1, x_2x_3)^{-1}].
\end{aligned}$$

Once again, the second of these two terms in brackets is a cocycle expression and is thus central in M. Therefore, we may permute the two terms in brackets to obtain

$$\begin{aligned}
& i(d^V y(x_1, x_2, x_3, x_4)) \\
& = \beta(x_1, x_2)\beta(x_1x_2, x_3)\beta(x_1x_2x_3, x_4)\beta(x_1x_2, x_3x_4)^{-1} \\
& \quad \beta(x_1, x_2)^{-1}x_1 \cdot (\beta(x_2, x_3x_4)\beta(x_2x_3, x_4)^{-1}) \\
& \quad [x_1 \cdot \beta(x_2x_3, x_4)\beta(x_1, x_2x_3x_4)\beta(x_1x_2x_3, x_4)^{-1}\beta(x_1, x_2x_3)^{-1}] \\
& \quad [\beta(x_1, x_2x_3)\beta(x_1x_2, x_3)^{-1}\beta(x_1x_2, x_3x_4)\beta(x_1, x_2x_3x_4)^{-1}x_1 \cdot \beta(x_2, x_3x_4)^{-1}] \\
& = \beta(x_1, x_2)\beta(x_1x_2, x_3)\beta(x_1x_2x_3, x_4)\beta(x_1x_2, x_3x_4)^{-1} \\
& \quad \beta(x_1, x_2)^{-1}x_1 \cdot \beta(x_2, x_3x_4)[\beta(x_1, x_2x_3x_4)\beta(x_1x_2x_3, x_4)^{-1}] \\
& \quad [\beta(x_1x_2, x_3)^{-1}\beta(x_1x_2, x_3x_4)\beta(x_1, x_2x_3x_4)^{-1}x_1 \cdot \beta(x_2, x_3x_4)^{-1}],
\end{aligned}$$

where we have once again reduced adjacent, mutually inverse terms. We rewrite this in terms of the conjugation $\mathrm{conj}_{x_1 \cdot \beta(x_2, x_3x_4)\beta(x_1, x_2x_3x_4)}$ as

$$\begin{aligned}
& i(d^V y(x_1, x_2, x_3, x_4)) \\
& = \beta(x_1, x_2)\beta(x_1x_2, x_3)\beta(x_1x_2x_3, x_4)\beta(x_1x_2, x_3x_4)^{-1} \\
& \quad \beta(x_1, x_2)^{-1}\mathrm{conj}_{x_1 \cdot \beta(x_2, x_3x_4)\beta(x_1, x_2x_3x_4)}[\beta(x_1x_2x_3, x_4)^{-1}\beta(x_1x_2, x_3)^{-1}\beta(x_1x_2, x_3x_4)].
\end{aligned}$$

Now use that $x_1 \cdot \beta(x_2, x_3x_4)\beta(x_1, x_2x_3x_4)\beta(x_1x_2, x_3x_4)^{-1}\beta(x_1, x_2)^{-1}$ is central in M. This implies that

$$\mathrm{conj}_{x_1 \cdot \beta(x_2, x_3x_4)\beta(x_1, x_2x_3x_4)} = \mathrm{conj}_{\beta(x_1, x_2)\beta(x_1x_2, x_3x_4)}.$$

Therefore, we obtain

$$\begin{aligned}
& i(d^V y(x_1, x_2, x_3, x_4)) \\
& = \beta(x_1, x_2)\beta(x_1x_2, x_3)\beta(x_1x_2x_3, x_4)\beta(x_1x_2, x_3x_4)^{-1} \\
& \quad \beta(x_1, x_2)^{-1}\mathrm{conj}_{\beta(x_1, x_2)\beta(x_1x_2, x_3x_4)}[\beta(x_1x_2x_3, x_4)^{-1}\beta(x_1x_2, x_3)^{-1}\beta(x_1x_2, x_3x_4)] \\
& = \beta(x_1, x_2)\beta(x_1x_2, x_3)\beta(x_1x_2x_3, x_4)\beta(x_1x_2, x_3x_4)^{-1} \\
& \quad \beta(x_1, x_2)^{-1}\beta(x_1, x_2)\beta(x_1x_2, x_3x_4)[\beta(x_1x_2x_3, x_4)^{-1}\beta(x_1x_2, x_3)^{-1}\beta(x_1x_2, x_3x_4)] \\
& \quad \beta(x_1x_2, x_3x_4)^{-1}\beta(x_1, x_2)^{-1} = 1.
\end{aligned}$$

This completes the proof. □

Lemma 1.2.4. *The class* $[y] \in H^3(G, V)$ *of the cocycle* y *does not depend on the choices involved in its definition (in particular, it does not depend on the choice of the section* ρ*).*

Proof.

(a) Let ρ and ρ' be two sections of $\pi : N \to G$. Denote by $\alpha(x,y)$, respectively, $\alpha'(x,y)$, $\beta(x,y)$, respectively, $\beta'(x,y)$, $\gamma(x,y,z)c$ respectively, $\gamma'(x,y,z)$, the elements of N, M and V constructed above with respect to ρ and ρ', respectively (for $x,y,z \in G$). By construction, we have

$$\rho'(x) = \delta(x)\rho(x)$$

for some map $\delta : G \to \mathrm{Ker}(\pi) \subset N$. But then α' may be written

$$\begin{aligned}\alpha'(x,y) &= \rho'(x)\rho'(y)\rho'(xy)^{-1}\\ &= \delta(x)\rho(x)\delta(y)\rho(y)(\delta(xy)\rho(xy))^{-1}\\ &= \delta(x)\mathrm{conj}_{\rho(x)}(\delta(y))\rho(x)\rho(y)\rho(xy)^{-1}\delta(xy)^{-1}\\ &= \delta(x)\mathrm{conj}_{\rho(x)}(\delta(y))\alpha(x,y)\delta(xy)^{-1}.\end{aligned}$$

This is a non-abelian way of saying that α and α' are cohomologuous. By the exactness argument, there exist $\beta(x,y)$ and $\beta'(x,y)$ in M such that $\mu(\beta(x,y)) = \alpha(x,y)$ and $\mu(\beta'(x,y)) = \alpha'(x,y)$. Furthermore, as $\delta : G \to \mathrm{Ker}(\pi) = \mathrm{Im}(\mu)$, there exists a map $\epsilon : G \to M$ such that $\mu \circ \epsilon = \delta$. We obtain from the above

$$\beta'(x,y) = \epsilon(x)\,{}^{\rho(x)}\epsilon(y)\beta(x,y)\epsilon(xy)^{-1}\zeta(x,y),$$

for some element $\zeta(x,y) \in \mathrm{Ker}(\mu)$. Now compute the coboundary of β'. For this, we use the action given by the section ρ'. It therefore reads

$$\begin{aligned}d^M\beta'(x,y,z) &= {}^{\rho'(x)}\beta'(y,z)\beta'(x,yz)\beta'(xy,z)^{-1}\beta'(x,y)^{-1}\\ &= {}^{\rho'(x)}\epsilon(y)\,{}^{\rho'(x)\rho(y)}\epsilon(z)\,{}^{\rho'(x)}\beta(y,z)\,{}^{\rho'(x)}\epsilon(yz)^{-1}\,{}^{\rho'(x)}\zeta(y,z)\\ &\quad \epsilon(x)\,{}^{\rho(x)}\epsilon(yz)\beta(x,yz)\epsilon(xyz)^{-1}\zeta(x,yz)\\ &\quad \zeta(xy,z)^{-1}\epsilon(xyz)\beta(xy,z)^{-1}\,{}^{\rho(xy)}\epsilon(z)^{-1}\epsilon(xy)^{-1}\\ &\quad \zeta(x,y)^{-1}\epsilon(xy)\beta(x,y)^{-1}\,{}^{\rho(x)}\epsilon(y)^{-1}\epsilon(x)^{-1}.\end{aligned}$$

In this expression, we can first of all collect all terms in ζ and put them together because all these terms have values in $\mathrm{Ker}(\mu) \subset Z(M)$. Then observe that replacing $\rho'(t) = \delta(t)\rho(t)$, the action by $\rho'(t)$ becomes an action by $\rho(t)$, conjugated by $\epsilon(t)$, because the action by $\delta(t) \in \mathrm{Im}(\mu)$ becomes conjugation by $\epsilon(t)$ by property (b) of a crossed module (and this for all $t \in G$). We obtain (canceling adjacent annihilating terms):

$$\begin{aligned}d^M\beta'(x,y,z) &= \epsilon(x)\,{}^{\rho(x)}[\epsilon(y)\,{}^{\rho(y)}\epsilon(z)\beta(y,z)\epsilon(yz)^{-1}]\epsilon(x)^{-1}\\ &\quad \epsilon(x)\,{}^{\rho(x)}\epsilon(yz)\beta(x,yz)\beta(xy,z)^{-1}\,{}^{\rho(xy)}\epsilon(z)^{-1}\\ &\quad \beta(x,y)^{-1}\,{}^{\rho(x)}\epsilon(y)^{-1}\epsilon(x)^{-1}\end{aligned}$$

$$^{\rho'(x)}\zeta(y,z)\zeta(x,yz)\zeta(xy,z)^{-1}\zeta(x,y)^{-1}.$$

Now there are two times two adjacent annihilating terms canceling each other. Furthermore, express the action by $\rho(x)\rho(y)$ as the action by $\alpha(x,y)\rho(xy)$ and obtain the action by $\rho(xy)$, conjugated by $\beta(x,y)$ as before. Also observe that the action by $\rho'(x)$ or $\rho(x)$ on $\zeta(y,z)$ gives the same result, because $\zeta(y,z) \in Z(M)$. We obtain

$$\begin{aligned} d^M\beta'(x,y,z) = {} & \epsilon(x)\,^{\rho(x)}\epsilon(y)\beta(x,y)\,^{\rho(xy)}\epsilon(z)[\beta(x,y)^{-1} \\ & ^{\rho(x)}\beta(y,z)\beta(x,yz)\beta(xy,z)^{-1}]\,^{\rho(xy)}\epsilon(z)^{-1}\beta(x,y)^{-1}\,^{\rho(x)} \\ & \epsilon(y)^{-1}\epsilon(x)^{-1}\,^{\rho(x)}\zeta(y,z)\zeta(x,yz)\zeta(xy,z)^{-1}\zeta(x,y)^{-1}. \end{aligned}$$

The term in brackets is a coboundary expression (up to conjugation) and lives therefore in $Z(M)$ (cf. Lemma 1.2.2). We can thus cancel the conjugation by $^{\rho(xy)}\epsilon(z)$. But then there is the conjugation by $\beta(x,y)$, then by $^{\rho(x)}\epsilon(y)$ and finally by $\epsilon(x)$, which cancel in the same way. One stays with

$$d^M\beta'(x,y,z) = d^M\beta(x,y,z)d\zeta(x,y,z),$$

and this means that the expressions for γ and γ' differ by a coboundary.

(b) Let us show that the class $[\gamma]$ does not depend on the choice of the preimage $\beta(x,y) \in M$ of $\alpha(x,y) \in N$. This means independence of the choice of a section $\mathrm{Im}(\mu) \to M$ of the map $\mu : M \to N$.
Indeed, two different choices $\beta(x,y)$ and $\beta'(x,y)$ with $\mu(\beta(x,y)) = \mu(\beta'(x,y)) = \alpha(x,y)$ for all $x, y \in G$ lead to a relation

$$\beta(x,y) = \beta'(x,y)\zeta(x,y),$$

with $\zeta(x,y) \in \mathrm{Ker}(\mu) = \mathrm{Im}(i)$. Now taking coboundaries on both sides implies that $d\beta$ and $d\beta'$ differ by a coboundary—observe that the ζ-terms commute with all other terms, because they take values in the center.

(c) There is no choice involved in the last step choosing a preimage under the map i because i is injective. □

Lemma 1.2.5. *Let $\mu : M \to N$ (with action η) and $\mu' : M' \to N'$ (with action η') be elementary equivalent crossed modules with* $\mathrm{Ker}(\mu) = \mathrm{Ker}(\mu') =: V$ *and* $\mathrm{Coker}(\mu) = \mathrm{Coker}(\mu') =: G$.

Then the corresponding cohomology classes $B([\mu]) = [\gamma]$ and $B([\mu']) = [\gamma']$ coincide in $H^3(G,V)$.

Proof. Denote by (φ,ψ) the morphism rendering the two crossed modules elementary equivalent; see Definition 1.1.6. The cocycle $\gamma \in Z^3(G,V)$ is defined using a section ρ of the map π. But then $\tilde{\rho} := \psi \circ \rho$ is a section of π'. Now Lemma 1.2.4 implies that the sections ρ' and $\tilde{\rho}$ both give rise to the same cohomology class $[\gamma']$. In the first step (computing the failure of the sections to be group homomorphisms), ρ' gives rise to $\alpha'(x,y)$ and $\tilde{\rho}$ gives rise to $\psi\circ\alpha(x,y)$ for $x,y\in G$ (because ψ is a group homomorphism).

In the second step, one lifts $\alpha'(x,y)$ and $\psi\circ\alpha(x,y)$ to M' to obtain $\beta'(x,y)$ on the one hand, and $\varphi\circ\beta(x,y)$ on the other. By construction,

$$\beta'(x,y)(\varphi\circ\beta(x,y))^{-1}\in \mathrm{Ker}(\mu')=\mathrm{Im}(i'),$$

thus there exists $\zeta(x,y)\in Z(M')$ such that

$$\beta'(x,y)(\varphi\circ\beta(x,y))^{-1}=\zeta(x,y).$$

As in the proof of Lemma 1.2.4, taking coboundary operators now shows that the classes γ' and γ differ by the coboundary $d\zeta$. □

In conclusion, we have constructed a well-defined map,

$$B:\mathrm{crmod}(G,V)\to H^3(G,V),$$

which associates to an equivalence class of crossed modules $[\mu:M\to N]$ the cohomology class $[\gamma]$ of the cocycle γ constructed above. Theorem 1.2.1 states that this map is an isomorphism of abelian groups.

1.3 Relation to third cohomology II

In this section, we look at a crossed module $\mu:M\to N$ in a different way. We follow here [83]. The constructions of the previous section were based on the four-term exact sequence

$$0\to \mathrm{Ker}(\mu)\to M\to N\to \mathrm{Coker}(\mu)\to 1$$

associated to the crossed module $\mu:M\to N$. Here, we change our point of view and look at a crossed module as the data of (an arbitrary) group extension

$$1\to L\overset{i}{\to}\widehat{G}\overset{\pi}{\to}G\to 1,$$

together with a central extension

$$0\to Z\to\widehat{L}\overset{p}{\to}L\to 1$$

of the normal subgroup L of the (first named) extension. The link to the previous section is obviously $Z=V=\mathrm{Ker}(\mu)$, $\widehat{L}=M$, $L=\mathrm{Im}(\mu)$ and $\widehat{G}=N$.

If we demand that a given group extension and a given central extension of its normal subgroup should form a crossed module $\mu:\widehat{L}\to G$ given by the composition

$$\widehat{L}\overset{p}{\to}L\overset{\pi}{\to}\widehat{G},$$

they have to be compatible in the sense that the conjugation action of L on $\widehat{L}$ extends to an action of $\widehat{G}$ on $\widehat{L}$ satisfying the requirements of Definition 1.1.1.

In the following, we will use this slightly different framework to derive once more the cohomology class $[\gamma]$ associated to a crossed module, that is, in the following, we will assume that the data of the two extensions together with the extended action form a crossed module. In particular, there is then a fixed action of G on Z.

The group product on the central extension $\widehat{L}$ can be described in a very explicit way (cf. Appendix A). Namely, set-theoretically $\widehat{L}$ is the product set $Z \times L$ and the multiplication reads in these *product coordinates*.

$$(z_1, l_1)(z_2, l_2) = (z_1 + z_2 + \omega(l_1, l_2), l_1 l_2),$$

where $\omega \in Z^2(L, Z)$ is the 2-cocycle defining the central extension. We will write in this situation $\widehat{L} = Z \times_\omega L$. Observe that the action of L on Z is trivial.

Now the $\widehat{G}$-action should extend the conjugation action of L on $\widehat{L}$. The conjugation in the central extension is given by

$$(z_1, l_1)(z_2, l_2)(z_1, l_1)^{-1} = (z_2 + \omega(l_1, l_2) - \omega(l_1 l_2 l_1^{-1}, l_1), l_1 l_2 l_1^{-1}).$$

The $\widehat{G}$-action may be written as

$$^{x}(z, l) = (^{x}z + \theta(x)(\mathrm{conj}_x(l)), \mathrm{conj}_x(l)) \tag{1.1}$$

for some map $\theta : \widehat{G} \times L \to Z$. In order to keep the number of notation to a minimum, we will also denote by θ the associated map $\theta : \widehat{G} \to C^1(L, Z)$. For a fixed $x \in \widehat{G}$, we will also write θ_x for $\theta(x)$.

Property (b) of Definition 1.1.1 implies that for $l_1 \in L \subset \widehat{G}$,

$$^{l_1}(z, l_2) = (0, l_1)(z, l_2)(0, l_1)^{-1} = (z + \omega(l_1, l_2) - \omega(\mathrm{conj}_{l_1}(l_2), l_1), \mathrm{conj}_{l_1}(l_2)), \tag{1.2}$$

meaning that the restriction $\theta|_{L\times L}$ to $L \times L \subset \widehat{G} \times L$, evaluated on (l_1, l_2), equals the negative of the expression

$$\tilde{\omega}(l_1)(l_2) := \omega(l_2, l_1) - \omega(l_1, \mathrm{conj}_{l_1^{-1}}(l_2)). \tag{1.3}$$

Indeed, it is enough to replace l_2 by $l_1 l_2 l_1^{-1} =: \tilde{l}_2$, and then interpret $\tilde{l}_2$ as the new l_2.

Lemma 1.3.1. *Fix an element $x \in \widehat{G}$.*

(a) *A map $\theta_x : L \to Z$ defines an automorphism*

$$\zeta(x) : \widehat{L} \to \widehat{L}, \quad (z, l) \mapsto (^{x}z + \theta_x(\mathrm{conj}_x(l)), \mathrm{conj}_x(l))$$

if and only if the cochain $\theta_x \in C^1(L, Z)$ satisfies

$$d\theta_x = x \cdot \omega - \omega,$$

where the $\widehat{G}$-action on $C^1(L,Z)$ given by

$$(x\cdot\alpha)(l) = {}^x\alpha(x^{-1}lx)$$

and the $\widehat{G}$-action on $\omega\in C^2(L,Z)$ is given by

$$(x\cdot\omega)(l_1,l_2) = {}^x\omega(\mathrm{conj}_{x^{-1}}(l_1),\mathrm{conj}_{x^{-1}}(l_2)).$$

(b) *Suppose that the map $\theta:\widehat{G}\to C^1(L,Z)$, $x\mapsto\theta_x$, satisfies (a). Then θ defines a representation of $\widehat{G}$ on $\widehat{L}$ via the formula*

$$^x(z,l) = ({}^xz+\theta_x(\mathrm{conj}_x(l)),\mathrm{conj}_x(l))$$

if and only if θ is a 1-cocycle with respect to the above $\widehat{G}$-action on $C^1(L,Z)$.

Proof.

(a) First of all, let $A: Z\times_\omega L\to Z\times_\omega L$, $A(z,l) = (\varphi(z)+h(\beta(l)),\beta(l))$ be an automorphism of the central extension $Z\times_\omega L$ which preserves Z, that is, $A(Z)\subset Z$, for some maps $\varphi: Z\to Z$, $\beta: L\to L$ and $h\in C^1(L,Z)$, and suppose that φ and β are automorphisms. We compute

$$\begin{aligned}A((z_1,l_1)(z_2,l_2)) &= A(z_1+z_2+\omega(l_1,l_2),l_1l_2)\\ &= (\varphi(z_1+z_2+\omega(l_1,l_2))+h(\beta(l_1l_2)),\beta(l_1l_2))\\ &= (\varphi(z_1)+\varphi(z_2)+\varphi(\omega(l_1,l_2))+h(\beta(l_1)\beta(l_2)),\beta(l_1l_2)).\end{aligned}$$

On the other hand, we obtain

$$\begin{aligned}A(z_1,l_1)A(z_2,l_2) &= (\varphi(z_1)+h(\beta(l_1)),\beta(l_1))(\varphi(z_2)+h(\beta(l_2)),\beta(l_2))\\ &= (\varphi(z_1)+h(\beta(l_1))+\varphi(z_2)+h(\beta(l_2))\\ &\quad+\omega(\beta(l_1),\beta(l_2)),\beta(l_1l_2))\end{aligned}$$

Therefore, A is an automorphism if and only if for all $l_1,l_2\in L$,

$$\varphi(\omega(l_1,l_2))+h(\beta(l_1)\beta(l_2)) = h(\beta(l_1))+h(\beta(l_2))+\omega(\beta(l_1),\beta(l_2)).$$

In order to link this to our setting, take $h=\theta_x$, $\beta=\mathrm{conj}_x$ and $\varphi(z) = {}^xz$. Then A is an automorphism if and only if for all $l_1,l_2\in L$,

$$\begin{aligned}&{}^x\omega(l_1,l_2)-\omega(\mathrm{conj}_x(l_1),\mathrm{conj}_x(l_2))\\ &\quad=\theta_x(\mathrm{conj}_x(l_2))-\theta_x(\mathrm{conj}_x(l_1l_2))+\theta_x(\mathrm{conj}_x(l_1)).\end{aligned}$$

This is exactly our claim—recall that l_1 acts trivially on $\theta_x(\mathrm{conj}_x(l_2))\in Z$.

(b) We have on the one hand,

$$^{xy}(z,l) = (^{xy}z + \theta_{xy}(\mathrm{conj}_{xy}(l)), \mathrm{conj}_{xy}(l)),$$

and on the other hand,

$$\begin{aligned} ^{x}(^{y}(z,l)) &= {}^{x}(^{y}z + \theta_y(\mathrm{conj}_y(l)), \mathrm{conj}_y(l)) \\ &= (^{x}(^{y}z) + {}^{x}\theta_y(\mathrm{conj}_y(l)) + \theta_x(\mathrm{conj}_x(\mathrm{conj}_y(l))), \mathrm{conj}_x(\mathrm{conj}_y(l))) \end{aligned}$$

Therefore, the above formula defines an action of $\widehat{G}$ on $\widehat{L}$ if and only if for all $x, y \in \widehat{G}$, all $z \in Z$ and all $l \in L$, we have

$$\theta_{xy}(\mathrm{conj}_{xy}(l)) = {}^{x}\theta_y(\mathrm{conj}_y(l)) + \theta_x(\mathrm{conj}_{xy}(l)).$$

Replacing $\mathrm{conj}_{xy}(l)$ by l, we transform this equation into

$$\theta_{xy}(l) = {}^{x}(\theta_y(\mathrm{conj}_{x^{-1}}(l))) + \theta_x(l) = ({}^{x}\theta_y)(l) + \theta_x(l).$$

This is the cocycle identity for the 1-cochain $\theta_x \in C^1(L,Z)$ for the above defined action of $\widehat{G}$ on $C^1(L,Z)$. □

This lemma expresses the conditions for the action of $\widehat{G}$ on $\widehat{L}$ given by the map $\theta : \widehat{G} \times L \to Z$ to be a representation of $\widehat{G}$ on $\widehat{L}$ acting by automorphisms.

In order to define in this context the 3-cocycle associated to the crossed module $\mu : \widehat{L} \to \widehat{G}$, we proceed as follows. Let us define the set D by

$$D := \{(\omega, \theta) \in Z^2(L,Z) \times Z^1(\widehat{G}, C^1(L,Z)) \mid d\theta_x = x \cdot \omega - \omega,\ \theta|_L = \tilde{\omega}\},$$

where we have used the notation defined in equation (1.3).

We want to construct a map $Q : D \to H^3(G,Z)$ such that $Q(\omega,\theta) = 0$ characterizes the extendability of ω to a cocycle $\bar{\omega} \in Z^2(\widehat{G},Z)$ (with respect to our fixed $\widehat{G}$-module structure on Z) such that $\bar{\omega}|_{\widehat{G}\times L} = \tilde{\omega}$. This is actually very close to what we did in the previous section. Namely, let $\rho : G \to \widehat{G}$ be a (set-theoretical) section (with $\rho(1) = 1$). Define then the lift of the outer action by

$$S := \eta \circ \rho : G \to \mathrm{Aut}(\widehat{L}),$$

where η denotes the action of $\widehat{G}$ on $\widehat{L}$, which is part of the data of a crossed module. For $x, y \in G$, let $\alpha(x,y)$ denote the failure of ρ to be a group homomorphism, $\alpha(x,y) = \rho(x)\rho(y)\rho(xy)^{-1}$. As $\pi(\alpha(x,y)) = 1$, there exists $\beta(x,y) \in M$ such that $\mu(\beta(x,y)) = \alpha(x,y)$. β may be chosen normalized, that is, vanishing as soon as $x = 1$ or $y = 1$.

The link to the framework of [83] is the following: $S = \eta \circ \rho : G \to \mathrm{Aut}(M)$ is an outer action of G on M. The failure δ_S of S to be a homomorphism is linked to α by

$$\delta_S = \eta \circ \alpha = \eta \circ \mu \circ \beta = \mathrm{conj} \circ \beta,$$

where conj is the conjugation action in M and where we have used property (b) of a crossed module for $\eta \circ \mu = \text{conj}$. Thus β plays the role of the 2-cochain $\omega \in C^2(G,M)$. It has two components when we write $M = \hat{L}$ in product coordinates as above: $\beta = (\beta_Z, \alpha) \in Z \times L$.

Then we have as before

Proposition 1.3.2. *The map*

$$\tilde{Q}(\omega,\theta,\rho,\beta) := d^M\beta : G^3 \to \hat{L},$$
$$(x,y,z) \mapsto S(x)(\beta(y,z))\beta(x,yz)\beta(xy,z)^{-1}\beta(x,y)^{-1}$$

has values in Z and is a 3-cocycle in $Z^3(G,Z)$. Its cohomology class $Q(\omega,\theta) := [d^M\beta] \in H^3(G,Z)$ does not depend on the choices of β and ρ.

Proof. This has already been proved in Lemmas 1.2.2, 1.2.3 and 1.2.4. □

1.4 Relation to general group extensions

This section is concerned with general group extensions, that is, extensions which are a priori more general than central or abelian extensions. To each general extension, one may associate an outer action of the group on the RHS on the group on the LHS. Our main point is that conversely, given an outer action, there exists a crossed module, which has zero class if and only if there exists a general extension realizing this outer action.

1.4.1 Outer actions and factor systems

Definition 1.4.1. A *general group extension* is a short exact sequence of groups of the form

$$1 \to L \xrightarrow{i} E \xrightarrow{\pi} G \to 1 \tag{1.4}$$

where the group L is not necessarily central in E, not even necessarily abelian.

Such a general extension defines an outer action (see Remark 1.1.4). Indeed, choosing a section $\rho : G \to E$ of π with $\rho(1) = 1$ (*normalized section*), one may lift elements of G to E and then act by the conjugation action in E on elements of L, which we included into E. This defines a map

$$\psi_\rho : G \to \text{Aut}(L). \quad g \mapsto \text{conj}_{\rho(g)}$$

As in Remark 1.1.4, two different sections lead to actions, which differ by the conjugation with some element, thus ψ_ρ induces a well-defined map

$$\psi : G \to \text{Out}(L).$$

This map ψ is called the *outer action* associated to the general extension (1.4).

Definition 1.4.2. Two general extensions

$$1 \to L \to E_1 \to G \to 1$$

and

$$1 \to L \to E_2 \to G \to 1$$

of the same group G by the same group L are called *equivalent* in case there exists a group homomorphism $\varphi : E_1 \to E_2$ such that the following diagram is commutative:

$$\begin{array}{ccccccccc} 1 & \longrightarrow & L & \longrightarrow & E_1 & \longrightarrow & G & \longrightarrow & 1 \\ & & \downarrow{\scriptstyle \mathrm{id}_L} & & \downarrow{\scriptstyle \varphi} & & \downarrow{\scriptstyle \mathrm{id}_G} & & \\ 1 & \longrightarrow & L & \longrightarrow & E_2 & \longrightarrow & G & \longrightarrow & 1 \end{array}$$

Observe that φ is automatically a group isomorphism by the Five Lemma. Denote by $\mathrm{Ext}(G, L)$ the set of equivalence classes of group extensions of G by L.

A general extension (1.4) is called *trivial* or *split* if there exists a group homomorphism $\rho : G \to E$ such that $\pi \circ \rho = \mathrm{id}_G$. In this case, the map

$$L \rtimes G \to E, (l, x) \mapsto l\rho(x)$$

is a group isomorphism, where the *semidirect product* $L \rtimes G$ is defined by the homomorphism $\psi_\rho : G \to \mathrm{Aut}(L)$, $\psi_\rho(x)(l) = \mathrm{conj}_{\rho(x)}(l)$ (where we have suppressed the reference to the inclusion $i : L \to E$ for better readability).

Remark 1.4.3. Given a general extension (1.4) and a (set-theoretical) section $\rho : G \to E$ (with $\rho(1) = 1$) of the projection π, one may associate to these data a map $\psi_\rho : G \to \mathrm{Aut}(L)$ and a map $\beta : G \times G \to L$ computing the failure of ρ to be a group homomorphism:

$$\alpha(x, y) = \rho(x)\rho(y)\rho(xy)^{-1},$$

for all $x, y \in G$. As $\pi \circ \alpha(x, y) = 1$, the map β is then defined by $i \circ \beta = \alpha$. Now the map

$$\varphi : L \times G \to E, \quad (l, x) \mapsto l\rho(x)$$

becomes a group isomorphism (see Lemma 1.4.1 below) when we endow $L \times G$ with the product

$$(l, x)(l', x') = (l\psi_\rho(x)(l')\beta(x, x'), xx'). \tag{1.5}$$

In this sense, the triple (ρ, ψ_ρ, β) determines the extension E up to isomorphism.

Associate to a map $S : G \to \mathrm{Aut}(L)$ its failure δ_S to be a group homomorphism, that is,

$$\delta_S(x,y) := S(x)S(y)S(xy)^{-1}.$$

Usually, we assume S to be normalized, that is, $S(1) = \mathrm{id}$, and in this case δ_S is normalized, that is, $\delta_S(x,y) = 1$ if at least one argument is 1.

Definition 1.4.4. A *factor system* is a pair (S,β) where $S : G \to \mathrm{Aut}(L)$ is a normalized map, $\beta : G \times G \to L$ is also normalized and satisfies $d^L\beta = 1$, and S is linked to β by $\delta_S = \mathrm{conj}_\beta$.

Here, d^L (also denoted d_S) is the formal non-abelian group coboundary operator with values in L using the map S as an "action," while S does not endow L with the structure of a G-module in general; cf. equation (A.1) in Appendix A. Denote by $Z^2(G,L)$ the set of these factor systems, also called *non-abelian 2-cocycles.*

Remark 1.4.5. Given a factor system (S,β), one may compose S with the canonical projection $q : \mathrm{Aut}(L) \to \mathrm{Aut}(L)/\mathrm{conj}_L =: \mathrm{Out}(L)$ to get an outer action $\psi := q \circ S$. This ψ is then a group homomorphism, because, up to conjugations by elements of L, S is a homomorphism. Up to this point, we did not use the condition $d^L\beta = 1$.

Conversely, a typical example of a factor system (S,β) is the factor system associated to an outer action $\psi : G \to \mathrm{Out}(L)$. Indeed, given ψ, there is on the one hand a (normalized) lift $S : G \to \mathrm{Aut}(L)$. On the other hand, as ψ is a group homomorphism, δ_S must be conjugation by some element, thus we may define β by the equation $\delta_S = \mathrm{conj}_\beta$.

Lemma 1.4.1. *Given a factor system (S,β), the set $L \times G$ becomes a group, denoted $L \times_{(S,\beta)} G$, using the product defined in equation (1.5) with S taking the place of ψ_ρ.*

Conversely, any general extension of G by L is isomorphic to some $L \times_{(S,\beta)} G$.

Proof. Let us compute

$$\begin{aligned}((l,x)(l',x'))(l'',x'') &= (lS(x)(l')\beta(x,x'),xx')(l'',x'') \\ &= (lS(x)(l')\beta(x,x')S(xx')(l'')\beta(xx',x''),(xx')x'').\end{aligned}$$

On the other hand,

$$\begin{aligned}(l,x)((l',x')(l'',x'')) &= (l,x)((l'S(x')(l'')\beta(x',x''),x'x'')) \\ &= (lS(x)(l'S(x')(l'')\beta(x',x''))\beta(x,x'x''),x(x'x'')).\end{aligned}$$

In order to have associativity, it remains then to show

$$\beta(x,x')S(xx')(l'')\beta(xx',x'') = S(x)(S(x')(l''))S(x)(\beta(x',x''))\beta(x,x'x'').$$

For this, we use that the hypothesis $\delta_S = \mathrm{conj}_\beta$ implies $S(xx') = \mathrm{conj}_{\beta(x,x')^{-1}} \circ S(x) \circ S(x')$ and obtain

$$\begin{aligned}&\beta(x,x')\beta(x,x')^{-1}S(x)(S(x')(l''))\beta(x,x')\beta(xx',x'')\\&= S(x)(S(x')(l''))S(x)(\beta(x',x''))\beta(x,x'x'').\end{aligned}$$

Canceling $S(x)(S(x')(l''))$, the remaining terms give the cocycle identity

$$\beta(x,x')\beta(xx',x'') = S(x)(\beta(x',x''))\beta(x,x'x''),$$

which is true by the hypothesis $d^L\beta = 1$.

The normalizing conditions $S(1) = \mathrm{id}$ and $\beta(1,x) = \beta(x,1) = 1$ imply that $(1,1)$ is an identity element for this product. For $(l,x) \in L \times G$, the element

$$(S(x)^{-1}(l^{-1}\beta(x,x^{-1})^{-1}), x^{-1})$$

is a right inverse to (l,x). Similarly,

$$(\beta(x^{-1},x)^{-1}S(x^{-1})(l^{-1}), x^{-1})$$

is a left inverse to (l,x). But we have already shown that the product is associative, so left inverse and right inverse must coincide and constitute thus a two-sided inverse to (l,x). In conclusion, we have shown that $L \times_{(S,\beta)} G$ is a group.

Conversely, let

$$1 \to L \xrightarrow{i} E \xrightarrow{\pi} G \to 1$$

be a general extension. As already sketched in Remark 1.4.3, choosing a normalized section ρ of π, we obtain a map ψ_ρ and a map β which form a factor system (ψ_ρ, β). It remains to show that the map

$$\varphi : L \times_{(\psi_\rho,\beta)} G \to E, \quad (l,x) \mapsto l\rho(x)$$

is a group isomorphism. Indeed, φ is a group homomorphism by

$$\begin{aligned}\varphi((l,x)(l',x')) &= \varphi(l\psi_\rho(x)(l')\beta(x,x'), xx') = l\psi_\rho(x)(l')\beta(x,x')\rho(xx')\\&= l\rho(x)l'\rho(x)^{-1}\rho(x)\rho(x') = l\rho(x)l'\rho(x') = \varphi(l,x)\varphi(l',x').\end{aligned}$$

But φ is clearly bijective, thus this shows our claim. □

In the preceding lemma, we reduced the study of general group extensions to standard extensions of the form $L \times_{(S,\beta)} G$ for some factor system (S,β). We will also call this description the *choice of product coordinates* on the extension. The following lemma now describes in how many equivalent ways one can define such a standard extension on the set $L \times G$.

In order to state the lemma, we need an action of the group $C^1(G,L)$ on the set of factor systems (S,β). Here, the set $C^1(G,L) := \mathrm{Map}(G,L)$ of all maps from G to L is considered as a (in general non-abelian!) group with respect to pointwise multiplication in L. The action is defined for $h \in C^1(G,L)$ by

$$h \cdot (S,\beta) = (h \cdot S, h *_S \beta),$$

where the terms are given by

$$h \cdot S := \mathrm{conj}_h \circ S,$$

the map which assigns to $x \in G$ the composition of the two automorphisms $\mathrm{conj}_{h(x)}$ and $S(x)$, and

$$(h *_S \beta)(x,x') := h(x)S(x)(h(x'))\beta(x,x')h(xx')^{-1}$$

for all $x, x' \in G$.

Lemma 1.4.2. *Let $(S,\beta), (S',\beta') \in Z^2(G,L)$ be two factor systems. Then the corresponding standard extensions $L \times_{(S,\beta)} G$ and $L \times_{(S',\beta')} G$ are equivalent if and only if there exists $h \in C^1(G,L)$ such that $h \cdot (S,\beta) = (S',\beta')$.*

If this condition is satisfied, then the map

$$\varphi : L \times_{(S,\beta)} G \to L \times_{(S',\beta')} G, \quad (l,x) \mapsto (lh(x),x)$$

is an equivalence of extensions, and all equivalences of extensions $L\times_{(S,\beta)} G \to L\times_{(S',\beta')} G$ are of this form.

Proof. Choosing product coordinates on the extensions $L \times_{(S,\beta)} G$ and $L \times_{(S',\beta')} G$, an equivalence $L \times_{(S,\beta)} G \to L \times_{(S',\beta')} G$ is necessarily of the form $(l,x) \mapsto (lh(x),x)$ for some $h \in C^1(G,L)$. We compute

$$\begin{aligned}
\varphi(l,x)\varphi(l',x') &= (lh(x),x)(l'h(x'),x') \\
&= (lh(x)S(x)(l'h(x'))\beta(x,x'),xx') \\
&= (l\mathrm{conj}_{h(x)}(S(x)(l'))h(x)S(x)(h(x'))\beta(x,x'),xx') \\
&= (l(h \cdot S)(x)(l')(h *_S \beta)(x,x')h(xx'),xx') \\
&= \varphi(l(h \cdot S)(x)(l')(h *_S \beta)(x,x'),xx').
\end{aligned}$$

Since φ is an equivalence of extensions, it is injective, and the product in $L \times_{(S',\beta')} G$ is therefore given by

$$(lS'(x)(l')\beta'(x,x'),xx') = (l,x)(l',x') = (l(h \cdot S)(x)(l')(h *_S \beta)(x,x'),xx').$$

This implies $\beta' = h *_S \beta$ (choosing $l = l' = 1$) and further $S' = h \cdot S$.

If conversely $\beta' = h *_S \beta$ and $S' = h \cdot S$, we define φ by the above formula, and the above computation shows that φ is indeed a group homomorphism. It is then clearly an equivalence of extensions. □

Definition 1.4.6. According to the previous lemma, let us introduce an equivalence relation. We call $S, S' : G \to \mathrm{Aut}(L)$ *equivalent*, denoted $S \sim S'$, if there exists $h \in C^1(G, L)$ such that $h \cdot S = S'$. The equivalence class of S is denoted $[S]$. Observe that it is the class $[S]$ of a normalized map $S : G \to \mathrm{Aut}(L)$ that corresponds to an outer action $\psi : G \xrightarrow{S} \mathrm{Aut}(L) \to \mathrm{Out}(L)$, because different representatives of the class differ at most by some conjugation. Therefore, one also calls the equivalence class $[S]$ an *outer action* of G on L, or a *G-kernel*, or the *structure of a G-kernel on L*. By abuse of language, we will sometimes call even a representative S of the class $[S]$ an outer action or a G-kernel.

We are now ready to describe the set of equivalence classes of L-extensions $\mathrm{Ext}(G, L)$ of G in terms of factor systems.

Corollary 1.4.3. *The map*

$$Z^2(G, L) \to \mathrm{Ext}(G, L), \quad (S, \beta) \mapsto [L \times_{(S,\beta)} G]$$

induces a bijection

$$H^2(G, L) := Z^2(G, L)/C^1(G, L) \to \mathrm{Ext}(G, L).$$

The preceding proposition implies in particular that $L \times_{(S,\beta)} G \sim L \times_{(S',\beta')} G$ implies $[S] = [S']$, that is, equivalent extensions correspond to the same G-kernel. In the following, we write $\mathrm{Ext}(G, L)_{[S]}$ for the set of equivalence classes of L-extensions of G corresponding to the G-kernel $[S]$. Similarly, the set of 2-cocycles is denoted $Z^2(G, Z(L))_{[S]}$, where the action of G on the (abelian!) center $Z(L)$ is given by the G-kernel $[S]$. $Z^2(G, Z(L))_{[S]}$ has the set of coboundaries $B^2(G, Z(L))_{[S]}$ as a subset, that is, the set of $\omega \in C^2(G, Z(L))$ such that there exists $\lambda \in C^1(G, Z(L))$ with $d_S\lambda = \omega$. This last equation means explicitly for all $x, x' \in G$,

$$S(x)(\lambda(x')) - \lambda(xx') + \lambda(x) = \omega(x, x').$$

Theorem 1.4.4. *Let S be an outer action of G on L with $\mathrm{Ext}(G, L)_{[S]} \neq \emptyset$. Then each extension class in $\mathrm{Ext}(G, L)_{[S]}$ can be represented by a group of the form $L \times_{(S,\beta)} G$. Any other extension $L \times_{(S,\beta')} G$ representing an element of $\mathrm{Ext}(G, L)_{[S]}$ satisfies*

$$\beta'\beta^{-1} \in Z^2(G, Z(L))_{[S]},$$

and the groups $L \times_{(S,\beta)} G$ and $L \times_{(S,\beta')} G$ define equivalent extensions if and only if

$$\beta'\beta^{-1} \in B^2(G, Z(L))_{[S]}.$$

Proof. By Lemma 1.4.1, each extension is equivalent to one of the form $L \times_{(S',\beta')} G$. If $[S] = [S']$ and $h \in C^1(G, L)$ satisfy $h \cdot S = S'$, then $h^{-1} \cdot (S', \beta') = (S, h^{-1} *_{S'} \beta')$, so that $\beta'' := h^{-1} *_{S'} \beta'$ satisfies $L \times_{(S',\beta')} G \cong L \times_{(S,\beta'')} G$. This means that each extension class in $\mathrm{Ext}(G, L)_{[S]}$ may be represented by a group of the form $L \times_{(S,\beta'')} G$.

If (S,β) and (S,β') are factor systems, then $\mathrm{conj}_\beta = \delta_S = \mathrm{conj}_{\beta'}$ implies $\epsilon := \beta'\beta^{-1} \in C^2(G,Z(L))$. Indeed, for all $l \in L$, $\mathrm{conj}_b(l) = \mathrm{conj}_{b'}(l)$ means $blb^{-1} = b'l(b')^{-1}$, that is, $(b(b')^{-1})l = l(b(b')^{-1})$ and, therefore, $b(b')^{-1} \in Z(L)$.

In view of Lemma 1.4.2, the equivalence of the extensions $L \times_{(S,\beta)} G$ and $L \times_{(S,\beta')} G$ is equivalent to the existence of an $h \in C^1(G,L)$ with

$$S = h \cdot S = \mathrm{conj}_h \circ S \quad \text{and} \quad \beta' = h *_S \beta.$$

But then $\mathrm{conj}_h = \mathrm{id}$ and, therefore, $h \in C^1(G,Z(L))$. This implies further $\beta' = h *_S \beta = (d_S h)\beta$, that is, $\beta'\beta \in B^2(G,Z(L))$. If, conversely, $\beta'\beta = d_S h$ for some $h \in C^1(G,Z(L))$, then $h \cdot S = S$ and $h *_S \beta = \beta'$. □

Corollary 1.4.5. *For a G-kernel* $[S]$ *such that* $\mathrm{Ext}(G,L)_{[S]} \neq \emptyset$, *the map,*

$$H^2(G,Z(L)) \times \mathrm{Ext}(G,L)_{[S]} \to \mathrm{Ext}(G,L)_{[S]}$$
$$(\epsilon, [L \times_{(S,\beta)} G]) \mapsto [L \times_{(S,\beta\epsilon)} G]$$

is a well-defined simply transitive action.

1.4.2 Significance of the zero class in $H^3(G,Z)$

We now go back to the framework of the previous section to show that $Q(\omega,\theta) = 0$ characterizes the extendability of ω to a cocycle $\tilde{\omega} \in Z^2(\widehat{G},Z)$. We include this theorem here, because we need for its proof factor systems and standard extensions.

Theorem 1.4.6. *Let* $\omega \in Z^2(L,Z)$ *define the central extension* $\widehat{L} = Z \times_\omega L$, *and* $\theta \in Z^1(\widehat{G},C^1(L,Z))$ *describe the action as in equation* (1.1) *such that* $\mu : \widehat{L} \to \widehat{G}$, $(z,l) \mapsto l$ *is a crossed module of groups. Further choose a section* p *and let* α *and* β *be defined as before. Then the following are equivalent:*

(1) $Q(\omega,\theta) = [d^M\beta] = 0$ *in* $H^3(G,Z)$.
(2) *There exists an extension*

$$0 \to Z \to \widehat{\widehat{G}} \xrightarrow{q} \widehat{G} \to 1$$

and a $\widehat{G}$*-equivariant equivalence* $\widehat{L} \to q^{-1}(L)$ *of central Z-extensions of L.*
(3) *There exists a cocycle* $\omega_{\widehat{G}} \in Z^2(\widehat{G},Z)$ *with* $\omega_{\widehat{G}}|_{L\times L} = \omega$ *and* $\tilde{\omega}_{\widehat{G}}|_{\widehat{G}\times L} = \theta$.

Proof. (1)⇒(2): Suppose that $[d^M\beta] = 0$, that is, that $d^M\beta = d^G\chi$ for some $\chi \in C^2(G,Z)$. Then $\beta' := \beta\chi^{-1}$ satisfies on the one hand $\mu \circ \beta' = \alpha$, because $\mu \circ \chi^{-1} = 1$ as χ has values in Z. On the other hand, $d^M\beta' = d^M\beta - d^G\chi = 0$. Denoting by S the outer action, the pair (S,β') permits to define the group $\widehat{\widehat{G}} = \widehat{L} \times_{(S,\beta')} G$ via the product

$$(l,g)(l',g') := (lS(g)(l')\beta'(g,g'), gg').$$

This group $\widehat{\widehat{G}}$ is an extension of G by $\widehat{L}$. Furthermore, we can view $(\mathrm{conj} \circ \rho, \alpha)$ as a factor system for G and L, which leads to the description of $\widehat{G}$ as $L \times G$, endowed with the product

$$(l,g)(l',g') = (l\mathrm{conj}_{\rho(g)}(l')\alpha(g,g'), gg').$$

Hence the map

$$q : \widehat{\widehat{G}} = \widehat{L} \times G = (Z \times_\omega L) \times G \to \widehat{G} = L \times G, ((z,l),g) \mapsto (l,g)$$

defines a group extension of $\widehat{G}$ by Z containing $\widehat{L}$. It remains to verify that the action of $\widehat{G}$ on $\widehat{L}$, induced by the conjugation action of $\widehat{\widehat{G}}$, for which Z acts trivially, coincides with the action given by θ.

It follows from the construction that the conjugation action of $\widehat{\widehat{G}}$ on $\widehat{L} = Z \times_\omega L$ is given by

$$\begin{aligned}
(l,g)(z,l')(l,g)^{-1} &= \mathrm{conj}_l \circ S(g)(z,l') \\
&= \mathrm{conj}_l({}^g z + \theta(g)(\mathrm{conj}_g(l')), \mathrm{conj}_g(l')) \\
&= ({}^g z + \theta(g)(\mathrm{conj}_g(l')) + \theta(l)(\mathrm{conj}_l \circ \mathrm{conj}_g(l')), \mathrm{conj}_l \circ \mathrm{conj}_g(l')) \\
&= ({}^g z + \theta(l,g)(\mathrm{conj}_l \circ \mathrm{conj}_g(l')), \mathrm{conj}_{(l,g)}(l')) \\
&= (l,g) \cdot (z,l').
\end{aligned}$$

This completes the proof of (2).

(2)⇒(3): Let $q : \widehat{\widehat{G}} \to \widehat{G}$ be an extension as in (2). We may without loss of generality assume that $\widehat{\widehat{G}} = Z \times_{\omega_{\widehat{G}}} \widehat{G}$ for some $\omega_{\widehat{G}} \in Z^2(\widehat{G}, Z)$ such that $\omega_{\widehat{G}}$ extends ω, that is, $\widehat{L} = Z \times_\omega L$ is contained in $\widehat{\widehat{G}}$. The condition that the inclusion $\widehat{L} \hookrightarrow \widehat{\widehat{G}}$ is $\widehat{G}$-equivariant means that $\tilde{\omega}_{\widehat{G}}|_{\widehat{G} \times L} = \theta$.

(3)⇒(1): Let $\omega_{\widehat{G}} \in Z^2(\widehat{G}, Z)$ be a cocycle as in (3) and let

$$q_{\widehat{G}} : \widehat{\widehat{G}} := Z \times_{\omega_{\widehat{G}}} \widehat{G} \to \widehat{G}$$

be the corresponding extension. The condition $\omega_{\widehat{G}}|_{L \times L} = \omega$ means that the group $\widehat{L} = Z \times_\omega L$ is a subgroup of $\widehat{G}$, and the second condition $\tilde{\omega}_{\widehat{G}}|_{\widehat{G} \times L} = \theta$ ensures that the conjugation action of $\widehat{\widehat{G}}$ on $\widehat{L}$ induces the action of $\widehat{G}$ on $\widehat{L}$ defined by θ.

Now take the section $\widehat{\rho} : G \to \widehat{G}$ defined by $g \mapsto (0, \rho(g))$. Then its corresponding failure to be a group homomorphism $\alpha_{\widehat{\rho}}$ satisfies $\mu \circ \alpha_{\widehat{\rho}} = \alpha$, where α is the failure for the section ρ. Therefore, $\alpha_{\widehat{\rho}}$ may be taken as β, and, observing that the outer action is just the composition of the conjugation action and $\widehat{\rho}$, we obtain

$$\begin{aligned}
d^M \alpha_{\widehat{\rho}}(x,y,z) &= \widehat{\rho}(x)\widehat{\rho}(y)\widehat{\rho}(z)\widehat{\rho}(yz)^{-1}\widehat{\rho}(x)^{-1}\widehat{\rho}(x)\widehat{\rho}(yz)\widehat{\rho}(xyz)^{-1}\widehat{\rho}(xyz)\widehat{\rho}(z)^{-1} \\
&\quad \widehat{\rho}(xy)^{-1}\widehat{\rho}(xy)\widehat{\rho}(y)^{-1}\widehat{\rho}(x)^{-1} \\
&= 1.
\end{aligned}$$

We deduce $\tilde{Q}(\omega, \theta, \rho, \alpha_{\widehat{\rho}}) = d^M \alpha_{\widehat{\rho}} = 1$, and thus $Q(\omega, \theta) = [d^M \alpha_{\widehat{\rho}}] = 0$. □

1.4.3 The obstruction crossed module

Definition 1.4.7. Let S be an outer action of G on L and define $\beta \in C^2(G,L)$ by $\delta_S = \mathrm{conj}_\beta$; cf. Remark 1.4.5. We then have $d_S\beta \in Z^3(G,Z(L))_{[S]}$; cf. Lemma 1.2.3. Indeed, the values of $d_S\beta$ are central because of the following computation:

$$\begin{aligned}
\mathrm{conj}_{d_S\beta(x,y,z)} &= \mathrm{conj}_{S(x)(\beta(y,z))\beta(x,yz)\beta(xy,z)^{-1}\beta(x,y)^{-1}}\\
&= S(x)\mathrm{conj}_{\beta(y,z)}S(x)^{-1}\mathrm{conj}_{\beta(x,yz)}\mathrm{conj}_{\beta(x,yz)}^{-1}\mathrm{conj}_{\beta(x,yz)}^{-1}\\
&= S(x)\delta_S(y,z)S(x)^{-1}\delta_S(x,yz)\delta_S(xy,z)^{-1}\delta_S(x,y)^{-1}\\
&= S(x)\delta_S(y,z)S(x)^{-1}S(x)S(yz)S(xyz)^{-1}S(xyz)S(z)^{-1}S(xy)^{-1}S(xy)S(y)^{-1}S(x)^{-1}\\
&= S(x)S(y)S(z)S(yz)^{-1}S(yz)S(z)^{-1}S(y)^{-1}S(x)^{-1}\\
&= 1.
\end{aligned}$$

This means that conjugation with $d_S\beta(x,y,z)$ is trivial for all $x,y,z \in G$ and, therefore, the values of $d_S\beta$ are central.

The corresponding cohomology class

$$[d_S\beta] \in H^3(G,Z(L))_{[S]}$$

is called the *obstruction class* of the outer action S.

We have already computed that taking a different β with $\delta_S = \mathrm{conj}_\beta$, we have $\epsilon := \beta'\beta^{-1} \in C^2(G,Z(L))$ and $d_S\beta$ and $d_S\beta'$ differ by the coboundary $d_S\epsilon$, thus the class $[d_S\beta]$ does not depend on the choice of β.

For equivalent S and S', that is, $S \sim S'$, there exists $h \in C^1(G,L)$ with $S' = h\cdot S$, and we may choose β and β' such that $h *_S \beta = \beta'$. It follows then from the computation in the proof of Lemma 1.2.4 that $d_S\beta = d_{S'}\beta'$. The upshot of this discussion is that the class $[d_S\beta]$ depends only on the equivalence class of S, that is, on the G-kernel S.

Up to now, we have studied the situation where a general extension of G by L already existed. Let us see now how a certain crossed module answers the *existence question*, that is, given an outer action, does there exist a general extension, which realizes this outer action? For this, we will associate to an outer action $S : G \to \mathrm{Aut}(L)$ a crossed module $\mu : L \to G^S$.

Indeed, let $S : G \to \mathrm{Aut}(L)$ be an outer action and let $\beta \in C^2(G,L)$ with $\delta_S = \mathrm{conj}_\beta$. Consider the normal subgroup $L \subset \mathrm{Aut}(L)$ of automorphisms given by conjugation with some element of L, and denote by $\mathrm{ad} : L \to \mathrm{conj}_L \subset \mathrm{Aut}(L)$ the quotient map $L \to L/Z(L) = \mathrm{conj}_L =: L_{\mathrm{ad}}$. On the product set,

$$G^S := L_{\mathrm{ad}} \times G,$$

we define the group product

$$\begin{aligned}
(\mathrm{conj}_l, x)(\mathrm{conj}_{l'}, x') &:= (\mathrm{conj}_{lS(x)(l')\beta(x,x')}, xx')\\
&= (\mathrm{conj}_l\mathrm{conj}_{S(x)(l')}\delta_S(x,x'), xx').
\end{aligned}$$

Note that the second form of the product shows that it does not depend on the choice of β. Define

$$S_1 : G \to \mathrm{Aut}(L_{\mathrm{ad}}), \quad S_1(g)(\mathrm{conj}_l) := \mathrm{conj}_{S(g)(l)}.$$

This defines an outer action on L_{ad} with $\delta_{S_1} = \mathrm{conj}_{\delta_S}$, if we consider δ_S as L_{ad}-valued.

Lemma 1.4.7. *The above defined product renders G^S a group with the following properties:*

(a) *The map $q_S : G^S \to G$, $(\mathrm{conj}_l, x) \mapsto x$ defines a group extension of G by L_{ad}. The equivalence class of this extension depends only on the G-kernel $[S]$.*
(b) *The map $\eta : G^S \to \mathrm{Aut}(L)$, $(\mathrm{conj}_l, x) \mapsto \mathrm{conj}_l \circ S(x)$ defines an action of G^S on L.*
(c) *The map $\mu : L \to G^S, l \mapsto (\mathrm{conj}_l, 1)$ defines a crossed module with $\mathrm{Ker}(\mu) = Z(L)$ and $\mathrm{Coker}(\mu) = G^S/L_{\mathrm{ad}} \cong G$, and S is the corresponding outer action of $G \cong G^S/L_{\mathrm{ad}}$ on L.*

Proof.
(a) Let us show that (S_1, δ_{S_1}) is a factor system for G and L_{ad}. We have clearly $\delta_{S_1} = \mathrm{conj}_{\delta_S}$. Let $x, x', x'' \in G$. Then compute

$$\begin{aligned} &S_1(x)(\delta_S(x', x''))\delta_S(x, x'x'') \\ &= S(x)(S(x')S(x'')S(x'x'')^{-1})S(x)^{-1}S(x)S(x'x'')S(xx'x'')^{-1} \\ &= S(x)S(x')S(x'')S(xx'x'')^{-1} \\ &= S(x)S(x')S(xx')^{-1}S(xx')S(x'')S(xx'x'')^{-1} \\ &= \delta_S(x, x')\delta_S(xx', x''). \end{aligned}$$

This shows that (S_1, δ_{S_1}) is a factor system and, therefore, by Lemma 1.4.1 that G^S is a group.
To see that the equivalence class of the extension $G^S = L_{\mathrm{ad}} \times_{(S_1, \delta_S)} G$ of G by L_{ad} depends only on the equivalence class $[S]$, let $\alpha \in C^1(G, L)$ and $S' := \mathrm{conj}_\alpha \circ S$. Then

$$S_1' := \mathrm{conj}_h \circ S_1 = h \cdot S_1$$

for $h := \mathrm{conj}_\alpha \in C^1(G, L_{\mathrm{ad}})$. We obtain

$$\delta_{S'}(x, x') = \mathrm{conj}_{\alpha(x)S(x)(\alpha(x'))} \circ \delta_S(x, x') \circ \mathrm{conj}_{\alpha(xx')^{-1}} = (h *_{S_1} \delta_S)(x, x').$$

According to Lemma 1.4.2, the map

$$\varphi : G^{S'} \to G^S, \quad (\mathrm{conj}_l, x) \mapsto (\mathrm{conj}_{l\alpha(x)}, x) = (\mathrm{conj}_l \circ h(x), x)$$

is an equivalence of extensions of G by L_{ad}.

(b) Let us show that η is a representation:

$$\begin{aligned}\eta(\mathrm{conj}_l, x)\eta(\mathrm{conj}_{l'}, x') &= \mathrm{conj}_l \circ S(x) \circ \mathrm{conj}_{l'} \circ S(x') \\ &= \mathrm{conj}_l \circ \mathrm{conj}_{S(x)(l')} \circ S(x) \circ S(x') \\ &= \mathrm{conj}_{lS(x)(l')} \circ \delta_S(x, x') \circ S(xx') \\ &= \eta(\mathrm{conj}_{lS(x)(l')} \circ \delta_S(x, x'), xx') \\ &= \eta((\mathrm{conj}_l, x)(\mathrm{conj}_{l'}, x')).\end{aligned}$$

(c) This follows directly from (a) and (b). □

Remark 1.4.8. On the one hand (see the lemma above), this construction gives the *obstruction crossed module* $\mu : L \to G^S$ as associated to the canonical central extension

$$0 \to Z(L) \to L \to L_{\mathrm{ad}} \to 1$$

of the normal subgroup L_{ad} in G^S, and on the other hand (see the lemma below), the crossed module $\mu : L \to G^S$ can also be seen as the *pullback crossed module* of the canonical crossed module

$$0 \to Z(L) \to L \to \mathrm{Aut}(L) \to \mathrm{Out}(L) \to 1$$

by the group homomorphism $\psi : G \to \mathrm{Out}(L)$ obtained by $\psi = q \circ S$ using the quotient map $q : \mathrm{Aut}(L) \to \mathrm{Out}(L)$.

Denote by $q_S : G^S \to G$ the projection.

Lemma 1.4.8. *The map*

$$\psi_S = (\eta, q_S) : G^S \to \mathrm{Aut}(L) \times G$$

is injective and yields an isomorphism of groups

$$G^S \cong \{(\varphi, x) \in \mathrm{Aut}(L) \times G \mid S(x) \in \mathrm{conj}_L \circ \varphi\}.$$

Here, $\mathrm{conj}_L = L_{\mathrm{ad}}$ *is the normal subgroup of* $\mathrm{Aut}(L)$ *of inner automorphisms of* L.

Note that the above action $\eta(\mathrm{conj}_l, x) = (l' \mapsto lS(x)(l')l^{-1})$ is given exactly by this precomposition of conj_l by the automorphism $S(x)$.

Proof. Since $\mathrm{Ker}(q_S) = L_{\mathrm{ad}}$ and $\mathrm{Ker}(\eta) \cap L_{\mathrm{ad}} = \{1\}$, the map ψ_S is an injective homomorphism of groups. In the following, we will therefore view G^S as a subgroup of the group product $\mathrm{Aut}(L) \times G$ and show that it is equal to

$$\{(\varphi, x) \in \mathrm{Aut}(L) \times G \mid S(x) \in \mathrm{conj}_L \circ \varphi\}.$$

For each element $(\mathrm{conj}_l, x) \in G^S$, we have

$$\psi_S(\mathrm{conj}_l, x) = (\mathrm{conj}_l \circ S(x), x),$$

which proves the left-to-right inclusion $\subset$. But for any pair $(\varphi, x) \in \mathrm{Aut}(L) \times G$ with $\varphi \in \mathrm{conj}_L \circ S(x)$, there exists an element $l \in L$ with $\varphi = \mathrm{conj}_l \circ S(x)$, which implies that (φ, x) is the ψ_S-image of (conj_l, x). □

The next lemma sheds some light on the passage from a given extension of G by L with outer action S to the corresponding extension $q_S : G^S \to G$ of G by $Z(L)$.

Lemma 1.4.9. *Let $q : \widehat{G} \to G$ be a group extension of G by L with associated G-kernel $[S]$ and denote (as usual) by* conj *the conjugation map* conj $: \widehat{G} \to \mathrm{Aut}(L)$. *Then $\widehat{G}/Z(L) \cong G^S$ and the map*

$$\psi = (\mathrm{conj} \circ (S \circ q)^{-1}, q) : \widehat{G} \to \mathrm{Aut}(L) \times G, \quad x \mapsto (\mathrm{conj}_x \circ S(q(x))^{-1}, q(x))$$

defines an extension

$$0 \to Z(L) \to \widehat{G} \xrightarrow{\psi} G^S \to 1.$$

This assignment has the following properties:

(a) *If $q_j : \widehat{G}_j \to G$, $j = 1, 2$, are equivalent extensions of G by L, then $\psi_j : \widehat{G}_j \to G^S$ are equivalent extensions of G^S by $Z(L)$. We thus obtain a map*

$$\tau : \mathrm{Ext}(G, L)_{[S]} \to \mathrm{Ext}(G^S, Z(L))_{[\eta]}.$$

(b) *An extension $\psi : \widehat{G^S} \to G^S$ of G^S by $Z(L)$ comes from an extension of G by L corresponding to $[S]$ if and only if there exists a G^S-equivariant equivalence*

$$\alpha : L \to \psi^{-1}(L_{\mathrm{ad}}) \subset \widehat{G^S}$$

of central extensions of L_{ad} by $Z(L)$. Note that $G^S \cong \widehat{G^S}/Z(L)$ acts on $\psi^{-1}(L_{\mathrm{ad}})$ because $Z(L)$ acts trivially on this group.

Proof. Since the extension $\widehat{G}$ corresponds to $[S]$, it is equivalent to an extension of the type $L \times_{(S,\beta)} G$ for some factor system (S, β) by Lemma 1.4.1. This means that

$$\mathrm{conj}_{(l,x)} = \mathrm{conj}_l \circ S(x) = \mathrm{conj}_l \circ S(q(l, x)),$$

so that $\psi(l, x) = (\mathrm{conj}_l, x)$. Now the explicit formulas for the multiplication in $L \times_{(S,\beta)} G$ and G^S imply that ψ is a surjective group homomorphism:

$$\begin{aligned}\psi((l, x)(l', x')) &= \psi(lS(x)(l')\beta(x, x'), xx') \\ &= (\mathrm{conj}_{lS(x)(l')\beta(x,x')}, xx')\end{aligned}$$

$$
\begin{aligned}
&= (\mathrm{conj}_{lS(x)(l')}\delta_S(x,x'),xx') \\
&= (\mathrm{conj}_l,x)(\mathrm{conj}_{l'},x') \\
&= \psi(l,x)\psi(l',x').
\end{aligned}
$$

Its kernel is $Z(L)$, and thus ψ defines an extension of G^S by $Z(L)$.

(a) If $\varphi : \widehat{G}_1 \to \widehat{G}_2$ is an equivalence of L-extensions of G, then we have for the conjugation actions conj^j of $\widehat{G}_j$ on L for $j = 1,2$ the relation $\mathrm{conj}^2 \circ \varphi = \mathrm{conj}^1$ because $\varphi|_L = \mathrm{id}_L$. Therefore, the quotient maps

$$\psi_j = (\mathrm{conj}^j \circ (S \circ q_j)^{-1}, q_j) : \widehat{G}_j \to G^S$$

satisfy

$$
\begin{aligned}
\psi_2 \circ \varphi &= ((\mathrm{conj}^2 \circ \varphi) \circ (S \circ q_2 \circ \varphi)^{-1}, q_2 \circ \varphi) \\
&= (\mathrm{conj}^1 \circ (S \circ q_1)^{-1}, q_1) = \psi_1.
\end{aligned}
$$

This means that $\varphi : \widehat{G}_1 \to \widehat{G}_2$ is also an equivalence of extensions of G^S by $Z(L)$.

(b) Suppose first that the extension $\psi : \widehat{G^S} \to G^S$ by $Z(L)$ comes from the L-extension $q : \widehat{G} \to G$ corresponding to $[S]$ in the sense that $\widehat{G} = \widehat{G^S}$ with $\widehat{G}/Z(L) \cong G^S$. We may assume that $\widehat{G} = Z(L) \times_{(S,\beta)} G$ by Lemma 1.4.1. Then Lemma 1.4.7 shows that $(l,x) \in \widehat{G}$ acts on L by $\mathrm{conj}_l \circ S(x) = \eta(\mathrm{conj}_l, x)$. Therefore, the inclusion $L \hookrightarrow \widehat{G}$ onto the subset $\psi^{-1}(L_{\mathrm{ad}})$ is equivariant with respect to the action of G^S and, therefore, in particular for the action of L_{ad}, so that it is an equivalence of central extensions of L_{ad} by $Z(L)$.

Suppose, conversely, that $\psi : \widehat{G^S} \to G^S$ is an extension of G^S by $Z(L)$ for which there exists a G^S-equivariant equivalence $\alpha : L \to \psi^{-1}(L_{\mathrm{ad}})$ of central extensions of L_{ad} by $Z(L)$. Then

$$\widehat{G^S}/\alpha(L) = \widehat{G^S}/\psi^{-1}(L_{\mathrm{ad}}) \cong G^S/L_{\mathrm{ad}} \cong G,$$

so that we obtain an extension of G by L using the quotient map $q = q_S \circ \psi : \widehat{G^S} \to G$ with kernel $\psi^{-1}(L_{\mathrm{ad}})$. The action of $G^S \cong \widehat{G^S}/Z(L)$ on L induced by the conjugation action of $\widehat{G^S}$ on L coincides with the given action

$$\eta : G^S \to \mathrm{Aut}(L), \quad (\mathrm{conj}_l, x) \mapsto \mathrm{conj}_l \circ S(x)$$

of G^S on L, because α is G^S-equivariant. Therefore, the outer action associated to the extension $q : \widehat{G^S} \to G$ is $[S]$.

For the conjugation action conj of $\widehat{G^S}$ on L, we have $\mathrm{conj} = \eta \circ \psi$ and $q = q_S \circ \psi$, so that the corresponding map $\widehat{G^S} \to G^S$ coincides with ψ. This means that $\psi : \widehat{G^S} \to G^S$ comes from the extension $q : \widehat{G^S} \to G$ in the sense described above. □

Let an outer action $[S]$ be given. Recall the obstruction crossed module $\mu : L \to G^S$. Its characteristic class is $[d_S\beta] \in H^3(G, Z(L))_{[S]}$.

The following theorem shows that the cohomology class $[d_S\beta] \in H^3(G, Z(L))_{[S]}$ associated to $[S]$ is exactly the obstruction against the existence of an extension of G by L realizing this outer action. The second part of the theorem concerns the realizability as an *abelian* extension (but of the more complicated group G^S).

Theorem 1.4.10. *Let S be an outer action of G on L.*

(a) *Then, concerning non-abelian extensions, we have*

$$[d_S\beta] = 1 \quad \Leftrightarrow \quad \mathrm{Ext}(G, L)_{[S]} \neq \emptyset.$$

(b) *Furthermore, the central extension*

$$0 \to Z(L) \to L \to L_{\mathrm{ad}} \to 1$$

can be embedded in a G^S-equivariant way into an abelian extension

$$0 \to Z(L) \to \widehat{G^S} \to G^S \to 1$$

if and only if $[d_S\beta] = 1$.

Proof.

(a) If there exists an extension corresponding to $[S]$, then we may assume by Lemma 1.4.1 that it is of the form $L \times_{(S,\beta)} G$ for some factor system (S, β). But then we have $d_S\beta = 1$, and thus $[d_S\beta] = 1$.
Suppose conversely that S is an outer action with $[d_S\beta] = 1$. Then there exists $\beta \in C^2(G, L)$ with $\delta_S = \mathrm{conj}_\beta$ and some $h \in C^2(G, Z(L))$ with $d_S\beta = d_S h^{-1}$ (because $[d_S\beta] = 1$), so that $\beta' := \beta \cdot h \in C^2(G, L)$ satisfies $d_S\beta' = d_S\beta d_S h = 1$ and $\delta_S = \mathrm{conj}_{\beta'}$ (because the factor h does not play a role in the conjugation). Hence (S, β') is a factor system and Lemma 1.4.1 implies the existence of a group extension $L \times_{(S,\beta')} G$ of G by L corresponding to $[S]$.

(b) We have seen in Lemma 1.4.7 that the map $\mu : L \to G^S$ defines a crossed module with outer action S. According to Theorem 1.4.6, the condition $[d_S\beta] = 1$ is equivalent to the possibility of embedding the central extension

$$0 \to Z(L) \to L \to L_{\mathrm{ad}} \to 1$$

in a G^S-equivariant way into an abelian extension

$$0 \to Z(L) \to \widehat{G^S} \to G^S \to 1. \qquad \square$$

1.5 Construction of crossed modules

1.5.1 The principal construction

In order to motivate the construction procedure, which we present below and which we will call the *principal construction* in the following, let us go back to the four-term exact sequence associated to a crossed module:

$$0 \to V \xrightarrow{i} M \xrightarrow{\mu} N \xrightarrow{\pi} G \to 1.$$

We have already seen how to derive from it two short exact sequences of groups:

$$1 \to M/i(V) \xrightarrow{\mu} N \xrightarrow{\pi} G \to 1,$$

and

$$0 \to V \xrightarrow{i} M \xrightarrow{\mu} \mathrm{Im}(\mu) \to 1.$$

Observe that $M/i(V) = M/\mathrm{Ker}(\mu) \cong \mathrm{Im}(\mu) =: L$ according to our standard notation.

Obviously, the crossed module $\mu : M \to N$ can be seen as the Yoneda product (i. e., the splicing together) of these two short exact sequences. The second sequence is a central extension, whereas the first is neither central, nor even an abelian extension in general. According to the terminology of the previous section, it is a *general group extension*.

Let us ask (some version of) the converse question: Given a group G, a short exact sequence of G-modules

$$0 \to V_1 \to V_2 \xrightarrow{p} V_3 \to 0 \tag{1.6}$$

(regarded as a short exact sequence of abelian groups) and an abelian extension E of G by the abelian group V_3,

$$0 \to V_3 \to E \to G \to 1, \tag{1.7}$$

is the Yoneda product of (1.6) and (1.7) a crossed module? In case (1.7) is given by a 2-cocycle α, we continue to use the notation $E = V_3 \times_\alpha G$. By convention, all our group cocycles are normalized, that is, for all $g \in G$, $\alpha(1, g) = \alpha(g, 1) = 0$ for the neutral element $1 \in G$.

Theorem 1.5.1. *In the above situation, the Yoneda product of (1.6) and (1.7) is a crossed module, the associated 3-cocycle of which is the image of the 2-cocycle defining the abelian extension (1.7) under the connecting homomorphism in the long exact cohomology sequence associated to (1.6).*

Proof. Splicing the sequences (1.6) and (1.7) together, one obtains a map

$$\mu : V_2 \to E, \tag{1.8}$$

given by definition as the composition of $p : V_2 \to V_3$ with the inclusion $V_3 \hookrightarrow E$. Writing $E = V_3 \times G$ in product coordinates (as a set), we have $\mu(v) = (v, 1)$ where $1 \in G$ is the neutral element. On the other hand, an E-action η on V_2 is induced by the action of G on V_2: $\eta(w,x)(v) := x \cdot v$, where $(w,x) \in E = V_3 \times G$, $v \in V_2$. With these structures, condition (b) for a crossed module is trivially true, while condition (a) is true by definition of the product in the abelian extension:

$$\mu(\eta(w,x)(v)) = (x \cdot v, 1) = (w,x)(v,1)(w,x)^{-1}.$$

Now let us discuss the second claim. The short exact sequence (2.9) induces a short exact sequence of complexes

$$0 \to C^*(G, V_1) \xrightarrow{i} C^*(G, V_2) \xrightarrow{\pi} C^*(G, V_3) \to 0,$$

where i is induced by the inclusion $V_1 \hookrightarrow V_2$ and π is induced by the map $p : V_2 \to V_3$. Take a cocycle $\alpha \in C^2(G, V_3)$, then the connecting homomorphism $\partial : C^2(G, V_3) \to C^3(G, V_1)$ is defined as follows:

$$\begin{array}{ccccc}
 & & \beta \in \pi^{-1}(\alpha) & \xmapsto{\pi} & \alpha \\
 & & \downarrow & & \downarrow \\
C^2(G, V_1) & \xrightarrow{i} & C^2(G, V_2) & \xrightarrow{\pi} & C^2(G, V_3) \\
\downarrow d^{V_1} & & \downarrow d^{V_2} & & \downarrow d^{V_3} \\
C^3(G, V_1) & \xrightarrow{i} & C^3(G, V_2) & \xrightarrow{\pi} & C^3(G, V_3) \\
 & & \downarrow & & \downarrow \\
\partial\alpha := i^{-1}(d^{V_2}\beta) & \xmapsto{i} & d^{V_2}\beta & \xmapsto{\pi} & 0 = d^{V_3}\alpha
\end{array}$$

Here, we wrote elements on top respectively on the bottom of the corresponding spaces, and denoted by d^{V_1}, d^{V_2} and d^{V_3} the group coboundary operators with values in V_1, V_2 and V_3, respectively. Summarizing, $\partial\alpha$ is constructed by choosing first an element β, preimage of α under π, then taking $d^{V_2}\beta$, and finally taking a preimage of $d^{V_2}\beta$ under i. It is obvious from Section 1.2 that this is exactly how we constructed the 3-cocycle γ corresponding to a crossed module. Observe that the map corresponding to π here is the map μ in (1.8) (modulo extending it trivially to the second factor), the map defining the crossed module. Having stated the coincidence of the two constructions, it remains to take for $\alpha \in C^2(G, V_3)$ the cocycle defining the abelian extension (1.7). □

1.5.2 End of the proof of Theorem 1.2.1

Here, we complete the proof of Theorem 1.2.1 using the previous construction. Using free crossed modules in the next subsection, we will see another proof of the surjectivity claim in Theorem 1.2.1.

Surjectivity of the map *B* from Theorem 1.2.1

We have to show that given a cohomology class $[\gamma] \in H^3(G, V)$, there is a crossed module whose associated class is $[\gamma]$. V is here some G-module. As the category of G-modules possesses enough injectives, there is an injective G-module I and a monomorphism $i : V \hookrightarrow I$. Consider now the short exact sequence of G-modules:

$$0 \to V \to I \to Q \to 0$$

where Q is the cokernel of i. As I is injective, the connecting homomorphism in the long exact sequence in cohomology is an isomorphism $H^2(G, Q) \cong H^3(G, V)$. Thus $[\gamma]$ corresponds under this isomorphism to a class $[\alpha] \in H^2(G, Q)$ (represented by a normalized 2-cocycle α), and the principal construction applied to the above short exact sequence of G-modules and the abelian extension of G by Q using the cocycle α gives a crossed module whose class is a preimage under B of $[\gamma]$ by Theorem 1.5.1.

Injectivity of the map *B* from Theorem 1.2.1

It is clear that B is a homomorphism of abelian groups. Thus in order to show the injectivity of B, it suffices to show that its kernel is trivial.

We have seen in Theorem 1.4.10 (a) that a crossed module

$$0 \to V \to M \xrightarrow{\mu} N \to G \to 1$$

corresponding to the trivial cohomology class gives rise to a general extension

$$1 \to M \xrightarrow{\varphi} E \xrightarrow{\psi} G \to 1,$$

which realizes the outer action S of G on M.

We now use this extension to construct a morphism of crossed modules. By the proof of Theorem 1.4.10(a), we have in product coordinates, $E = M \times_{(S,\beta)} G$, where S is the outer action associated to the crossed module $\mu : M \to N$. By equation (1.5), the group product in E reads

$$(l, x)(l', x') = (lS(x)(l')\beta(x, x'), xx'). \tag{1.9}$$

Let us also use product coordinates in order to describe the group product in N, using the section $\rho : G \to N$, which defines the outer action S.

Lemma 1.5.2. *The map $m : E \to N$, induced by μ and ρ on the two factors of the cartesian product set $E = M \times G$ in a description of the extension E in product coordinates, induces a morphism of short exact sequences,*

$$\begin{array}{ccccccccccc} & & 1 & \longrightarrow & M & \longrightarrow & E & \longrightarrow & G & \longrightarrow & 1 \\ & & & & \downarrow{\scriptstyle \mathrm{id}_M} & & \downarrow{\scriptstyle m} & & \downarrow{\scriptstyle \mathrm{id}_G} & & \\ 0 & \longrightarrow & V & \xrightarrow{i} & M & \xrightarrow{\mu} & N & \xrightarrow{\pi} & G & \longrightarrow & 1 \end{array}$$

Proof. The squares commute by definition of m. It remains to show that m is a group homomorphism. This follows from the fact that the group product in N can be described in the same way as we described the group product for E above.

Indeed, N is identified with $\mu(M) \times G$ via the maps $n \mapsto (n\rho(\pi(n))^{-1}, \pi(n))$ and $(\mu(m), x) \mapsto \mu(m)\rho(x)$, using a set-theoretical section $\rho : G \to N$ of $\pi : N \to G$. This is the meaning of a description of N in product coordinates. The group product is written in product coordinates as

$$\begin{aligned} (\mu(m), x)(\mu(m'), x') &\mapsto \mu(m)\rho(x)\mu(m')\rho(x') \\ &= \mu(m)\rho(x)\mu(m')\rho(x)^{-1}\rho(x)\rho(x') \\ &= \mu(m)S(x)(\mu(m'))\alpha(x, x')\rho(xx') \\ &\mapsto (\mu(m)S(x)(\mu(m'))\alpha(x, x'), xx'). \end{aligned}$$

Comparing this to the group product in E, that is, equation (1.9), together with $\mu(\beta(x, x')) = \alpha(x, x')$, shows our claim. □

Lemma 1.5.3. *If a crossed module*

$$0 \to V \to M \xrightarrow{\mu} N \to G \to 1$$

admits a morphism

$$\begin{array}{ccccccccccc} & & 1 & \longrightarrow & M & \xrightarrow{\varphi} & E & \xrightarrow{\psi} & G & \longrightarrow & 1 \\ & & & & \downarrow{\scriptstyle \mathrm{id}_M} & & \downarrow{\scriptstyle m} & & \downarrow{\scriptstyle \mathrm{id}_G} & & \\ 0 & \longrightarrow & V & \xrightarrow{i} & M & \xrightarrow{\mu} & N & \xrightarrow{\pi} & G & \longrightarrow & 1 \end{array}$$

then it represents the zero class.

Proof. Indeed, in case there is a morphism as indicated, we have a commutative diagram

$$\begin{array}{ccccccccccc}
0 & \longrightarrow & V & \xrightarrow{\mathrm{id}_V} & V & \xrightarrow{1} & G & \xrightarrow{\mathrm{id}_G} & G & \longrightarrow & 1 \\
 & & \uparrow{\scriptstyle \mathrm{id}_V} & & \uparrow{\scriptstyle \mathrm{proj}_1} & & \uparrow{\scriptstyle \psi} & & \uparrow{\scriptstyle \mathrm{id}_G} & & \\
0 & \longrightarrow & V & \xrightarrow{\mathrm{incl}_1} & V \times M & \xrightarrow{(1,\varphi)} & E & \xrightarrow{\psi} & G & \longrightarrow & 1 \\
 & & \downarrow{\scriptstyle \mathrm{id}_V} & & \downarrow{\scriptstyle i\cdot\mathrm{id}_M} & & \downarrow{\scriptstyle m} & & \downarrow{\scriptstyle \mathrm{id}_G} & & \\
0 & \longrightarrow & V & \xrightarrow{i} & M & \xrightarrow{\mu} & N & \xrightarrow{\pi} & G & \longrightarrow & 1
\end{array}$$

Here, we used the map $i \cdot \mathrm{id}_M : V \times M \to M$ defined by $(v, m) \mapsto i(v)m$ (the product in M), which is a group homomorphism because $i(V)$ is central in M. We denoted by incl_1 and proj_1 once again the standard inclusion and projection maps.

The diagram shows that the crossed module $\mu : M \to N$ is equivalent to the zero crossed module, that is, represents the zero element in $\mathrm{crmod}(G, V)$. This concludes the proof of the Lemma, of the injectivity of the map B, and thus the proof of Theorem 1.2.1. □

Remark 1.5.1. In fact, the implication of the preceding lemma is an equivalence.

Corollary 1.5.4. *Every crossed module is equivalent to one coming from the principal construction.*

1.5.3 Constructions involving free groups

The main idea of the constructions in this subsection is kind of dual to the idea we used in the previous subsection: Instead of resolving the *module* by an injective module, we resolve here the *group* by a free group.

Let G be a group, V be a G-module. Suppose given a presentation

$$1 \to R \to F \xrightarrow{\nu} G \to 1$$

of G as the quotient of a free group F by a normal subgroup R, called the *subgroup of relations*. Consider the construction of the 3-class γ in some crossed module

$$0 \to V \xrightarrow{i} M \xrightarrow{\mu} F \to G \to 1,$$

where $R = \mathrm{Im}(\mu)$. The image $\mathrm{Im}(\mu) = R \subset F$ is a free group (as subgroup of the free group F), hence M is the direct product of $i(V)$ with some group isomorphic to R. Indeed, one may construct a section of the quotient map $\mu : M \to \mu(M) = R$ by choosing preimages for the elements of a basis of R. We may therefore choose these representatives $v(r) \in M$ such that $\mu(v(r)) = r$ and such that v is a group homomorphism. In this way, M is presented as the direct product $M = i(V) \times v(R)$.

Denote by $u : G \to F$ a set-theoretical section of the quotient map $v : F \to G$ such that $u(1) = 1$. By axiom (a) of a crossed module, we have

$$\mu(\eta(u(x))(v(r))) = u(x)ru(x)^{-1}.$$

Lifting to M, this means that there exists $m(x, r) \in V$ such that

$$\eta(u(x))(v(r)) = i(m(x, r))v(u(x)ru(x)^{-1}). \tag{1.10}$$

As the action η is by automorphisms and v is a group homomorphism, we have

$$m(x, r) + m(x, r') = m(x, rr'), \tag{1.11}$$

that is, we obtain a map $m \in C^1(G, \mathrm{Hom}(R, V))$, which satisfies $m(x, 1) = 0$. With respect to the multiplication in G, we have on the other hand,

$$\eta(u(x))\eta(u(y)) = \eta(\alpha(x, y))\eta(u(xy))$$

for all $x, y \in G$ (cf. Remark 1.1.4 (c)), where here $\alpha(x, y) := u(x)u(y)u(xy)^{-1}$.

Lemma 1.5.5. *We have*

$$x \cdot m(y, r) + m(x, u(y)ru(y)^{-1}) = m(xy, r)$$

for all $x, y \in G$ and $r \in R$.

Proof. We start from the condition

$$\eta(u(x))\eta(u(y)) = \eta(\alpha(x, y))\eta(u(xy)),$$

where the products on both sides are taken in the automorphism group of M. Apply both sides to an element $v(r) \in M$. The left-hand side gives then

$$\begin{aligned}\eta(u(x))\eta(u(y))(v(r)) &= \eta(u(x))(\eta(u(y))(v(r)))\\ &= \eta(u(x))(i(m(y, r))v(u(y)ru(y)^{-1}))\\ &= \eta(u(x))(i(m(y, r)))\eta(u(x))(v(u(y)ru(y)^{-1}))\\ &= i(x \cdot m(y, r))i(m(x, u(y)ru(y)^{-1}))v(u(x)u(y)ru(y)^{-1}u(x)^{-1}),\end{aligned}$$

where we defined the action of G on V via the section u, that is,

$$i(x \cdot m(y, r)) := \eta(u(x))(i(m(y, r))).$$

The right-hand side gives

$$\eta(\alpha(x, y))\eta(u(xy))(v(r)) = \eta(\alpha(x, y))(\eta(u(xy))(v(r)))$$

$$\begin{aligned}
&= \eta(\alpha(x,y))(i(m(xy,r))v(u(xy)ru(xy^{-1})))\\
&= \eta(\mu(\beta(x,y)))(i(m(xy,r)))\eta(\mu(\beta(x,y)))(v(u(xy)ru(xy)^{-1}))\\
&= \beta(x,y)i(m(xy,r))\beta(x,y)^{-1}\eta(\mu(\beta(x,y)))(v(u(xy)ru(xy)^{-1}))\\
&= i(m(xy,r))\eta(\mu(\beta(x,y)))(v(u(xy)ru(xy)^{-1}))\\
&= i(m(xy,r))\beta(x,y)v(u(xy)ru(xy)^{-1})\beta(x,y)^{-1}\\
&= i(m(xy,r))v(\alpha(x,y)u(xy)ru(xy)^{-1}\alpha(x,y)^{-1})\\
&= i(m(xy,r))v(u(x)u(y)ru(y)^{-1}u(x)^{-1}).
\end{aligned}$$

Here, we have used property (b) of a crossed module, that $i(V)$ is central in M, and the fact that v is a section of μ.

Canceling the term $v(u(x)u(y)ru(y)^{-1}u(x)^{-1})$ on both sides and using the injectivity of i shows the claim. □

The equation from Lemma 1.5.5 is a cocycle condition for $m \in C^1(G, \mathrm{Hom}(R, V))$ letting G act on $\mathrm{Hom}(R, V)$ via

$$(x \cdot m)(y, r) := x \cdot m(y, r) - m(x, u(y)ru(y)^{-1})$$

for all $x, y \in G$ and $r \in R$.

Definition 1.5.2. The class $[m] \in H^1(G, \mathrm{Hom}(R, V))$ is called the *deviation class* associated to the crossed module

$$0 \to V \xrightarrow{i} M \xrightarrow{\mu} F \to G \to 1.$$

For a given class $[m] \in H^1(G, \mathrm{Hom}(R, V))$, we will now construct a crossed module whose deviation class is $[m]$.

Theorem 1.5.6. *Let G be a group, V be a G-module and $v : F \to G$ be a free presentation of G with relation group R. For any class $[m] \in H^1(G, \mathrm{Hom}(R, V))$, there exists a crossed module*

$$0 \to V \to M \to F \xrightarrow{v} G \to 1$$

with deviation class $[m]$.

Proof. Define M to be the direct product $M = V \times R$, and let F act on M via the group homomorphism $\eta : F \to \mathrm{Aut}(M)$ defined by

$$(h, r) \mapsto \eta(a)(h, r) := (v(a) \cdot h + m(v(a), r), ara^{-1})$$

for all $a \in F$, $h \in V$ and $r \in R$; cf. equation (1.1). Observe that the conjugation in the r-variable of m is trivial due to the homomorphism property (1.11). It follows from

Lemma 1.3.1 that η is an action. The map $\mu : M \to F$ is given by the inclusion of the projection to the second factor. This leads to a crossed module

$$0 \to V \to M \xrightarrow{\mu} F \xrightarrow{\nu} G \to 1,$$

as one verifies easily conditions (a) and (b): $\mu(\eta(a)(h,r)) = ara^{-1} = a\mu(h,r)a^{-1}$ for all $a \in F$, $r \in R$ and $h \in V$ gives (a), and $\eta(r)(h,r') = (1 \cdot h + m(1,r'), rr'r^{-1}) = (0,r)(h,r')(0,r)^{-1}$ for all $r,r' \in R$ and all $h \in V$ gives (b), because Lemma 1.5.5 implies $m(1,r') = 0$ for all $r' \in R$. By construction, the deviation class of the crossed module is represented by m. □

Next, we link $H^1(G, \mathrm{Hom}(R,V))$ to $H^3(G,V)$ using cup-products: Let as before $u : G \to F$ be a set-theoretical section of $\nu : F \to G$ with $u(1) = 1$, and let $\alpha(x,y) \in \mathrm{Im}(\mu) = R$ denote the failure of u to be a group homomorphism for all $x,y \in F$. A cocycle $m \in Z^1(G, \mathrm{Hom}(R,V))$ determines a 3-cochain $\Lambda(m) \in C^3(G,V)$ via the formula

$$\Lambda(m)(x,y,z) := m(x, \alpha(y,z))$$

for all $x,y,z \in G$.

Lemma 1.5.7. *The map $\Lambda(m) \in C^3(G,V)$ is a 3-cocycle.*

Proof. In order to simplify notation, let us write here Λ for $\Lambda(m)$. We have for all $x_1, x_2, x_3, x_4 \in G$:

$$\begin{aligned}
d\Lambda(x_1,x_2,x_3,x_4) &= x_1 \cdot \Lambda(x_2,x_3,x_4) - \Lambda(x_1x_2,x_3,x_4) + \Lambda(x_1,x_2x_3,x_4) \\
&\quad - \Lambda(x_1,x_2,x_3x_4) + \Lambda(x_1,x_2,x_3) \\
&= x_1 \cdot m(x_2,\alpha(x_3,x_4)) - m(x_1x_2,\alpha(x_3,x_4)) \\
&\quad + m(x_1,\alpha(x_2x_3,x_4)) - m(x_1,\alpha(x_2,x_3x_4)) + m(x_1,\alpha(x_2,x_3)) \\
&= m(x_1x_2,\alpha(x_3,x_4)) - m(x_1,u(x_2)\alpha(x_3,x_4)u(x_2)^{-1}) \\
&\quad - m(x_1x_2,\alpha(x_3,x_4)) + m(x_1,\alpha(x_2x_3,x_4)) \\
&\quad - m(x_1,\alpha(x_2,x_3x_4)) + m(x_1,\alpha(x_2,x_3)) \\
&= -m(x_1,u(x_2)\alpha(x_3,x_4)u(x_2)^{-1}) + m(x_1,\alpha(x_2x_3,x_4)) \\
&\quad - m(x_1,\alpha(x_2,x_3x_4)) + m(x_1,\alpha(x_2,x_3)) \\
&= -m(x_1,u(x_2)\alpha(x_3,x_4)u(x_2)^{-1}) + m(x_1,\alpha(x_2,x_3)) \\
&\quad + m(x_1,\alpha(x_2x_3,x_4)) - m(x_1,\alpha(x_2,x_3x_4)) \\
&= m(x_1,u(x_2)\alpha(x_3,x_4)^{-1}u(x_2)^{-1}\alpha(x_2,x_3)\alpha(x_2x_3,x_4)\alpha(x_2,x_3x_4)^{-1}) \\
&= m(x_1,1) = 0.
\end{aligned}$$

Here, we used the cocycle identity in order to express the term $x_1 \cdot m(x_2,\alpha(x_3,x_4))$, then we used that $m(x,-)$ are homomorphisms, then we rearranged terms in the abelian group V, and finally we expressed $\alpha(x,y) = u(x)u(y)u(xy)^{-1}$. □

In a similar way, one shows that for a 1-coboundary $m \in B^1(G, \mathrm{Hom}(R, V))$, the corresponding Λ is a 3-coboundary. The upshot is a well-defined map

$$\Lambda : H^1(G, \mathrm{Hom}(R, V)) \to H^3(G, V).$$

Theorem 1.5.8. *The map Λ is an isomorphism.*

Proof. This is Theorem 11.2 in [69]. □

All this preparation served to show the following theorem.

Theorem 1.5.9. *Let G be a group, equipped with a free presentation $\nu : F \to G$ and relation group R, let V be a G-module and $[\gamma] \in H^3(G, V)$ a 3-class.*

Then there exists a crossed module

$$0 \to V \to M \to F \to G \to 1$$

representing the class $[\gamma]$.

Proof. Given $[\gamma] \in H^3(G, V)$, we use the previous theorem to associate to γ an element $m \in Z^1(G, \mathrm{Hom}(R, V))$ such that $\Lambda(m) = \gamma$. By Theorem 1.5.6, m gives rise to a crossed module

$$0 \to V \to M \to F \to G \to 1$$

representing the deviation class $[m] \in H^1(G, \mathrm{Hom}(R, V))$, where F is some free group. It remains to show that the Λ carries the deviation class of a crossed module of the form

$$0 \to V \to M \to F \to G \to 1$$

to the obstruction class $[d_S\beta]$ of the crossed module.

Let u and v be the maps used in the construction of the deviation class. For the construction of the obstruction class, we may use as $\beta(x, y)$ the expression $v(\alpha(x, y))$ for all $x, y \in G$. Then we have for all $x, y, z \in G$,

$$\begin{aligned} \eta(u(x))(v(\alpha(y,z)))v(\alpha(x,yz)) &= i(m(x,\alpha(y,z)))v(u(x)\alpha(y,z)u(x)^{-1}\alpha(x,yz)) \\ &= i(m(x,\alpha(y,z)))v(\alpha(x,y))v(\alpha(xy,z)), \end{aligned}$$

where we have used the identity

$$u(x)\alpha(y,z)u(x)^{-1}\alpha(x,yz) = \alpha(x,y)\alpha(xy,z)$$

(which follows simply by expressing $\alpha(x, y) = u(x)u(y)u(xy)^{-1}$ as before), the definition of the deviation m in equation (1.10), and the fact that v is a group homomorphism.

Associated to the extension $L \subset N$, there is a Lyndon–Hochschild–Serre spectral sequence. By Hochschild–Serre's theorem, its E_2-term reads

$$E_2^{p,q} = H^p(G, H^q(L, V)).$$

On the other hand, the central extension $\widehat{L}$ is described (up to equivalence) by a 2-cohomology class $[\omega] \in H^2(L, V)$. Observe that the first condition to have a crossed module $\widehat{L} \to N$ is that the L-action on $\widehat{L}$ extends to an N-action on $\widehat{L}$ by automorphisms. This is condition (a) in Lemma 1.3.1. We claim that this condition is equivalent to the condition that the class $[\omega] \in H^2(L, V)$ is N-invariant. Indeed, Lemma 1.3.1 states this condition (a) as for all $x \in N$, $x \cdot \omega - \omega = d\theta_x$, which means precisely that the class $[\omega] \in H^2(L, V)$ is N-invariant. Recall that the group L acts trivially on its cohomology $H^2(L, V)$, therefore the quotient group $G = N/L$ acts on $H^2(L, V)$. We thus obtain in particular that $[\omega] \in H^2(L, V)$ is G-invariant. From the point of view of the spectral sequence, this condition means that

$$[\omega] \in H^0(G, H^2(L, V)) = E_2^{0,2}.$$

By the very position of $E_2^{0,2}$ in the spectral sequence, no nonzero differential has its image in $E_2^{0,2}$. On the other hand, there are only two possibly nonzero differentials starting from $E_2^{0,2}$. Consider the differentials $d_2^{0,2} : E_2^{0,2} \to E_2^{2,1}$ and $d_3^{0,2} : E_3^{0,2} \to E_3^{3,0}$. The situation may be depicted as follows:

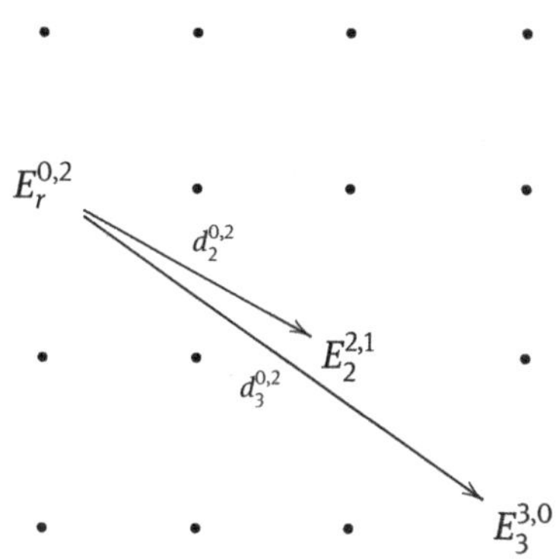

for $r = 2, 3$.

Definition 1.6.1. An element $e \in E_2^{0,2}$ is called *transgressive* in case $d_2^{0,2}e = 0$. Given a transgressive element $e \in E_2^{0,2}$, the *transgression* of e is the image $d_3^{0,2}e$ in

$$E_3^{3,0} \cong H^3(G, V)/d_2^{1,1}(H^1(G, H^1(L, V))).$$

Now we can state the main theorem of this section.

Theorem 1.6.2. *Suppose that $V = \mathbb{C}^\times$ or S^1 and that the action of N on V is trivial.*

Given the data which we listed above, the conditions for $\hat{L} \to N$ to give a crossed module are equivalent to the conditions that the class $[\omega]$ of the central extension $\hat{L}$ is N-invariant and transgressive.

Moreover, if the conditions hold true, the characteristic class of the crossed module $\hat{L} \to N$ is then the transgression of $[\omega]$.

Proof. Denote the conjugation action of L on $\hat{L}$ as usual by conj. Any lift of $l \in L$ to $\tilde{l} \in \hat{L}$ permits to put unambiguously

$$\mathrm{conj}_l(\tilde{l}') := \tilde{l}\tilde{l}'(\tilde{l})^{-1},$$

for all $\tilde{l}' \in \hat{L}$, because the extension is central. On the other hand, N acts on its normal subgroup L by conjugation, denote this action by ρ:

$$\rho_n(l) = nln^{-1},$$

for all $n \in N$ and all $l \in L$. The action ρ induces an action of N on the cohomology of L, and thus N acts on set $H^2(L, V)$ of classes of V-central extensions, in particular on $[\omega] \in H^2(L, V)$. Denote by $\hat{L}_x$ the image of the V-central extension $\hat{L}$ under the map ρ_x.

We have already seen in the discussion before the statement of the theorem that the N-invariance of the class $[\omega]$ is exactly condition (a) in Lemma 1.3.1. This condition (a) means that for all $x \in N$, the automorphism $\rho_x \in \mathrm{Aut}(L)$ lifts to an automorphism $\tilde{\rho}_x \in \mathrm{Aut}(\hat{L})$. More precisely, $\tilde{\rho}_x$ sends $\hat{L}$ first to $\hat{L}_x$, which is then sent to $\hat{L}$ by equivalence of central extensions. The different liftings correspond to automorphisms of a fixed central extension and, therefore, form a torsor under $H^1(L, V)$.

Concerning the transgressivity of $[\omega]$, consider now this collection of lifted automorphisms $\tilde{\rho}_x$ on $\hat{L}$, parametrized by $x \in N$. Choose the lifting $\tilde{\rho}_x$ such that

$$\tilde{\rho}_1 = 1, \quad \tilde{\rho}_{xl} = \tilde{\rho}_x \,\mathrm{conj}_l \quad \text{for all } l \in L, x \in N. \tag{1.12}$$

For any two $n, n' \in N$, both $\widetilde{\rho_{nn'}}$ and $\widetilde{\rho_n} \circ \widetilde{\rho_{n'}}$ are liftings of $\rho_{nn'}$. By the preceding discussion, there exists then a unique $\bar{b}(n, n') \in H^1(L, V)$ sending $\widetilde{\rho_n} \circ \widetilde{\rho_{n'}}$ to $\widetilde{\rho_{nn'}}$. This element $\bar{b}(n, n')$ is the failure of $\tilde{\rho}$ to be a genuine action, that is, to have

$$\widetilde{\rho_n} \circ \widetilde{\rho_{n'}} = \widetilde{\rho_{nn'}}.$$

In other words, $\bar{b}(n, n')$ (seen as a map $\hat{L} \to \hat{L}$) makes the following diagram:

$$\begin{array}{ccc} \hat{L} & \xrightarrow{\bar{b}(n,n')} & \hat{L} \\ \Big\downarrow{\scriptstyle \widetilde{\rho_{nn'}}} & & \Big\downarrow{\scriptstyle \widetilde{\rho_{n'}}} \\ \hat{L}_{nn'} & \xleftarrow[\widetilde{\rho_n}]{} & \hat{L}_{n'} \end{array}$$

commutative. This construction gives rise to $\bar{b} \in C^2(N, H^1(L, V))$. One can check that

(a) $\bar{b}(ln, n') = \bar{b}(n, n'l) = \bar{b}(n, n')$ for all $l \in L$ and all $n, n' \in N$. Therefore, $\bar{b}$ is in fact in $C^2(G, H^1(L, V))$.
(b) Define $b(n, n') := \bar{b}((n')^{-1}, n^{-1})$. Then b is a cocycle. Choosing different liftings $\widetilde{\rho_n}$ satisfying (1.12), the corresponding cocycles b differ by a coboundary. Therefore, $[b]$ is a well-defined cohomology class in $H^2(G, H^1(L, V))$.
(c) If $[b] = 0$ in $H^2(G, H^1(L, V))$, we can choose a compatible family of liftings $\widetilde{\rho_n}$ such that the map $\tilde{\rho}$, defined by $\tilde{\rho}_n := \widetilde{\rho_n}$, lifts ρ to a genuine action.
(d) All the liftings form a torsor under $Z^1(G, H^1(L, V))$.
(e) $[b] = d_2([\omega])$.

We leave the verification of the properties (a) to (d) to the reader. A proof may be found in [32], proof of Proposition 5.2. Property (c) together with property (e) shows our second claim, namely, $d_2([\omega]) = 0$ implies the existence of a lifting, which is an *action* of N on $\hat{L}$. Let us prove property (e).

We choose for any $n \in N$ a lifting $\widetilde{\rho_n} : \hat{L} \to \hat{L}$, which satisfies (1.12). Choose a set-theoretical section $s : G \to N$ of the quotient map $\pi : N \to N/L = G$ with $s(1) = 1$. Then any element of N can be written as $s(x)l$ for $x \in G$ and $l \in L$. For $x, x' \in G$, we denote by

$$t(x, x') := s(xx')^{-1} s(x) s(x') \in L$$

the failure of s to be a homomorphism.

Remark 1.6.2. This is not the same "failure to be a group homomorphism," which we used before (e. g., to define $\alpha(x, y) = s(x)s(y)s(xy)^{-1}$). Let us show how to transform t into α. Let $\mathrm{inv} : G \to G$ and $\mathrm{inv} : N \to N$ denote the inversion maps. Instead of s, we take the section (!) $\tilde{s} := \mathrm{inv} \circ s \circ \mathrm{inv}$. With respect to $\tilde{s}$, t reads

$$t_{\tilde{s}}(x, y) = s((xy)^{-1}) s(x^{-1})^{-1} s(y^{-1})^{-1} = (s(y^{-1}) s(x^{-1}) s(y^{-1}x^{-1})^{-1})^{-1}.$$

We obtain thus

$$(t_{\tilde{s}}(y^{-1}, x^{-1}))^{-1} = s(x)s(y)s(xy)^{-1} = \alpha(x, y).$$

We also choose a section $L \to \hat{L}$. To every element $l \in L$, we thus associate $\tilde{l} \in \hat{L}$ which we also write $\tilde{l} = (0, l)$ in product coordinates for the central extension. Now define $\tilde{\omega} \in C^2(N, V)$ (see [32]) by

$$\tilde{\omega}(s(x)l, s(x')l') := \frac{\widetilde{t(x, x')}\widetilde{\rho_{s(x')^{-1}}}(\tilde{l})\tilde{l}'}{(0, t(x, x')\rho_{s(x')^{-1}}(l)l')}.$$

By this formula, we mean the "difference" between the product of the lifted terms and the lifting of the product. This "difference" is necessarily in V—recall that we assumed that V is $\mathbb{C}^\times$ or S^1. It is clear that the restriction of $\tilde{\omega}$ to L is a cocycle representing the class $[\omega]$. This is the key point for computing $d_2([\omega])$. Observe further that $\tilde{\omega}(s(x), s(x')) = 1$ and also

$$\tilde{\omega}(l, s(x')) = \frac{\widetilde{\rho_{s(x')^{-1}}}(\tilde{l})}{(0, \rho_{s(x')^{-1}}(l))}.$$

It is easy to check that $d\tilde{\omega} \in F^2C^3(N, V)$ (see [32]). Now by definition of the differential d_2, a cocycle representing $d_2([\omega]) \in H^2(G, H^1(L, V))$ can be chosen as $d\tilde{\omega}(l, s(x), s(x'))$ where $l \in L$ and $x, x' \in G$. Let us compute this cocycle:

$$\begin{aligned}
d\tilde{\omega}(l, s(x), s(x')) &= \frac{\tilde{\omega}(s(x), s(x'))\tilde{\omega}(l, s(x)s(x'))}{\tilde{\omega}(ls(x), s(x'))\tilde{\omega}(l, s(x))} \\
&= \frac{\widetilde{\rho_{s(xx')^{-1}}}(\tilde{l})\widetilde{t(x,x')}}{(0, t(x,x'))\widetilde{\rho_{s(x')^{-1}}}(0, \rho_{s(x)^{-1}}(l))} \frac{(0, t(x,x')\rho_{s(x')^{-1}}(\rho_{s(x)^{-1}}(l)))}{(0, \rho_{s(xx')^{-1}}(l)t(x,x'))} \\
&\quad \frac{(0, \rho_{s(x)^{-1}}(l))}{\widetilde{\rho_{s(x)^{-1}}}(\tilde{l})} \\
&= \frac{\widetilde{\rho_{s(xx')^{-1}}}(\tilde{l})\widetilde{t(x,x')}}{(0, t(x,x'))\widetilde{\rho_{s(x')^{-1}}}(0, \rho_{s(x)^{-1}}(l))} \frac{(0, \rho_{s(x)^{-1}}(l))}{\widetilde{\rho_{s(x)^{-1}}}(\tilde{l})} \\
&= \frac{\widetilde{\rho_{s(xx')^{-1}}}(\tilde{l})\widetilde{t(x,x')}}{\widetilde{t(x,x')}\widetilde{\rho_{s(x')^{-1}}}(0, \rho_{s(x)^{-1}}(l))} \widetilde{\rho_{s(x')^{-1}}}\left(\frac{(0, \rho_{s(x)^{-1}}(l))}{\widetilde{\rho_{s(x)^{-1}}}(\tilde{l})}\right) \\
&= \frac{\widetilde{\rho_{s(xx')^{-1}}}(\tilde{l})\widetilde{t(x,x')}}{\widetilde{t(x,x')}\widetilde{\rho_{s(x')^{-1}}}(\widetilde{\rho_{s(x)^{-1}}}(\tilde{l}))} \\
&= \widetilde{\rho_{s(x')^{-1}}}(\widetilde{\rho_{s(x)^{-1}}}(b(x,x')(l))) \\
&= b(x,x')(l).
\end{aligned}$$

Here, we used that L acts trivially on V, and we wrote the terms multiplicatively, as all the computation takes place in $V = \mathbb{C}^\times$ or S^1. From the first to the second line, we used that $s(x)s(x') = s(xx')t(x,x')$ in order to write the $s(x)s(x')$ in the form $s(y)l$, which permits to apply the definition of $\tilde{\omega}$. We also used $ls(x) = s(x)(s(x)^{-1}ls(x))$ in order to write $ls(x)$ in the form $s(y)l'$. Then we used $s(x)^{-1}ls(x) = \rho_{s(x)^{-1}}(l)$ by definition of ρ.

From the second to the fourth line, we used

$$\rho_{s(xx')^{-1}}(l)t(x,x') = t(x,x')\rho_{s(x')^{-1}}(\rho_{s(x)^{-1}}(l)),$$

which follows from expressing all factors in terms of s.

This concludes the proof of the claim that the existence of a lift of the action of N on L to $\hat{L}$ is equivalent to the N-invariance and transgressivity of $[\omega]$. It remains now to show that $d_3^{0,2}([\omega])$ equals the class $[d_S\beta] \in H^3(G, V)$.

For this, let as before $s : G \to N$ be a section (with $s(1) = 1$), $\alpha(x,y) = s(x)s(y)s(xy)^{-1}$ be the failure of s to be a group homomorphism, and $\beta(x,y)$ such that $\mu \circ \beta = \alpha$. The cocycle $d_S\beta$ is then by definition

$$d_S\beta(x,y,z) = S(x)(\beta(y,z))\beta(x,yz)\beta(xy,z)^{-1}\beta(x,y)^{-1},$$

where $x,y,z \in G$ and $S = \tilde{\rho}$ is the outer action defined via the section s and the action of N on $\hat{L}$. Using a different lifting $\tilde{\rho}'$ of the action ρ, we know that $\tilde{\rho}'_n = \tilde{\rho}_n u(n^{-1})$ for some $u \in Z^1(G, H^1(L,V))$. Therefore, we have

$$d_{\tilde{\rho}'}\beta(x,y,z)\, u(s(x))(\alpha(y,z)) = d_{\tilde{\rho}}\beta(x,y,z).$$

Applying the same method as above, one can show that $u(s(x))(\alpha(y,z))$ is a cocycle representing $d_2^{1,1}([u])$. This means that the image of $d_S\beta$ in

$$E_3^{3,0} \cong H^3(G,V)/d_2^{1,1}(H^1(G,H^1(L,V)))$$

is independent of the choice of the lifting. In order to show that this image is $d_3^{0,2}([\omega])$, we use once again a cochain similar to $\tilde{\omega}$, the difference being that now we have a uniform lift $\tilde{\rho}$ of the action ρ. Our new cochain $\tilde{\tilde{\omega}}$ reads

$$\tilde{\tilde{\omega}}(s(x)l, s(x')l') = \frac{\widetilde{\tilde{\rho}_{s(x)^{-1}}(\tilde{l'})\tilde{l}t(x,x')}}{(0, \rho_{s(x)^{-1}}(l')l\,t(x,x'))}.$$

$\tilde{\tilde{\omega}}$ has the same properties as $\tilde{\omega}$. Let us finally show that

$$d\tilde{\tilde{\omega}}(s(x),s(y),s(z)) = d_S\beta(x,y,z).$$

Indeed, noting that $\tilde{\tilde{\omega}}(s(r),s(s)) = 1$ for all $r,s \in G$, we obtain

$$\begin{aligned} d\tilde{\tilde{\omega}}(s(x),s(y),s(z)) &= \frac{s(x)\cdot\tilde{\tilde{\omega}}(s(y),s(z))\tilde{\tilde{\omega}}(s(x),s(y)s(z))}{\tilde{\tilde{\omega}}(s(x)s(y),s(z))\tilde{\tilde{\omega}}(s(y),s(z))} \\ &= \frac{\tilde{\tilde{\omega}}(s(x),s(y)s(z))}{\tilde{\tilde{\omega}}(s(x)s(y),s(z))} \\ &= \frac{\rho_{s(x)^{-1}}(\widetilde{t(y,z)})\widetilde{t(x,yz)}}{\widetilde{t(x,y)}\widetilde{t(xy,z)}} \\ &= \rho_{s(x)^{-1}}(\widetilde{t(y,z)})\widetilde{t(x,yz)}\widetilde{t(xy,z)}^{-1}\widetilde{t(x,y)}^{-1} \\ &= (d_{\tilde{\rho}}\tilde{t})(x,y,z). \end{aligned}$$

Here, we used from the second to the third line that the values are already in V *a fortiori* and that therefore all terms of the type $(0,l)$ can be ignored.

In order to pass to $d_S\beta$, it remains to remark that the passage from t to α amounts to changing the section s, a procedure which does not change the cohomology class. □

Remark 1.6.3. In conclusion, given a 3-cohomology class in $H^3(G, V)$, we can associate to it a crossed module and then take the corresponding short exact sequence

$$1 \to L \to N \to G \to 1$$

and its associated Lyndon–Hochschild–Serre filtration and spectral sequence. The theorem above then shows that all these classes will land in the factor $E_\infty^{3,0}$ in

$$H^3(N, V) \cong \bigoplus_{i+j=3} E_\infty^{i,j},$$

as they should.

Note that results similar to those of this section have been obtained by Ratcliffe [91] and Huebschmann [54].

1.7 Examples

We will present here some examples illustrating the topics discussed in the preceding sections. Among other things, we will construct in this section crossed modules which are related to two famous 3-cohomology classes, namely to the string class in $H^3(G, \mathbb{R})$ (de Rham cohomology group) for a connected, simply connected, finite dimensional, simple compact Lie group G, and a crossed module representing the Godbillon–Vey class in $H^3(\widetilde{\mathrm{Diff}(S^1)}, \mathbb{R})$. Both are linked to the corresponding Lie algebra cohomology classes, and we suggest to read the passage on the crossed modules for the Lie algebra cocycles first.

The *string group* arises in the following context. Recall that for a connected, simply connected, finite dimensional, semisimple compact Lie group G, $\pi_3(G)$ counts the number of simple factors of G. Starting with an arbitrary topological group K, the connected component of $1 \in K$, denoted K_1, has $\pi_i(K) = \pi_i(K_1)$ for all $i > 0$, but $\pi_0(K_1) = 0$. In the same way (in case K is, e. g., locally contractible and connected), the universal covering $\tilde{K}$ of K, satisfies $\pi_i(\tilde{K}) = \pi_i(K)$ for all $i > 1$, but $\pi_1(\tilde{K}) = \pi_0(\tilde{K}) = 0$. All finite dimensional Lie groups K have $\pi_2(K) = 0$. The string group $\mathrm{Str}(G)$ of G is defined in the same way as being a group S such that $\pi_i(S) = \pi_i(G)$ for all $i > 3$, but $\pi_3(S) = \pi_2(S) = \pi_1(S) = \pi_0(S) = 0$. By what we remarked before, the group S cannot be a finite dimensional Lie group, because its maximal compact subgroup (which is homotopy equivalent to S) should contain a simple factor. The string group $S(G)$ for a connected, simply connected, finite dimensional, semisimple compact Lie group G has many realizations ("models") coming from homotopy theory. We will here discuss two of them from the point of view of crossed modules. As these require many deep mathematical results, our exposition is necessarily somewhat sketchy.

1.7.1 The string group (after Stolz–Teichner)

The goal of this section is to sketch the construction of the string group model of Stolz–Teichner, exposed in [96], Section 5, as an example of a nontrivial general extension. For more details and some explicit computations, we refer to [96].

Recall the notion of a von Neumann algebra. For further information about these matters, we refer to [20]. Let H be a (infinite dimensional, separable, complex) Hilbert space, and let $\mathcal{L}(H)$ be the C^*-algebra of bounded operators $H \to H$. Let

$$\mathcal{L}(H)_* := \{\rho \in \mathcal{L}(H) \mid \mathrm{Tr}|\rho| < \infty\} \subset \mathcal{L}(H)^*$$

be the *pre-dual* of $\mathcal{L}(H)$. $\mathcal{L}(H)_*$ is seen as a subspace of the full dual $\mathcal{L}(H)^*$ by regarding a $\rho \in \mathcal{L}(H)_*$ as the linear form $T \mapsto \mathrm{Tr}|\rho T|$ for all $T \in \mathcal{L}(H)$. $\mathcal{L}(H)_*$ is the unique subspace of $\mathcal{L}(H)^*$ whose dual is $\mathcal{L}(H)$. $\mathcal{L}(H)_*$ is Banach and separable and behaves in many respects better than $\mathcal{L}(H)$ itself. Give $\mathcal{L}(H)_*$ the *ultraweak topology* $\sigma(\mathcal{L}(H), \mathcal{L}(H)_*)$, that is, the coarsest topology on $\mathcal{L}(H)_*$ such that the evaluation maps $T \mapsto \mathrm{Tr}(\rho T)$ are continuous for all $T \in \mathcal{L}(H)$ and all $\rho \in \mathcal{L}(H)_*$.

Recall further the notion of the *commutant*: For a subspace S of $\mathcal{L}(H)$, the commutant S' is defined as

$$S' := \{T \in \mathcal{L}(H) \mid TA = AT \text{ for all } A \in S\}.$$

The following double commutant theorem of von Neumann relates a topological condition and an algebraic condition on a $\star$-subalgebra M of $\mathcal{L}(H)$.

Theorem 1.7.1 (Double commutant theorem of von Neumann). *The following are equivalent for a $\star$-subalgebra M of $\mathcal{L}(H)$:*
(a) *M is $\sigma(\mathcal{L}(H), \mathcal{L}(H)_*)$-closed*
(b) $M = (M')'$

Definition 1.7.1. A *von Neumann algebra* on H is a $\star$-subalgebra M of $\mathcal{L}(H)$, containing $1 \in \mathcal{L}(H)$, such that conditions (a) and (b) of the above theorem hold for M.

Taking the double commutant of an arbitrary $\star$-subalgebra M, one obtains the von Neumann algebra generated by M. By the above theorem, this von Neumann algebra is also the $\sigma(\mathcal{L}(H), \mathcal{L}(H)_*)$-closure of M.

Recall now the *loop group* LG of a finite dimensional Lie group G. Most of the time, G is here considered to be connected, simply connected, compact and simple. The most important example of G in relation to the string group is the group $G = Spin(n)$, the universal covering group of $SO(n, \mathbb{R})$.

By definition $LG = C^\infty(S^1, G)$, the space of smooth maps from the unit circle S^1 to the group G. It is a group by pointwise multiplication of group-valued loops, and becomes a topological group (even an infinite-dimensional Lie group) in the

C^∞-topology. There is a vast literature on loop groups, their structure and representations, see [88] and references therein.

Let ρ be a projective unitary representation of LG, that is, a continuous homomorphism $\rho : LG \to PU(H)$ from LG to the projective unitary group $PU(H) := U(H)/\mathbb{T}$ of some (infinite dimensional separable) complex Hilbert space H. $\mathbb{T}$ denotes still the unit circle, but here seen as homothety operators on H. The group $PU(H)$ carries the quotient topology of the weak operator topology on the group of unitary operators $U(H)$ on H. We will suppose that ρ is defined even for all piecewise smooth loops on G. $PU(H)$ has a canonical central extension

$$1 \to \mathbb{T} \to U(H) \to PU(H) \to 1.$$

Pulling it back to LG via ρ, we obtain a central extension

$$1 \to \mathbb{T} \to \tilde{L}G \to LG \to 1$$

and a *unitary* representation $\tilde{\rho} : \tilde{L}G \to U(H)$ of this central extension.

Let $I \subset S^1$ be the upper semicircle consisting of all $z \in S^1$ with nonnegative imaginary part. Let $L_IG \subset LG$ be the subgroup consisting of those loops $\gamma : S^1 \to G$ with support in I, that is, $\gamma(z) = 1 \in G$ for all $z \notin I$. Let $\tilde{L}_IG \subset \tilde{L}G$ be the subgroup corresponding to L_IG. Define

$$A_\rho := \tilde{\rho}(\tilde{L}_IG)'' \subset \mathcal{L}(H)$$

to be the von Neumann algebra generated by the operators $\tilde{\rho}(\gamma)$ with $\gamma \in \tilde{L}_IG$. By what we have recalled above, this is just the weak operator closure of linear combinations of group elements $\tilde{\rho}(\gamma)$. In fact, A_ρ is a so-called hyperfinite III_1 factor in the classification of subfactors of von Neumann algebras.

In order to construct the Stolz–Teichner model of the string group (see [96]), we start from the group extension

$$1 \to L_IG \overset{\mathrm{res}_I}{\longrightarrow} P_1^IG \overset{\mathrm{ev}_{-1}}{\longrightarrow} G \to 1,$$

where $P_1^IG = \{\gamma : I \to G \mid \gamma(1) = 1 \in G\}$ and where res_I denotes the restriction to $I \subset S^1$ and ev_{-1} denotes the evaluation in $-1 \in I$.

The idea for the construction of the string group is to modify this extension by replacing the normal subgroup L_IG by the projective unitary group $PU(A_\rho)$ of the von Neumann algebra A_ρ. Here, the unitary group $U(A_\rho) \subset A_\rho$ is given by those $a \in A_\rho$ such that $aa^* = a^*a = 1$, and the projective unitary group $PU(A_\rho)$ is its quotient by its subgroup of multiples of $1 \in A_\rho$. The replacement uses the map

$$\rho : L_IG \to PU(A_\rho),$$

which is given by restricting the representation ρ to $L_IG \subset LG$. Note that by definition of A_ρ, we have $\rho(L_IG) \subset PU(A_\rho) \subset PU(H)$.

In order to realize this idea, observe that the conjugation action of $P_1^I G$ on its normal subgroup $L_I G$ extends to $PU(A_\rho)$. Indeed, $\gamma \in P_1^I G$ acts on a class $[a] \in PU(A_\rho)$ (with representative $a \in U(A_\rho)$) as follows: Lift γ first to a piecewise smooth loop $\gamma : S^1 \to G$ and then to $\tilde{\gamma} \in \tilde{L}G$. Then define

$$\gamma \cdot [a] := [\tilde{\rho}(\tilde{\gamma}) a \tilde{\rho}(\tilde{\gamma}^{-1})].$$

It is clear that $\tilde{\rho}(\tilde{\gamma}) a \tilde{\rho}(\tilde{\gamma}^{-1})$ is again a unitary element of $\mathcal{L}(H)$. It is shown in *loc. cit.* that it is in fact an element of A_ρ and independent of the extension to a piecewise smooth loop $S^1 \to G$.

Lemma 1.7.2. *With the above action of $P_1^I G$ on $PU(A_\rho)$, the representation $\rho : L_I G \to PU(A_\rho)$ is $P_1^I G$-equivariant. Therefore, there is a well-defined monomorphism*

$$r : L_I G \to PU(A_\rho) \rtimes P_1^I G, \quad r(\gamma) := \rho(\gamma^{-1}, \gamma)$$

into the semidirect product, whose image is a normal subgroup.

Proof. This is Lemma 5.26 in *loc. cit.* □

Definition 1.7.2. The (Stolz–Teichner model of the) *string group* G_ρ is by definition the group

$$G_\rho := PU(A_\rho) \rtimes P_1^I G / r(L_I G).$$

We therefore have a general extension

$$1 \to PU(A_\rho) \to G_\rho \to G \to 1,$$

where the projection onto G is given by $[(u, \gamma)] \mapsto \gamma(-1)$. Observe that this extension is not central, nor abelian.

Remark 1.7.3. It is the homotopy theoretical content of the above group extension which is most important. Indeed, the contractibility of $U(A_\rho)$ (a deep theorem of Wassermann for type III_1 subfactors; cf. Theorem 5.17 in *loc. cit.*) shows that $PU(A_\rho)$ is an Eilenberg–Mac Lane space $K(\mathbb{Z}, 2)$. The main result of Stolz–Teichner is that the above constructed group extension induces in the long exact sequence of homotopy groups a connecting homomorphism

$$\pi_3(G) \to \pi_2(PU(A_\rho)) \cong \mathbb{Z},$$

which corresponds as an element of $\mathrm{Hom}(\pi_3(G), \mathbb{Z}) \cong H^4(BG)$ to a generator, where $H^4(BG)$ is the singular cohomology of the classifying space BG of G.

It is shown in [53] that the continuous group cohomology of a compact connected group G is trivial, and thus (real valued) singular cohomology and Lie algebra cohomology are isomorphic. This explains why we find the string class from Lie algebra cohomology here in the homotopy of the topological space G, rather than in the group cohomology.

In order to understand the relation of Stolz–Teichner's model to string group models by crossed modules and to 3-cohomology, we refer to [94]. Schommer–Pries explains within other things that extensions of G by a $K(\mathbb{Z},2)$ are classified by the Segal–Mitchison 3-cohomology space $H^3_{SM}(G,S^1)$ (see Theorem 100). This is consistent with the preceding because $H^4(BG) \cong H^3_{SM}(G,S^1)$, as Schommer–Pries shows in Corollary 99.

1.7.2 The string group (after Baez–Crans–Schreiber–Stevenson)

The construction of the string group model of Baez–Crans–Schreiber–Stevenson [5] is geometrically inspired by a certain gerbe construction. The point is here that the Cartan cocycle

$$\theta(x,y,z) := \kappa([x,y],z),$$

constructed from the Killing form κ and the bracket $[,]$ and which generates $H^3(\mathfrak{g},\mathbb{R})$ for a real simple Lie algebra, may be seen as a biinvariant closed differential 3-form (still denoted) θ on the corresponding simple, simply connected, compact Lie group G and generates then the de Rham cohomology group $H^3(G,\mathbb{R})$. The Cartan cocycle will be studied with more detail in the chapter on crossed modules of Lie algebras; we will see there that θ is intimately linked to the *Kac–Moody cocycle* ω on the loop algebra $\Omega\mathfrak{g} = C^\infty(S^1,\mathfrak{g})$ defined by

$$\omega(f,g) = 2\int_{S^1} \kappa(f,dg).$$

In the following, we will also regard ω (by abuse of notation) as a (left-invariant) differential form on the corresponding loop group ΩG.

One of the crossed modules, which we will construct in the next chapter on Lie algebras, is in fact the derived crossed module of the crossed module of groups, which we construct here.

We begin with the construction of the Kac–Moody group $\widehat{\Omega G}$ according to Murray and Stevenson [79]. As before, most of the time G will denote a connected, simply connected, compact, simple Lie group.

Let $\mathcal{G}$ be a Lie group. Denote by $P_0\mathcal{G}$ the space of smooth based paths in $\mathcal{G}$:

$$P_0\mathcal{G} = \{f \in C^\infty([0,2\pi],\mathcal{G}) \mid f(0) = 1 \in \mathcal{G}\}.$$

$P_0\mathcal{G}$ is a group under pointwise multiplication of paths. The map $\pi : P_0\mathcal{G} \to \mathcal{G}$, which evaluates a paths f at its endpoint $f(2\pi)$ is a group homomorphism. The kernel of π is

$$\Omega\mathcal{G} = \{f \in C^\infty([0,2\pi],\mathcal{G}) \mid f(0) = f(2\pi) = 1\}.$$

It is thus a normal subgroup of $P_0\mathcal{G}$, and we have an exact sequence of groups

$$1 \to \Omega\mathcal{G} \to P_0\mathcal{G} \xrightarrow{\pi} \mathcal{G} \to 1,$$

which is some version of the path-loop fibration of the topological space $\mathcal{G}$. Note that $\Omega\mathcal{G}$ is rather different from the previously defined loop group LG: The group $\Omega\mathcal{G}$ consists of loops $f : S^1 \to \mathcal{G}$, which are smooth anywhere except the base point. At the basepoint, left and right derivatives exist to all orders, but do not necessarily agree.

We will apply the above to the (infinite dimensional) Lie group $\mathcal{G} = \Omega G$. Then an element of $P_0\mathcal{G}$ is a map $f : [0, 2\pi] \times S^1 \to G$ with $f(0, \theta) = 1$ for all $\theta \in S^1$ and $f(t, 0) = 1$ for all $t \in [0, 2\pi]$. It is shown in *loc. cit.* that the map

$$\zeta : P_0\Omega G \times P_0\Omega G \to U(1)$$

defined by

$$\zeta(f, g) = \exp\left(2i \int_0^{2\pi}\int_0^{2\pi} \kappa(f(t)^{-1}f'(t), g'(\theta)g(\theta)^{-1})d\theta dt\right)$$

is a group 2-cocycle. We therefore obtain a central extension $U(1) \times_\zeta P_0\Omega G$, which carries the multiplication

$$(z_1, f_1) \cdot (z_2, f_2) = (z_1 z_2 \zeta(f_1, f_2), f_1, f_2).$$

Let N be the subset of $U(1) \times_\zeta P_0\Omega G$ consisting of pairs (z, γ) where $z \in U(1)$ and $\gamma : [0, 2\pi] \to \Omega G$ is a loop based at $1 \in \Omega G$ such that

$$z = \exp\left(-i \int_{D_\gamma} \omega\right),$$

where D_γ is any disk in ΩG which γ bounds (and where we regard ω as a differential 2-form on ΩG). It is shown in *loc. cit.*, using [78], that N is a normal subgroup of the group $U(1) \times_\zeta P_0\Omega G$. To construct now $\widehat{\Omega G}$, we form the quotient group $(U(1) \times_\zeta P_0\Omega G)/N$. In [78], [79], it is shown that the resulting central extension is equivalent to the standard Kac–Moody central extension. This construction has the advantage of describing the Kac–Moody central extension, up to the normal subgroup N, as a topologically trivial fiber bundle.

Notice that the space of based paths P_0G acts on ΩG by conjugation. This action lifts to an action on $\widehat{\Omega G}$ by Proposition 24 in [5].

The upshot of this construction is (see the proof of Proposition 25 in [5]) the following.

Theorem 1.7.3. *The map $\mu : \widehat{\Omega G} \to P_0G$ obtained from splicing together the central extension*

$$1 \to U(1) \to \widehat{\Omega G} \to \Omega G \to 1$$

and the "path-loop fibration"

$$1 \to \Omega G \to P_0G \to G \to 1$$

gives a crossed module of groups.

Proof. Details of the computations are given in Proposition 25 of [5]. □

1.7.3 A crossed module of diffeomorphisms

In this subsection, we will use the construction of a Lie algebra crossed module representing the Godbillon–Vey cocycle in $H^3(\text{Vect}(S^1), \mathbb{R})$ in order to construct a group crossed module of diffeomorphism groups representing the Godbillon–Vey cocycle in group cohomology. We advise the reader to go over the Lie algebra construction first. Concerning continuous group cohomology, see Appendix A.

Let $\text{Diff}(S^1)$ denote the group of orientation preserving diffeomorphisms of the circle $S^1 = \mathbb{R}/\mathbb{Z}$. It is a Fréchet manifold, homotopy equivalent to S^1, and has as its universal covering group the group $\widetilde{\text{Diff}(S^1)} = \text{Diff}_{\mathbb{Z}}(\mathbb{R})$ of orientation preserving $\mathbb{Z}$-equivariant diffeomorphisms of $\mathbb{R}$. The universal covering sequence reads

$$0 \to \mathbb{Z} \to \widetilde{\text{Diff}(S^1)} \xrightarrow{\pi} \text{Diff}(S^1) \to 1.$$

For these basic properties of the group of diffeomorphisms, we refer to [40].

By the example of Section 2.8.3, we have a construction which corresponds to the Godbillon–Vey cocycle for W_1, the Lie algebra of vector fields on the line $\mathbb{R}$. As we already remarked in that section, one cannot transpose it directly to $\text{Vect}(S^1)$. Denote by $\mathcal{F}_\lambda$ the λ-density $\text{Vect}(S^1)$-module consisting as a vector space of smooth functions on S^1. Now the corresponding sequence of λ-density modules reads

$$0 \to \mathbb{R} \to \mathcal{F}_0 \to \mathcal{F}_1 \to H^1(S^1, \mathbb{R}) = \mathbb{R} \to 0.$$

Obviously, the last term prevents us from using the previous construction.

The trick is to pass in an appropriate manner to the universal covering. Lifting elements of $\text{Vect}(S^1)$ to 1-periodic functions on $\mathbb{R}$, we can make them act on F_0 and F_1, the modules of densities on $\mathbb{R}$.

Now let us pass to group level. The relation of continuous group cohomology, cohomology as a topological space and Lie algebra cohomology of the Lie algebra of a

Lie group is given by the *Hochschild–Mostow spectral sequence* (see [35] p. 295 Theorem 3.4.1). One can refine this sequence to start from the cohomology of the homogeneous space of the group by some compact subgroup, abutting then to the relative Lie algebra cohomology modulo this subgroup (see [35] p. 297 Theorem 3.4.2). An isomorphism between continuous group and relative Lie algebra cohomology stemming from acyclicity of the homogeneous space is then known as *van Est's theorem* (cf. [35] p. 298 Corollary).

The diffeomorphism group version of van Est's theorem ([12] Theorem 1 for the version with trivial coefficients which can be adapted to nontrivial coefficients; cf. [86] p. 286) with nontrivial coefficients implies the following.

Theorem 1.7.4.

$$H_c^*(\mathrm{Diff}(S^1); \underline{\mathcal{F}_\lambda}) \cong H^*(\mathrm{Vect}(S^1), SO(2); \mathcal{F}_\lambda).$$

Here, $H_c^*(\mathrm{Diff}(S^1); \underline{M})$ denotes group cohomology with continuous cochains and M is the corresponding Lie algebra module for a Lie group module $\underline{M}$. The $\mathrm{Vect}(S^1)$-module $\mathcal{F}_\lambda$ is the module of λ-densities; see, for example, [86]. The isomorphism is functorial.

By the version with trivial coefficients (cf. the corollary on p. 298 in [35]), we have an isomorphism

$$H^*(\mathrm{Diff}(S^1), \mathbb{R}) \cong H^*(\mathrm{Vect}(S^1), SO(2); \mathbb{R})$$

and applied to $\widetilde{\mathrm{Diff}(S^1)}$, we get an isomorphism

$$H^*(\widetilde{\mathrm{Diff}(S^1)}, \mathbb{R}) \cong H^*(\mathrm{Vect}(S^1); \mathbb{R}).$$

The last statement follows from the Hochschild–Mostow spectral sequence (see [35] p. 295 Theorem 3.4.1), using the fact that $\widetilde{\mathrm{Diff}(S^1)}$ is contractible.

Therefore, the Godbillon–Vey cocycle $\theta(0)$ arises in $H^*(\widetilde{\mathrm{Diff}(S^1)}, \mathbb{R})$, but not in $H^*(\mathrm{Diff}(S^1), \mathbb{R})$. But its class is sent to the Bott–Thurston class $[\mu]$ in the Gysin sequence

$$H^3(\widetilde{\mathrm{Diff}(S^1)}, \mathbb{R}) \to H^2(\mathrm{Diff}(S^1), \mathbb{R}), \quad [\theta(0)] \mapsto [\mu].$$

We have seen in Section 1.5 that in group cohomology crossed modules can be constructed by what we have called there the principal construction. This is completely analogous to the construction mechanism for Lie algebra crossed modules, which we introduced in Section 2.5. Let us thus consider the crossed module

$$0 \to \mathbb{R} \to \underline{F_0} \to \underline{F_1} \times_{\pi^* B_1} \widetilde{\mathrm{Diff}(S^1)} \to \widetilde{\mathrm{Diff}(S^1)} \to 1. \tag{1.13}$$

Here, $\pi^* : H^*(\mathrm{Diff}(S^1), \underline{F_1}) \to H^*(\widetilde{\mathrm{Diff}(S^1)}, \underline{F_1})$ is the map induced in cohomology by the covering projection and B_1 is the generator of $H^2(\mathrm{Diff}(S^1), \underline{F_1})$ given by Ovsienko and Roger in [86]. The modules are still modules of λ-densities on group level, but now over the real line instead of the circle S^1. B_1 corresponds to the cocycle we named α in Section 2.8.3.

The upshot of the discussion is the following theorem.

Theorem 1.7.5. *The crossed module* (1.13) *represents the Gobillon–Vey cocycle in* $H^3(\widetilde{\mathrm{Diff}(S^1)}, \mathbb{R})$.

Proof. The proof of the theorem follows from functoriality of the van Est isomorphism and Theorem 1.5.1. Indeed, by these two facts, we have a commutative diagram:

$$
\begin{array}{ccc}
[\alpha] & \longmapsto & [\pi^* B_1] \\
\top & & \top \\
H^2(\mathrm{Vect}(S^1), F_1) & \xrightarrow{\cong} & H^2(\widetilde{\mathrm{Diff}(S^1)}, \underline{F_1}) \\
\downarrow \partial & & \downarrow \partial \\
H^3(\mathrm{Vect}(S^1), \mathbb{R}) & \xrightarrow{\cong} & H^3(\widetilde{\mathrm{Diff}(S^1)}, \underline{\mathbb{R}}) \\
\downarrow & & \downarrow \\
[\partial\alpha] = [\theta(0)] & \longmapsto & [\partial\pi^* B_1]
\end{array}
$$

This shows the claim. □

1.8 Relation to relative cohomology

In this section, we report on a different approach to the relation between crossed modules and cohomology. Namely, when one looks for the data to fix in a crossed module $\mu : M \to N$ in order to make it correspond to a *relative* cohomology class, it comes out that this is not only the kernel $\mathrm{Ker}(\mu) = V$ and the cokernel $\mathrm{Coker}(\mu) = G$, but one has to fix the whole quotient morphism $\pi : N \to G$. Then the theory becomes much simpler, because fixing π, the set $M \approx V \times \mathrm{Im}(\mu)$ is already fixed. This approach is due to Loday in [63].

For the sake of this section only, we introduce a new notion of equivalence of crossed modules. Fix a group epimorphism $\pi : N \to G$ and a G-module V. A *relative crossed module* $\mu : M \to N$ is an ordinary crossed module where the epimorphism $\pi : N \to G$ is supposed to be fixed. The difference between relative and ordinary crossed modules becomes visible only in the equivalence relation.

Definition 1.8.1. Two (relative) crossed modules $\mu_i : M_i \to N$ for $i = 1,2$ are called *relatively equivalent* if there exists a group homomorphism $\varphi : M_1 \to M_2$ such that the diagram

$$\begin{array}{ccccccccc} 0 & \longrightarrow & V & \xrightarrow{i_1} & M_1 & \xrightarrow{\mu_1} & N & \xrightarrow{\pi} & G & \longrightarrow & 0 \\ & & \downarrow{\scriptstyle \mathrm{id}_V} & & \downarrow{\scriptstyle \varphi} & & \downarrow{\scriptstyle \mathrm{id}_N} & & \downarrow{\scriptstyle \mathrm{id}_G} & & \\ 0 & \longrightarrow & V & \xrightarrow{i_2} & M_2 & \xrightarrow{\mu_2} & N & \xrightarrow{\pi} & G & \longrightarrow & 0 \end{array}$$

is commutative and such that φ is equivariant for the actions η_i of N on M_i for $i = 1,2$.

Remark 1.8.2. By the Five Lemma, the homomorphism φ is necessarily an isomorphism. Therefore, it is clear that relative equivalence is an equivalence relation.

The main theorem of this section reads then the following.

Theorem 1.8.1. *The map associating to a relative crossed module the relative 3-cocycle $\kappa^* g$ (to be defined below) induces a natural isomorphism*

$$\mathrm{crmod}(G,N,V) \cong H^3(G,N,V)$$

between the abelian group of relative equivalence classes of crossed modules $\mu : M \to N$ with fixed quotient morphism $\pi : N \to G$ and fixed kernel $V = \mathrm{Ker}(\mu)$, and the third relative cohomology group $H^3(G,N,V)$.

The proof of this theorem will be complete at the end of this section. In order to prove this theorem, associate first of all to a crossed module $\mu : M \to N$ a relative cohomology class.

Choosing a normalized, set-theoretical section $\rho : G \to N$, $\rho(1) = 1$, the failure of ρ to be a group homomorphism is measured by

$$\alpha(x,y) = \rho(x)\rho(y)\rho(xy)^{-1}$$

for all $x,y \in G$. As $\pi \circ \alpha = 1$, there exists a normalized cochain $\beta : G \times G \to M$ with $\mu \circ \beta = \alpha$.

On the other hand, we have $\pi \circ \rho \circ \pi = \pi$. Therefore, for any $n \in N$, the elements $\rho(\pi(n))$ and n differ at most by an element in $\mathrm{Im}(\mu)$. Thus there exists a map $\psi : N \to M$, normalized by the requirement $\psi(1) = 1$, such that

$$\mu(\psi(n))\rho(\pi(n)) = n$$

for all $n \in N$.

Lemma 1.8.2. *There exists a unique normalized 2-cochain $g \in C^2(N,V)$ such that*

$$\eta(n)(\psi(n'))\psi(n)\beta(\pi(n),\pi(n')) = i(g(n,n'))\psi(nn'), \tag{1.14}$$

where $i : V \to M$ denotes the inclusion of $\mathrm{Ker}(\mu) = V$.

Proof. It is enough to show

$$\mu(\eta(n)(\psi(n'))\psi(n)\beta(\pi(n),\pi(n'))) = \mu(\psi(nn')).$$

Put $\bar{n} := \rho(\pi(n))^{-1}$. The defining equation for ψ then reads $\mu(\psi(n)) = n\bar{n}$. Property (a) of a crossed module reads in turn

$$\mu(\eta(n)(\psi(n'))) = n\mu(\psi(n'))n^{-1} = nn'\overline{n'}n^{-1}.$$

Put all together, we obtain

$$\begin{aligned}\mu(\eta(n)(\psi(n'))\psi(n)\beta(\pi(n),\pi(n'))) &= nn'\overline{n'}n^{-1}n\bar{n}\alpha(\pi(n),\pi(n'))\\ &= nn'\overline{n'}n^{-1}n\bar{n}\bar{n}^{-1}\overline{n'}^{-1}\overline{nn'}\\ &= \mu(\psi(nn')).\end{aligned}$$ □

This lemma leads to the definition of the relative cocycle associated to a crossed module. Recall the framework of relative group cohomology (all cochains we consider are normalized):

The epimorphism $\pi : N \to G$ induces a monomorphism of complexes $\pi^* : C^*(G,V) \to C^*(N,V)$, and the *complex of relative group cohomology* $C^*(G,N,V)$ is by definition the cokernel of π^*, that is, we have an exact sequence of complexes

$$0 \to C^*(G,V) \xrightarrow{\pi^*} C^*(N,V) \xrightarrow{\kappa^*} C^*(G,N,V) \to 0. \tag{1.15}$$

In order to have that crossed modules correspond to H^3, there is a degree shift while passing to cohomology: The cohomology of the relative complex at $C^k(G,N,V)$ is denoted $H^{k+1}(G,N,V)$.

Theorem 1.8.3. *The image $\kappa^*g \in C^2(G,N,V)$ of the cochain g is a cocycle whose cohomology class $[\kappa^*g] \in H^3(G,N,V)$ only depends on the relative equivalence class of the crossed module $\mu : M \to N$.*

Definition 1.8.3. The cohomology class $[\kappa^*g] \in H^3(G,N,V)$ is called the *characteristic class* of the relative crossed module $\mu : M \to N$. It is denoted by $\xi(M,\mu)$.

Before proving the theorem, we will establish the link between the characteristic class $\xi(M,\mu)$ and the previously defined class $[\gamma]$, which we have associated to the crossed module $\mu : M \to N$.

Lemma 1.8.4. *The equality $dg = \pi^*\gamma$ holds in $C^3(N,V)$.*

Proof. The element $i(dg)^{-1}$ is given by the formula

$$ig(n,n')ig(nn',n'') = i(dg(n,n',n'')^{-1})i(\eta(n)(g(n',n'')))ig(n,n'n'')$$

for all $n, n', n'' \in N$. Rewrite the above relation using the definition of g, that is, equation (1.14), and the relation

$$(\eta(n)(\psi(n')))\psi(n) = \psi(n)(\eta(\rho(\pi(n)))(\psi(n'))), \tag{1.16}$$

which follows directly from property (b) of a crossed module (by multiplying by $\psi(n)^{-1}$ from left and then applying (b)). To simplify expressions, we will write $n \cdot m$ for $\eta(n)(m)$ for all $n \in N$ and all $m \in M$. One obtains

$$\begin{aligned}
&(n \cdot \psi(n'))\psi(n)\beta(\pi(n), \pi(n'))\psi(nn')^{-1}\psi(nn')(\rho\pi(nn') \cdot \psi(n''))\\
&\quad \beta(\pi(nn'), \pi(n''))\psi(nn'n'')^{-1}\\
&\quad = idg(n, n', n'')^{-1} n \cdot (\psi(n')(\rho\pi(n') \cdot \psi(n''))\beta(\pi(n'), \pi(n''))\psi(n'n'')^{-1})\\
&\quad (n \cdot \psi(n'n''))\psi(n)\beta(\pi(n), \pi(n'n''))\psi(nn'n'')^{-1}.
\end{aligned}$$

Observe that in the same way, one may also replace $n \cdot m$ by $\psi(n)(\rho\pi(n) \cdot m)\psi(n)^{-1}$. Apply this in the expression after the equality sign. We obtain after simplification, using that $idg(n, n', n'')^{-1}$ is central in M,

$$\begin{aligned}
&idg(n, n', n'')\psi(n)\beta(\pi(n), \pi(n'))(\rho\pi(nn') \cdot \psi(n''))\beta(\pi(nn'), \pi(n''))\\
&\quad = \psi(n)((\rho\pi(n)\rho\pi(n') \cdot \psi(n''))(\rho\pi(n) \cdot \beta(\pi(n'), \pi(n''))))\\
&\quad \psi(n)^{-1}\psi(n)\beta(\pi(n), \pi(n'n'')).
\end{aligned}$$

Observe further that $\rho\pi(n)\rho\pi(n') \cdot \psi(n'')$ equals

$$\beta(\pi(n), \pi(n'))(\rho\pi(nn') \cdot \psi(n''))\beta(\pi(n), \pi(n'))^{-1}$$

by using as above property (b) of a crossed module together with $\mu\beta = \alpha$ and the definition of α. Replacing this in the second line, we obtain (again simplifying some terms)

$$\begin{aligned}
&idg(n, n', n'')\beta(\pi(n), \pi(n'))(\rho\pi(nn') \cdot \psi(n''))\beta(\pi(nn'), \pi(n''))\\
&\quad = \beta(\pi(n), \pi(n'))(\rho\pi(nn') \cdot \psi(n''))\beta(\pi(n), \pi(n'))^{-1}\\
&\quad (\rho\pi(n) \cdot \beta(\pi(n'), \pi(n'')))\beta(\pi(n), \pi(n'n'')).
\end{aligned}$$

This may be simplified (using again that $idg(n, n', n'')^{-1}$ is central in M) to give

$$\begin{aligned}
&\beta(\pi(n), \pi(n'))\beta(\pi(nn'), \pi(n''))\\
&\quad = idg(n, n', n'')^{-1}(\rho\pi(n) \cdot \beta(\pi(n'), \pi(n'')))\beta(\pi(n), \pi(n'n'')).
\end{aligned}$$

This means $dg = \pi^* \gamma$, because $i\gamma = d_{\eta\rho}\beta$ and i is injective. □

Proof of the theorem. Thanks to the lemma, it is clear that $\kappa^* g$ is a cocycle. Indeed,

$$d\kappa^* g = \kappa^* dg = \kappa^* \pi^* \gamma = 0.$$

We will show in the lemmas below the independence of the class $[\kappa^* g]$ of the choices of β, ψ and ρ.

Lemma 1.8.5. *Given ρ and ψ, a modification in the choice of β does not change $\kappa^* g$.*

Proof. Let β' be a normalized cochain with $\mu \circ \beta' = \alpha$. As the difference of β and β' is in the kernel of μ, there exists $h : G \times G \to V$ with $\beta' = \beta\, ih$ (pointwise multiplication in M). Passing to the corresponding cochains g respectively g', it follows from equation (1.14) that we have the relation $g' = g + \pi^* h$. Thus

$$\kappa^* g' = \kappa^* g + \kappa^* \pi^* h = \kappa^* g. \qquad \square$$

Lemma 1.8.6. *Given ρ and β, a different choice of ψ modifies $\kappa^* g$ by adding a coboundary.*

Proof. Let ψ' be a normalized cochain with $\mu\psi'(n)\rho\pi(n) = n$ for all $n \in N$. As above, there exists a map $\theta : N \to V$ with $\psi' = \psi\, i\theta$. Passing to the corresponding cochains g respectively g', it follows from equation (1.14) that we have the relation $g' = g + d\theta$. Thus

$$\kappa^* g' = \kappa^* g + d\kappa^* \theta. \qquad \square$$

Lemma 1.8.7. *Each different choice of ρ may be followed by a modification in the choice of β and ψ in order to keep the cochain g unchanged.*

Proof. Let ρ' be a different section of π with $\rho'(1) = 1$. Then there exists a map $q : G \to M$ with $\rho' = \mu q\, \rho$ (pointwise multiplication in N). Modify now β to β' defined by

$$\beta'(x, y) = q(x)(\eta(\rho x)(q(y)))\beta(x, y)q(xy)^{-1}$$

for all $x, y \in G$, and modify ψ to ψ' defined by

$$\psi'(n) = \psi(n)q\pi(n)^{-1}$$

for all $n \in N$. Denote by g' the 2-cochain associated to (ρ', β', ψ'), and compute $g'(n, n')$ using the notation $\eta(n)(m) = n \cdot m$ for all $n \in N$ and all $m \in M$ and $\widehat{n} = q(\pi(n))$ for all $n \in N$.

By definition (see equation (1.14) and the relation (1.16)), we have

$$\psi'(n)(\rho'\pi(n) \cdot \psi'(n'))\beta'(\pi(n), \pi(n')) = ig'(n, n')\psi'(nn')$$

and, therefore,

$$\psi(n)\widehat{n}^{-1}(\mu(\widehat{n})\rho\pi(n) \cdot (\psi(n')\widehat{n'}^{-1}))\widehat{n}(\rho\pi(n) \cdot \widehat{n'})\beta(\pi(n), \pi(n'))\widehat{nn'}^{-1} = ig'(n, n')\psi(nn')\widehat{nn'}^{-1}.$$

Use property (b) of a crossed module to transform the action by $\mu(\widehat{n})$ into conjugation, which then simplifies. We obtain, simplifying also the right-hand terms $\widehat{nn'}^{-1}$,

$$\psi(n)(\rho\pi(n) \cdot \psi(n'))(\rho\pi(n) \cdot \widehat{n'}^{-1})(\rho\pi(n) \cdot \widehat{n'})\beta(\pi(n), \pi(n')) = ig'(n, n')\psi(nn')$$

which simplifies further to

$$\psi(n)(\rho\pi(n)\cdot\psi(n'))\beta(\pi(n),\pi(n')) = ig'(n,n')\psi(nn').$$

Equation (1.14) implies then $ig = ig'$. □

In order to conclude the proof of Theorem 1.8.3, it remains to prove that two relatively equivalent crossed modules have the same characteristic class. But if φ gives a relative equivalence between $\mu : M \to N$ and $\mu' : M' \to N$ and the 2-cochain g associated with $\mu : M \to N$ has been defined using ρ, β and ψ, then the 2-cochain g' associated with $\mu' : M' \to N$ may be defined using $\rho' = \rho$, $\beta' = \varphi\beta$ and $\psi' = \varphi\psi$. Equation (1.14) then implies that $(\varphi \circ i)g = i'g'$, and thus finally $g = g'$. □

Let us also record the relation to the characteristic class $[d_S\beta]$.

Proposition 1.8.8. *The image of $\xi(M,\mu)$ under the connecting homomorphism $\partial : H^3(G,N,V) \to H^3(G,V)$ is the characteristic class $[\gamma]$ of the crossed module $\mu : M \to N$.*

Proof. We have seen $dg = \pi^*\gamma$ in Lemma 1.8.4. According to the definition of the connecting homomorphism ∂, we must first lift κ^*g to $C^2(N,V)$, then take its coboundary d, and finally take $(\pi^*)^{-1}$. Choosing g as a lift to $C^2(N,V)$, the equation $dg = \pi^*\gamma$ tells us that $\partial[\kappa^*g] = [\gamma]$, as claimed. □

Up to now, we have constructed a well-defined map

$$\xi : \mathrm{crmod}(G,N,V) \to H^3(G,N,V).$$

We will not go into the details (see some details in [63]), which permit to construct an abelian group structure on $\mathrm{crmod}(G,N,V)$, to show that ξ is additive and to characterize the zero class in $\mathrm{crmod}(G,N,V)$ as being represented by crossed modules $\mu : M \to N$ such that the map $\mu : M \to \mathrm{Ker}(\pi)$ admits an N-equivariant section. In the following, we will denote $\mathrm{Ker}(\pi) = \mathrm{Im}(\mu)$ by L.

Let us just show that ξ is injective and surjective. For the injectivity, let $\mu : M \to N$ be a relative crossed module with associated 2-cochain g. Suppose that $\xi(M,\mu) = 0$, that is, there exists $c \in C^1(G,N,V)$ with $\kappa^*g = dc$. By exactness of the sequence of complexes, there exists $b \in C^1(N,V)$ with $\kappa^*b = c$. One obtains that $db - g \in \mathrm{Ker}(\kappa^*)$, that is, there exists $a \in C^2(G,V)$ such that $g = db + \pi^*a$. In conclusion, $\xi(M,\mu) = 0$ means that there exist $b \in C^1(N,V)$ and $a \in C^2(G,V)$ such that $g = db + \pi^*a$.

Put $s(n) := \psi(n)\, ib(n)^{-1}$ for all $n \in N$. Let us show that the restriction of s to L is an N-equivariant section. By what we have said above, this will show that the equivalence class of $\mu : M \to N$ is the zero class. Apply equation (1.14) to a pair of elements $l, l' \in L$, using $g = db + \pi^*a$:

$$\begin{aligned}(l\cdot\psi(l'))\psi(l)\beta(\pi(l),\pi(l')) &= idb(l,l')i\pi^*a(l,l')\psi(ll')\\ &= i((l\cdot b(l')) - b(ll') + b(l'))ia(\pi(l),\pi(l'))\psi(ll').\end{aligned}$$

Use that l, l' are μ-images, and thus $\pi(l) = \pi(l') = 1$. This takes care of the β-term and the $\pi^* a$-term. Then we obtain

$$(l \cdot \psi(l'))\psi(l) = i((l \cdot b(l')) - b(ll') + b(l'))\psi(ll').$$

Moving as before $\psi(l)$ to the left on the LHS (see relation (1.16)), we obtain

$$\psi(l)(\rho\pi(l) \cdot \psi(l')) = i((l \cdot b(l')) - b(ll') + b(l'))\psi(ll').$$

Finally, we get because of $\pi(l) = 1$ and again the triviality of the action of $\mathrm{Im}(\mu)$ on V:

$$\psi(l)\psi(l') = i(b(l') - b(ll') + b(l))\psi(ll').$$

This translates into $s(ll') = s(l)s(l')$. Therefore, the restriction of s to L is a homomorphism. Now apply again equation (1.14) to a pair $(n, l) \in N \times L$, and then to a pair $(n \cdot l, n)$ for $n \in N$ and $l \in L$. We find in the same way as before $(\eta(n)(s(l)))s(n) = s(nl)$, and $s(n \cdot l)s(n) = s(nl)$. Taking these two together, we obtain

$$\eta(n)(s(l)) = s(n \cdot l).$$

Therefore, the restriction of s to L is N-equivariant. In order to show that s is a section, it suffices to write down $\mu s(l) = \mu\psi(l)\, \mu i b(l)^{-1} = l$. In conclusion, we have shown that $\xi(M, \mu) = 0$ implies that the equivalence class of $\mu : M \to N$ is zero.

Let us finally show the surjectivity of the map ξ. Given $g \in C^2(N, V)$ such that $\kappa^* g$ is a relative cocycle, we have to construct a relative crossed module $\mu : M \to N$ (with the fixed quotient morphism $\pi : N \to G$ and the fixed G-module $V = \mathrm{Ker}(\mu)$) such that $\xi(M, \mu) = [\kappa^* g]$. As N acts on V via $\pi : N \to G$, L acts trivially on V.

The condition that $\kappa^* g$ is a cocycle is equivalent to the existence of $k \in C^3(G, V)$ such that $dg = \pi^* k$. Indeed, $d\kappa^* g = \kappa^* dg = 0$ implies $dg \in \mathrm{Ker}(\kappa^*) = \mathrm{Im}(\pi^*)$. We will use the property $dg = \pi^* k$ in the following form: If one of the elements $n, n', n'' \in N$ is in L, then $dg(n, n', n'') = 0$. The image of g in $C^2(L, V)$ is thus a 2-cocycle (which we still note as g) defining an extension $M := V \times_g L$. In product coordinates, an element of M is written (v, l), and the product reads (as the action of L on V is trivial)

$$(v, l)(v', l') = (v + v' + g(l, l'), ll').$$

We put $\mu(v, l) = l$ and define an action η of N on M by

$$\eta(n)(v, l) := (\pi(n) \cdot v + g(n, l) - g(nln^{-1}, n), nln^{-1}).$$

Observe that this is just equation (1.2) in our context here. We want to show that $\mu : M \to N$ with this action η is indeed a crossed module. Let us do this here again explicitly in our context without referring to the preceding sections. First, for all $n \in N$, $\eta(n)$ is a group homomorphism. As G acts on V, it suffices for this to show that

$\eta(n)(0,l)\eta(n)(0,l') = \eta(g(l,l'),ll')$ for all $l,l' \in L$ and all $n \in N$. This boils down to showing that

$$g(n,l) - g(nln^{-1},n) + g(n,l') - g(n(l')n^{-1},n) + g(nln^{-1},n(l')n^{-1})$$
$$= \pi(n)\cdot g(l,l') + g(n,ll') - g(n(ll')n^{-1},n).$$

The equation is true due to the relation

$$-dg(n,l,l') + dg(nln^{-1},n,l') - dg(nln^{-1},n(l')n^{-1},n) = 0,$$

where we use as explained before that if one of the elements $n,n',n'' \in N$ is in L, then $dg(n,n',n'') = 0$.

Furthermore, the map $n \mapsto \eta(n)$ is a group homomorphism. For this, it suffices to verify that $\eta(nn')(0,l) = \eta(n)(\eta(n')(0,l))$ for all $n,n' \in N$ and all $l \in L$, which reduces to

$$g(nn',l) - g(nn'l(nn')^{-1},nn')$$
$$= \pi(n)\cdot g(n',l) - \pi(n)\cdot g(n'l(n')^{-1},n') + g(n,n'l(n')^{-1}) - g((nn')l(nn')^{-1},n).$$

This follows from the equality

$$dg(n,n',l) - dg(n,n'l(n')^{-1},n') + dg((nn')l(nn')^{-1},n,n') = 0.$$

The property (a) of a crossed module is trivially true. For property (b), it suffices to show

$$(v,l)(v',l')(v,l)^{-1} = \eta(l)(v',l').$$

This translates into

$$v + v' + g(l,l') = v' + g(l,l') - g(l(l')l^{-1},l) + v + g(l(l')l^{-1},l),$$

which is true. In conclusion, $\mu : M \to N$ is a crossed module satisfying the given requirements.

We show that the (relative) characteristic class of the (relative) crossed module $\mu : M \to N$ (which we have constructed from the given class $[\kappa^* g]$) is the class $[\kappa^* g]$ we started with. This will show surjectivity of the map ξ and end the proof of Theorem 1.8.1.

Let ρ be a set-theoretical section of π with $\rho(1) = 1$. Put $\psi(n) = (g(n,\bar{n}),n\bar{n}) \in M$ with $n \in N$ and $\bar{n} = \rho\pi(n)^{-1}$. With respect to the product in $M = V \times_g L$, the inverse of $\psi(nn')$ reads

$$\psi(nn')^{-1} = (-g(nn',\overline{nn'}) - g(nn'\overline{nn'},\overline{nn'}^{-1}(n\bar{n})^{-1}),\overline{nn'}^{-1}(nn')^{-1}).$$

Define further

$$\beta(\pi(n),\pi(n')) = (-\pi(nn')\cdot g(\bar{n'},\bar{n}) - \pi(nn')\cdot g(\bar{n'}\bar{n},\bar{n}^{-1}\bar{n'}^{-1}\overline{nn'}),\bar{n}^{-1}\bar{n'}^{-1}\overline{nn'}).$$

Call g' the 2-cochain on N associated to this choice of ρ, ψ and β via the formula (1.14). One computes using the formula of the product in $M = V \times_g L$:

$$\begin{aligned} g'(n,n') = {}& \pi(n)\cdot g(n',\bar{n'}) + g(n,n'\bar{n'}) - g(nn'\bar{n'}n^{-1},n) + g(n,\bar{n}) \\ & + g(nn'\bar{n'}n^{-1},n\bar{n}) - \pi(nn')\cdot g(\bar{n'},\bar{n}) - \pi(nn')\cdot g(\bar{n'}\bar{n},\bar{n}^{-1}\bar{n'}^{-1}\overline{nn'}) \\ & + g(nn'\bar{n'}\bar{n},\bar{n}^{-1}\bar{n'}^{-1}\overline{nn'}) - g(nn',\overline{nn'}) - g(nn'\overline{nn'},\overline{nn'}^{-1}(n\bar{n})^{-1}) \\ & + g(nn'\overline{nn'},\overline{nn'}^{-1}(n\bar{n})^{-1}). \end{aligned}$$

Explanation of this formula: Take formula (1.14) with the term $\psi(nn')$ brought to the LHS. The term $\eta(n)(\psi(n'))$ comprises the first three terms, the fourth term corresponds to $\psi(n)$, the fifth term comes from the product in $M = V \times_g L$, sixth and seventh term correspond to $\beta(\pi(n),\pi(n'))$, term eight comes again from the product in $M = V \times_g L$, terms nine and ten correspond to $\psi(nn')^{-1}$ and the last term comes again from the product in $M = V \times_g L$.

From here, one computes that

$$\begin{aligned} (g'+g)(n,n') = {}& dg(n,n',\bar{n'}) - dg(nn',\bar{n'},\bar{n}) \\ & - dg(nn',\bar{n'}\bar{n},\bar{n}^{-1}\bar{n'}^{-1}\overline{nn'}) + dg(nn'\bar{n'}n^{-1},n,\bar{n}). \end{aligned}$$

As we have $dg = \pi^* k$, we obtain $\kappa^* dg = \kappa^*\pi^* k = 0$, and thus the above formula implies $\kappa^* g = -\kappa^* g'$. Thus the above crossed modules realizes the class of $-\kappa^* g$, showing the surjectivity of ξ (modulo the compatibility with the vector space structure which we did not show in detail).

Remark 1.8.4. Exactly as in Remark 2.9.5 in the context of Lie algebras, one can show that the relative Gerstenhaber Theorem 1.8.1 implies the absolute Gerstenhaber Theorem 1.2.1. We leave the details to the interested reader.

1.9 Strict 2-groups

We have seen interpretations of crossed modules in terms of cohomology classes or as obstructions to the existence of extensions or relative crossed modules as the obstruction to the existence of equivariant sections. In this section, we will see another important interpretation of crossed modules, namely, in terms of categorified groups, or 2-groups. For more ample details on this matter, we refer to [73], Chapter XII, [64] and [87].

In recent times, *categorification*, that is, the passage to categorified algebraic structures, plays a growing role in algebra and geometry. Here, categorification means the replacement of the underlying sets in some algebraic structure by categories and maps between these sets by functors. For example, instead of regarding a group in Sets, the category of sets (which is the ordinary concept of a group), one considers a group object in the category Cat of (small) categories. For example, a categorified group, or 2-group, is a group object in the category of categories. Amazingly, this is the same as a category object in the category of groups.

Proposition 1.9.1. *A group object in the category of (small) categories* Cat *is the same as a category object in the category of groups* Grps.

Proof. A category object in Grps is the data of two groups G_0, the group of objects, and G_1, the group of arrows, together with group homomorphisms $s, t : G_1 \to G_0$, source and target, $i : G_0 \to G_1$, inclusion of identities and $m : G_1 \times_{G_0} G_1 \to G_1$, the categorical composition (of arrows), which satisfy the usual axioms of a category (see pp. 267–268 in [73]). Denote by $\mu_1 : G_1 \times G_1 \to G_1$ the multiplication of G_1, and by μ_0 the multiplication of G_0.

Let us turn the above category C (with objects G_0, arrows G_1, source s and target t etc) into a group in Cat with multiplication $\mu : C \times C \to C$ by setting $\mu = \mu_0$ on objects and $\mu = \mu_1$ on arrows. Then μ defines a functor $\mu : C \times C \to C$. The unit for C is the functor $u : 1 \to C$, where 1 is the category with one object $*$, sending id_* to $i(u(*))$, the inversion is the functor $\iota : C \to C$, sending an object $g \in G_0$ to the object $g^{-1} \in G_0$ and similarly for the arrows. It is straightforward to show that C is a group. It is thus a group object in Cat.

Conversely, given a group object C in Cat, restriction of the multiplication $\mu : C \times C \to C$ yields multiplications μ_0 and μ_1 on the set of objects and on the set of arrows. Functoriality of μ translates into associativity of μ_0 and μ_1. The maps s, t, i and m come from the underlying category. □

Definition 1.9.1. A *2-group* is a category object in the category Grp, that is, it is the data of two groups G_0, the group of objects, and G_1, the group of arrows, together with group homomorphisms $s, t : G_1 \to G_0$, source and target, $i : G_0 \to G_1$, inclusion of identities, and $m : G_1 \times_{G_0} G_1 \to G_1$, the categorical composition (of arrows), which satisfy the usual axioms of a category.

We should emphasize, however, that the 2-groups introduced here are *strict* 2-groups, that is, in the categorified version all laws are verified as equations. In other categorifications, one relaxes some or all laws to hold only up to natural transformation, transformations, which should then satisfy coherence conditions. This leads to much more general notions (like coherent 2-groups, weak 2-groups; see [3]), but does not concern us here.

Now we come to the notion of a morphism between strict 2-groups.

Definition 1.9.2. A *morphism* $F : (G_0, G_1) \to (H_0, H_1)$ between 2-groups (G_0, G_1) and (H_0, H_1) is a functor internal to the category Grp, that is, the data of morphisms of groups $F_0 : G_0 \to H_0$ and $F_1 : G_1 \to H_1$ such that the following diagrams commute:

$$\begin{array}{ccc} G_1 \times_{G_0} G_1 & \xrightarrow{F_1 \times F_1} & H_1 \times_{H_0} H_1 \\ \downarrow{\scriptstyle m_G} & & \downarrow{\scriptstyle m_H} \\ G_1 & \xrightarrow{F_1} & H_1 \end{array} \qquad \begin{array}{ccccc} G_1 & \xrightarrow{s,t} & G_0 & \xrightarrow{i} & G_1 \\ \downarrow{\scriptstyle F_1} & & \downarrow{\scriptstyle F_0} & & \downarrow{\scriptstyle F_1} \\ H_1 & \xrightarrow{s,t} & H_0 & \xrightarrow{i} & H_1 \end{array}$$

With this notion of morphisms, strict 2-groups form a category. 2-groups also form a 2-category (cf. [87], [73]), but we will stick to the easiest categorical framework.

Crossed modules of groups also form a category. They also form a 2-category, as explained, for example, in [87]. The following theorem can be found in [73], Chapter XII, in [64] and in the 2-categorical version with all details in [87].

Theorem 1.9.2. *The categories of 2-groups and of crossed modules of groups are equivalent.*

Remark 1.9.3. One interesting point about this theorem is that the categorical composition $m : G_1 \times_{G_0} G_1 \to G_1$ does not represent additional structure, but is already encoded in the group law of G_1, namely, one has

$$g \circ f := m(f, g) = f\big(i(b)\big)^{-1} g,$$

where $t(f) = b$; this formula involves only the group multiplication in G_1 on the RHS. Thus the data of two groups G_0, G_1 and morphisms $s, t : G_1 \to G_0$ and $i : G_0 \to G_1$ satisfying the usual axioms of source, target and object inclusion in a category is already equivalent to the data of a crossed module.

Proof. Suppose given a 2-group, that is, the data of two groups G_0 and G_1 together with group homomorphisms $s, t : G_1 \to G_0$, $i : G_0 \to G_1$ and $m : G_1 \times_{G_0} G_1 \to G_1$. Let us first of all show the remark, that is, that the categorical composition $\circ := m : G_1 \times_{G_0} G_1 \to G_1$ does not represent additional structure, but is already encoded in the group law of G_1. Indeed, the fact that the composition is a group homomorphism reads as

$$(g_1 g_2) \circ (f_1 f_2) = (g_1 \circ f_1)(g_2 \circ f_2).$$

This is the *middle four exchange law*. Here, we supposed that the elements f_1, f_2, g_1, g_2 are composable in the above way, that is, $f_1, f_2 : a \to b$ and $g_1, g_2 : b \to c$ for some $a, b, c \in G_0$. Applying this law two times using the identity arrow 1_b in b, we obtain

$$\begin{aligned} g \circ f &= (1_b 1_b^{-1} g) \circ (f 1_b^{-1} 1_b) \\ &= (1_b \circ f)\big((1_b^{-1} g) \circ (1_b^{-1} 1_b)\big) \\ &= (1_b \circ f)(1_b^{-1} \circ 1_b^{-1})(g \circ 1_b) \\ &= f(1_b^{-1} \circ 1_b^{-1}) g. \end{aligned}$$

Observe that 1_b^{-1} is the inverse of 1_b with respect to the group product, it does not have properties with respect to composition, a priori. This formula already expresses the composition $f \circ g$ in terms of the product in G_2, but in order to better understand the middle term $1_b^{-1} \circ 1_b^{-1}$, let us apply the formula which we obtained to $f = g = 1_b$:

$$1_b = 1_b(1_b^{-1} \circ 1_b^{-1})1_b.$$

One deduces from here

$$(1_b^{-1} \circ 1_b^{-1}) = 1_b^{-1}.$$

Inserting this in the above formula, we obtain

$$g \circ f = f\, 1_b^{-1}\, g,$$

which is the formula we alluded to in the above remark. Similarly, one obtains

$$\begin{aligned} g \circ f &= (g1_b^{-1}1_b) \circ (1_b 1_b^{-1} f) \\ &= (g \circ 1_b)(1_b^{-1} \circ 1_b^{-1})(1_b \circ f) \\ &= g\, 1_b^{-1} f. \end{aligned}$$

Putting these two equations together, we get

$$f\, 1_b^{-1} g = g\, 1_b^{-1} f.$$

In particular, if $b = 1 \in G_0$ is the unit in the group G_0, then $1_b = i(b)$ must be the unit in G_1, because i is a group homomorphism. In this situation, we have $fg = gf$. But if $b = 1$, the source of g is 1 and the target of f is 1, that is, $g \in \mathrm{Ker}(s)$ and $f \in \mathrm{Ker}(t)$. We thus obtain as a corollary to the above formula the following.

Corollary 1.9.3.

$$[\mathrm{Ker}(s), \mathrm{Ker}(t)] = 1.$$

Let us now come to the crossed module $\mu : M \to N$, which we associate to the given 2-group. It is defined by $N := G_0$, $M := \mathrm{Ker}(s)$ and $\mu := t|_{\mathrm{Ker}(s)}$. The general extension

$$1 \to \mathrm{Ker}(s) \to G_1 \xrightarrow{s} G_0 \to 1$$

splits by $s \circ i = \mathrm{id}_{G_0}$, and determines a genuine action of G_0 on $\mathrm{Ker}(s)$ by conjugation. More precisely, $a \in G_0$ acts on $f \in \mathrm{Ker}(s)$ by

$$a \cdot f := i(a) f\, i(a)^{-1} = 1_a f\, 1_{a^{-1}}.$$

Property (a) of a crossed module reads now

$$\mu(a\cdot f) = t(1_a f 1_{a^{-1}}) = t(1_a)t(f)t(1_{a^{-1}}) = at(f)a^{-1} = a\mu(f)a^{-1}.$$

Property (b) of a crossed module reads

$$\mu(f)\cdot g = 1_{t(f)}\, g\, 1_{t(f)^{-1}} = fgf^{-1}.$$

The last equality is true, because f and $1_{t(f)}$ differ only by an element of $\mathrm{Ker}(t)$:

$$t(f^{-1}1_{t(f)}) = t(f)^{-1}t(f) = 1.$$

Thus, as by assumption $g \in \mathrm{Ker}(s)$, the conjugation of g by f and $1_{t(f)}$ coincide, because $\mathrm{Ker}(s)$ and $\mathrm{Ker}(t)$ commute. In conclusion, we have constructed a crossed module from a strict 2-group.

Conversely, let a crossed module $\mu : M \to N$ be given. One defines a strict 2-group by $G_0 := N$ and $G_1 := M \rtimes N$, where the semidirect product uses the given action of N on M. The maps are defined as follows: $i : G_0 \to G_1$, $i(n) := (1, n)$, $s : G_1 \to G_0$, $s(m, n) := n$ and $t : G_1 \to G_0$, $t(m, n) := \mu(m)n$ for all $n \in N$ and all $m \in M$. It is clear that i and s are group homomorphisms. We compute

$$\begin{aligned} t((m,n)(m',n')) &= t(m(n\cdot m'), nn') = \mu(m(n\cdot m'))nn' \\ &= \mu(m)n\mu(m')n^{-1}nn' = \mu(m)n\mu(m')n' = t(m,n)t(m',n'), \end{aligned}$$

where we used property (a) of the crossed module $\mu : M \to N$ in the third equality. The categorical composition $m : G_1 \times_{G_0} G_1 \to G_1$ is defined by the formula

$$m(f,g) := f(i(b))^{-1}g.$$

Written in terms of elements in N and $M \rtimes N$, this reads

$$(m', \mu(m)n) \circ (m, n) = (m'm, n).$$

A direct computation shows that m is a group homomorphism, that is, the middle four exchange law:

$$\begin{aligned} &((m',\mu(m)n)(\tilde{m}',\mu(\tilde{m})\tilde{n})) \circ ((m,n)(\tilde{m},\tilde{n})) \\ &\quad = ((m',\mu(m)n)\circ(m,n))((\tilde{m}',\mu(\tilde{m})\tilde{n})\circ(\tilde{m},\tilde{n})). \end{aligned}$$

On the one side, we obtain

$$\begin{aligned} &((m',\mu(m)n)(\tilde{m}',\mu(\tilde{m})\tilde{n})) \circ ((m,n)(\tilde{m},\tilde{n})) \\ &\quad = (m'((\mu(m)n)\cdot\tilde{m}'), \mu(m)n\mu(\tilde{m})\tilde{n}) \circ (m(n\cdot\tilde{m}), n\tilde{n}) \\ &\quad = (m'((\mu(m)n)\cdot\tilde{m}')m(n\cdot\tilde{m}), n\tilde{n}) \end{aligned}$$

$$= (m'm(n \cdot \tilde{m}')m^{-1}m(n \cdot \tilde{m}), n\tilde{n})$$
$$= (m'm(n \cdot \tilde{m}'\tilde{m}), n\tilde{n}).$$

On the other side, we obtain

$$\begin{aligned}&((m', \mu(m)n) \circ (m, n))((\tilde{m}', \mu(\tilde{m})\tilde{n}) \circ (\tilde{m}, \tilde{n}))\\ &\quad = (m'm, n)(\tilde{m}'\tilde{m}, \tilde{n})\\ &\quad = (m'm(n \cdot \tilde{m}'\tilde{m}), n\tilde{n}).\end{aligned}$$

This shows our claim. □

We now extract from the proof a new definition (see [64]).

Definition 1.9.4. A cat^1-*group* is the data of a group G together with a subgroup N and two homomorphisms $s, t : G \to N$ such that:
(a) $s|_N = t|_N = \mathrm{id}_N$
(b) $[\mathrm{Ker}(t), \mathrm{Ker}(s)] = 1$

It is obvious how to define morphisms $\varphi : (G, N, s, t) \to (G', N', s', t')$ between cat^1-groups, namely as morphisms $\varphi : G \to G'$ with $\varphi(N) \subset N'$ such that $s' \circ \varphi = \varphi|_N \circ s$ and $t' \circ \varphi = \varphi|_N \circ t$. It is clear that cat^1-groups form a category with these morphisms.

Theorem 1.9.4. *The category of* cat^1*-groups and the category of crossed modules of groups are equivalent.*

Proof. We will restrict to the relation between objects of the two categories in order to keep technicalities to a minimum.

We have already seen how to associate to a crossed module $\mu : M \to N$ a cat^1-group: Namely, associate to it first its strict 2-group $s, t : M \rtimes N \to N$ and take $G := M \rtimes N$ together with N, s and t as defined. By the above lemma, (G, N, s, t) is a cat^1-group.

Conversely, we have seen that the data of a cat^1-group is enough to reconstruct the 2-group, thus we also get a crossed module. □

Remark 1.9.5. In conclusion, the four notions of crossed module of groups, group object in the category of categories, category object in the category of groups and cat^1-group are equivalent; see [64]. Moreover, they are all equivalent to the notion of a *simplicial group with Moore complex of length one*; see [64].

The main interest for Loday in the notion of a cat^1-group is that this is a flexible enough notion to generalize to the corresponding n-fold notion: cat^n-groups, which are then related to homotopy $n+1$-types, generalizing the relation of crossed modules or cat^1-groups to homotopy 2-types (see Section 1.10.3) or which are related to homology $n+1$-classes, generalizing the relation of crossed modules to 3-classes.

1.10 Motivation: Classification of homotopy 2-types

In this last section, we will come back to the motivations for introducing crossed modules of groups. Crossed modules were introduced by J. H. C. Whitehead in 1949 [104] for the needs of homotopy theory. Namely, Mac Lane and Whitehead [70] classified 2-homotopy types using crossed modules. We will sketch this development here. Along the way, we will introduce Whitehead's free crossed modules and report on the proof in [90] that a 2-cell attachment corresponds to a free crossed module.

Instead of considering arbitrary topological spaces, we will most of the time restrict to CW complexes for pedagogical reasons.

1.10.1 The crossed module $\pi_2 \to \pi_1$

Let us introduce in this subsection homotopy groups of topological spaces and illustrate how the first and second homotopy groups $\pi_1(A)$ and $\pi_2(X, A)$ of a topological pair (X, A) fit together such that the connecting homomorphism

$$\partial : \pi_2(X, A) \to \pi_1(A)$$

defines a crossed module of groups. Our basic reference in [105] where the presented material appears in Chapters III and (the beginning of) IV.

Let $I = [0, 1]$ be the interval and X be a topological space with base point $x_0 \in X$. Recall that the *homotopy groups* of X are defined as the sets of based homotopy classes

$$\pi_n(X, x_0) = [(S^n, *), (X, x_0)] = [(I^n, \partial I^n), (X, x_0)]$$

for $n \geq 1$. These sets inherit a group structure from the cogroup structure of S^n. More explicitly, for $f, g : I^n \to X$ taking ∂I^n to x_0, the product of the classes of f and g is represented by the map

$$(f * g)(t_1, \dots, t_n) = \begin{cases} f(t_1, \dots, t_{n-1}, 2t_n) & \text{for } t_n \leq \frac{1}{2}, \\ g(t_1, \dots, t_{n-1}, 2t_n - 1) & \text{for } t_n \geq \frac{1}{2}. \end{cases}$$

The fact that we perform the group product along the last variable is immaterial; different choices leads to the same product (between classes). As usual, we will feel free to suppress the base point x_0 in the notation and write equally well $\pi_n(X, x_0)$ and $\pi_n(X)$ for the homotopy groups.

Now we define an action of $\pi_1(X)$ on the higher $\pi_n(X)$: Let X be a topological space and $x_0 \in X$ be a nondegenerate base point. This means by definition that the pair (X, A) with $A = \{x_0\}$ has the homotopy extension property for all spaces Y, that is, given a continuous map $f : X \to Y$ and a homotopy $h : I \times A \to Y$ from $f|_A$ to some $f' : A \to Y$, there exists $H : I \times X \to Y$, a homotopy from f to an extension of f', restricting to h, f and f'.

Consider now the particular situation where $f : (X, x_0) \to (Y, y_0)$ and $u : (I, 0, 1) \to (Y, y_0, y_1)$ are continuous. The homotopy extension property assures the existence of a map $g : (X, x_0) \to (Y, y_1)$, which is homotopic to f relative to u (i. e., there exists a continuous $H : I \times X \to Y$ with $H(t, x_0) = u(t)$, $H(0, x) = f(x)$ and $H(1, x) = g(x)$). In fact, the homotopy class of g depends only on the homotopy classes α and ξ of f and u, respectively (see, e. g., [105] Chapter III, (1.5), (1.6) p. 99). Denote by $\tau_\xi(\alpha)$ the homotopy class of g and by $\pi_1(Y, y_1, y_0)$ the set of homotopy classes of paths $I \to Y$ from y_0 to y_1.

Lemma 1.10.1. *If $\alpha \in [(X, x_0), (Y, y_0)]$, $\xi \in \pi_1(Y, y_1, y_0)$, $\eta \in \pi_1(Y, y_2, y_1)$, then*

$$\tau_{\eta\xi}(\alpha) = \tau_\eta(\tau_\xi(\alpha)).$$

If $\xi \in \pi_1(Y, y_0, y_0)$ is the homotopy class of the constant path, then $\tau_\xi(\alpha) = \alpha$.

Proof. The concatenation of the homotopies defining $\tau_\xi(\alpha)$ and $\tau_\eta(\tau_\xi(\alpha))$ (with rescaling in order to make them fit into I) defines a homotopy defining $\tau_{\eta\xi}(\alpha)$. By independence of the choice of the homotopy, we obtain the formula. The second assumption is shown in the same way. □

Introducing the set $M(y) := [(X, x_0), (Y, y)]$, this lemma implies that $\pi_1(Y, y_1, y_0)$ acts on the sets $M(y)$, sending $M(y_0)$ to $M(y_1)$ by "translating the base point." In particular, the fundamental group $\pi_1(Y, y)$ acts on $M(y)$. In particular, $\pi_1(Y, y)$ acts in this way on all higher homotopy groups $\pi_n(Y)$, $n \geq 1$. For $n = 1$, this action is the canonical conjugation action. In case Y possesses a universal covering, the action of $\pi_1(Y)$ on $\pi_n(Y)$ for $n \geq 2$ may also be expressed in terms of deck transformations.

Observe further that this whole discussion may be done in a relative framework for a pair (X, A). This brings us to the definition of the *relative homotopy groups* $\pi_n(X, A)$ for $n \geq 1$ as the sets of homotopy classes of maps from $(\mathbb{E}^{n+1}, S^n)$, the $n+1$ dimensional unit ball relative to its boundary S^n, to (X, A). The group structure is induced by the same formula as above. The above action of $\pi_1(X)$ on $\pi_n(X)$, $n \geq 1$, also extends to the relative framework. Observe that the base point must stay in A, thus we obtain then an action of $\pi_1(A)$ on $\pi_n(X, A)$ for $n \geq 1$.

We recall now the *long exact sequence in homotopy*.

Let $f : (\mathbb{E}^{n+1}, S^n) \to (X, A)$ represent a homotopy class in $\pi_n(X, A)$. The connecting homomorphism $\partial_* : \pi_n(X, A) \to \pi_{n-1}(A)$ sends by definition the class of f to the class represented by $f|_{S^n}$. It is a group homomorphism.

Proposition 1.10.2. *The sequence of homotopy groups*

$$\dots \to \pi_{n+1}(X, A) \xrightarrow{\partial_*} \pi_n(A) \xrightarrow{i_*} \pi_n(X) \xrightarrow{j_*} \pi_n(X, A) \to \dots \tag{1.17}$$

is exact. Here, j_* is induced by the inclusion $j : (X, *) \hookrightarrow (X, A)$.

Proof. This is Theorem (2.4) in Chapter IV, page 162, in [105]. □

By the above, $\pi_1(A)$ operates on the homotopy groups $\pi_n(A)$ and on the relative homotopy groups $\pi_n(X,A)$. Using $\imath_* : \pi_1(A) \to \pi_1(X)$, $\pi_1(A)$ also acts on $\pi_n(X)$, for all $n \geq 1$. Therefore, $\pi_1(A)$ acts on each term in the homotopy sequence.

Proposition 1.10.3. *For a pair (X,A), the homotopy group $\pi_1(A)$ acts on the homotopy sequence, that is, it acts on each term and the maps ∂_*, i_* and j_* are $\pi_1(A)$-linear.*

Proof. This is Theorem (3.1) in Chapter IV, page 164, in [105]. □

The operation of $\pi_1(A)$ on $\pi_n(A)$ and on $\pi_n(X,A)$ are consistent with the group operation on the latter sets.

Proposition 1.10.4. *If $\xi \in \pi_1(A)$, then $\tau_\xi : \pi_n(A) \to \pi_n(A)$ and $\tau_\xi : \pi_{n+1}(X,A) \to \pi_{n+1}(X,A)$ are group homomorphisms for $n \geq 1$.*

Proof. This is Theorem 3.2 in Chapter IV of [105]. □

Proposition 1.10.5. *Let $\alpha, \beta \in \pi_2(X,A)$, $\xi = \partial_*\beta \in \pi_1(A)$. Then*

$$\tau_\xi(\alpha) = \operatorname{conj}_\beta(\alpha) = \beta\alpha\beta^{-1}.$$

Proof. Let $h : (\mathbb{E}^2 \vee \mathbb{E}^2, S^1 \vee S^1) \to (X,A)$ be a map such that $h \circ j_1 : (\mathbb{E}^2, S^1) \to (X,A)$ represents α and $h \circ j_2 : (\mathbb{E}^2, S^1) \to (X,A)$ represents β. Define elements $\iota_1, \iota_2 \in \pi_2(\mathbb{E}^2 \vee \mathbb{E}^2, S^1 \vee S^1)$ as homotopy classes of the maps $j_1, j_2 : (\mathbb{E}^2, S^1) \to (\mathbb{E}^2 \vee \mathbb{E}^2, S^1 \vee S^1)$. ι_1 and ι_2 have the property that $h_*(\iota_1) = \alpha$ and $h_*(\iota_2) = \beta$. Further, let $\lambda = \partial_*\iota_2$ and obtain

$$h_*(\lambda) = h_*(\partial_*\iota_2) = \partial_* h_*(\iota_2) = \xi \in \pi_1(A),$$

by functoriality of ∂_*. Therefore,

$$h_*\tau_\lambda(i_1) = \tau_{h_*(\lambda)}(h_*(i_1)) = \tau_\xi(\alpha),$$

by functoriality of τ. Thus, if our claim holds for the elements $\iota_1, \iota_2 \in \pi_2(\mathbb{E}^2 \vee \mathbb{E}^2, S^1 \vee S^1)$, that is, $\tau_\lambda(\iota_1) = \operatorname{conj}_{\iota_2}(\iota_1)$, then it will hold for α and β:

$$\begin{aligned} h_*\tau_\lambda(\iota_1) &= h_*(\operatorname{conj}_{\iota_2}(\iota_1)) \\ &= \operatorname{conj}_{h_*(\iota_2)}(h_*(\iota_1)) \\ &= \operatorname{conj}_\beta(\alpha), \end{aligned}$$

by naturality of the group product. It therefore suffices to show the relation $\tau_\lambda(\iota_1) = \operatorname{conj}_{\iota_2}(\iota_1)$. Let us compute

$$\begin{aligned} \partial_*\tau_\lambda(\iota_1) &= \tau_\lambda(\partial_*\iota_1) \\ &= \lambda(\partial_*\iota_1)\lambda^{-1} \\ &= (\partial_*\iota_2)(\partial_*\iota_1)(\partial_*\iota_2)^{-1} \end{aligned}$$

$$= \partial_* \mathrm{conj}_{\iota_2}(\iota_1),$$

where we have used $\pi_1(A)$-equivariance of ∂_* in the first line, the fact that the $\pi_1(A)$-action identifies to conjugation on $\pi_1(A)$ in the second line, the definition of λ in the third line, and finally the fact that ∂_* is a group homomorphism in the last line.

In order to conclude, observe that $\mathbb{E}^2 \vee \mathbb{E}^2$ is contractible, and the connecting homomorphism

$$\partial_* : \pi_2(\mathbb{E}^2 \vee \mathbb{E}^2, S^1 \vee S^1) \to \pi_1(S^1 \vee S^1)$$

is thus an isomorphism by exactness of the homotopy sequence of the pair (X, A). Therefore, we may conclude from the above that $\tau_\lambda(\iota_1) = \mathrm{conj}_{\iota_2}(\iota_1)$ as claimed. □

Corollary 1.10.6. *Let (X, A) be a pair of topological spaces.*

Then the map $\partial_* : \pi_2(X, A) \to \pi_1(A)$, together with the action of $\pi_1(A)$ on $\pi_2(X, A)$, defines a crossed module of groups.

Proof. We have seen above that $\pi_1(A)$ acts on $\pi_2(X, A)$ by group homomorphisms, and that with respect to this action, ∂_* is equivariant. This is property (a) of a crossed module. Proposition 1.10.5 is exactly property (b), that is, the Peiffer identity, for a crossed module. □

Remark 1.10.1. It is clear that this corollary gives a vast supply of crossed modules; cf. Exercise 1.11.11.

1.10.2 Free crossed modules

In this subsection, we show a result due to J. H. C. Whitehead [104]. Namely, we show that in the special case where the pair of spaces (X, A) is given by a space X obtained from A by attaching some 2-cells, the crossed module $\partial_* : \pi_2(X, A) \to \pi_1(A)$ is a so-called *free crossed module*. We will follow here [90].

Definition 1.10.2. Let $\mu : M \to N$ be a crossed module and $\{m_\alpha\}_{\alpha \in A}$ be an indexed set of elements of M. The crossed module $\mu : M \to N$ is called *free with basis* $\{m_\alpha\}_{\alpha \in A}$ if it satisfies the following universal property:

For every crossed module $\mu' : M' \to N'$, every indexed set $\{m'_\alpha\}_{\alpha \in A}$ of elements of M' and every homomorphism $h : N \to N'$ with $h(\mu(m_\alpha)) = \mu'(m'_\alpha)$ for all $\alpha \in A$, there exists a unique homomorphism $\phi : M \to M'$ such that $\phi(m_\alpha) = m'_\alpha$ for all $\alpha \in A$ and such that (ϕ, h) defines a morphism of crossed modules from $\mu : M \to N$ to $\mu' : M' \to N'$.

The following construction of crossed modules in due to J. H. C. Whitehead [104]. Let N be some group and $\{n_\alpha\}_{\alpha \in A}$ some indexed set of elements of N. Let E be the free group on $N \times A$. Let M_W be the quotient of E by the normal subgroup generated by

$$W = \{(x,\alpha)(y,\beta)(x,\alpha)^{-1}(xn_\alpha x^{-1}y,\beta)^{-1} \mid x,y \in N, \text{ and } \alpha,\beta \in A\}.$$

Observe that the relations in W imply that we have in the quotient space M_W,

$$\mathrm{conj}_{(x,\alpha)}(y,\beta) = (xn_\alpha x^{-1}y,\beta).$$

Lemma 1.10.7. *The map $\mu : E \to N$, $(x,\alpha) \mapsto xn_\alpha x^{-1}$, induces a map $\mu_W : M_W \to N$. The action of N on E defined by $n \cdot (x,\alpha) = (nx,\alpha)$ induces an action of N on M_W. The map $\mu_W : M_W \to N$, together with this action of N on M_W, forms a crossed module.*

Proof. Since

$$\mu_W(xn_\alpha x^{-1}y,\beta) = xn_\alpha x^{-1}yn_\beta y^{-1}xn_\alpha^{-1}x^{-1} = \mu_W(x,\alpha)\mu_W(y,\beta)\mu_W(x,\alpha)^{-1},$$

μ_W is well-defined on the quotient. Since we have in the quotient

$$\mathrm{conj}_{(zx,\alpha)}(zy,\beta) = (zxn_\alpha x^{-1}z^{-1}zy,\beta) = (zxn_\alpha x^{-1}y,\beta),$$

the map $(x,\alpha) \mapsto (zx,\alpha)$ gives for each $z \in N$ a well-defined endomorphism of M_W. As $1 \in N$ gives the identity endomorphism and $z'(z(x,\alpha)) = z'z(x,\alpha)$ for all $z, z' \in N$, N acts on M_W by group homomorphisms. We have property (a) of a crossed module by

$$\mu_W(n \cdot (x,\alpha)) = \mu_W(nx,\alpha) = nxn_\alpha x^{-1}n^{-1} = n\mu_W(x,\alpha)n^{-1}.$$

We have property (b) of a crossed module by

$$\mu_W(x,\alpha) \cdot (y,\beta) = xn_\alpha x^{-1} \cdot (y,\beta) = (xn_\alpha x^{-1}y,\beta) = \mathrm{conj}_{(x,\alpha)}(y,\beta),$$

where we used a relation in W in the last step. □

Let $p : E \to M_W$ denote the natural projection and set $m_\alpha = p(1,\alpha)$ for all $\alpha \in A$. The following proposition is Lemma 2 in [104].

Proposition 1.10.8. *The above defined crossed module $\mu_W : M_W \to N$ is free with basis $\{m_\alpha\}_{\alpha \in A}$.*

Proof. We have to show the universal mapping property. For this, suppose given a crossed module $\mu' : M' \to N'$, some indexed set $\{m'_\alpha\}_{\alpha \in A}$ of elements of M' and some homomorphism $h : N \to N'$ with $h(\mu_W(m_\alpha)) = \mu'(m'_\alpha)$ for all $\alpha \in A$. We have to show that there exists a unique homomorphism $\phi : M_W \to M'$ such that $\phi(m_\alpha) = m'_\alpha$ for all $\alpha \in A$ and such that (ϕ, h) defines a morphism of crossed modules from $\mu_W : M_W \to N$ to $\mu' : M' \to N'$.

We have $\mu'(m'_\alpha) = h(\mu_W(m_\alpha)) = h(\mu_W(1,\alpha)) = h(n_\alpha)$, and it follows from property (b) in $\mu' : M' \to N'$ that for all $x, y \in N$ and all $\alpha, \beta \in A$,

$$\mathrm{conj}_{h(x)\cdot m'_\alpha}(h(y) \cdot m'_\beta) = \mu'(h(x) \cdot m'_\alpha) \cdot (h(y) \cdot m'_\beta)$$

$$
\begin{aligned}
&= (h(x)\mu'(m'_\alpha)h(x)^{-1}) \cdot (h(y) \cdot m'_\beta) \\
&= h(xn_\alpha x^{-1}y) \cdot m'_\beta.
\end{aligned}
$$

Therefore, the map $(x, \alpha) \mapsto h(x) \cdot m'_\alpha$ determines a well-defined homomorphism $\phi : M_W \to M'$. Indeed, we get by the universal property of the free group a group homomorphism $E \to M'$, which by what we have shown passes to the quotient of E with respect to W. This homomorphism ϕ is obviously equivariant with respect to the actions of N on M_W and on M' via h. We have also $\phi(m_\alpha) = \phi(1, \alpha) = m'_\alpha$ and $\mu'(\phi(m_\alpha)) = \mu'(m'_\alpha) = h(\mu_W(m_\alpha))$. We thus obtain from property (a) in $\mu' : M' \to N'$ that

$$
\begin{aligned}
\mu' \circ \phi(x, \alpha) &= \mu'(h(x) \cdot m'_\alpha) \\
&= h(x)\mu'(m'_\alpha)h(x)^{-1} \\
&= h(x)h(\mu_W(m_\alpha))h(x)^{-1} \\
&= h(xn_\alpha x^{-1}) \\
&= h \circ \mu_W(x, \alpha).
\end{aligned}
$$

As the elements (x, α) generate M_W, it follows that $\mu' \circ \phi = h \circ \mu_W$. In a similar way, one shows that ϕ is uniquely determined by the requirement $\phi(m_\alpha) = m'_\alpha$. □

Remark 1.10.3. It follows from Whitehead's construction and the universal property of a free crossed module that a free crossed module $\mu : M \to N$ with basis $\{m_\alpha\}_{\alpha \in A}$ depends only on the indexed set $\{\mu(m_\alpha)\}_{\alpha \in A}$ of N up to a isomorphism of crossed modules inducing the identity on N.

Let us consider the structure of the free crossed module $\mu_W : M_W \to N$. The elements of $I := \operatorname{Ker}(\mu_W)$ (viewed in E !) are called *identities among the images n_α*. A typical element $\Pi_{i=1}^n (x_i, \alpha_i)^{\epsilon_i}$ of E is an identity if and only if $\Pi_{i=1}^n x_i n_{\alpha_i}^{\epsilon_i} x_i^{-1} = 1$ in N. Observe that $\operatorname{Ker}(p : E \to M_W) \subset I = \operatorname{Ker}(\mu_W)$ (viewed as elements of E !) as has been shown the proof of Lemma 1.10.7. Thus all elements of W are identities. The elements of $P = \operatorname{Ker}(p)$ are called *Peiffer identities* among the images n_α.

Lemma 1.10.9.

$$P \cap [E, E] = [E, I].$$

Proof. This is Lemma 2.1 in [90]. □

The following theorem characterizes free crossed modules.

Theorem 1.10.10. *Let $\mu : M \to N$ be a crossed module, $L = \operatorname{Im}(\mu)$, $G = N/L$ and $\{m_\alpha\}_{\alpha \in A}$ be an indexed subset of M. Then $\mu : M \to N$ is a free crossed module with basis $\{m_\alpha\}_{\alpha \in A}$ if and only if:*

(a) $M_{ab} = M/[M, M]$ *is a free G-module with basis $\{\bar{m}_\alpha\}_{\alpha \in A}$;*

(b) *L is the normal subgroup generated by $\{\mu(m_\alpha)\}_{\alpha\in A}$ in N;*
(c) *$\partial_* : H_2(M) \to H_2(L)$ is trivial.*

Proof. Given a free crossed module $\mu : M \to N$ with basis $\{m_\alpha\}_{\alpha\in A}$, we may build Whitehead's free crossed module $\mu_W : M_W \to N$ on $(N, \{\mu(m_\alpha)\}_{\alpha\in A})$. By the universal properties of the two free crossed modules, $\mu : M \to N$ and $\mu_W : M_W \to N$ are isomorphic via an isomorphism of crossed modules inducing the identity on N. But the above three conditions are clearly transmitted by this kind of equivalence, thus we may suppose in the following that $(\mu : M \to N) = (\mu_W : M_W \to N)$ is the free crossed module constructed by Whitehead.

Let us show that $(M_W)_{\mathrm{ab}} = M_W/[M_W, M_W]$ is a free G-module with basis $\{\bar{m}_\alpha\}_{\alpha\in A}$. First of all, property (b) of a crossed module tells us that the image $\mu(M_W)$ acts trivially on $(M_W)_{\mathrm{ab}}$, and thus $(M_W)_{\mathrm{ab}}$ becomes a G-module. Now, given any G-module V with generating set $\{v_\alpha\}_{\alpha\in A}$, the trivial homomorphism $* : V \to G$ is a crossed module, and by the universal property of $\mu_W : M_W \to N$, there is a morphism of crossed modules $(\phi, \psi) : (\mu_W : M_W \to N) \to (V \to G)$. The N-morphism $\phi : M_W \to V$ induces a G-morphism $\bar{\phi} : (M_W)_{\mathrm{ab}} \to V$ with $\bar{\phi}(m_\alpha) = v_\alpha$. This shows that V is a quotient G-module of $(M_W)_{\mathrm{ab}}$, meaning that $(M_W)_{\mathrm{ab}}$ is free.

Let us show condition (b). The normal subgroup L' generated by the $\{\mu(m_\alpha)\}_{\alpha\in A}$ is clearly included in L. On the other hand, the very definition of μ_W shows that L is included in L', meaning $L = L'$.

In order to show condition (c), note that we have a central extension

$$0 \to A \to M_W \to L \to 1,$$

where $A = \mathrm{Ker}(\mu_W)$. This extension induces a 5-term exact sequence (as the 5-term exact sequence of the homological Lyndon–Hochschild–Serre spectral sequence, see Appendix A), which reads

$$H_2(M_W) \stackrel{(\mu_W)_*}{\to} H_2(L) \to A \to H_1(M_W) \stackrel{(\mu_W)_*}{\to} H_1(L) \to 0.$$

Observe that we have a free presentation

$$1 \to I \to E \to L \to 1$$

for L and in the same way a free presentation

$$1 \to P \to E \to M_W \to 1$$

for M_W. According to Hopf's formula (see [103], Theorem 6.8.8, p. 198), the H_2 of a group may be computed via a free presentation as $H_2(L) = \frac{I\cap[E,E]}{[E,I]}$ and similarly for M_W. We obtain thus

$$\frac{P \cap [E,E]}{[E,P]} \stackrel{(\mu_W)_*}{\to} \frac{I \cap [E,E]}{[E,I]} \to \frac{I}{P} \to (M_W)_{\mathrm{ab}} \to L_{\mathrm{ab}} \to 0,$$

where the first two homomorphisms are induced by inclusion. Now Lemma 1.10.9 shows that the first homomorphism is trivial, that is, $(\mu_W)_* : H_2(M_W) \to H_2(L)$ is trivial, as we have claimed.

Suppose conversely that $(\mu : M \to N, \{m_\alpha\}_{\alpha \in A})$ satisfies the conditions (a), (b) and (c). On the other hand, let $\mu_W : M_W \to N$ be a free crossed module with basis $\{m'_\alpha\}_{\alpha \in A}$ such that $\mu_W(m'_\alpha) = \mu(m_\alpha)$ for all α. By condition (b), we have $\mathrm{Im}(\mu_W) = L$. By the universal property of the free crossed module, there exists an N-equivariant group homomorphism $\phi : M_W \to M$ such that $\phi(m'_\alpha) = m_\alpha$ for all α. As $\mu_W = \mu \circ \phi$ (the morphism of crossed modules induces the identity on N !), ϕ induces a homomorphism $\phi_0 : A' := \mathrm{Ker}(\mu_W) \to A := \mathrm{Ker}(\mu)$.

By the naturality of the 5-term exact sequence, we obtain a diagram

$$\begin{array}{ccccccccc}
0 & \longrightarrow & H_2(L) & \longrightarrow & A' & \longrightarrow & (M_W)_{\mathrm{ab}} & \xrightarrow{\bar{\mu}_W} & L_{\mathrm{ab}} & \longrightarrow & 0 \\
 & & \downarrow{\scriptstyle \mathrm{id}_{H_2(L)}} & & \downarrow{\scriptstyle \phi_0} & & \downarrow{\scriptstyle \bar{\phi}} & & \downarrow{\scriptstyle \mathrm{id}_{L_{\mathrm{ab}}}} & & \\
0 & \longrightarrow & H_2(L) & \longrightarrow & A & \longrightarrow & M_{\mathrm{ab}} & \xrightarrow{\bar{\mu}} & L_{\mathrm{ab}} & \longrightarrow & 0
\end{array}$$

Observe that both rows are exact. Observe that both G-modules $(M_W)_{\mathrm{ab}}$ and M_{ab} are free and $\bar{\phi}$ sends a basis onto a basis, thus $\bar{\phi}$ is an isomorphism. By the Five Lemma, ϕ_0 is then an isomorphism.

Now pass to the diagram

$$\begin{array}{ccccccccc}
0 & \longrightarrow & A' & \longrightarrow & M_W & \xrightarrow{\mu_W} & L & \longrightarrow & 1 \\
 & & \downarrow{\scriptstyle \phi_0} & & \downarrow{\scriptstyle \phi} & & \downarrow{\scriptstyle \mathrm{id}_L} & & \\
0 & \longrightarrow & A & \longrightarrow & M & \xrightarrow{\mu} & L & \longrightarrow & 1
\end{array}$$

It is still commutative with exact rows, and ϕ_0 is an isomorphism, thus ϕ is finally an isomorphism by the Five Lemma. It follows that both crossed modules are isomorphic, and in particular, $\mu : M \to N$ is free with basis $\{m_\alpha\}_{\alpha \in A}$. □

We now come to Whitehead's main theorem on this subject, namely the fact that the crossed module $\partial_* : \pi_2(X, A) \to \pi_1(A)$ where X is obtained from A by attaching some 2-cells is a free crossed module.

Theorem 1.10.11. *Let A be a connected CW complex and X be obtained from A by attaching 2-cells, then $\partial_* : \pi_2(X, A) \to \pi_1(A)$ is a free crossed module with basis corresponding to the attached 2-cells.*

Proof. Homotopy computations for cell attachments are resumed in Chapter V, Section 1 of [105]. Our space X is in the terminology of that section a 2-cellular extension of A. We follow closely that source in the first part of the proof.

Let $\{E_\alpha \mid \alpha \in C\}$ be the 2-cells which are to be attached to A (we denote the indexing set by C in order to avoid confusion with the connected CW complex A), let $h_\alpha : (\Delta^2, \dot{\Delta}^2) \to (X, A)$ be the attaching maps, fix for each α a homotopy class u_α of paths in A from the base point $*$ to $h_\alpha(*)$, denote $\epsilon_\alpha = h_{\alpha,*}(\epsilon) \in \pi_2(X, A, h_\alpha(*))$ the image of a generator $\epsilon \in \pi_2(\Delta^2, \dot{\Delta}^2) \cong \mathbb{Z}$, and denote finally by $\epsilon'_\alpha = \tau_{u_\alpha}(\epsilon_\alpha)$ the image of ϵ_α under action of u_α on it (cf. previous subsection).

Let us show that the elements $\epsilon'_\alpha \in \pi_2(X, A)$, together with their *translates*, that is, their images under the action of $\pi_1(A)$ on these elements, generate $\pi_2(X, A)$. We will show this claim successively marking the following steps:

(a) the claim holds for $(X, A) = (K_2, K_1)$ for some simplicial subdivision K of the simplex Δ^2 (with 1-skeleton K_1 and 2-skeleton K_2);
(b) the claim holds for (X, A) having only one cell;
(c) the claim holds for (X, A) having only a finite number of cells;
(d) the claim holds in general.

In case (a), $\partial_* : \pi_2(X, A) \to \pi_1(A)$ is an isomorphism and we must show that $\pi_1(A)$ is generated, as a normal subgroup, by the elements $\partial_* \epsilon'_\alpha$. The elements $\partial_* \epsilon'_\alpha$ depend on certain choices, but these choices only have the effect to replace some $\partial_* \epsilon'_\alpha$ by their conjugates, thus the generated normal subgroup remains the same.

Recall how to compute the fundamental group of a connected simplicial complex K. Order the vertices and choose a maximal tree T in K. For each vertex v of K, let ξ_v be the unique homotopy class of paths in T from the base point $*$ to v. For each 1-simplex σ of $K \setminus T$ with vertices $a < b$, let $\eta_{a,b}$ be the unique homotopy class of paths in σ from a to b. The elements $\zeta_{a,b} := \xi_a \eta_{a,b} \xi_b^{-1}$ form a set of free generators of the fundamental group $\pi_1(K_1)$ of the 1-skeleton K_1 of $K = K_2$. Let R be defined by

$$1 \to R \to \pi_1(K_1) \to \pi_1(K_2) \to 1.$$

To find R, set $\omega_\alpha := \zeta_{a,b}\zeta_{b,c}\zeta_{a,c}^{-1}$ for each 2-simplex α of K with vertices $a < b < c$. We agree that $\zeta_{x,y} = 1$ if $x < y$ are the vertices of a 1-simplex in T. Then R is generated, as a normal subgroup, by the elements ω_α. In our case K, as a subdivision of Δ^2, is 1-connected, so that $\pi_1(K_1)$ is generated, as a normal subgroup, by the elements ω_α. If we choose h_α to be the map of Δ^2 into α which sends the standard vertices e_0, e_1, e_2 into a, b, c and if we choose $u_\alpha = \xi_a$, we see that $\omega_\alpha = \partial_* \epsilon'_\alpha$. This proves our claim in case (a).

Case (b) is done using the following lemma.

Lemma 1.10.12. *Let X be a 2-cellular extension of A having just one cell E with attaching map $h : (\Delta^2, \dot{\Delta}^2) \to (X, A)$. Let $f : (\Delta^2, \dot{\Delta}^2, e_0) \to (X, A, *)$ be a map. Then there exists a subdivision K of Δ^2 and a map $f_2 \simeq f$ such that for each 2-simplex σ of K, either $f_2(\sigma) \subset A$ or $f_2|_\sigma = h \circ f_\sigma$ for some map $f_\sigma : (\sigma, \dot{\sigma}) \to (\Delta^2, \dot{\Delta}^2)$.*

Proof. This is Lemma (1.4) in [105] page 214. □

Now let ι be the element of $\pi_2(K_2, K_1, e_0)$ represented by the identity map of $\triangle^2$. For each 2-simplex σ of K, let ξ_σ be a homotopy class of paths in K_2 from e_0 to a vertex e_σ of σ, and let $\iota_\sigma \in \pi_2(K_2, K_1, e_0)$ be the element represented by an attaching map for σ. By case (a), ι is a sum of terms, each of which is a translate of $\pm\iota_\sigma$.

Coming back to the situation in the lemma, we obtain now that $\beta = f_*(\iota) = f_{2,*}(\iota)$ is a sum of translates of $\pm f_{2,*}(\iota_\sigma)$. If $f_2(\sigma) \subset A$, then $f_{2,*}(\iota_\sigma) = 0$. Otherwise $f_2|_\sigma = h \circ f_\sigma$ with $f_\sigma : (\sigma, \dot{\sigma}) \to (\triangle^2, \dot{\triangle}^2)$. Then $f_{\sigma,*}(\iota_\sigma)$ is a translate of $n_\sigma \epsilon_2$ for some integer n_σ. Thus $f_{2,*}(\iota_\sigma)$ is a translate of $n_\sigma h_*(\epsilon_2)$ and, therefore, of $n_\sigma \epsilon$. Hence β is a sum of integral multiples of translates of ϵ'. This shows our claim in case (b).

Case (c) is shown by induction on the finite number of cells r. $r = 1$ is case (b), initializing our induction. Suppose our claim holds for (X', A) with $X' = E_2 \cup \cdots \cup E_r \cup A$. Then there is an exact sequence (part of the sequence of the triple (X, X', A)),

$$\pi_2(X', A) \xrightarrow{i} \pi_2(X, A) \xrightarrow{j} \pi_2(X, X') \to 0,$$

where i and j are induced by the natural injections. The first two groups have operators in $\pi_1(A)$, while the third has operators in $\pi_1(X')$. Since $\pi_1(A) \to \pi_1(X')$ is an epimorphism and i and j commute with the operators, the exact sequence may be regarded as an exact sequence with action of $\pi_1(A)$ (i. e., i and j commute with the action of $\pi_1(A)$). As $\pi_2(X', A)$ has a set of generators, which are mapped by i into $\epsilon_2', \dots, \epsilon_r'$ and as $\pi_2(X, X')$ has one generator $j(\epsilon_1')$, $\pi_2(X, A)$ is generated by $\epsilon_1', \dots, \epsilon_r'$, showing our claim in case (c).

Case (d) follows from case (c) by a direct limit argument. This ends the proof of the claim that the elements $\epsilon_\alpha' \in \pi_2(X, A)$, together with their translates generate $\pi_2(X, A)$. Observe that this shows condition (b) from Theorem 1.10.10 to be a free crossed module.

In order to verify condition (a) of Theorem 1.10.10, observe that that X has a universal covering space $p : \tilde{X} \to X$. Let $\tilde{A} = p^{-1}(A)$. Then the Hurewicz map induces a natural isomorphism $\pi_2(X, A)_{\mathrm{ab}} \to H_2(\tilde{X}, \tilde{A})$ and, therefore, an isomorphism of $\pi_1(X)$-modules. But $H_2(\tilde{X}, \tilde{A})$ is a free $\pi_1(X)$-module (it is a free $\pi_1(X)$-module with basis given by the 2-cells, which are to be attached; see, for example, the discussion on pages 211–212 in [105]) and condition (a) follows.

In order to verify condition (c) of Theorem 1.10.10, let us consider the central extension

$$0 \to \mathrm{Ker}(\partial) \to \pi_2(\tilde{X}, \tilde{A}) \xrightarrow{\partial} \pi_1(\tilde{A}) \to 1$$

and (part of) its associated exact homology sequence

$$H_2(\pi_2(\tilde{X}, \tilde{A})) \xrightarrow{\partial_*} H_2(\pi_1(\tilde{A})) \xrightarrow{\delta} \mathrm{Ker}(\partial).$$

It suffices to show that the connecting homomorphism δ is a monomorphism.

Consider the following portion of the homotopy–homology ladder for the pair $(\tilde{X}, \tilde{A})$:

$$\begin{array}{ccccccccc} \pi_2(\tilde{A}) & \xrightarrow{i_*} & \pi_2(\tilde{X}) & \xrightarrow{j_*} & \pi_2(\tilde{X},\tilde{A}) & \xrightarrow{\partial} & \pi_1(\tilde{A}) & \longrightarrow & 1 \\ \downarrow{\scriptstyle \Phi_1} & & \downarrow{\scriptstyle \Phi_2} & & \downarrow{\scriptstyle \Phi_3} & & \downarrow{\scriptstyle \Phi_4} & & \\ H_2(\tilde{A}) & \xrightarrow{i_*} & H_2(\tilde{X}) & \xrightarrow{j_*} & H_2(\tilde{X},\tilde{A}) & \xrightarrow{\partial} & H_1(\tilde{A}) & \longrightarrow & 1 \end{array}$$

Note that the Hurewicz homomorphisms Φ_3 and Φ_4 are epimorphisms with kernels equal to the respective commutator subgroups. Φ_2 is an isomorphism and we have $\mathrm{Coker}(\Phi_1) \cong H_2(\pi_1(\tilde{A}))$. This last fact is the statement that the quotient of $H_2(Y)$ by the subgroup of *spherical cycles* $\Sigma_2(Y)$, that is, the image of the Hurewicz homomorphism (in degree 2), is the group theoretical H_2 of the fundamental group $\pi_1(Y)$:

$$H_2(Y)/\Sigma_2(Y) \cong H_2(\pi_1(Y)).$$

This result is due to Hopf [52] for the case of connected polyhedra, but enlarged in its scope and explained more thoroughly in [25].

Consider the homomorphism $j_* \circ \Phi_2^{-1} \circ i_* : H_2(\tilde{A}) \to \pi_2(\tilde{X},\tilde{A})$. By exactness of the upper line, $\mathrm{Im}(\Phi_1)$ is in the kernel, so $j_* \circ \Phi_2^{-1} \circ i_*$ induces a homomorphism d : $\mathrm{Coker}(\Phi_1) \to \mathrm{Ker}(\partial)$ (again by exactness). In [25], there is defined a canonical isomorphism $\kappa : \mathrm{Coker}(\Phi_1) \to H_2(\pi_1(\tilde{A}))$. Define $\triangle : H_2(\pi_1(\tilde{A})) \to \mathrm{Ker}(\partial)$ by $\triangle = d \circ \kappa^{-1}$. By looking closely at the definitions, one sees that $\triangle = \delta$ up to a sign.

Note that $i_* : H_2(\tilde{A}) \to H_2(\tilde{X})$ is a monomorphism, since $H_3(\tilde{X},\tilde{A}) = 0$. This implies that d is a monomorphism, hence δ is also. This shows that condition (c) of Theorem 1.10.10 is also satisfied. □

Remark 1.10.4. Let us come back one moment to the canonical isomorphism κ : $\mathrm{Coker}(\Phi_1) \to H_2(\pi_1(\tilde{A}))$ constructed in [25]. It comes from a chain transformation

$$\kappa : S_1(X) \to K(\pi_1(X))$$

from a "based version" $S_1(X)$ of the singular chain complex $S(X)$ (i. e., all vertices of standard simplexes are mapped to the base point) to the group cohomological chain complex $K(\pi_1(X))$ of the group $\pi_1(X)$. As all vertices are mapped to the base point, edges are mapped to based loops in X and give elements of $\pi_1(X)$. In this way, a based singular chain in X is mapped to a group cohomological chain of $\pi_1(X)$. The main property of this morphism of complexes is that spherical cycles are mapped to zero (see Lemma 8.2 in *loc. cit.*).

1.10.3 Classifying 2-homotopy types

In this subsection, we use the crossed module $\partial_* : \pi_2(X,A) \to \pi_1(A)$ to classify 2-homotopy types. We base our exposition on the article [70].

Let K be a connected CW complex, and denote by K_n its n-skeleton. The CW complexes K and K' are called of the same $n-1$-*(homotopy)-type* in case there exist maps $\phi : K_n \to K'_n$ and $\phi' : K'_n \to K_n$ and homotopies

$$\phi' \circ \phi|_{K_{n-1}} \simeq i : K_{n-1} \to K_n,$$

and

$$\phi \circ \phi'|_{K'_{n-1}} \simeq i' : K'_{n-1} \to K'_n,$$

where i and i' are the inclusion maps of the $n-1$-skeleton. Let us denote this situation as $\phi : K_n \equiv_{n-1} K'_n$. Observe that any two (connected) CW complexes have the same 0-type, so this relation is meaningful only for $n > 1$.

The classification of CW complexes according to their 1-type is equivalent to the classification of groups by the relation of isomorphism. This can be seen by using the functors $K \mapsto \pi_1(K)$ and $G \mapsto BG$. The goal of this subsection is to deal with 2-types and to sketch the proof that their classification is given by crossed modules.

Since the $n-1$-type of a CW complex K depends only on its n-skeleton K_n, we may in our study always replace K by K_3. Therefore, we may (and will) assume in the following that K is at most 3-dimensional.

Definition 1.10.5. By an *algebraic 2-type*, we mean a triple $T = (\pi_1, \pi_2, \kappa)$ which consists of:
(a) an arbitrary (multiplicative) group π_1;
(b) an additively written abelian group π_2 affording the structure of a π_1-module;
(c) a degree 3 cohomology class $\kappa \in H^3(\pi_1, \pi_2)$.

Associate to a CW complex K an algebraic 2-type $T(K)$: Given K, we have constructed the crossed module $\partial_* : \pi_2(K, K_1) \to \pi_1(K_1)$ in Corollary 1.10.6. Note that by the long homotopy sequence (1.17), we have $\mathrm{Coker}(\partial_*) = \pi_1(K)$ and $\mathrm{Ker}(\partial_*) = \pi_2(K)$. The crossed module defines a cohomology class $[\gamma(K)] \in H^3(\pi_1(K), \pi_2(K))$. Now put $T(K) := (\pi_1(K), \pi_2(K), [\gamma(K)])$.

Definition 1.10.6. Let $T = (\pi_1, \pi_2, \kappa)$ and $T' = (\pi'_1, \pi'_2, \kappa')$ be two algebraic 2-types. By a *homomorphism* $\theta = (\theta_1, \theta_2) : T \to T'$, we mean a pair of a group homomorphism $\theta_1 : \pi_1 \to \pi'_1$ and a homomorphism $\theta_2 : \pi_2 \to \pi'_2$ of π_1-modules from the π_1-module π_2 to the π_1-module π'_2, seen as a π_1-module via θ_1. If k (resp., k') is a normalized 3-cocycle with $\kappa = [k]$ (resp., $\kappa' = [k']$), we require furthermore that $\theta_2 k(x, y, z)$ is cohomologuous to $k'(\theta_1(x), \theta_1(y), \theta_1(z))$ for all $x, y, z \in \pi_1$.

The homomorphism θ is called an *isomorphism* in case both θ_1 and θ_2 are isomorphisms.

The isomorphism relation is clearly an equivalence relation on the class of algebraic homotopy types.

We shall say that a given algebraic 2-type T is *realized* by a complex K in case there exists a CW complex K such that $T \cong T(K)$. Let CW complexes $K = K_3$ and $K' = K_3'$ be given and suppose $\phi : K \to K'$ is a map of CW complexes. Let $\pi_n = \pi_n(K)$ and $\pi_n' = \pi_n(K')$ for $n = 1, 2$. Then the induced homomorphisms $\phi_1 : \pi_1 \to \pi_1'$ and $\phi_2 : \pi_2 \to \pi_2'$ satisfy the requirements of a homomorphism of algebraic 2-types (the fact that $\theta_2 k(x, y, z)$ is cohomologuous to $k'(\theta_1(x), \theta_1(y), \theta_1(z))$ follows from the explicit definition of k and k'). Thus ϕ induces a homomorphism $\phi : T(K) \to T(K')$. Conversely, given a homomorphism $\phi : T(K) \to T(K')$, a map of CW complexes $\phi : K \to K'$ inducing it will be called a *realization* of ϕ.

The main results of [69] are the following.

Theorem 1.10.13. *The CW complexes K and K' have the same 2-type if and only if $T(K) \cong T(K')$.*

Theorem 1.10.14. *Any algebraic 2-type may be realized by some CW complex.*

Theorem 1.10.15. *For CW complexes K and K', a given homomorphism $T(K) \to T(K')$ has a realization $\phi : K \to K'$ provided* $\dim(K) \leq 3$.

We will concentrate here on (and sketch the proof of) Theorem 1.10.14 and give no hints about the other two theorems.

Suppose given an algebraic 2-type $T = (\pi_1, \pi_2, \kappa)$. Our goal is to construct a CW complex K with $T(K) = (\pi_1(K), \pi_2(K), [\gamma(K)]) = T$. For this, we represent first the group π_1 as a quotient of a free group $F =: X$ by a relation subgroup R. The class $\kappa \in H^3(\pi_1, \pi_2)$ can then be translated into a deviation class using Theorem 1.5.8, and for this deviation class, there exists by Theorem 1.5.6 a crossed module

$$0 \to \pi_2 \to M \stackrel{\mu}{\to} F \to \pi_1 \to 1,$$

which represents the deviation class. Denote X by ρ_1.

In a second step, choose a set of generators $\{m_\alpha\}_{\alpha \in A}$ for M, and construct the free crossed module $\mu_W : M_W \to X$ with symbolic generators (x, m) for all $x \in X$ and $m_\alpha \in M$ and with μ_W determined by $\mu_W(x, m_\alpha) = x\mu(m_\alpha)x^{-1}$; see Lemma 1.10.7. Denote M_W by ρ_2. We have seen that the crossed module $\mu_W : M_W \to X$ is a free crossed module (see Proposition 1.10.8) and, therefore, there exists a morphism of crossed modules $\omega : (\mu_W : M_W \to X) \to (\mu : M \to X)$ induced by $\omega : M_W \to M$, which is defined by $(1, m_\alpha) \mapsto m_\alpha$. Visually,

$$\begin{array}{ccccccccccc}
0 & \longrightarrow & V & \longrightarrow & M_W = \rho_2 & \stackrel{\mu_W}{\longrightarrow} & X = \rho_1 & \longrightarrow & \pi_1 & \longrightarrow & 1 \\
 & & \downarrow & & \downarrow{\scriptstyle \omega} & & \downarrow{\scriptstyle \mathrm{id}_X} & & \downarrow{\scriptstyle \mathrm{id}_{\pi_1}} & & \\
0 & \longrightarrow & \pi_2 & \longrightarrow & M & \stackrel{\mu}{\longrightarrow} & X = \rho_1 & \longrightarrow & \pi_1 & \longrightarrow & 1
\end{array}$$

This presents our crossed module $\mu : M \to X$ as

$$0 \to V/\omega^{-1}(0) \to \rho_2/\omega^{-1}(0) \to \rho_1 \to \pi_1 \to 1.$$

The last step is to replace the π_1-module $\pi_2 \cong V/\omega^{-1}(0)$ by a free π_1-module. For this, let ρ_3 be a free π_1-module such that $\omega^{-1}(0)$ is its quotient (by some relation submodule). We obtain a (not necessarily exact) sequence

$$0 \to \rho_3 \to \rho_2 \to \rho_1 \to \pi_1 \to 1.$$

This is a homotopy system of dimension 3 in the sense of [70]. As such, it admits a realization by a CW complex K by Theorem 2 in [104]. The resulting CW complex K is the solution of our realization problem. It is obtained from attaching k-cells successively to the $k-1$ skeleton, starting from the base point e_0. For this, the freeness of the terms ρ_i is important. The attaching maps are chosen using according to a choice of generators and the isomorphisms involved in the construction. More details are found in the proof of Theorem 2 in [104].

1.11 Exercises

Exercise 1.11.1. Consider the usual nonsplit exact sequence

$$0 \to \mathbb{Z}_2 \to \mathbb{Z}_4 \to \mathbb{Z}_2 \to 0$$

as a central extension and obtain a 2-cocycle defining it.

Exercise 1.11.2. Verify the assertions in (b), (c) and (d) of Remark 1.1.5.

Compute the group of outer automorphisms and write down the corresponding crossed module for the symmetric group S_6 and for $\mathbb{Z}_2 \oplus \mathbb{Z}_2$. What are the associated 3-classes?

Exercise 1.11.3. Suppose that the base field $\mathbb{K}$ is of characteristic zero. Then all cohomology spaces $H^k(G, \mathbb{K})$ for $k \geq 1$ and for a finite group G are trivial. This is due to the existence of a norm map.

On the other hand, for a nontrivial p-group G over a field $\mathbb{K}$ of prime characteristic p, the cohomology spaces $H^k(G, \mathbb{K})$ of G with values in the trivial module $\mathbb{K}$ are nonzero for all $k \geq 0$. Show this for $p = 2$ and $G = C_2$, the cyclic group with two elements, by gluing together an infinite number of sequences

$$0 \to \mathbb{K}[G]^+ \to \mathbb{K}[G] \to \mathbb{K} \to 0,$$

where $\mathbb{K}[G]^+$ is the augmentation ideal of the group algebra $\mathbb{K}[G]$ of G.

Thus in particular $H^3(C_2, \mathbb{K}) \neq 0$. Determine a crossed module representing a generator of this cohomology space.

Exercise 1.11.4. Compare the equivalence relation between crossed modules to morphisms in the derived category of an abelian category. Is it always possible to reduce the succession of morphisms between two crossed modules going in opposition directions to a single roof?

Exercise 1.11.5. Complete the proof of Lemma 1.1.1 by examining the functoriality of $\mathrm{crmod}(G,-)$ and by showing the additivity of the functor $\mathrm{crmod}(G,-)$.

Exercise 1.11.6. Related to the proof of Lemma 1.2.3, show that for $m, \tilde{m} \in M$ with $m\tilde{m} \in Z(M)$, we have

$$\mathrm{conj}_m = \mathrm{conj}_{\tilde{m}}.$$

Exercise 1.11.7. Show Remark 1.5.1.

Exercise 1.11.8. Related to the proof of Theorem 1.6.2, show that the different liftings of an automorphism of L to an automorphism of a fixed central extension $\hat{L}$ of L by V form a torsor under the cohomology group $H^1(L, V)$.

Exercise 1.11.9. Consider the following diagram for a given epimorphism $\pi : N \to G$ and a given action of G on V (which induces an action of N on V via π):

$$\begin{array}{ccc}
\mathrm{Ext}(N, V) & \xrightarrow{\cong} & H^2(N, V) \\
\downarrow{\scriptstyle \varphi} & & \downarrow{\scriptstyle \kappa^*} \\
\mathrm{crmod}(G, N, V) & \xrightarrow{\xi} & H^3(G, N, V) \\
\downarrow{\scriptstyle \text{forget}} & & \downarrow{\scriptstyle \partial} \\
\mathrm{crmod}(G, V) & \xrightarrow{\cong} & H^3(G, V)
\end{array}$$

The isomorphisms in the diagram are induced by the interpretation of abelian extensions in terms of 2-cocycles and by the Gerstenhaber map B (from Theorem 1.2.1). The right-hand column is exact and induced from the short exact sequence of complexes (1.15). In the left-hand column, the map forget sends a relative crossed module to the underlying crossed module. Note that this is well-defined on equivalence classes. The map φ is defined as follows: Given an abelian extension $V \times_\omega N$, the 2-cocycle $\omega \in Z^2(N, V)$ is sent to $i^*\omega \in Z^2(L, V)$ via the map $i : L \to N$ for $L = \mathrm{Ker}(\pi)$. Then exactly as in the proof of Theorem 1.8.1 at the end of Section 1.8, $i^*\omega$ permits to construct a relative crossed module

$$0 \to V \to V \times_{i^*\omega} L \to N \to G \to 0.$$

We put $\varphi([\omega]) = [\mu : V \times_{i^*\omega} L \to N]$. Note that this is also well-defined.

Show that the two squares in the above diagram are commutative.

Exercise 1.11.10. Provide the details for Remark 1.8.4, that is, show in detail that the relative Gerstenhaber Theorem 1.8.1 implies the absolute Gerstenhaber Theorem 1.2.1. A way how to proceed is indicated in Remark 2.9.5 in the context of Lie algebras.

Exercise 1.11.11. Use Theorem 1.10.14 to devise a proof of the surjectivity claim in Theorem 1.2.1. How to recover the entire content of Theorem 1.2.1 from the homotopy theoretic context?

1.12 Bibliographical notes

The definition of a crossed module of groups has been coined by J. H. C. Whitehead in 1949 [104]. The procedure of associating a 3-cocycle to a crossed module is due to Eilenberg–Mac Lane [26] according to Gerstenhaber [37]. Theorem 1.2.1 has been stated for the first time by Gerstenhaber in [37], according to the historical note by Mac Lane [71]. For the proof, Gerstenhaber refers to [36] where he proves in fact that splicing together long exact sequences of R-modules with abelian extensions yields a cohomology theory, which is again $\mathrm{Ext}^*_R(M,N)$. The proof is by stating that both theories satisfy the axioms and vanish on injectives (without further details).

The lemmas in Section 1.2 are either new or taken from Neeb's article [83]. Sections 1.3 and 1.4 are entirely due to Neeb [83], where he adapts Eilenberg–Mac Lane's theory [26] to the case of infinite dimensional Lie groups. It depicts crossed modules as obstructions to the existence of extensions.

Section 1.5 is the adaptation of my approach in [101] to the setting of groups. Lemma 1.5.3 is due to Gerstenhaber; see [37]. Section 1.5.3 is due to Mac Lane [69]. Section 1.6 is taken from [32], but note that results similar to those of this section have been obtained by Ratcliffe and Huebschmann; see [91] and [54].

The first example in Section 1.7.1 is due to Stolz–Teichner [96], while the second example in Section 1.7.2 is due to Baez–Crans–Schreiber–Stevenson [5]. The example of Section 1.7.3 is from [101]. Section 1.8 is due to Loday [63]. I understood only from Casas [16] and private communication with M. Ladra that the relative framework implies the absolute framework and is thus completely equivalent; see Remark 1.8.4.

I learned about the interpretation of crossed modules in terms of categorified groups from Baez–Lauda [3]. An early reference for these matters is Loday's article [64].

Section 1.10 is the origin of the theory of crossed modules and taken from J. H. C. Whitehead's article [104] and his article with S. Mac Lane [70]. The introductory material is taken from the book of G. W. Whitehead [105]. Ratcliffe reworked Whitehead's theory of free crossed modules and generalized it to projective crossed modules [90]. In the material from [70], we applied a shift in degree in order to come back to homotopy 2-types (which Mac Lane and Whitehead called homotopy 3-types).

2 Crossed modules of Lie algebras

In this chapter, we discuss crossed modules of Lie algebras. In order to emphasize the similarities with Chapter 1, we proceed in the same order, namely after giving definitions in Section 2.1, we discuss crossed modules in relation to third cohomology (Sections 2.2 to 2.5). In Section 2.6, we report on Hochschild's theory of Lie algebra kernels. Section 2.7 covers the relation with the Hochschild–Serre spectral sequence. In Section 2.8, we list some less elementary examples of crossed modules of Lie algebras, related to the string Lie algebra and the Godbillon–Vey cocycle. Lie 2-algebras, a categorized version of Lie algebras, appear in Section 2.10. The chapter closes with exercises in Section 2.11. Please see Appendix B for a definition of Lie algebra cohomology and the Hochschild–Serre spectral sequence.

All Lie algebras in this chapter are Lie algebras over a fixed field $\mathbb{K}$ of characteristic zero.

2.1 Definitions

Definition 2.1.1. A *crossed module of Lie algebras* is the data of a homomorphism of Lie algebras $\mu : \mathfrak{m} \to \mathfrak{n}$ together with an action η of $\mathfrak{n}$ on $\mathfrak{m}$ by derivations, denoted $\eta : \mathfrak{n} \to \mathfrak{Der}(\mathfrak{m})$ or sometimes simply $m \mapsto n \cdot m$ for all $m \in \mathfrak{m}$ and all $n \in \mathfrak{n}$, such that:
(a) $\mu(n \cdot m) = [n, \mu(m)]$ for all $n \in \mathfrak{n}$ and all $m \in \mathfrak{m}$;
(b) $\mu(m) \cdot m' = [m, m']$ for all $m, m' \in \mathfrak{m}$.

Remark 2.1.2. Property (a) means that the morphism μ is equivariant with respect to the $\mathfrak{n}$-action via η on $\mathfrak{m}$ and the adjoint action on $\mathfrak{n}$. Property (b) is called *Peiffer identity*.

Remark 2.1.3. To each crossed module of Lie algebras $\mu : \mathfrak{m} \to \mathfrak{n}$, one associates a four-term exact sequence

$$0 \to V \xrightarrow{i} \mathfrak{m} \xrightarrow{\mu} \mathfrak{n} \xrightarrow{\pi} \mathfrak{g} \to 0,$$

where $\mathrm{Ker}(\mu) =: V$ and $\mathfrak{g} := \mathrm{Coker}(\mu)$.

Remark 2.1.4.
(a) By property (a), $\mathrm{Im}(\mu)$ is an ideal, and thus $\mathfrak{g}$ is a Lie algebra.
(b) By property (b), V is a central ideal of $\mathfrak{m}$, and in particular abelian.
(c) Lifting elements of $\mathfrak{g}$ to $\mathfrak{n}$, the action of $\mathfrak{n}$ on $\mathfrak{m}$ induces an *outer action* of $\mathfrak{g}$ on $\mathfrak{m}$. Given linear sections ρ and ρ' of π and an element $x \in \mathfrak{g}$, $\eta(\rho(x))$ and $\eta(\rho'(x))$ differ by the inner derivation $\mathrm{ad}_{m'}$ for some $m' \in \mathfrak{m}$. Indeed,

$$\eta(\rho(x))(m) - \eta(\rho'(x))(m) = \eta((\rho - \rho')(x))(m)$$

https://doi.org/10.1515/9783110750959-002

$$= \eta(\mu(m'))(m)$$
$$= [m', m].$$

Here, we used property (b) in the last line, and m' exists by exactness of the four-term sequence, because $(\rho - \rho')(x) \in \operatorname{Ker}(\pi) = \operatorname{Im}(\mu)$. This means that the expression $\eta(\rho(x))$ is well-defined up to inner isomorphism.

Moreover, $\eta \circ \rho$ satisfies the requirements of an action also up to inner derivations. Indeed, denote for all $x, y \in \mathfrak{g}$ by

$$\alpha(x,y) := [\rho(x), \rho(y)] - \rho([x,y]),$$

the failure of ρ to be a morphism of Lie algebras. Then $\alpha(x,y) \in \operatorname{Ker}(\pi)$, because π is a morphism and ρ a section of π. Therefore, by exactness of the four-term sequence, there exists $\beta(x,y) \in \mathfrak{m}$ such that

$$\mu(\beta(x,y)) = \alpha(x,y).$$

Then to show that $\eta \circ \rho$ is an outer action, we have to consider $\eta([\rho(x), \rho(y)] - \rho([x,y]))(m)$ for some $m \in \mathfrak{m}$. But this gives

$$\eta(\alpha(x,y))(m) = \eta(\mu(\beta(x,y)))(m) = [\beta(x,y), m]$$

by property (b), and in this sense, an outer action is an action up to inner derivations.

Now property (a) implies that the restriction of this outer action to V induces the structure of a $\mathfrak{g}$-module on V, because as V is central, the inner automorphisms of $\mathfrak{m}$ act trivially on V.

(d) Note that by this lifting procedure, the action η (resp., the adjoint action) does not in general render $\mathfrak{m}$ (resp., $\mathfrak{n}$) a $\mathfrak{g}$-module.

Remark 2.1.5. Let us list some elementary examples of crossed modules:

(a) Each central extension is a crossed module. Indeed, central extensions correspond exactly to the case where the map μ is surjective.

(b) Each (inclusion of an) ideal in a Lie algebra constitutes a crossed module. Indeed, an (inclusion of an) ideal correspond exactly to the case where the map μ is injective.

(c) For each Lie algebra $\mathfrak{l}$, there is a canonical crossed module

$$\mu : \mathfrak{l} \to \mathfrak{der}(\mathfrak{l}),$$

where μ sends an element $x \in \mathfrak{l}$ to the inner derivation ad_x defined by $\mathrm{ad}_x(y) = [x,y]$ for all $y \in \mathfrak{l}$. The action of $\mathfrak{der}(\mathfrak{l})$ on $\mathfrak{l}$ is the usual action as a derivation. The kernel of μ is the center $\mathfrak{z}(\mathfrak{l})$ of $\mathfrak{l}$ and the cokernel of μ is the Lie algebra of outer derivations

$$\mathfrak{out}(\mathfrak{l}) := \mathfrak{der}(\mathfrak{l})/\mathrm{ad}(\mathfrak{l}).$$

Definition 2.1.6. Two crossed modules $\mu : \mathfrak{m} \to \mathfrak{n}$ (with action η) and $\mu' : \mathfrak{m}' \to \mathfrak{n}'$ (with action η') such that $\mathrm{Ker}(\mu) = \mathrm{Ker}(\mu') =: V$ and $\mathrm{Coker}(\mu) = \mathrm{Coker}(\mu') =: \mathfrak{g}$ are called *elementary equivalent* if there are morphisms of Lie algebras $\varphi : \mathfrak{m} \to \mathfrak{m}'$ and $\psi : \mathfrak{n} \to \mathfrak{n}'$, which are compatible with the actions, meaning

$$\varphi(\eta(n)(m)) = \eta'(\psi(n))(\varphi(m)),$$

for all $n \in \mathfrak{n}$ and all $m \in \mathfrak{m}$, and such that the following diagram is commutative:

$$\begin{array}{ccccccccc}
0 & \longrightarrow & V & \xrightarrow{i} & \mathfrak{m} & \xrightarrow{\mu} & \mathfrak{n} & \xrightarrow{\pi} & \mathfrak{g} & \longrightarrow & 0 \\
 & & \downarrow{\scriptstyle \mathrm{id}_V} & & \downarrow{\scriptstyle \varphi} & & \downarrow{\scriptstyle \psi} & & \downarrow{\scriptstyle \mathrm{id}_{\mathfrak{g}}} & & \\
0 & \longrightarrow & V & \xrightarrow{i'} & \mathfrak{m}' & \xrightarrow{\mu'} & \mathfrak{n}' & \xrightarrow{\pi'} & \mathfrak{g} & \longrightarrow & 0
\end{array}$$

We call *equivalence of crossed modules* the equivalence relation generated by elementary equivalence. One easily sees that two crossed modules are equivalent in case there exists a zig-zag of elementary equivalences going from one to the other (which are not necessarily going all in the same direction).

Let us denote by $\mathrm{crmod}(\mathfrak{g}, V)$ the set of equivalence classes of Lie algebra crossed modules with respect to fixed kernel V and fixed cokernel $\mathfrak{g}$.

Remark 2.1.7. Compare this equivalence relation to the equivalence of two abelian extensions; cf. Appendix B. In the framework of extensions, equivalence imposes the underlying vector space of the extension up to isomorphism. For crossed modules, this is not the case, and leads thus to much more different representatives of the equivalence class of a crossed module.

Recall the definition of the *sum of two crossed modules*; cf. [36].

Consider two crossed modules $\mu : \mathfrak{m} \to \mathfrak{n}$ and $\mu' : \mathfrak{m}' \to \mathfrak{n}'$ with isomorphic kernel and cokernel and their corresponding four-term exact sequences

$$0 \to V \xrightarrow{i} \mathfrak{m} \xrightarrow{\mu} \mathfrak{n} \xrightarrow{\pi} \mathfrak{g} \to 0$$

and

$$0 \to V \xrightarrow{i'} \mathfrak{m}' \xrightarrow{\mu'} \mathfrak{n}' \xrightarrow{\pi'} \mathfrak{g} \to 0.$$

Denote by $K := \{(v, -v) \in V \oplus V\}$ the kernel of the addition map $V \oplus V \to V$. Notice that the diagonal $\triangle : V \to V \oplus V$ followed by the quotient map $V \oplus V \to (V \oplus V)/K$ identifies V and $(V \oplus V)/K$. The space K can be considered as a subspace in $\mathfrak{m} \oplus \mathfrak{m}'$ via $i \oplus i'$. As V is central in $\mathfrak{m}$ and $\mathfrak{m}'$, K is an ideal of $\mathfrak{m} \oplus \mathfrak{m}'$.

Denote by $\mathfrak{n} \oplus_{\mathfrak{g}} \mathfrak{n}'$ the pullback associated to the maps $\pi : \mathfrak{n} \to \mathfrak{g}$ and $\pi' : \mathfrak{n}' \to \mathfrak{g}$. More explicitly,

$$\mathfrak{n} \oplus_{\mathfrak{g}} \mathfrak{n}' = \{(n, n') \in \mathfrak{n} \oplus \mathfrak{n}' \mid \pi(n) = \pi'(n')\}.$$

Notice that the two maps $\pi : (\mathfrak{n} \oplus_{\mathfrak{g}} \mathfrak{n}') \to \mathfrak{g}$ and $\frac{1}{2}(\pi + \pi') : (\mathfrak{n} \oplus_{\mathfrak{g}} \mathfrak{n}') \to \mathfrak{g}$ coincide (because in our base field $\mathbb{K}$, 2 is invertible).

With these preparations, we have the following definition:

Definition 2.1.8. The *sum* of two crossed modules $\mu : \mathfrak{m} \to \mathfrak{n}$ and $\mu' : \mathfrak{m}' \to \mathfrak{n}'$ such that $\mathrm{Ker}(\mu) = \mathrm{Ker}(\mu') = V$ and $\mathrm{Coker}(\mu) = \mathrm{Coker}(\mu) = \mathfrak{g}$ is by definition the crossed module

$$0 \to V \xrightarrow{(i\oplus i')\circ\Delta} (\mathfrak{m} \oplus \mathfrak{m}')/K \xrightarrow{\mu\oplus\mu'} (\mathfrak{n} \oplus_{\mathfrak{g}} \mathfrak{n}') \xrightarrow{\pi} \mathfrak{g} \to 0.$$

The action of $\mathfrak{n}\oplus_{\mathfrak{g}}\mathfrak{n}'$ on $(\mathfrak{m}\oplus\mathfrak{m}')/K$ by derivations is induced from the sum of actions on the two summands. The compatibility relations (a) and (b) of definition 2.1.1 are true in the direct sum, thus are true for the crossed module sum.

Lemma 2.1.1. *The sum of crossed modules defines an abelian group structure on the set of equivalence classes of crossed modules* $\mathrm{crmod}(\mathfrak{g}, V)$ *with given kernel* V *and cokernel* $\mathfrak{g}$.

Proof. It is clear that the sum of crossed modules is associative and commutative up to equivalence as it is induced by the direct sum. It is equally clear that the sum is compatible with the equivalence relation as we can sum the maps giving the equivalences.

We have to show that there is a zero element and an inverse to every crossed module. These definitions are taken from [36], page 4: We define the zero crossed module with given kernel V and cokernel $\mathfrak{g}$ to be

$$0 \to V \overset{\mathrm{id}_V}{\to} V \overset{0}{\to} \mathfrak{g} \overset{\mathrm{id}_{\mathfrak{g}}}{\to} \mathfrak{g} \to 0$$

and the inverse of a crossed module

$$0 \to V \overset{i}{\to} \mathfrak{m} \to \mathfrak{n} \to \mathfrak{g} \to 0$$

to be

$$0 \to V \overset{-i}{\to} \mathfrak{m} \to \mathfrak{n} \to \mathfrak{g} \to 0.$$

In order to show that this crossed module is inverse to the given one, notice that $\mathrm{crmod}(\mathfrak{g}, -)$ is an additive functor. Thus we have for an equivalence class $[\mu : \mathfrak{m} \to \mathfrak{n}] \in \mathrm{crmod}(\mathfrak{g}, V)$,

$$(\alpha_1 + \alpha_2)[\mu : \mathfrak{m} \to \mathfrak{n}] = \alpha_1[\mu : \mathfrak{m} \to \mathfrak{n}] + \alpha_2[\mu : \mathfrak{m} \to \mathfrak{n}],$$

where $\alpha_i : V \to V'$ ($i = 1, 2$) are two $\mathfrak{g}$-module morphisms. We refer to [68] Chapter III, Section 5, for more information. This reduces the proof to showing that pushforward by the zero map $0 : V \to V$ gives the zero class. Now, pushforward by the zero map splits up a direct factor V in $(V \oplus \mathfrak{m})/(0 \oplus (-i(V)))$ and we have then a commutative diagram

$$\begin{array}{ccccccccccc}
0 & \longrightarrow & V & \xrightarrow{\quad i \quad} & \mathfrak{m} & \xrightarrow{\quad \mu \quad} & \mathfrak{n} & \xrightarrow{\pi} & \mathfrak{g} & \longrightarrow & 0 \\
 & & \downarrow{\scriptstyle 0} & & \downarrow{\scriptstyle \mathrm{incl}_2} & & \downarrow{\scriptstyle \mathrm{id}_{\mathfrak{n}}} & & \downarrow{\scriptstyle \mathrm{id}_{\mathfrak{g}}} & & \\
0 & \longrightarrow & V & \longrightarrow & (V \oplus \mathfrak{m})/(0 \oplus (-i(V))) & \xrightarrow{0 \oplus \mu} & \mathfrak{n} & \xrightarrow{\pi} & \mathfrak{g} & \longrightarrow & 0 \\
 & & \downarrow{\scriptstyle \mathrm{id}_V} & & \downarrow{\scriptstyle \mathrm{proj}_1} & & \downarrow{\scriptstyle \pi} & & \downarrow{\scriptstyle \mathrm{id}_{\mathfrak{g}}} & & \\
0 & \longrightarrow & V & \xrightarrow{\quad \mathrm{id}_V \quad} & V & \xrightarrow{\quad 0 \quad} & \mathfrak{g} & \xrightarrow{\mathrm{id}_{\mathfrak{g}}} & \mathfrak{g} & \longrightarrow & 0
\end{array}$$

Here, incl_2 and proj_2 are the standard inclusions and projections to/from the direct sum. This shows that the class $0[\mu : \mathfrak{m} \to \mathfrak{n}]$ is the zero class. □

2.2 Relation to third cohomology I

2.2.1 Gerstenhaber's theorem

Crossed modules may serve as explicit representatives of third cohomology classes; this is the essence of the following theorem, which Mac Lane [67] attributes to Gerstenhaber.

Theorem 2.2.1. *The map, which associates to a crossed module its 3-cocycle γ (to be defined below), induces an isomorphism of abelian groups*

$$b : \mathrm{crmod}(\mathfrak{g}, V) \cong H^3(\mathfrak{g}, V).$$

The proof of Theorem 2.2.1 will be completed in Section 2.5. In a first step, we will only discuss in this section how to associate a cohomology class in $H^3(\mathfrak{g}, V)$ to a given crossed module.

We shall use the notation $b([\mu : \mathfrak{m} \to \mathfrak{n}])$ or simply $b([\mu])$ for b applied to a crossed module $\mu : \mathfrak{m} \to \mathfrak{n}$ representing an equivalence class $[\mu : \mathfrak{m} \to \mathfrak{n}]$.

Let us show how to associate to a crossed module a 3-cocycle of $\mathfrak{g}$ with values in V. For this, recall the exact sequence from Remark 2.1.3:

$$0 \to V \xrightarrow{i} \mathfrak{m} \xrightarrow{\mu} \mathfrak{n} \xrightarrow{\pi} \mathfrak{g} \to 0.$$

The first step is to take a linear section ρ of π and to compute the failure of ρ to be a Lie algebra homomorphism, that is,

$$\alpha(x_1, x_2) = [\rho(x_1), \rho(x_2)] - \rho([x_1, x_2]).$$

Here, $x_1, x_2 \in \mathfrak{g}$. The map α is bilinear and skewsymmetric in x_1, x_2. We have obviously $\pi(\alpha(x_1, x_2)) = 0$, because π is a Lie algebra homomorphism, so $\alpha(x_1, x_2) \in \mathrm{Im}(\mu) = \mathrm{Ker}(\pi)$. This means by exactness that there exists $\beta(x_1, x_2) \in \mathfrak{m}$ such that

$$\mu(\beta(x_1, x_2)) = \alpha(x_1, x_2).$$

Choosing a linear section σ on $\mathrm{Im}(\mu)$, one can choose β as

$$\beta(x_1, x_2) = \sigma(\alpha(x_1, x_2)) \tag{2.1}$$

showing that we can suppose β bilinear and skewsymmetric in x_1, x_2.

Now, $\mu(d^{\mathfrak{m}}\beta(x_1, x_2, x_3)) = 0$ by the following lemma. The coboundary operator $d^{\mathfrak{m}}$ is here formally defined as in (B.1); see Appendix B.

Lemma 2.2.2. $\mu(d^{\mathfrak{m}}\beta(x_1, x_2, x_3)) = 0$ *where* $x_1, x_2, x_3 \in \mathfrak{g}$ *and where* $d^{\mathfrak{m}}$ *is the formal expression of the Lie algebra cohomology boundary operator corresponding to cohomology of* $\mathfrak{g}$ *with values in* $\mathfrak{m}$ *(cf. Formula* (B.1) *in Appendix* B*).*

Remark 2.2.1. For general information about the cohomology of Lie algebras, please refer to Appendix B.

By cohomology of $\mathfrak{g}$ with values in $\mathfrak{m}$, we mean that $\mathfrak{g}$ "acts" on $\mathfrak{m}$ by $\eta \circ \rho$. This gives just formally the expression of a Lie algebra coboundary operator, because the map $\eta \circ \rho$ is not a Lie algebra action by derivations in general.

Proof.

$$\begin{aligned}
\mu(d^{\mathfrak{m}}\beta(x_1, x_2, x_3)) &= \mu\Big(\sum_{\text{cycl.}} \beta([x_1, x_2], x_3) - \sum_{\text{cycl.}} \eta(\rho(x_1)) \cdot \beta(x_2, x_3)\Big) \\
&= \sum_{\text{cycl.}} \alpha([x_1, x_2], x_3) - \sum_{\text{cycl.}} [\rho(x_1), \mu(\beta(x_2, x_3))] \\
&= \sum_{\text{cycl.}} \alpha([x_1, x_2], x_3) - \sum_{\text{cycl.}} [\rho(x_1), \alpha(x_2, x_3)] \\
&= \sum_{\text{cycl.}} ([\rho([x_1, x_2]), \rho(x_3)] - \rho([[x_1, x_2], x_3])) \\
&\quad - \sum_{\text{cycl.}} ([\rho(x_1), [\rho(x_2), \rho(x_3)]] + [\rho(x_1), \rho([x_2, x_3])]) = 0.
\end{aligned}$$

In the last two lines, we observed that the second and third cyclic sums vanish individually due to the Jacobi identity, while the first and last cyclic sum cancel. □

This means that $d^{\mathfrak{m}}\beta(x_1, x_2, x_3) \in \mathrm{Ker}(\mu) = \mathrm{Im}(i) = i(V)$, that is, there exists $\gamma(x_1, x_2, x_3) \in V$ such that $d^{\mathfrak{m}}\beta(x_1, x_2, x_3) = i(\gamma(x_1, x_2, x_3))$. The explicit formula of $d^{\mathfrak{m}}\beta(x_1, x_2, x_3)$ shows trilinearity and skew symmetry in the three variables x_1, x_2 and x_3. Choosing a linear section τ on $i(V) = \mathrm{Ker}(\mu)$, one can choose γ to be $\tau \circ d^{\mathfrak{m}}\beta$ (in the obvious sense) gaining that γ is also trilinear and skew symmetric in x_1, x_2 and x_3. By the following lemma, γ is a 3-cocycle of $\mathfrak{g}$ with values in V.

Lemma 2.2.3. *The map γ is a 3-cocycle of $\mathfrak{g}$ with values in V.*

Proof. We have to show that $d^V\gamma(x_1,x_2,x_3,x_4) = 0$, denoting by x_1, x_2, x_3, x_4 four elements of $\mathfrak{g}$ and by d^V as before the Lie algebra coboundary operator of $\mathfrak{g}$ with values in V. The expression for $d^V\gamma(x_1,x_2,x_3,x_4)$ is the sum of "action terms" and "bracket terms". It is enough to show $i(d^V\gamma(x_1,x_2,x_3,x_4)) = 0$:

$$\begin{aligned} i(d^V\gamma(x_1,x_2,x_3,x_4)) &= d^{\mathfrak{m}} i(\gamma(x_1,x_2,x_3,x_4)) \\ &= d^{\mathfrak{m}} d^{\mathfrak{m}} \beta(x_1,x_2,x_3,x_4). \end{aligned}$$

One has to be careful because $d^{\mathfrak{m}}d^{\mathfrak{m}}$ is not automatically zero, because of the fact that $\eta\circ\rho$ is not an action of $\mathfrak{g}$ on $\mathfrak{m}$ in general. Let us display here only some terms of it, while the sum of the other terms vanishes as usual. The terms we choose are all "action terms of the action terms" and some "action terms of the bracket terms."

$$\begin{aligned} &i(d^V\gamma(x_1,x_2,x_3,x_4)) \\ &\quad= \sum_{1\le i<j\le 4} (-1)^{i+j}\eta(\rho([x_i,x_j]))\beta(x_1,\dots,\widehat{x_i},\dots,\widehat{x_j},\dots,x_4) \\ &\quad\quad - \sum_{i=1}^{4}(-1)^i\eta(\rho(x_i))\sum_{l=1}^{3}(-1)^l\eta(\rho(z_l))\beta(z_1,\dots,\widehat{z_l},\dots,z_3). \end{aligned}$$

Here, we denote by z_1, z_2, z_3 the three remaining x_r after having chosen x_i from the list. Now, the difference of acting by $\eta(\rho([x,y]))$ and acting by $\eta(\rho(x))\eta(\rho(y)) - \eta(\rho(y))\eta(\rho(x))$ is just the action by $\eta(\alpha(x,y))$. Calculating the differences of the actions by the bracket and the action of the single elements gives thus terms of the form

$$\eta(\alpha(x,y))\beta(u,v) = \eta(\mu(\beta(x,y)))\beta(u,v) = [\beta(x,y),\beta(u,v)],$$

where the last equality follows from property (b) in Definition 2.1.1. Then the sum reads

$$\begin{aligned} &[\beta(x_1,x_2),\beta(x_3,x_4)] - [\beta(x_2,x_3),\beta(x_4,x_1)] + [\beta(x_3,x_4),\beta(x_1,x_2)] \\ &\quad - [\beta(x_4,x_1),\beta(x_2,x_3)] + [\beta(x_4,x_2),\beta(x_1,x_3)] - [\beta(x_1,x_3),\beta(x_2,x_4)] \end{aligned}$$

and vanishes. □

Lemma 2.2.4. *The class of the cocycle γ in $C^3(\mathfrak{g},V)$ does not depend on the choice of the sections ρ, σ and τ.*

Proof. (1) Let ρ and ρ' be two sections of π. Denote by $\alpha(x,y)$ respectively $\alpha'(x,y)$, $\beta(x,y)$ respectively $\beta'(x,y)$, $\gamma(x,y,z)$ respectively $\gamma'(x,y,z)$, the elements of $\mathfrak{n}$, $\mathfrak{m}$ and V constructed above with respect to ρ and ρ'. Here, $x,y,z\in\mathfrak{g}$. By construction, we have

$$\rho'(x) = \rho(x) + \delta(x),$$

for some linear map $\delta : \mathfrak{g} \to \mathrm{Ker}(\pi) \subset \mathfrak{n}$. But then α' may be written

$$\begin{aligned}\alpha'(x,y) &= [\rho'(x),\rho'(y)] - \rho'([x,y]) \\ &= [(\rho+\delta)(x),(\rho+\delta)(y)] - (\rho+\delta)([x,y]) \\ &= \alpha(x,y) + [\rho(x),\delta(y)] + [\delta(x),\rho(y)] + [\delta(x),\delta(y)] - \delta([x,y]). \end{aligned} \tag{2.2}$$

Let $\sigma : \mathrm{Im}(\mu) \to \mathfrak{m}$ be a section of μ. As δ has values in $\mathrm{Ker}(\pi) = \mathrm{Im}(\mu)$, $\sigma\delta(x)$ is a preimage of $\delta(x)$ with respect to μ. We can thus transform the terms in the following way using property (b) of a crossed module and the fact that μ is a morphism:

$$\begin{aligned}&[\rho(x),\delta(y)] + [\delta(x),\rho(y)] + [\delta(x),\delta(y)] - \delta([x,y]) \\ &\quad = \mu(\rho(x)\cdot\sigma\delta(y) - \rho(y)\cdot\sigma\delta(x) + [\sigma\delta(x),\sigma\delta(y)] - \sigma\delta[x,y]).\end{aligned}$$

The whole equation (2.2) becomes thus

$$\begin{aligned}&\mu(\beta'(x,y) - \beta(x,y)) \\ &\quad = \mu(\rho(x)\cdot\sigma\delta(y) - \rho(y)\cdot\sigma\delta(x) + [\sigma\delta(x),\sigma\delta(y)] - \sigma\delta[x,y]).\end{aligned}$$

Therefore, there exists $\zeta(x,y) \in V$ such that

$$\begin{aligned}&\beta'(x,y) - \beta(x,y) \\ &\quad = i\zeta(x,y) + \rho(x)\cdot\sigma\delta(y) - \rho(y)\cdot\sigma\delta(x) + [\sigma\delta(x),\sigma\delta(y)] - \sigma\delta[x,y].\end{aligned}$$

When applying in the next step d^{m} to all terms, the term in ζ will give a coboundary term, because $d^{\mathrm{m}}(i\zeta(x,y,z)) = id^V\zeta(x,y,z)$. Let us show that by applying d^{m} to all terms, we obtain $d^{\mathrm{m}}\beta'(x,y,z) - d^{\mathrm{m}}\beta(x,y,z) = id^V\zeta(x,y,z)$. This implies that the associated cocycles $\gamma(x,y,z)$ and $\gamma'(x,y,z)$ differ only by a coboundary.

Observe that we want the "action" in $d^{\mathrm{m}}\beta'(x,y,z)$ to be given by ρ', while at first, we take the cyclic sum of the terms $\beta'([x,y],z)$ minus the cyclic sum of the terms $\rho(x)\cdot\beta'(y,z)$, thus the action terms appear with the wrong "action" ρ. The additional terms in the equation will help transform the action terms in ρ into action terms in ρ'.

Namely, we claim that

$$\begin{aligned}&\sum_{\text{cycl.}} \rho'[x,y]\cdot\sigma\delta(z) - \rho'(x)\cdot(\rho'(y)\cdot\sigma\delta(z)) + \rho'(y)\cdot(\rho'(x)\cdot\sigma\delta(z)) \\ &\quad = \sum_{\text{cycl.}} \rho[x,y]\cdot\sigma\delta(z) - \rho(x)\cdot(\rho(y)\cdot\sigma\delta(z)) + \delta[x,y]\cdot\sigma\delta(z) + \rho(y)\cdot(\rho(x)\cdot\sigma\delta(z)) \\ &\quad\quad - \rho(x)\cdot(\delta(y)\cdot\sigma\delta(z)).\end{aligned}$$

This will show what we asserted, because the cyclic sum d^{m} applied (with the use of ρ) to

$$\rho(x)\cdot\sigma\delta(y) - \rho(y)\cdot\sigma\delta(x) + [\sigma\delta(x),\sigma\delta(y)] - \sigma\delta[x,y]$$

leads to

$$\sum_{\text{cycl.}} \rho[x,y]\cdot\sigma\delta(z) - \rho(y)\cdot\sigma\delta[x,y] + [\sigma\delta[x,y],\sigma\delta(z)] - \sigma\delta[[x,y],z]$$
$$- \sum_{\text{cycl.}} \rho(x)\cdot(\rho(y)\cdot\sigma\delta(z) - \rho(z)\cdot\sigma\delta(y) + [\sigma\delta(y),\sigma\delta(z)] - \sigma\delta[y,z]),$$

and here the last term in the first sum is zero due to the Jacobi identity, while the second term in the first sum and the last term in the second sum cancel. The remaining terms are exactly those after the equality sign in our claim (up to using again property (b) of a crossed module).

Moreover, the terms

$$\sum_{\text{cycl.}} \rho'[x,y]\cdot\sigma\delta(z) - \rho'(x)\cdot(\rho'(y)\cdot\sigma\delta(z)) + \rho'(y)\cdot(\rho'(x)\cdot\sigma\delta(z))$$
$$= -\sum_{\text{cycl.}} \alpha'(x,y)\cdot\sigma\delta(z) = \sum_{\text{cycl.}} \delta(z)\cdot\beta'(x,y)$$

are exactly the terms which permit to pass from $d^{\mathrm{m}}\beta'(x,y,z)$ with the action by ρ to $d^{\mathrm{m}}\beta'(x,y,z)$ with the correct action by ρ' (recall that the action terms come in with an overall minus sign!).

Thus we are reduced to showing the claim. When computing the first line of the claim by replacing $\rho'(u) = \rho(u) + \delta(u)$, we obtain the terms after the equality sign plus following the terms:

$$\sum_{\text{cycl.}} -\delta(x)\cdot(\rho(y)\cdot\sigma\delta(z)) - \delta(x)\cdot(\delta(y)\cdot\sigma\delta(z))$$
$$+ \rho(y)\cdot(\delta(x)\cdot\sigma\delta(z)) + \delta(y)\cdot(\rho(x)\cdot\sigma\delta(z)) + \delta(y)\cdot(\delta(x)\cdot\sigma\delta(z)).$$

It remains thus to show that this cyclic sum is zero. Transforming $\delta(x)\cdot(\delta(y)\cdot\sigma\delta(z))$ into $[\sigma\delta(x),[\sigma\delta(y),\sigma\delta(z)]]$ thanks to property (b) again, it is clear that the terms with three δ are zero by the Jacobi identity. For the still remaining terms, we compute (where we use the cyclicity of the sum and that the action is a derivation of the bracket!)

$$\begin{aligned}
&- \delta(z)\cdot(\rho(x)\cdot\sigma\delta(y)) + \delta(y)\cdot(\rho(x)\cdot\sigma\delta(z))\\
&\quad = -[\sigma\delta(z),\rho(x)\cdot\sigma\delta(y)] + [\sigma\delta(y),\rho(x)\cdot\sigma\delta(z)]\\
&\quad = -\rho(x)\cdot[\sigma\delta(z),\sigma\delta(y)] + [\rho(x)\cdot\sigma\delta(z),\sigma\delta(y)] + [\sigma\delta(y),\rho(x)\cdot\sigma\delta(z)]\\
&\quad = -\rho(x)\cdot(\delta(z)\cdot\sigma\delta(y)).
\end{aligned}$$

This shows that the first and fourth term in the above cyclic sum cancel the third term. In conclusion, we have shown the independence of the 3-cocycle up to a coboundary when using a different section ρ.

(2) Now suppose that we chose two different linear sections σ and σ' of μ on $\mathrm{Im}(\mu)$, leading to different lifts $\beta = \sigma\alpha$ and $\beta' = \sigma'\alpha$. We have then $\beta - \beta' \in \mathrm{Ker}(\mu) = \mathrm{Im}(i)$, and it is easily seen that this leads to the corresponding γ and γ' differing by a coboundary.

(3) Two "different" sections τ and τ' of i have to be the same; they are both inverses of the isomorphism i on its image. □

Lemma 2.2.5. *Let $\mu : \mathfrak{m} \to \mathfrak{n}$ (with action η) and $\mu' : \mathfrak{m}' \to \mathfrak{n}'$ (with action η') such that* $\mathrm{Ker}(\mu) = \mathrm{Ker}(\mu') =: V$ *and* $\mathrm{Coker}(\mu) = \mathrm{Coker}(\mu') =: \mathfrak{g}$ *be two elementary equivalent crossed modules.*

Then the corresponding cohomology classes $b([\mu]) = [\gamma]$ and $b([\mu']) = [\gamma']$ coincide in $H^3(\mathfrak{g}, V)$.

Proof. Denote by (φ, ψ) the morphism rendering the two crossed modules elementary equivalent; see Definition 2.1.6. Denote further by γ (resp., γ') the cocycle associated to $\mu : \mathfrak{m} \to \mathfrak{n}$ (resp., to $\mu' : \mathfrak{m}' \to \mathfrak{n}'$) by choosing sections ρ, σ and τ (resp., ρ', σ' and τ') as in the preceding lemma.

As ρ is a section of π, $\tilde{\rho}' := \psi \circ \rho$ is a section of π'. Lemma 2.2.4 shows that α' (defined by the section ρ') and $\tilde{\alpha}' := \psi \circ \alpha$ (defined by the section $\tilde{\rho}'$) give rise to the same cohomology class $[\gamma']$ represented by γ'.

Now let us compute

$$(\tilde{d}^{\mathfrak{m}'}(\sigma' \circ \psi \circ \alpha) - \tilde{d}^{\mathfrak{m}'}(\varphi \circ \beta))(x, y, z)$$

for $x, y, z \in \mathfrak{g}$ where $\tilde{d}^{\mathfrak{m}'}$ is the formal Lie algebra coboundary operator (cf. Formula (B.1)) with values in $\mathfrak{m}'$ and with the formal action $\eta' \circ \psi \circ \rho$. Let us call $\tilde{\beta}' := \sigma' \circ \psi \circ \alpha$.

First, we remark that $(\tilde{\beta}' - \varphi \circ \beta)(x, y) \in \mathrm{Ker}(\mu')$, thus we introduce $v(x, y) \in V$ such that $(\tilde{\beta}' - \varphi \circ \beta)(x, y) = i'v(x, y)$ for all $x, y \in \mathfrak{g}$. A computation using that (φ, ψ) satisfies the conditions of Definition 2.1.6 shows that

$$\tilde{d}^{\mathfrak{m}'}(\varphi \circ \beta)(x, y, z) = \varphi(d^{\mathfrak{m}}\beta(x, y, z)).$$

Another computation shows that $(\tilde{d}^{\mathfrak{m}'} i'v)(x, y, z) = i'(d^V v(x, y, z))$. We thus obtain by injectivity of i'

$$(\gamma' - \gamma)(x, y, z) = d^V v(x, y, z),$$

showing that the two elementary equivalent crossed modules have the same cohomology class. □

In conclusion, we have constructed a well-defined map b associating to an equivalence class of crossed modules with kernel V and cokernel $\mathfrak{g}$ a cohomology class in $H^3(\mathfrak{g}, V)$. Theorem 2.2.1 states that this map is an isomorphism of abelian groups.

2.3 Relation to third cohomology II

In this section, we look at a crossed module of Lie algebras $\mu : \mathfrak{m} \to \mathfrak{n}$ in a different way. Most of the previous section was based on the four-term exact sequence

$$0 \to \mathrm{Ker}(\mu) \to \mathfrak{m} \to \mathfrak{n} \to \mathrm{Coker}(\mu) \to 0$$

associated to the crossed module. Here, we rather look at a crossed module of Lie algebras as the data of a general extension of Lie algebras

$$0 \to \mathfrak{l} \xrightarrow{\iota} \widehat{\mathfrak{g}} \xrightarrow{\pi} \mathfrak{g} \to 0, \tag{2.3}$$

together with a central extension

$$0 \to \mathfrak{z} \to \widehat{\mathfrak{l}} \xrightarrow{p} \mathfrak{l} \to 0$$

of the ideal $\mathfrak{l}$ of the extension (2.3).

If we want that the given general extension and central extension of $\mathfrak{l}$ form a crossed module

$$\mu := \widehat{\mathfrak{l}} \xrightarrow{p} \mathfrak{l} \xrightarrow{\iota} \widehat{\mathfrak{g}},$$

they have to be compatible in the sense that the action of $\mathfrak{l}$ on $\widehat{\mathfrak{l}}$ extends to an action of $\widehat{\mathfrak{g}}$ on $\widehat{\mathfrak{l}}$ satisfying the requirements of Definition 2.1.1.

In the following, we will use this slightly different framework to derive once again the cohomology class $[\gamma]$ associated to a crossed module, that is, in the following, we assume the data of the two extensions together with the extension of the action to form a crossed module. In particular, there is a fixed action of $\mathfrak{g}$ on $\mathfrak{z}$.

The Lie bracket on the central extension $\widehat{\mathfrak{l}}$ can be given in a very explicit way (cf. Appendix B). Namely, we may assume $\widehat{\mathfrak{l}} = \mathfrak{z} \oplus_\omega \mathfrak{l}$, meaning that the bracket reads as

$$[(z_1, l_1), (z_2, l_2)] = (\omega(l_1, l_2), [l_1, l_2]).$$

Here, $\omega \in Z^2(\mathfrak{l}, \mathfrak{z})$ is the 2-cocycle defining the central extension. Note that $\mathfrak{l}$ acts trivially on $\mathfrak{z}$, and the given $\mathfrak{g}$-action on $\mathfrak{z}$ extends trivially to a $\widehat{\mathfrak{g}}$-action on $\mathfrak{z}$, still denoted $z \mapsto x \cdot z$.

Now this $\widehat{\mathfrak{g}}$-action on $\mathfrak{z}$ is supposed to extend to an action on $\widehat{\mathfrak{l}}$, denoted by

$$x \cdot (z, l) = (x \cdot z + \theta(x, l), [x, l]) \tag{2.4}$$

with some linear map $\theta : \widehat{\mathfrak{g}} \oplus \mathfrak{l} \to \mathfrak{z}$. Property (b) of Definition 2.1.1 implies for $l_1 \in \mathfrak{l} \subset \widehat{\mathfrak{g}}$

$$l_1 \cdot (z, l_2) = [(0, l_1), (z, l_2)] = (\omega(l_1, l_2), [l_1, l_2]),$$

meaning that $\theta|_{\mathfrak{l} \times \mathfrak{l}} = \omega$.

Lemma 2.3.1. *Fix an element $x \in \hat{\mathfrak{g}}$.*

(a) *A linear map $\theta_x \in \mathrm{Hom}(\mathfrak{l}, \mathfrak{z})$ defines a derivation*

$$\zeta(x) : \hat{\mathfrak{l}} \to \hat{\mathfrak{l}}, \quad (z, l) \mapsto (x \cdot z + \theta_x(l), [x, l])$$

if and only if $d_{\mathfrak{l}}\theta_x = x \cdot \omega$, where $d_{\mathfrak{l}}$ is the Lie algebra coboundary operator of $\mathfrak{l}$ with values in $\mathrm{Hom}(\mathfrak{l}, \mathfrak{z})$ and ω is seen as $\omega \in C^1(\mathfrak{l}, \mathrm{Hom}(\mathfrak{l}, \mathfrak{z}))$.
The condition $d_{\mathfrak{l}}\theta_x = x \cdot \omega$ means explicitly for all $l, l' \in \mathfrak{l}$,

$$x \cdot \omega(l, l') - \omega([x, l], l') - \omega([l', x], l) + \theta_x([l, l']) = 0.$$

(b) *Suppose that the linear map $\theta : \hat{\mathfrak{g}} \to \mathrm{Hom}(\mathfrak{l}, \mathfrak{z})$, $x \mapsto \theta_x$ satisfies condition (a). Then θ defines a representation of $\hat{\mathfrak{g}}$ on $\hat{\mathfrak{l}}$ via the formula*

$$x \cdot (z, l) = (x \cdot z + \theta_x(l), [x, l])$$

if and only if θ is a 1-cocycle with respect to the natural $\hat{\mathfrak{g}}$-structure on $C^1(\mathfrak{l}, \mathfrak{z}) = \mathrm{Hom}(\mathfrak{l}, \mathfrak{z})$.
This condition means explicitly for $x, x' \in \hat{\mathfrak{g}}$ and $l \in \mathfrak{l}$

$$x \cdot \theta(x', l) - x' \cdot \theta(x, l) - \theta([x, x'], l) + \theta(x, [x', l]) + \theta(x', [l, x]) = 0.$$

Proof.

(a) First of all, let $D : \mathfrak{z} \oplus_\omega \mathfrak{l} \to \mathfrak{z} \oplus_\omega \mathfrak{l}$, $D(z, l) = (\alpha(z) + \gamma(l), \beta(l))$ be a derivation of the central extension which preserves $\mathfrak{z}$, that is, $D(\mathfrak{z}) \subset \mathfrak{z}$. It is clear that β should be a derivation of $\mathfrak{l}$. Then

$$\begin{aligned} D([(z_1, l_1), (z_2, l_2)]) &= D(\omega(l_1, l_2), [l_1, l_2]) \\ &= (\alpha(\omega(l_1, l_2)) + \gamma([l_1, l_2]), \beta([l_1, l_2])) \end{aligned}$$

on the one hand, and on the other hand,

$$\begin{aligned} [D(z_1, l_1), (z_2, l_2)] &= [(\alpha(z_1) + \gamma(l_1), \beta(l_1)), (z_2, l_2)] \\ &= (\omega(\beta(l_1), l_2), [\beta(l_1), l_2]), \end{aligned}$$

which implies then

$$\begin{aligned} &[D(z_1, l_1), (z_2, l_2)] + [(z_1, l_1), D(z_2, l_2)] \\ &= (\omega(\beta(l_1), l_2) + \omega(l_1, \beta(l_2)), [\beta(l_1), l_2] + [l_1, \beta(l_2)]). \end{aligned}$$

It follows that

$$\alpha(\omega(l_1, l_2)) + \gamma([l_1, l_2]) = \omega(\beta(l_1), l_2) + \omega(l_1, \beta(l_2)).$$

Taking $\alpha(z) = x \cdot z$, $\beta(l) = [x, l]$ and $\gamma(l) = \theta_x(l)$, this shows our claim.

(b) Let $x, y \in \hat{\mathfrak{g}}$. Then we have on the one hand

$$[x,y]\cdot(z,l) = ([x,y]\cdot z + \theta_{[x,y]}(l), [[x,y],l]),$$

and on the other hand

$$x\cdot(y\cdot(z,l)) = (x\cdot(y\cdot z + \theta_y(l)) + \theta_x([y,l]), [x,[y,l]]),$$

leading to

$$\begin{aligned} &x\cdot(y\cdot(z,l)) - y\cdot(x\cdot(z,l)) \\ &\quad = (x\cdot(y\cdot z + \theta_y(l)) + \theta_x([y,l]) \\ &\quad\quad - (y\cdot(x\cdot z + \theta_x(l)) + \theta_y([x,l]), [x,[y,l]] - [y,[x,l]])). \end{aligned}$$

Using the Jacobi identity and the fact that $\hat{\mathfrak{g}}$ acts on $\mathfrak{z}$, the requirement reduces to

$$x\cdot\theta_y(l) + \theta_x([y,l]) - y\cdot\theta_x(l) + \theta_y([x,l]) - \theta_{[x,y]}(l) = 0.$$

Once again, this is exactly our claim. □

We saw that the action of $\hat{\mathfrak{g}}$ on $\hat{\mathfrak{l}}$ is determined by the map $\theta : \hat{\mathfrak{g}} \times \mathfrak{l} \to \mathfrak{z}$. The preceding lemma states conditions on θ coming from the fact that $\hat{\mathfrak{g}} \to \mathrm{End}(\hat{\mathfrak{l}})$ is a representation (condition (b)) and that it is a representations by derivations, that is, in fact $\hat{\mathfrak{g}} \to \mathfrak{der}(\hat{\mathfrak{l}})$ (condition (a)).

Now we define the 3-cocycle associated in this context to the crossed modules we started with. $\theta|_{\mathfrak{l}\times\mathfrak{l}}$ is alternating, thus we can extend $\theta : \hat{\mathfrak{g}} \times \mathfrak{l} \to \mathfrak{z}$ by skew symmetry to a bilinear map

$$\tilde{\theta} : \hat{\mathfrak{g}} \times \hat{\mathfrak{g}} \to \mathfrak{z}. \tag{2.5}$$

Then $d_{\hat{\mathfrak{g}}}\tilde{\theta} \in Z^3(\hat{\mathfrak{g}}, \mathfrak{z})$ is a 3-cocycle, which vanishes on $\mathfrak{l} \times \hat{\mathfrak{g}}^2$ by condition (b) of the lemma. This means that as soon as the cocycle is evaluated on a triple of elements where at least one comes from $\mathfrak{l}$, it vanishes. Therefore, it passes to the quotient $\mathfrak{g} = \hat{\mathfrak{g}}/\mathfrak{l}$ to a 3-cocycle $\chi \in Z^3(\mathfrak{g}, \mathfrak{z})$ with $d_{\hat{\mathfrak{g}}}\tilde{\theta} = \pi^*\chi$.

Remark 2.3.1. Observe that one advantage of this approach is that here the coboundary operator $d_{\hat{\mathfrak{g}}}$ refers to a genuine action of $\hat{\mathfrak{g}}$ on $\mathfrak{z}$, and we need not deal with formal expressions of coboundary operators.

Lemma 2.3.2. *The 3-cocycle χ defines a class $[\chi] \in H^3(\mathfrak{g}, \mathfrak{z})$, which does not depend on the choice of the cocycle ω, neither on the choice of the extension $\tilde{\theta}$ of θ.*

Proof. If $\tilde{\theta}$ and $\tilde{\theta}'$ are both extensions of θ, then $\tilde{\theta} - \tilde{\theta}'$ vanishes on $\hat{\mathfrak{g}} \times \mathfrak{l}$, and thus can be written $\pi^*\lambda$ for some $\lambda \in C^2(\mathfrak{g}, \mathfrak{z})$. We therefore get

$$d_{\hat{\mathfrak{g}}}\tilde{\theta} - d_{\hat{\mathfrak{g}}}\tilde{\theta}' = d_{\hat{\mathfrak{g}}}\pi^*\lambda = \pi^*(d_{\hat{\mathfrak{g}}}\lambda).$$

Passing to χ respectively χ', this means that both χ and χ' define the same cohomology class, because π^* is injective on the cochain level.

Now let $\omega' := \omega + d_{\mathfrak{l}}\zeta$ be a cohomologous cocycle. Then the two central extensions $\mathfrak{z} \oplus_\omega \mathfrak{l}$ and $\mathfrak{z} \oplus_{\omega'} \mathfrak{l}$ are equivalent via the morphism

$$\varphi : \mathfrak{z} \oplus_{\omega'} \mathfrak{l} \to \mathfrak{z} \oplus_\omega \mathfrak{l}, \quad (z, l) \mapsto (z + \zeta(l), l).$$

If the action of $\widehat{\mathfrak{l}}$ on $\mathfrak{z} \oplus_{\omega'} \mathfrak{l}$ is given by θ', then the map φ should become $\widehat{\mathfrak{g}}$-equivariant. With

$$\varphi(x \cdot (z, l)) = (x \cdot z + \theta'(x, l) + \zeta([x, l]), [x, l])$$

and

$$x \cdot \varphi(z, l) = (x \cdot z + \theta(x, l) + x \cdot \zeta(l), [x, l]),$$

the equivariance implies that $\theta' = \theta + x \cdot \zeta$, where the action of $\widehat{\mathfrak{g}}$ on $C^1(\mathfrak{l}, \mathfrak{z}) = \mathrm{Hom}(\mathfrak{l}, \mathfrak{z})$ is given by $(x \cdot \zeta)(l) = x \cdot \zeta(l) - \zeta([x, l])$.

Now extend ζ to $\tilde{\zeta} \in C^1(\widehat{\mathfrak{g}}, \mathfrak{z})$. For $x \in \widehat{\mathfrak{g}}$ and $l \in \mathfrak{l}$, we then have

$$(d_{\widehat{\mathfrak{g}}}\tilde{\zeta})(x, l) = x \cdot \tilde{\zeta}(l) - l \cdot \tilde{\zeta}(x) - \tilde{\zeta}([x, l]) = x \cdot \zeta(l) - \zeta([x, l]) = (x \cdot \zeta)(l),$$

so that $\tilde{\theta} + d_{\widehat{\mathfrak{g}}}\tilde{\zeta}$ is an alternating continuous extension of θ'. In view of the first part of the proof, we may use this extension to compute the cohomology class associated to θ', which is therefore given by factorization of $d_{\widehat{\mathfrak{g}}}\tilde{\theta} + (d_{\widehat{\mathfrak{g}}})^2\tilde{\zeta} = d_{\widehat{\mathfrak{g}}}\tilde{\theta}$ to $\mathfrak{g}$, and thus equal to the class defined by θ. □

Let us come back to the formula for the characteristic class of a crossed module which we defined in the previous section.

Proposition 2.3.3. *Let $\mu : \mathfrak{m} \to \mathfrak{n}$ be a crossed module. Let*

$$0 \to V \xrightarrow{i} \mathfrak{m} \xrightarrow{\mu} \mathfrak{n} \xrightarrow{\pi} \mathfrak{g} \to 0$$

be the associated four-term exact sequence and denote $\mathfrak{l} = \mathrm{Im}(\mu)$.

Then the classes $[\gamma]$, where $\gamma \in Z^3(\mathfrak{g}, V)$ is obtained as in Lemma 2.2.3 by using a section $\rho : \mathfrak{g} \to \mathfrak{n}$ in the four-term exact sequence, and $[\chi]$, where $\chi = d_{\mathfrak{m}}\tilde{\theta}$ is defined in Lemma 2.3.2 (with $\mathfrak{z} = V$), coincide in $H^3(\mathfrak{g}, V)$.

Proof. First choose $\omega \in Z^2(\mathfrak{l}, V)$ with $\mathfrak{m} =: \widehat{\mathfrak{l}} = V \oplus_\omega \mathfrak{l}$ and a linear section $\rho : \mathfrak{g} \to \mathfrak{n}$. Associated to ρ, we have for $x, y, z \in \mathfrak{g}$,

$$\alpha(x, y) = [\rho(x), \rho(y)] - \rho([x, y]) \in \mathfrak{n},$$

and $\beta(x, y) \in \mathfrak{m}$ with

$$\mu\beta(x, y) = \alpha(x, y).$$

We therefore get $d^{\mathfrak{m}}\beta(x,y,z)$ and $\gamma(x,y,z) \in V$ such that $i\gamma(x,y,z) = d^{\mathfrak{m}}\beta(x,y,z)$ as before. Now as $\beta(x,y) \in \widehat{\mathfrak{l}} = V \oplus_\omega \mathfrak{l}$, write

$$\beta = (\beta_V, \alpha) \quad \text{with } \beta_V \in C^2(\mathfrak{g}, V).$$

On the other hand, write the action of $\mathfrak{m}$ on $\widehat{\mathfrak{l}}$ as $x \cdot (z,l) = (x \cdot z + \theta(x,l), [x,l])$ with a bilinear map $\theta : \mathfrak{n} \times \mathfrak{l} \to V$. Then

$$\begin{aligned} d^{\mathfrak{m}}\beta(x,y,z) &= -\sum_{\text{cycl.}} (\rho(x) \cdot (\beta_V(y,z), \alpha(y,z)) - (\beta_V([x,y],z), \alpha([x,y],z))) \\ &= -\sum_{\text{cycl.}} ((x \cdot \beta_V(y,z) + \theta(\rho(x), \alpha(y,z)), [\rho(x), \alpha(y,z)]) \\ &\quad - (\beta_V([x,y],z), \alpha([x,y],z))). \end{aligned}$$

As $d^{\mathfrak{m}}\beta(x,y,z) \in \operatorname{Ker}(\mu)$, the $\mathfrak{l}$-component of this expression must be zero. This may be seen as an abstract Bianchi identity; cf. [77]. We thus obtain

$$d^{\mathfrak{m}}\beta(x,y,z) = \left(d^V\beta_V(x,y,z) + \sum_{\text{cycl.}} \theta(\rho(x), \alpha(y,z)), 0\right) = i\gamma(x,y,z).$$

We must now compare this to $\chi(x,y,z)$. For this, let $\tilde{\theta} \in C^2(\mathfrak{n}, V)$ be an alternating extension of θ. We may assume that $\rho^*\tilde{\theta} = \beta_V$. We show now that $\pi^*(d^{\mathfrak{m}}\beta) = d_{\mathfrak{n}}\tilde{\theta}$, so that $[\chi] = [d^{\mathfrak{m}}\beta] \in H^3(\mathfrak{g}, V)$ as claimed:

$$\begin{aligned} d_{\mathfrak{n}}\tilde{\theta}(\rho(x),\rho(y),\rho(z)) &= -\sum_{\text{cycl.}} (\rho(x) \cdot \tilde{\theta}(\rho(y),\rho(z)) - \tilde{\theta}([\rho(x),\rho(y)],\rho(z))) \\ &= -\sum_{\text{cycl.}} (\rho(x) \cdot \tilde{\theta}(\rho(y),\rho(z)) - \tilde{\theta}(\rho([x,y]) + \alpha(x,y),\rho(z))) \\ &= -\sum_{\text{cycl.}} (x \cdot \beta_V(y,z) - \beta_V([x,y],z) - \tilde{\theta}(\alpha(x,y),\rho(z))) \\ &= -\sum_{\text{cycl.}} (x \cdot \beta_V(y,z) - \beta_V([x,y],z) + \theta(\rho(x),\alpha(y,z))) \\ &= d^V\beta_V(x,y,z) - \sum_{\text{cycl.}} \theta(\rho(x),\alpha(y,z)) \\ &= \operatorname{proj}_1(d^{\mathfrak{m}}\beta(x,y,z)), \end{aligned}$$

where $\operatorname{proj}_1 : V \times \mathfrak{l} \to V$ is the projection onto the first factor. We will suppress this projection in the following.

In this computation, we used the definition of α passing from the first to the second line, then the assumption $\rho^*\tilde{\theta} = \beta_V$, then the fact that $\tilde{\theta}$ reduces to θ in case one variable is in $\mathfrak{l}$, and finally the relation between $d^{\mathfrak{m}}\beta$ and $d^{\mathfrak{m}}\beta_V$ we obtained earlier. Notice also that we have shown

$$d_{\mathfrak{n}}\tilde{\theta}(\rho(x),\rho(y),\rho(z)) = d^{\mathfrak{m}}\beta(x,y,z)$$

for all $x, y, z \in \mathfrak{g}$. This implies the wanted relation $\pi^*(d^{\mathfrak{m}}\beta) = d_{\mathfrak{n}}\tilde{\theta}$ by noting that $d_{\mathfrak{n}}\tilde{\theta}$ is zero as soon as one argument is in $\mathrm{Im}(\mu)$ and, therefore, zero on the difference between $\rho(\pi(n))$ and $n \in \mathfrak{n}$. □

Let us conclude this section with a characterization of the zero class in the present framework.

Theorem 2.3.4. *Let $\mu : \mathfrak{m} \to \mathfrak{n}$ be a crossed module with associated $\mathfrak{n}$-module $\mathrm{Ker}(\mu) =: \mathfrak{z}$, with $\mathrm{Coker}(\mu) =: \mathfrak{g}$ and $\mathrm{Im}(\mu) =: \mathfrak{l}$. The following assertions are equivalent:*
(a) *$[\chi] = 0$ in $H^3(\mathfrak{g}, \mathfrak{z})$.*
(b) *If $\mathfrak{m} = \mathfrak{z} \times_\omega \mathfrak{l} =: \hat{\mathfrak{l}}$ for some 2-cocycle $\omega \in Z^2(\mathfrak{l}, \mathfrak{z})$ and the action is written $x \cdot (z, l) = (x \cdot z + \theta(x, l), [x, l])$, then θ extends to a cocycle in $Z^2(\hat{\mathfrak{l}}, \mathfrak{z})$.*
(c) *There exists an abelian extension*

$$0 \to \mathfrak{z} \to \hat{\mathfrak{n}} \xrightarrow{\pi} \mathfrak{n} \to 0$$

and an $\mathfrak{n}$-equivariant equivalence $\mathfrak{m} \to \pi^{-1}(\mathfrak{l})$ of $\mathfrak{z}$-extensions of $\mathfrak{l}$.

Proof. *(a) ⇒ (b)*: If $[\chi] = 0$ and $\tilde{\theta} \in C^2(\mathfrak{n}, \mathfrak{z})$ is an alternating extension of θ, then there exists a $\chi \in C^2(\mathfrak{g}, \mathfrak{z})$ with $d_{\mathfrak{n}}\tilde{\theta} = \pi^*(d_{\mathfrak{g}}\chi) = d_{\mathfrak{n}}(\pi^*\chi)$, so that $\tilde{\theta} - \pi^*\chi$ is an extension of θ to a cocycle of $\mathfrak{n} = \hat{\mathfrak{l}}$.

(b) ⇒ (c): Identifying $\mathfrak{m} = \mathfrak{z} \times_\omega \mathfrak{l}$ means that the map $\mu : \mathfrak{m} \to \mathfrak{n}$ becomes the projection onto the second component. Let $\tilde{\theta} \in Z^2(\mathfrak{m}, \mathfrak{z})$ be a cocycle extending θ and $\hat{\mathfrak{m}} = \mathfrak{z} \times_{\tilde{\theta}} \mathfrak{m}$ the corresponding abelian extension. Then the inclusion $\hat{\mathfrak{l}} = \mathfrak{z} \times_\omega \mathfrak{l} \hookrightarrow \hat{\mathfrak{m}}$, $(z, l) \mapsto (z, l)$, induces an $\mathfrak{n}$-equivariant equivalence $\mathfrak{m} \to \pi^{-1}(\mathfrak{l})$.

(c) ⇒ (a): Suppose that we have an $\mathfrak{n}$-equivariant equivalence of $\mathfrak{z}$-extensions

$$\mathfrak{m} \to \pi^{-1}(\mathfrak{l}) \subset \hat{\mathfrak{n}}.$$

Write $\hat{\mathfrak{n}} = \mathfrak{z} \times_{\tilde{\theta}} \mathfrak{n}$ for some $\tilde{\theta} \in Z^2(\mathfrak{n}, \mathfrak{z})$. Our assertion means that $\mathfrak{m} \cong \mathfrak{z} \times_\theta \mathfrak{n}$ for $\theta := \tilde{\theta}|_{\mathfrak{l} \times \mathfrak{l}}$, so that we may identify $\mathfrak{m}$ with the subspace $\mathfrak{z} \times \mathfrak{l} \subset \hat{\mathfrak{n}}$, and that the representation of $\mathfrak{n}$ on this subspace $\mathfrak{m}$ is given by

$$x \cdot (z, l) = (x \cdot z + \theta(x, l), [x, l]).$$

Then $\tilde{\theta}$ is an extension of θ, so that (a) follows from the definition of χ using furthermore $d_{\mathfrak{n}}\tilde{\theta} = 0$. □

2.4 Relation to general extensions

This section is concerned with general extensions of Lie algebras. To each extension of Lie algebras, one may associate an outer action of the right-hand Lie algebra on the Lie algebra on the left. The main point is that given an outer action, there is a crossed module which has zero class if and only if there exists an extension realizing this outer action.

2.4.1 Outer actions and factor systems

Definition 2.4.1. A *general extension* of Lie algebras is a short exact sequence of Lie algebras of the form

$$0 \to \mathfrak{l} \xrightarrow{i} \mathfrak{e} \xrightarrow{\pi} \mathfrak{g} \to 0, \tag{2.6}$$

where the Lie algebra $\mathfrak{l}$ is not necessarily central in $\mathfrak{e}$, nor even abelian.

Such a general extension defines an outer action (for the definition, see Remark 2.1.4). Indeed, lifting elements of $\mathfrak{g}$ to $\mathfrak{e}$ via a linear section ρ of the projection π, one may act on elements of $\mathfrak{l}$ by the adjoint action. This defines a linear map

$$\psi_\rho : \mathfrak{g} \to \mathfrak{der}(\mathfrak{l}).$$

As in Remark 2.1.4, two different sections lead to actions, which differ by an inner derivation, thus ψ induces a well-defined homomorphism of Lie algebras

$$\psi : \mathfrak{g} \to \mathfrak{out}(\mathfrak{l}) = \mathfrak{der}(\mathfrak{l})/\mathrm{ad}(\mathfrak{l}).$$

This homomorphism ψ is called the *outer action* associated to the general extension (2.6).

Definition 2.4.2. Two general extensions of Lie algebras

$$0 \to \mathfrak{l} \to \mathfrak{e}_1 \to \mathfrak{g} \to 0$$

and

$$0 \to \mathfrak{l} \to \mathfrak{e}_2 \to \mathfrak{g} \to 0$$

of the same Lie algebra $\mathfrak{g}$ by the same Lie algebra $\mathfrak{l}$ are called *equivalent* in case there exists a homomorphism $\varphi : \mathfrak{e}_1 \to \mathfrak{e}_2$ such that the following diagram is commutative:

$$\begin{array}{ccccccccc}
0 & \longrightarrow & \mathfrak{l} & \longrightarrow & \mathfrak{e}_1 & \longrightarrow & \mathfrak{g} & \longrightarrow & 0 \\
 & & \downarrow{\scriptstyle \mathrm{id}_{\mathfrak{l}}} & & \downarrow{\scriptstyle \varphi} & & \downarrow{\scriptstyle \mathrm{id}_{\mathfrak{g}}} & & \\
0 & \longrightarrow & \mathfrak{l} & \longrightarrow & \mathfrak{e}_2 & \longrightarrow & \mathfrak{g} & \longrightarrow & 0
\end{array}$$

Observe that φ is automatically an isomorphism of Lie algebras by the Five Lemma. Denote by $\mathrm{Ext}(\mathfrak{g}, \mathfrak{l})$ the set of equivalence classes of general extensions of $\mathfrak{g}$ by $\mathfrak{l}$.

A general extension (2.6) is called *trivial* or *split*, if there exists a homomorphism of Lie algebras $\rho : \mathfrak{g} \to \mathfrak{e}$ such that $\pi \circ \rho = \mathrm{id}_{\mathfrak{g}}$. In this case, the map

$$\mathfrak{l} \rtimes \mathfrak{g} \to \mathfrak{e}, \quad (l, x) \mapsto l + \rho(x)$$

is an isomorphism, where the semi-direct product is defined by the homomorphism $\psi : \mathfrak{g} \to \mathfrak{der}(\mathfrak{l})$, $\psi(x)(l) = [\rho(x), l]$.

Now we analyze the structure of a general extension.

Remark 2.4.3. Given a general extension (2.6) and a linear section $\rho : \mathfrak{g} \to \mathfrak{e}$ of the projection π, one may associate to these data a linear map $\psi_\rho : \mathfrak{g} \to \mathfrak{Der}(\mathfrak{l})$ and a bilinear map $\beta : \mathfrak{g} \times \mathfrak{g} \to \mathfrak{l}$ computing the failure of ρ to be a homomorphism of Lie algebras

$$\alpha(x,y) = [\rho(x), \rho(y)] - \rho([x,y]),$$

for $x, y \in \mathfrak{g}$. As $\pi \circ \alpha = 0$, the map β is then defined by $i \circ \beta = \alpha$. Now the map

$$\varphi : \mathfrak{l} \oplus \mathfrak{g} \to \mathfrak{e}, \quad (l,x) \mapsto l + \rho(x)$$

becomes an isomorphism of Lie algebras when we endow $\mathfrak{l} \oplus \mathfrak{g}$ with the bracket

$$[(l,x),(l',x')] = ([l,l'] + \psi_\rho(x)(l') - \psi_\rho(x')(l) + \beta(x,x'), [x,x']). \tag{2.7}$$

In this sense, the triple (ρ, ψ_ρ, β) determines the extension $\mathfrak{e}$.

Associate to a linear map $S : \mathfrak{g} \to \mathfrak{Der}(\mathfrak{l})$ its failure R_S to be a homomorphism of Lie algebras, that is,

$$R_S(x,y) := [S(x), S(y)] - S([x,y]).$$

R_S is a bilinear map $R_S : \mathfrak{g} \times \mathfrak{g} \to \mathfrak{Der}(\mathfrak{l})$.

Definition 2.4.4. A *factor system* is a pair (S, ω) where $S : \mathfrak{g} \to \mathfrak{Der}(\mathfrak{l})$ is a linear map, $\omega : \mathfrak{g} \times \mathfrak{g} \to \mathfrak{l}$ is bilinear and satisfies $d^{\mathfrak{l}}\omega = 0$, and S is linked to ω by imposing $R_S = \mathrm{ad} \circ \omega$.

Here, $d^{\mathfrak{l}}$ (also denoted d_S) denotes the formal expression of the Lie algebra coboundary operator (cf. Formula (B.1)) with values in $\mathfrak{l}$ using the map S, while S does not necessarily endow $\mathfrak{l}$ with the structure of a $\mathfrak{g}$-module. Denote by $Z^2(\mathfrak{g}, \mathfrak{l})$ the set of these factor systems.

Remark 2.4.5. Given a factor system (S, ω), one may compose S with the canonical projection $Q : \mathfrak{Der}(\mathfrak{l}) \to \mathfrak{Der}(\mathfrak{l})/\mathrm{ad}(\mathfrak{l}) =: \mathfrak{out}(\mathfrak{l})$ to get an outer action $\psi := Q \circ S$. This ψ is then a homomorphism, because, up to inner derivations, $R_S = 0$, that is, up to inner derivations, S is a homomorphism. In this construction, we did not use the condition $d^{\mathfrak{l}}\omega = 0$.

Conversely, a typical example of a pair (S, ω) (without necessarily having $d^{\mathfrak{l}}\omega = 0$) is the one associated to an outer action $\psi : \mathfrak{g} \to \mathfrak{out}(\mathfrak{l})$. Indeed, given ψ, on the one hand, there is a lift $S : \mathfrak{g} \to \mathfrak{Der}(\mathfrak{l})$. On the other hand, as ψ is a homomorphism, R_S must be an inner derivation, thus we may define ω by the equation $R_S = \mathrm{ad} \circ \omega$.

Lemma 2.4.1. *Given a factor system (S, ω), the vector space $\mathfrak{l} \oplus \mathfrak{g}$ becomes a Lie algebra, denoted $\mathfrak{l} \times_{(S,\omega)} \mathfrak{g}$, using the bracket given in equation (2.7).*

Conversely, any extension of $\mathfrak{g}$ by $\mathfrak{l}$ is isomorphic to some $\mathfrak{l} \times_{(S,\omega)} \mathfrak{g}$.

Proof. It is clear that the bracket given in equation (2.7) is alternating. Let us form its *Jacobiator J* defined by

$$J((l,x),(l',x'),(l'',x'')) = \sum_{\text{cycl.}} [[(l,x),(l',x')],(l'',x'')].$$

J is an alternating trilinear map, and it suffices to show that J vanishes on triples of the form (l,l',l''), (x,l',l''), (x,x',l'') and (x,x',x''), where $l,l',l'' \in \mathfrak{l}$ and $x,x',x'' \in \mathfrak{g}$. We will write simply $J(l,x',x'')$ for $J((l,0),(0,x'),(0,x''))$ for example. As the inclusion map $\mathfrak{l} \hookrightarrow \mathfrak{l} \times_{(S,\omega)} \mathfrak{g}$ preserves the bracket, we have $J(l,l',l'') = 0$.

It is clear that $[\mathfrak{l} \times_{(S,\omega)} \mathfrak{g}, \mathfrak{l}] \subset \mathfrak{l}$. Therefore, $J(x,l',l'') = 0$ follows from $S(x) \in \mathfrak{der}(\mathfrak{l})$ for all $x \in \mathfrak{g}$.

The vanishing of the expressions $J(x,x',l'')$ means that

$$\begin{aligned} &S(x)\cdot(S(x')\cdot l'') - S(x')\cdot(S(x)\cdot l'') \\ &\quad = [(0,x),(0,x')]\cdot l'' = [\omega(x,x'),l''] + S([x,x'])\cdot l'', \end{aligned}$$

which is $R_S = \operatorname{ad}\circ\,\omega$.

Finally,

$$\begin{aligned} [[(0,x),(0,x')],(0,x'')] &= [(\omega(x,x'),[x,x']),(0,x'')] \\ &= (-S(x'')\cdot\omega(x,x') + \omega([x,x'],x''),[[x,x'],x'']) \end{aligned}$$

implies that $J(x,x',x'') = (-d^{\mathfrak{l}}\omega(x,x',x''),0) = (0,0)$, where as usual $d^{\mathfrak{l}}\omega$ denotes the formal Lie algebra coboundary operator with values in $\mathfrak{l}$ as if $\mathfrak{l}$ was a $\mathfrak{g}$-module via S (which, in general, is not the case).

Conversely, given a general extension

$$0 \to \mathfrak{l} \to \mathfrak{e} \to \mathfrak{g} \to 0,$$

one chooses a linear section $\rho : \mathfrak{g} \to \mathfrak{e}$ of the quotient map and defines S and ω as ψ_ρ and β in Remark 2.4.3. We claim that (S,ω) is a factor system. Indeed,

$$\begin{aligned} R_S(x,y) &= [S(x),S(y)] - S([x,y]) = [\psi_\rho(x),\psi_\rho(y)] - \psi_\rho([x,y]) \\ &= [\operatorname{ad}_{\rho(x)},\operatorname{ad}_{\rho(y)}] - \operatorname{ad}_{\rho([x,y])} = \operatorname{ad}_{\alpha(x,y)}, \end{aligned}$$

and, therefore, we get $\operatorname{ad}\circ\,\omega = R_S$ where we have identified α, β and ω, viewing $\mathfrak{l}$ as a subspace of $\mathfrak{e}$ and the map i as the natural inclusion.

On the other hand, one has $d^{\mathfrak{l}}\omega = 0$ thanks to the abstract Bianchi identity, which we mentioned in the proof of Proposition 2.3.3; see Proposition I.8 in [77]. One concludes that (S,ω) is a factor system. Remark 2.4.3 shows that the extension $\mathfrak{e}$ is equivalent to $\mathfrak{l} \times_{(S,\omega)} \mathfrak{g}$. □

In the preceding lemma, we reduced the study of general Lie algebra extensions to standard extensions associated to a factor system (S,ω). The following lemma now describes in how many inequivalent ways one can define such a standard extension on the product space $\mathfrak{l}\oplus\mathfrak{g}$.

Lemma 2.4.2. *Let $(\zeta,\xi)\in \mathrm{Aut}(\mathfrak{l})\times\mathrm{Aut}(\mathfrak{g})$ and $\epsilon\in C^1(\mathfrak{g},\mathfrak{l})$. Then the map*

$$\varphi:\mathfrak{l}\times\mathfrak{g}\to\mathfrak{l}\times\mathfrak{g},\quad (l,x)\mapsto(\zeta(l)+\epsilon(\xi(x)),\xi(x))$$

is an isomorphism of Lie algebras $\varphi:\mathfrak{l}\times_{(S,\omega)}\mathfrak{g}\to\mathfrak{l}\times_{(S',\omega')}\mathfrak{g}$ if and only if

$$(\zeta,\xi)\cdot S=S'+\mathrm{ad}\circ\epsilon\quad\text{and}\quad(\zeta,\xi)\cdot\omega:=\zeta\circ\omega\circ(\xi\times\xi)^{-1}=\omega'+d\epsilon+\frac{1}{2}[\epsilon,\epsilon],$$

which can be written in a compact form as $(\zeta,\xi)\cdot(S,\omega)=\epsilon(S',\omega')$. Here,

$$((\zeta,\xi)\cdot S)(x):=\zeta\circ(S(\xi^{-1}(x)))\circ\zeta^{-1}.$$

Proof. We have

$$\begin{aligned}\varphi([(l,x),(l',x')])&=\varphi([l,l']+S(x)\cdot l'-S(x')\cdot l+\omega(x,x'),[x,x'])\\&=(\zeta([l,l'])+\zeta(S(x)\cdot l')\\&\quad-\zeta(S(x')\cdot l)+\zeta\omega(x,x')+\epsilon(\xi([x,x'])),\xi([x,x']))\end{aligned}$$

and

$$\begin{aligned}[\varphi(l,x),\varphi(l',x')]&=[(\zeta(l)+\epsilon(\xi(x)),\xi(x)),(\zeta(l')+\epsilon(\xi(x')),\xi(x'))]\\&=([\zeta(l)+\epsilon(\xi(x)),\zeta(l')+\epsilon(\xi(x'))]+S'(\xi(x))\cdot(\zeta(l')+\epsilon(\xi(x')))\\&\quad-S'(\xi(x'))\cdot(\zeta(l)+\epsilon(\xi(x)))+\omega'(\xi(x),\xi(x')),\xi([x,x'])).\end{aligned}$$

Therefore, the requirement that φ is a homomorphism of Lie algebras is equivalent to the two conditions

$$S'(\xi(x))\circ\zeta+\mathrm{ad}_{\epsilon(\xi(x))}\circ\zeta=\zeta\circ S(x)\quad\text{for } x\in\mathfrak{g},$$

and

$$\begin{aligned}\zeta\omega(x,x')+\epsilon(\xi([x,x']))&=\omega'(\xi(x),\xi(x'))+S'(\xi(x))\cdot\epsilon(\xi(x'))\\&\quad-S'(\xi(x'))\cdot\epsilon(\xi(x))+[\epsilon(\xi(x)),\epsilon(\xi(x'))]\quad\text{for } x,x'\in\mathfrak{g}.\end{aligned}$$

The first condition can be rewritten as

$$S'(x)+\mathrm{ad}_{\epsilon(x)}=\zeta\circ S(\xi^{-1}(x))\circ\zeta^{-1}=((\zeta,\xi)\cdot S)(x)\quad\text{for } x\in\mathfrak{g}.$$

Similarly, the second condition can be rewritten as

$$(\zeta,\xi)\cdot\omega=\omega'+d\epsilon+\frac{1}{2}[\epsilon,\epsilon],$$

where the coboundary term $d\epsilon = d_{S'}\epsilon$ is understood to involve the map S' and by definition

$$\frac{1}{2}[\epsilon,\epsilon](x,x')=[\epsilon(\xi(x)),\epsilon(\xi(x'))].$$

□

Remark 2.4.6. Suppose that $S = S' + \mathrm{ad}_{\mathfrak{l}}\circ\epsilon$ for some $\epsilon\in C^1(\mathfrak{g},\mathfrak{l})$. In the Lie algebra $\mathfrak{l}\times_{(S,\omega)}\mathfrak{g}$, we can replace the section $\rho:\mathfrak{g}\to\mathfrak{l}\times_{(S,\omega)}\mathfrak{g}$, $x\mapsto(0,x)$, by $\rho' := \rho+\epsilon$. Then $\omega = R_\rho$, $S=\mathrm{ad}_{\mathfrak{l}}\circ\rho$ and $S'=\mathrm{ad}_{\mathfrak{l}}\circ\rho'$, so that one computes

$$\begin{aligned}R_{\rho'}(x,y)&=[\rho'(x),\rho'(y)]-\rho'([x,y])\\&=[(\rho+\epsilon)(x),(\rho+\epsilon)(y)]-(\rho+\epsilon)([x,y])\\&=R_\rho(x,y)+[\epsilon(x),\epsilon(y)]-\epsilon([x,y])+[\rho(x),\epsilon(y)]+[\epsilon(x),\rho(y)]\\&=\left(R_\rho+\frac{1}{2}[\epsilon,\epsilon]+d_S\epsilon\right)(x,y)\end{aligned}$$

for all $x,y\in\mathfrak{g}$. Therefore, the passage from a pair (S,ω) to the corresponding pair obtained from changing the section by adding ϵ is given by the action described in the previous lemma.

Definition 2.4.7. According to the discussion above, let us introduce an equivalence relation. We call $S,S':\mathfrak{g}\to\mathfrak{der}(\mathfrak{l})$ *equivalent*, denoted by $S\sim S'$, if there exists $\epsilon\in C^1(\mathfrak{g},\mathfrak{l})$ such that $S=S'+\mathrm{ad}_{\mathfrak{l}}\circ\epsilon$. The equivalence class of a given S is denoted by $[S]$. Observe that it is the class $[S]$ of a linear map $S:\mathfrak{g}\to\mathfrak{der}(\mathfrak{l})$ that corresponds to an outer action (in particular, a Lie algebra homomorphism) $\psi:\mathfrak{g}\xrightarrow{S}\mathfrak{der}(\mathfrak{l})\to\mathfrak{out}(\mathfrak{l})$. The class $[S]$ is therefore also called an *outer action* of $\mathfrak{g}$ on $\mathfrak{l}$ or a $\mathfrak{g}$-*kernel*. By language abuse, we will sometimes call even a representative S of the class $[S]$ an outer action or a $\mathfrak{g}$-kernel.

Now we are ready to describe the set of equivalence classes of extensions $\mathrm{Ext}(\mathfrak{g},\mathfrak{l})$ by factor systems.

Theorem 2.4.3. *The map*

$$\Gamma:Z^2(\mathfrak{g},\mathfrak{l})\to\mathrm{Ext}(\mathfrak{g},\mathfrak{l}),\quad (S,\omega)\mapsto[\mathfrak{l}\times_{(S,\omega)}\mathfrak{g}]$$

factors through the action of $C^1(\mathfrak{g},\mathfrak{l})$ *to a bijection*

$$\bar{\Gamma}:Z^2(\mathfrak{g},\mathfrak{l})/C^1(\mathfrak{g},\mathfrak{l})\to\mathrm{Ext}(\mathfrak{g},\mathfrak{l}).$$

For every outer action S of $\mathfrak{g}$ *on* $\mathfrak{l}$ *such that* $\mathrm{Ext}(\mathfrak{g},\mathfrak{l})$ *is nonempty, the map*

$$Z^2(\mathfrak{g},\mathfrak{l})_S\to\mathrm{Ext}(\mathfrak{g},\mathfrak{l})_{[S]},\quad \omega\mapsto[\mathfrak{l}\times_{(S,\omega)}\mathfrak{g}]$$

is surjective and its fibers are the cosets of $B^2(\mathfrak{g},\mathfrak{z}(\mathfrak{l}))_S$ *in the affine space* $Z^2(\mathfrak{g},\mathfrak{l})_S$ *with translation group* $Z^2(\mathfrak{g},\mathfrak{z}(\mathfrak{l}))_S$. *Thus* $\mathrm{Ext}(\mathfrak{g},\mathfrak{l})_{[S]}$ *inherits the structure of an affine space with translation group* $Z^2(\mathfrak{g},\mathfrak{z}(\mathfrak{l}))_S$.

Proof. Thanks to Lemma 2.4.1, the map Γ is surjective. According to Lemma 2.4.2, an equivalence of standard extensions $\varphi : \mathfrak{l}\times_{(S,\omega)}\mathfrak{g} \to \mathfrak{l}\times_{(S',\omega')}\mathfrak{g}$ has the form $\varphi(l,x) = (l+\epsilon(x),x)$ with $\epsilon \in C^1(\mathfrak{g},\mathfrak{l})$ satisfying $(S,\omega) = \epsilon\cdot(S',\omega')$. Indeed, being an equivalence implies that the induced automorphisms on $\mathfrak{g}$ and $\mathfrak{l}$ are the identity, thus ζ and ξ drop out. This implies that the fibers of Γ are the $C^1(\mathfrak{g},\mathfrak{l})$-orbits, so that Γ factors through the bijection $\bar{\Gamma}$.

If $\mathrm{Ext}(\mathfrak{g},\mathfrak{l})_{[S]}$ is nonempty, then it follows from the preceding paragraph that each extension $\mathfrak{l}\times_{(S',\omega')}\mathfrak{g}$ with $S' = S+\mathrm{ad}_{\mathfrak{l}}\epsilon$ is equivalent to one of the form $\mathfrak{l}\times_{(S,\omega)}\mathfrak{g}$ for some $\omega\in Z^2(\mathfrak{g},\mathfrak{l})_S$ (indeed, take ω to be $\omega = \omega' + d_{S'}\epsilon + \frac{1}{2}[\epsilon,\epsilon]$).

All other extensions corresponding to $[S]$ are equivalent to Lie algebras of the form $\mathfrak{l}\times_{(S,\omega')}\mathfrak{g}$. The requirement $\mathrm{ad}\circ\omega = R_S = \mathrm{ad}\circ\omega'$ implies then $\beta := \omega' - \omega \in C^2(\mathfrak{g},\mathfrak{z}(\mathfrak{l}))$. Therefore,

$$0 = d_S\omega' = d_S\omega + d_S\beta = d_S\beta,$$

and thus $b\in Z^2(\mathfrak{g},\mathfrak{z}(\mathfrak{l}))_S$. This means that $\omega'\in\omega + Z^2(\mathfrak{g},\mathfrak{z}(\mathfrak{l}))_S$.

According to Lemma 2.4.2, an equivalence $\varphi : \mathfrak{l}\times_{(S,\omega)}\mathfrak{g}\to\mathfrak{l}\times_{(S,\omega')}\mathfrak{g}$ has the form $\phi(l,x) = ((l+\epsilon'x),x)$ with $\epsilon\in C^1(\mathfrak{g},\mathfrak{z}(\mathfrak{l}))$ satisfying $\omega-\omega' = d_S\epsilon$. This completes the proof. □

Corollary 2.4.4. *For a given* $\mathfrak{g}$*-kernel* $[S]$, *the map*

$$H^2(\mathfrak{g},\mathfrak{z}(\mathfrak{l}))_S\times\mathrm{Ext}(\mathfrak{g},\mathfrak{l})_{[S]}\to\mathrm{Ext}(\mathfrak{g},\mathfrak{l})_{[S]},\quad (\beta,[\mathfrak{l}\times_{(S,\omega)}\mathfrak{g}])\mapsto[\mathfrak{l}\times_{(S,\omega+\beta)}\mathfrak{g}]$$

is a well-defined simply transitive action, so that $\mathrm{Ext}(\mathfrak{g},\mathfrak{l})_{[S]}$ *carries the structure of an affine space with translation group* $H^2(\mathfrak{g},\mathfrak{z}(\mathfrak{l}))_S$.

2.4.2 The obstruction crossed module

Definition 2.4.8. Let S be an outer action of $\mathfrak{g}$ on $\mathfrak{l}$ and define $\omega\in C^2(\mathfrak{g},\mathfrak{l})$ by $R_S = \mathrm{ad}\circ\omega$ (cf. Remark 2.4.5). We have $d_S\omega\in Z^3(\mathfrak{g},\mathfrak{z}(\mathfrak{l}))_S$. Indeed, for $S_1 := \mathrm{ad}\circ S : \mathfrak{g}\to\mathrm{End}(\mathfrak{der}(\mathfrak{l}))$, we have

$$\mathrm{ad}\circ(d_S\omega) = d_{S_1}(\mathrm{ad}\circ\omega) = d_{S_1}R_S,$$

and this last expression is zero by the abstract Bianchi identity, which we mentioned in the proof of Proposition 2.3.3; see Proposition I.8 in [77]. This proves that $d_S\omega$ has values in $\mathfrak{z}(\mathfrak{l})$. One computes that $d_S\omega$ is a cocycle, that is, $d_Sd_S\omega = 0$; cf. Lemma 2.2.3.

The corresponding cohomology class

$$[\chi(S)] = [d_S\omega] \in H^3(\mathfrak{g}, \mathfrak{z}(\mathfrak{l}))_S$$

is called the *obstruction class* of the outer action S.

If $\omega' \in C^2(\mathfrak{g}, \mathfrak{l})$ also satisfies $R_S = \mathrm{ad} \circ \omega'$, then $\beta := \omega' - \omega \in C^2(\mathfrak{g}, \mathfrak{z}(\mathfrak{l}))$ satisfies

$$d_S\omega' = d_S\omega + d_S\beta,$$

and, therefore, $[d_S\omega'] = [d_S\omega]$ does not depend on the choice of ω.

Moreover, for $S \sim S'$, that is, $S' = S + \mathrm{ad}_{\mathfrak{l}}\epsilon$, we have $\omega = \omega' + d_{S'}\epsilon + \frac{1}{2}[\epsilon, \epsilon]$ for the corresponding ω and ω'. One computes then

$$\begin{aligned}
d_{S'}\omega'(x, x', x'') &= d_S\omega'(x, x', x'') + \sum_{\text{cycl.}} \mathrm{ad}_{\epsilon(x)}\omega'(x', x'') \\
&= d_S\omega(x, x', x'') + d_S\Big(d_S\epsilon + \frac{1}{2}[\epsilon, \epsilon]\Big)(x, x', x'') + [\epsilon, \omega'](x, x', x'') \\
&= (d_S\omega + [\omega, \epsilon] + [d_S\epsilon, \epsilon] - [\omega', \epsilon])(x, x', x'') \\
&= (d_S\omega + [\omega + d_S\epsilon - \omega', \epsilon])(x, x', x'') \\
&= \Big(d_S\omega - \frac{1}{2}[[\epsilon, \epsilon], \epsilon]\Big)(x, x', x'') \\
&= d_S\omega(x, x', x'').
\end{aligned}$$

Here, we used the bracket of cochains $[\epsilon, \omega']$ defined by

$$[\epsilon, \omega'](x, x', x'') := \sum_{\text{cycl.}} [\epsilon(x), \omega'(x', x'')],$$

we used that d_S is a derivation of this bracket, that $d_S d_S\epsilon = [\omega, \epsilon]$, and finally that $[[\epsilon, \epsilon], \epsilon] = 0$.

The upshot of this computation is that the obstruction class $[\chi(S)] = [d_S\omega]$ depends only on the equivalence class of S.

Up to now, we studied the situation where a general extension of $\mathfrak{g}$ by $\mathfrak{l}$ already existed. Let us now see how a certain crossed module answers the existence question. For this, we will associate to a given outer action $S : \mathfrak{g} \to \mathfrak{der}(\mathfrak{l})$ a crossed module $\mu : \mathfrak{l} \to \mathfrak{g}^S$.

Indeed, let $S : \mathfrak{g} \to \mathfrak{der}(\mathfrak{l})$ be an outer action and $\omega \in C^2(\mathfrak{g}, \mathfrak{l})$ with $R_S = \mathrm{ad} \circ \omega$. Consider the Lie subalgebra $\mathrm{ad}\,\mathfrak{l} \subset \mathfrak{der}(\mathfrak{l})$ of inner derivations, and consider $\mathrm{ad} : \mathfrak{l} \to \mathfrak{l}_{\mathrm{ad}}$ for $\mathfrak{l}_{\mathrm{ad}} := \mathfrak{l}/\mathfrak{z}(\mathfrak{l})$ as the quotient map. Define on the space

$$\mathfrak{g}^S := \mathfrak{l}_{\mathrm{ad}} \times \mathfrak{g}$$

an alternating bilinear map by

$$[(\mathrm{ad}\,l, x), (\mathrm{ad}\,l', x')] := (\mathrm{ad}([l, l'] + S(x) \cdot l' - S(x') \cdot l + \omega(x, x')), [x, x'])$$

$$= ([\operatorname{ad} l, \operatorname{ad} l'] + [S(x), \operatorname{ad} l'] - [S(x'), \operatorname{ad} l] + R_S(x, x'), [x, x']).$$

Note that the second form of the bracket implies that it does not depend on ω. We observe that

$$S_1 := \operatorname{ad} \circ S : \mathfrak{g} \to \mathfrak{der}(\mathfrak{l}_{\mathrm{ad}})$$

is a linear map with $R_{S_1} = \operatorname{ad} \circ R_S$.

Lemma 2.4.5. *$\mathfrak{g}^S$ is a Lie algebra with the following properties:*

(a) *The map $\pi_S : \mathfrak{g}^S \to \mathfrak{g}$, $(\operatorname{ad} l, x) \mapsto x$ is an extension of $\mathfrak{g}$ by $\mathfrak{l}_{\mathrm{ad}}$, which depends only on the class $[S]$, up to equivalence of extensions.*
(b) *The map $\eta : \mathfrak{g}^S \to \mathfrak{der}(\mathfrak{l})$, $(\operatorname{ad} l, x) \mapsto \operatorname{ad} l + S(x)$ defines a $\mathfrak{g}$-module structure on $\mathfrak{l}$.*
(c) *The map $\mu : \mathfrak{l} \to \mathfrak{g}^S$, $l \mapsto (\operatorname{ad} l, 0)$, defines a crossed module with $\operatorname{Ker}(\mu) = \mathfrak{z}(\mathfrak{l})$, the center of $\mathfrak{l}$, and $\operatorname{Coker}(\mu) = \mathfrak{g}$.*

Proof. The bracket is a special case of the bracket in equation (2.7). $\mathfrak{g}^S$ is therefore a Lie algebra isomorphic to $\mathfrak{l}_{\mathrm{ad}} \times_{(S_1, R_S)} \mathfrak{g}$.

(a) To see that the Lie algebra $\mathfrak{g}^S$ depends, as an extension of $\mathfrak{g}$ by $\mathfrak{l}_{\mathrm{ad}}$, only on the equivalence class $[S]$ of S, let $\epsilon : \mathfrak{g} \to \mathfrak{l}$ be a linear map and $S' := S - \operatorname{ad}_{\mathfrak{l}} \circ \epsilon$. Then $(S_1, R_S) = (\operatorname{ad}_{\mathfrak{l}} \circ \epsilon) \cdot (S_1', R_{S'})$ holds in $Z^2(\mathfrak{g}, \mathfrak{l}_{\mathrm{ad}})$ and Lemma 2.4.2 shows that

$$\varphi : \mathfrak{g}^S \to \mathfrak{g}^{S'}, \quad (\operatorname{ad} l, x) \mapsto (\operatorname{ad}(l + \epsilon(x)), x)$$

is an equivalence of extensions of $\mathfrak{g}$ by $\mathfrak{l}_{\mathrm{ad}}$.

(b) Let us show that η is a homomorphism of Lie algebras:

$$\begin{aligned}
&\eta([(\operatorname{ad} l, x), (\operatorname{ad} l', x')]) \\
&\quad = [\operatorname{ad} l, \operatorname{ad} l'] + [S(x), \operatorname{ad} l'] - [S(x'), \operatorname{ad} l] + R_S(x, x') + S([x, x']) \\
&\quad = [\operatorname{ad} l, \operatorname{ad} l'] + [S(x), \operatorname{ad} l'] - [S(x'), \operatorname{ad} l] + [S(x), S(x')] \\
&\quad = [\eta(\operatorname{ad} l, x), \eta(\operatorname{ad} l', x')].
\end{aligned}$$

(c) Follows from (a) and (b). □

Remark 2.4.9. On the one hand (see the Lemma above), this construction gives the *obstruction crossed module* $\mu : \mathfrak{l} \to \mathfrak{g}^S$ as the canonical central extension

$$0 \to \mathfrak{z}(\mathfrak{l}) \to \mathfrak{l} \to \mathfrak{l}_{\mathrm{ad}} \to 0$$

of the ideal $\mathfrak{l}_{\mathrm{ad}}$ in $\mathfrak{g}^S$; on the other hand (see the lemma below), the crossed module $\mu : \mathfrak{l} \to \mathfrak{g}^S$ can be seen as the pullback crossed module of the canonical crossed module

$$0 \to \mathfrak{z}(\mathfrak{l}) \to \mathfrak{l} \stackrel{\operatorname{ad}}{\to} \mathfrak{der}(\mathfrak{l}) \to \mathfrak{out}(\mathfrak{l}) \to 0$$

by the Lie algebra morphism $[S] : \mathfrak{g} \to \mathfrak{out}(\mathfrak{l})$.

Lemma 2.4.6. *The map* $\varphi = (\eta, \pi_S) : \mathfrak{g}^S \to \mathfrak{der}(\mathfrak{l}) \times \mathfrak{g}$ *is injective and yields an isomorphism of Lie algebras*

$$\mathfrak{g}^S \cong \{(d, x) \in \mathfrak{der}(\mathfrak{l}) \times \mathfrak{g} \mid S(x) \in d + \operatorname{ad} \mathfrak{l}\}.$$

Proof. Since $\operatorname{Ker}(\pi_S) = \mathfrak{l}_{\mathrm{ad}}$ and $\operatorname{Ker}(\eta) \cap \mathfrak{l}_{\mathrm{ad}} = \{0\}$, the map φ is a monomorphism of Lie algebras.

For each element $(\operatorname{ad} l, x) \in \mathfrak{g}^S$, we have $\varphi(\operatorname{ad} l, x) = (\operatorname{ad} l + S(x), x)$, which proves the inclusion $\operatorname{Im}(\varphi) \subset \{(d, x) \mid S(x) \in d + \operatorname{ad} \mathfrak{l}\}$. On the other hand, for any pair $(d, x) \in \mathfrak{der}(\mathfrak{l}) \times \mathfrak{g}$ with $d \in S(x) + \operatorname{ad} \mathfrak{l}$, we find an element $l \in \mathfrak{l}$ with $d = S(x) + \operatorname{ad} l$, which means that $(d, x) = \varphi(\operatorname{ad} l, x)$. This proves the lemma. □

The next lemma sheds some light on the passage from a given general extension of $\mathfrak{g}$ by $\mathfrak{l}$ to the corresponding abelian extension $\mathfrak{g}^S$ by $\mathfrak{z}(\mathfrak{l})$.

Lemma 2.4.7. *Let* $\pi : \mathfrak{e} \to \mathfrak{g}$ *be an extension of* $\mathfrak{g}$ *by* $\mathfrak{l}$ *corresponding to the* $\mathfrak{g}$*-kernel* $[S]$ *and* $\operatorname{ad}_{\mathfrak{l}}$ *the corresponding representation of* $\mathfrak{e}$ *on* $\mathfrak{l}$*. Then the map*

$$\gamma = (\operatorname{ad}_{\mathfrak{l}} - S \circ \pi, \pi) : \mathfrak{e} \to \mathfrak{der}(\mathfrak{l}) \times \mathfrak{g}, \quad x \mapsto (\operatorname{ad}_{\mathfrak{l}}(x) - S(\pi(x)), \pi(x))$$

defines an extension

$$0 \to \mathfrak{z}(\mathfrak{l}) \to \mathfrak{e} \xrightarrow{\gamma} \mathfrak{g}^S \to 0.$$

This assignment has the following properties:

(a) *If* $\pi_j : \mathfrak{e}_j \to \mathfrak{g}$, $j = 1, 2$, *are equivalent extensions of* $\mathfrak{g}$ *by* $\mathfrak{l}$*, then* $\gamma_j : \mathfrak{e}_j \to \mathfrak{g}^S$ *are equivalent extensions of* $\mathfrak{g}^S$ *by the* $\mathfrak{g}^S$*-module* $\mathfrak{z}(\mathfrak{l})$*. We thus obtain a map*

$$\tau : \operatorname{Ext}(\mathfrak{g}, \mathfrak{l})_{[S]} \to \operatorname{Ext}(\mathfrak{g}^S, \mathfrak{z}(\mathfrak{l})).$$

(b) *An extension* $\gamma : \mathfrak{e} \to \mathfrak{g}^S$ *of* $\mathfrak{g}^S$ *by* $\mathfrak{z}(\mathfrak{l})$ *comes from an extension of* $\mathfrak{g}$ *by* $\mathfrak{l}$ *if and only if there exists a* $\mathfrak{g}^S$*-equivariant equivalence* $\alpha : \mathfrak{l} \to \gamma^{-1}(\mathfrak{l}_{\mathrm{ad}})$ *of central extensions of* $\mathfrak{l}_{\mathrm{ad}}$ *by* $\mathfrak{z}(\mathfrak{l})$*.*

Proof. That $\mathfrak{e}$ corresponds to $[S]$ means that it is equivalent to a Lie algebra of the form $\mathfrak{l} \times_{(S,\omega)} \mathfrak{g}$ where (S, ω) is a factor system. This in turn means that

$$\operatorname{ad}_{\mathfrak{l}}(l, x) = \operatorname{ad}(l) + S(x) = \operatorname{ad}(l) + S(\pi(l, x)), \tag{2.8}$$

so that $\gamma(l, x) = (\operatorname{ad} l, x)$. Now the explicit formula for the bracket in $\mathfrak{l} \times_{(S,\omega)} \mathfrak{g}$ and $\mathfrak{g}^S$ implies that γ is an epimorphism. Its kernel is $\mathfrak{z}(\mathfrak{l})$, so that γ defines an extension of $\mathfrak{g}^S$ by $\mathfrak{z}(\mathfrak{l})$ as claimed.

(a) If $\varphi : \mathfrak{e}_1 \to \mathfrak{e}_2$ is an equivalence of $\mathfrak{l}$-extensions of $\mathfrak{g}$, then the representations $\mathrm{ad}^j_{\mathfrak{l}}$, $j = 1, 2$, of $\mathfrak{e}_j$ on $\mathfrak{l}$ satisfy $\mathrm{ad}^2_{\mathfrak{l}} \circ \varphi = \mathrm{ad}^1_{\mathfrak{l}}$ because $\varphi|_{\mathfrak{l}} = \mathrm{id}$. Therefore, the quotient maps

$$\gamma_j = (\mathrm{ad}^j_{\mathfrak{l}} - S \circ \pi_j, \pi_j) : \mathfrak{e}_j \to \mathfrak{g}^S$$

satisfy $\gamma_2 \circ \varphi = (\mathrm{ad}^2_{\mathfrak{l}} \circ \varphi - S \circ \pi_2 \circ \varphi, \pi_2 \circ \varphi) = (\mathrm{ad}^1_{\mathfrak{l}} - S \circ \pi_1, \pi_1) = \gamma_1$. This means that $\varphi : \mathfrak{e}_1 \to \mathfrak{e}_2$ is an equivalence of extensions of $\mathfrak{g}^S$ by $\mathfrak{z}(\mathfrak{l})$.

(b) Suppose first that the extension $\gamma : \mathfrak{e} \to \mathfrak{g}^S$ by $\mathfrak{z}(\mathfrak{l})$ comes from the $\mathfrak{l}$-extension $\pi : \mathfrak{e} \to \mathfrak{g}$ corresponding to $[S]$. We may assume that $\mathfrak{e} = \mathfrak{l} \times_{(S,\omega)} \mathfrak{g}$ (Lemma 2.4.2). Then equation (2.8) shows that $(l, x) \in \mathfrak{e}$ acts on $\mathfrak{l}$ by $\mathrm{ad}\, l + S(x) = \eta(\mathrm{ad}\, l, x)$. Therefore, the inclusion $\mathfrak{l} \hookrightarrow \mathfrak{e}$ onto the subspace $\gamma^{-1}(\mathfrak{l}_{\mathrm{ad}})$ is equivariant with respect to the action of $\mathfrak{g}^S$ and, therefore, in particular for the action of $\mathfrak{l}_{\mathrm{ad}}$, so that it is an equivalence of central extensions of $\mathfrak{l}_{\mathrm{ad}}$.

Suppose, conversely, that $\gamma : \mathfrak{e} \to \mathfrak{g}^S$ is an extension of $\mathfrak{g}^S$ by $\mathfrak{z}(\mathfrak{l})$ for which there exists a $\mathfrak{g}^S$-equivariant equivalence $\alpha : \mathfrak{l} \to \gamma^{-1}(\mathfrak{l}_{\mathrm{ad}})$ of central extensions of $\mathfrak{l}_{\mathrm{ad}}$ by $\mathfrak{z}(\mathfrak{l})$. Then

$$\mathfrak{e}/\alpha(\mathfrak{l}) = \mathfrak{e}/\gamma^{-1}(\mathfrak{l}_{\mathrm{ad}}) \cong \mathfrak{g}^S/\mathfrak{l}_{\mathrm{ad}} \cong \mathfrak{g},$$

so that we obtain by the quotient map $\pi : \mathfrak{e} \to \mathfrak{g}$ an extension of $\mathfrak{g}$ by $\mathfrak{l}$. The action of $\mathfrak{g}^S \cong \mathfrak{e}/\mathfrak{z}(\mathfrak{l})$ on $\mathfrak{l}$, induced by the adjoint representation of $\mathfrak{e}$ on $\mathfrak{l}$, coincides with the given action

$$\eta : \mathfrak{g}^S \to \mathfrak{der}(\mathfrak{l}), \quad (\mathrm{ad}\, l, x) \mapsto \mathrm{ad}\, l + S(x)$$

of $\mathfrak{g}^S$ on $\mathfrak{l}$, because α is $\mathfrak{g}^S$-equivariant. This implies that the $\mathfrak{g}$-kernel of the extension $\pi : \mathfrak{e} \to \mathfrak{g}$ is $[S]$.

For the adjoint representation $\mathrm{ad}_{\mathfrak{l}}$ of $\mathfrak{e}$ on $\mathfrak{l}$, we have $\mathrm{ad}_{\mathfrak{m}} = \eta \circ \gamma$ and $\pi = \pi_S \circ \gamma$, so that the corresponding map $\mathfrak{e} \to \mathfrak{g}^S$ coincides with γ. This means that $\gamma : \mathfrak{e} \to \mathfrak{g}^S$ is associated to the extension $\pi : \mathfrak{e} \to \mathfrak{g}$ by the process described above. □

The following theorem shows that the cohomology class $[d_S\omega]$ of an outer action $[S]$ is exactly the obstruction against the existence of an extension of $\mathfrak{g}$ by $\mathfrak{l}$ realizing this outer action.

We will assume $\mathfrak{l} = \mathfrak{z}(\mathfrak{l}) \times_f \mathrm{ad}\, \mathfrak{l}$ for some 2-cocycle $f \in Z^2(\mathrm{ad}\, \mathfrak{l}, \mathfrak{z}(\mathfrak{l}))$. Then the action of $\mathfrak{g}^S$ on $\mathfrak{l}$ is described by a bilinear map $\theta : \mathfrak{g}^S \times \mathfrak{l} \to \mathfrak{z}(\mathfrak{l})$ via

$$(\mathrm{ad}\, l, x) \cdot (z, \mathrm{ad}\, l') = (x \cdot z + \theta((\mathrm{ad}\, l, x), \mathrm{ad}\, l'), [\mathrm{ad}\, l + S(x), \mathrm{ad}\, l']).$$

In the following, we write

$$Z^2(\mathfrak{g}^S, \mathfrak{z}(\mathfrak{l}))_\theta := \{\tilde{\theta} \in Z^2(\mathfrak{g}^S, \mathfrak{z}(\mathfrak{l})) \mid \tilde{\theta}|_{\mathfrak{g}^S \times \mathfrak{l}} = \theta\}$$

for the set of all $\mathfrak{z}(\mathfrak{l})$-valued 2-cocycles extending θ.

Theorem 2.4.8.

(a) *For the* $\mathfrak{g}$*-kernel* $[S]$*, the cohomology class* $[\chi_S] = [d_S\omega] \in H^3(\mathfrak{g}, \mathfrak{z}(\mathfrak{l}))$ *vanishes if and only if* $\operatorname{Ext}(\mathfrak{g}, \mathfrak{l}) \neq 0$.

(b) *If* $[d_S\omega] = 0$*, then each* $\mathfrak{l}$*-extension of* $\mathfrak{g}$ *corresponding to* $[S]$ *is equivalent to an extension of the form*

$$\pi : \mathfrak{z}(\mathfrak{l}) \times_{\tilde{\theta}} \mathfrak{g}^S \to \mathfrak{g}, \quad (z, x) \mapsto \pi_S(x), \quad \tilde{\theta} \in Z^2(\mathfrak{g}^S, \mathfrak{z}(\mathfrak{l}))_\theta.$$

The set $Z^2(\mathfrak{g}^S, \mathfrak{z}(\mathfrak{l}))_\theta$ *is an affine space on which the vector space* $Z^2(\mathfrak{g}, \mathfrak{z}(\mathfrak{l}))_S$ *acts simply transitively by* $\omega \cdot \tilde{\theta} := \tilde{\theta} + \pi_S^*\omega$*. Two* $\mathfrak{l}$*-extensions of* $\mathfrak{g}$ *corresponding to* $\tilde{\theta}_j \in Z^2(\mathfrak{g}^S, \mathfrak{z}(\mathfrak{l}))_\theta$ *for* $j = 1, 2$ *are equivalent if and only if* $\tilde{\theta}_2 - \tilde{\theta}_1 \in \pi_S^* B^2(\mathfrak{g}, \mathfrak{z}(\mathfrak{l}))$.

Proof.

(a) In view of Proposition 2.3.3, the characteristic class $[\chi_\mu] \in H^3(\mathfrak{g}, \mathfrak{z}(\mathfrak{l}))$ of the crossed module $\mu : \mathfrak{l} \to \mathfrak{g}^S$ is represented by the cocycle $d_\rho\omega$, where $\rho : \mathfrak{g} \to \mathfrak{g}^S$, $x \mapsto (0, x)$, is the canonical section and $\omega : \mathfrak{g} \times \mathfrak{g} \to \mathfrak{l}$ is an alternating map with $R_S = R_\rho = \mu \circ \omega = \operatorname{ad} \circ \omega$. Hence

$$\begin{aligned} d_\rho\omega(x, x', x'') &= \sum_{\text{cycl.}} \rho(x) \cdot \omega(x', x'') - \omega([x, x'], x'') \\ &= \sum_{\text{cycl.}} S(x) \cdot \omega(x', x'') - \omega([x, x'], x'') = d_S\omega(x, x', x''). \end{aligned}$$

In view of Theorem 2.3.4, $[\chi_\mu]$ vanishes is and only if θ extends to a cocycle on $\mathfrak{g}^S$, that is, if and only if $Z^2(\mathfrak{g}^S, \mathfrak{z}(\mathfrak{l}))_\theta \neq \emptyset$. Again by Theorem 2.3.4, $[\chi_\mu] = 0$ implies that the condition of Lemma 2.4.7 (b) is fulfilled, and we have thus $\operatorname{Ext}(\mathfrak{g}, \mathfrak{l})_{[S]} \neq \emptyset$.
On the other hand, in case $\operatorname{Ext}(\mathfrak{g}, \mathfrak{l})_{[S]} \neq \emptyset$, the map τ of Lemma 2.4.7 (a) gives rise to an extension of $\mathfrak{g}^S$ by $\mathfrak{z}(\mathfrak{l})$, which implies that $Z^2(\mathfrak{g}^S, \mathfrak{z}(\mathfrak{l}))_\theta \neq \emptyset$, and thus $[\chi_\mu] = 0$ as above.

(b) For two cocycles $\tilde{\theta}_j \in Z^2(\mathfrak{g}^S, \mathfrak{z}(\mathfrak{l}))_\theta$, $j = 1, 2$, the difference $\tilde{\theta}_2 - \tilde{\theta}_1$ vanishes on $\mathfrak{g}^S \times \mathfrak{l}_{\mathrm{ad}}$, hence can be written as $\pi_S^*\omega$ for some $\omega \in Z^2(\mathfrak{g}, \mathfrak{z}(\mathfrak{l}))$. Conversely, for $\tilde{\theta} \in Z^2(\mathfrak{g}^S, \mathfrak{z}(\mathfrak{l}))_\theta$ and $\omega \in Z^2(\mathfrak{g}, \mathfrak{z}(\mathfrak{l}))$, the cocycle $\tilde{\theta} + \pi_S^*\omega$ is also contained in $Z^2(\mathfrak{g}^S, \mathfrak{z}(\mathfrak{l}))_\theta$, because $\pi_S^*\omega$ vanishes on $\mathfrak{g}^S \times \mathfrak{l}_{\mathrm{ad}}$. As the map $\pi_S^* : Z^2(\mathfrak{g}, \mathfrak{z}(\mathfrak{l})) \to Z^2(\mathfrak{g}^S, \mathfrak{z}(\mathfrak{l}))$ is injective, $Z^2(\mathfrak{g}^S, \mathfrak{z}(\mathfrak{l}))_\theta$ is an affine space with translation group $Z^2(\mathfrak{g}, \mathfrak{z}(\mathfrak{l}))$ acting by $\omega \cdot \tilde{\theta} = \tilde{\theta} + \pi_S^*\omega$.
Let $\tilde{\theta}_j \in Z^2(\mathfrak{g}^S, \mathfrak{z}(\mathfrak{l}))_\theta$, $j = 1, 2$ and $\pi_j : \mathfrak{e}_j \to \mathfrak{g}$ be the corresponding $\mathfrak{l}$-extensions of $\mathfrak{g}$. If $\varphi : \mathfrak{e}_1 \to \mathfrak{e}_2$ is an equivalence of $\mathfrak{l}$-extensions of $\mathfrak{g}$, then Lemma 2.4.7(a) implies that φ is also an equivalence of $\mathfrak{z}(\mathfrak{l})$-extensions of $\mathfrak{g}^S$. Therefore, it can be written in the form

$$\varphi : \mathfrak{e}_1 = \mathfrak{z}(\mathfrak{l}) \times_{\tilde{\theta}_1} \mathfrak{g}^S \to \mathfrak{e}_2 = \mathfrak{z}(\mathfrak{l}) \times_{\tilde{\theta}_2} \mathfrak{g}^S, \quad (z, l) \mapsto (z + \beta(x), x),$$

where $\beta : \mathfrak{g}^S \to \mathfrak{z}(\mathfrak{l})$ satisfies $d_{\mathfrak{g}^S}\beta = \tilde{\theta}_1 - \tilde{\theta}_2$. Since φ fixes $\mathfrak{l} = \mathfrak{z}(\mathfrak{l}) \times_f \mathfrak{l}_{\mathrm{ad}} \subset \mathfrak{e}_j$ pointwise, we have $\mathfrak{l}_{\mathrm{ad}} \subset \operatorname{Ker}(\beta)$, so that $\beta = \pi_S^*\bar{\beta}$ for some $\bar{\beta} \in C^1(\mathfrak{g}, \mathfrak{z}(\mathfrak{l}))$. This means that $\tilde{\theta}_1 - \tilde{\theta}_2 \in \pi_S^* B^2(\mathfrak{g}, \mathfrak{z}(\mathfrak{l}))$.

If, conversely, $\tilde{\theta}_1 - \tilde{\theta}_2 = d_{\mathfrak{g}^S}(\pi_S^*\bar{\beta})$ for some $\bar{\beta} \in C^1(\mathfrak{g}, \mathfrak{z}(\mathfrak{l}))$, then

$$\varphi : \mathfrak{z}(\mathfrak{l}) \times_{\tilde{\theta}_1} \mathfrak{g}^S \to \mathfrak{e}_2 = \mathfrak{z}(\mathfrak{l}) \times_{\tilde{\theta}_2} \mathfrak{g}^S, \quad (z, x) \mapsto (z + \bar{\beta}(\pi_S(x)), x)$$

is an equivalence of $\mathfrak{l}$-extensions of $\mathfrak{g}$. □

2.5 Construction of crossed modules

2.5.1 The principal construction

In order to motivate the construction procedure, which we present below and which we will call the *principal construction* in the following, let us go back to the four-term exact sequence associated to a crossed module:

$$0 \to V \xrightarrow{i} \mathfrak{m} \xrightarrow{\mu} \mathfrak{n} \xrightarrow{\pi} \mathfrak{g} \to 0.$$

We have already seen that this is an exact sequence of Lie algebras. We derive from it now two short exact sequences of Lie algebras:

$$0 \to \mathfrak{m}/i(V) \xrightarrow{\mu} \mathfrak{n} \xrightarrow{\pi} \mathfrak{g} \to 0,$$

and

$$0 \to V \xrightarrow{i} \mathfrak{m} \xrightarrow{\mu} \mathrm{Im}(\mu) \to 0.$$

Obviously, the crossed module $\mu : \mathfrak{m} \to \mathfrak{n}$ can be seen as a Yoneda product (i. e., the splicing together) of these two short exact sequences. The second sequence is a central extension, whereas the first is not a central extension, not even an abelian extension in general. According to the terminology of the previous section, it is a *general extension* of Lie algebras.

Let us ask (some version of) the converse question. Given a Lie algebra $\mathfrak{g}$, a short exact sequence of $\mathfrak{g}$-modules

$$0 \to V_1 \to V_2 \to V_3 \to 0 \tag{2.9}$$

(regarded as a short exact sequence of abelian Lie algebras) and an abelian extension $\mathfrak{e}$ of $\mathfrak{g}$ by the abelian Lie algebra V_3

$$0 \to V_3 \to \mathfrak{e} \to \mathfrak{g} \to 0, \tag{2.10}$$

is in this situation the Yoneda product of the sequences in equations (2.9) and (2.10) a crossed module? In case the sequence (2.10) is given by a 2-cocycle α, we continue to use the notation $\mathfrak{e} = V_3 \times_\alpha \mathfrak{g}$.

Theorem 2.5.1. *In the above situation, the Yoneda product of sequences* (2.9) *and* (2.10) *is a crossed module, the associated* 3*-cocycle of which is the image of the* 2*-cocycle defining the central extension* (2.10) *under the connecting homomorphism in the long exact cohomology sequence associated to the coefficient sequence* (2.9).

Proof. Splicing the sequences in equations (2.9) and (2.10) together, one obtains a map

$$\mu : V_2 \to \mathfrak{e}. \tag{2.11}$$

Writing $\mathfrak{e} = V_2 \oplus \mathfrak{g}$ as vector spaces, we have $\mu(v) = (v, 0)$. On the other hand, the $\mathfrak{e}$-action η on V_2 is induced by the action of $\mathfrak{g}$ on V_2: $\eta(w,x)(v) := x \cdot v$, where $(w,x) \in \mathfrak{e}$, $v \in V_2$. With these structures, condition (b) for a Lie algebra crossed module is trivially true, while condition (a) is true by definition of the bracket in the abelian extension: $\mu(\eta(w,x)(v)) = (x \cdot v, 0) = [(w,x),(v,0)]$.

Now let us discuss the second claim. The short exact sequence (2.9) induces a short exact sequence of complexes

$$0 \to C^*(\mathfrak{g}, V_1) \xrightarrow{i} C^*(\mathfrak{g}, V_2) \xrightarrow{\pi} C^*(\mathfrak{g}, V_3) \to 0.$$

Take a cocycle $\alpha \in C^2(\mathfrak{g}, V_3)$, then the connecting homomorphism $\partial : C^2(\mathfrak{g}, V_3) \to C^3(\mathfrak{g}, V_1)$ is defined as follows:

$$\begin{array}{ccccc}
 & & \beta \in \pi^{-1}(\alpha) & \stackrel{\pi}{\longmapsto} & \alpha \\
 & & \top & & \top \\
 & & | & & | \\
C^2(\mathfrak{g}, V_1) & \stackrel{i}{\rightarrowtail} & C^2(\mathfrak{g}, V_2) & \stackrel{\pi}{\twoheadrightarrow} & C^2(\mathfrak{g}, V_3) \\
\downarrow{\scriptstyle d^{V_1}} & & \downarrow{\scriptstyle d^{V_2}} & & \downarrow{\scriptstyle d^{V_3}} \\
C^3(\mathfrak{g}, V_1) & \stackrel{i}{\rightarrowtail} & C^3(\mathfrak{g}, V_2) & \stackrel{\pi}{\twoheadrightarrow} & C^3(\mathfrak{g}, V_3) \\
 & & \downarrow & & \downarrow \\
\partial\alpha := i^{-1}(d^{V_2}\beta) & \stackrel{i}{\longmapsto} & d^{V_2}\beta & \stackrel{\pi}{\longmapsto} & 0 = d^{V_3}\alpha
\end{array}$$

Here, we wrote elements on top resp. on bottom of the corresponding spaces, and denoted by d^{V_1}, d^{V_2} and d^{V_3} the Lie algebra coboundaries with values in V_1, V_2 and V_3, respectively. Summarizing, $\partial\alpha$ is constructed by choosing an element β preimage of α under π, taking $d^{V_2}\beta$ and taking an preimage of $d^{V_2}\beta$ under i. It is obvious that this is exactly how we constructed the 3-cocycle γ corresponding to a crossed module. Having stated the coincidence of the two constructions, it remains to take for $\alpha \in C^2(\mathfrak{g}, V_3)$ the cocycle defining the abelian extension (2.10). □

2.5.2 End of the proof of Theorem 2.2.1

Here, we end the proof of Theorem 2.2.1 using the previous construction. It is amazing that the surjectivity is completely formal, while the injectivity relies on general extension theory. This is in contrast to the theory of Lie algebra kernels of Hochschild [43], where the injectivity is formal, while the surjectivity is involved; see Section 2.6.1.

Surjectivity of the map *b* from Theorem 2.2.1

We have to show that given a cohomology class $[\gamma] \in H^3(\mathfrak{g}, V)$, there is a crossed module whose associated class is $[\gamma]$. V is here some $\mathfrak{g}$-module. As the category of $\mathfrak{g}$-modules possesses enough injectives, there is an injective $\mathfrak{g}$-module I and a monomorphism $i : V \hookrightarrow I$. Consider now the short exact sequence of $\mathfrak{g}$-modules:

$$0 \to V \to I \to Q \to 0,$$

where Q is the cokernel of i. As I is injective, the long exact sequence in cohomology gives $H^2(\mathfrak{g}, Q) \cong H^3(\mathfrak{g}, V)$. Thus $[\gamma]$ corresponds under the isomorphism (which is induced by the connecting homomorphism) to a class $[\alpha] \in H^2(\mathfrak{g}, Q)$. Now the principal construction in Theorem 2.5.1 applied to the above short exact sequence of $\mathfrak{g}$-modules and the abelian extension of $\mathfrak{g}$ by Q using the cocycle α gives a crossed module whose class is a preimage under b of $[\gamma]$.

Injectivity of the map *b* from Theorem 2.2.1

It is clear that b is a homomorphism of abelian groups. Thus in order to show the injectivity of b, it suffices to show that its kernel is trivial.

(a) A crossed module

$$0 \to V \xrightarrow{i} \mathfrak{m} \xrightarrow{\mu} \mathfrak{n} \xrightarrow{\pi} \mathfrak{g} \to 0$$

corresponding to the trivial cohomology class gives rise to a general extension

$$0 \to \mathfrak{m} \to \mathfrak{e} \to \mathfrak{g} \to 0.$$

Indeed, if $[\gamma] = 0$, there exists $\omega \in C^2(\mathfrak{g}, V)$ such that $d\omega = \gamma$. Set $\zeta := \beta(x, y) - i(\omega(x, y))$ for $x, y \in \mathfrak{g}$, where $\beta \in C^2(\mathfrak{g}, \mathfrak{m})$ is the 2-cochain defining $[\gamma]$ as usual by $d\beta = i\gamma$ with respect to a section ρ of π. Then

$$d\zeta = d\beta - di\omega = d\beta - i(d\omega) = d\beta - d\beta = 0,$$

thus the 2-cochain $\zeta \in C^2(\mathfrak{g}, \mathfrak{m})$ is a cocycle. Observe that ρ gives rise to an outer action S of $\mathfrak{g}$ on $\mathfrak{m}$ and that (S, ζ) forms a factor system, because $R_S = R_\rho = \mathrm{ad}_\beta = \mathrm{ad}_\zeta$; see Remark 2.1.4. There exists thus an extension of $\mathfrak{g}$ by $\mathfrak{m}$ using the cocycle $\zeta \in Z^2(\mathfrak{g}, \mathfrak{m})$ by Lemma 2.4.1.

(b) We now use this extension to construct a morphism of crossed modules. As a vector space, $\mathfrak{e} = \mathfrak{m} \oplus \mathfrak{g}$. We fix such a decomposition of $\mathfrak{e}$ and call it product coordinates. The bracket in product coordinates is given in equation (2.7).

Lemma 2.5.2. *The map m written in product coordinates as $m : \mathfrak{m} \oplus \mathfrak{g} \to \mathrm{Im}(\mu) \oplus \mathfrak{g}$, $(a,x) \mapsto m(a,x) = (\mu(a), x)$ induces a morphism of crossed modules*

$$\begin{array}{ccccccccccc} & & 0 & \longrightarrow & \mathfrak{m} & \longrightarrow & \mathfrak{e} & \longrightarrow & \mathfrak{g} & \longrightarrow & 0 \\ & & & & \downarrow{\scriptstyle \mathrm{id}_{\mathfrak{m}}} & & \downarrow{\scriptstyle m} & & \downarrow{\scriptstyle \mathrm{id}_{\mathfrak{g}}} & & \\ 0 & \longrightarrow & V & \xrightarrow{i} & \mathfrak{m} & \xrightarrow{\mu} & \mathfrak{n} & \xrightarrow{\pi} & \mathfrak{g} & \longrightarrow & 0 \end{array}$$

Proof. According to equation (2.7), the bracket in $\mathfrak{n}$ reads in product coordinates

$$[(a,x),(b,y)] = ([a,b] + x \cdot b - y \cdot a + \alpha(x,y), [x,y]),$$

and the one in $\mathfrak{e}$ reads

$$[(a,x),(b,y)] = ([a,b] + x \cdot b - y \cdot a + \beta(x,y), [x,y]).$$

Now, applying μ to the first component of the lower bracket gives the first component of the upper bracket thanks to the axioms of a crossed module and to $\mu\beta = \alpha$. The map is also compatible with the actions. □

(c)

Lemma 2.5.3. *If a crossed module*

$$0 \to V \to \mathfrak{m} \xrightarrow{\mu} \mathfrak{n} \to \mathfrak{g} \to 0$$

admits a morphism

$$\begin{array}{ccccccccccc} & & 0 & \longrightarrow & \mathfrak{m} & \xrightarrow{\varphi} & \mathfrak{e} & \xrightarrow{\psi} & \mathfrak{g} & \longrightarrow & 0 \\ & & & & \downarrow{\scriptstyle \mathrm{id}_{\mathfrak{m}}} & & \downarrow{\scriptstyle m} & & \downarrow{\scriptstyle \mathrm{id}_{\mathfrak{g}}} & & \\ 0 & \longrightarrow & V & \xrightarrow{i} & \mathfrak{m} & \xrightarrow{\mu} & \mathfrak{n} & \xrightarrow{\pi} & \mathfrak{g} & \longrightarrow & 0, \end{array}$$

then it represents the zero equivalence class.

Proof. Indeed, in case there is a morphism as indicated, we have a commutative diagram

$$\begin{array}{ccccccccccc} 0 & \longrightarrow & V & \xrightarrow{\mathrm{id}_V} & V & \xrightarrow{0} & \mathfrak{g} & \xrightarrow{\mathrm{id}_{\mathfrak{g}}} & \mathfrak{g} & \longrightarrow & 0 \\ & & \uparrow{\scriptstyle \mathrm{id}_V} & & \uparrow{\scriptstyle \mathrm{proj}_1} & & \uparrow{\scriptstyle \psi} & & \uparrow{\scriptstyle \mathrm{id}_{\mathfrak{g}}} & & \\ 0 & \longrightarrow & V & \xrightarrow{\mathrm{incl}_1} & V \oplus \mathfrak{m} & \xrightarrow{0+\varphi} & \mathfrak{e} & \xrightarrow{\psi} & \mathfrak{g} & \longrightarrow & 0 \\ & & \downarrow{\scriptstyle \mathrm{id}_V} & & \downarrow{\scriptstyle i+\mathrm{proj}_2} & & \downarrow{\scriptstyle m} & & \downarrow{\scriptstyle \mathrm{id}_{\mathfrak{g}}} & & \\ 0 & \longrightarrow & V & \xrightarrow{i} & \mathfrak{m} & \xrightarrow{\mu} & \mathfrak{n} & \xrightarrow{\pi} & \mathfrak{g} & \longrightarrow & 0 \end{array}$$

Thus $\mu : \mathfrak{m} \to \mathfrak{n}$ represents the zero class (as equivalence class of crossed modules). We denoted by incl_1 and proj_2 once again the standard inclusion and projection maps. This concludes the proof of the lemma, the injectivity of the map b, and thus the proof of Theorem 2.2.1. □

Corollary 2.5.4. *Every crossed module is equivalent to one coming from the principal construction.*

Remark 2.5.1. Now we can use the preceding arguments to characterize once again the equivalence class of crossed modules corresponding to the zero class in $H^3(\mathfrak{g}; V_1)$. This characterization is different from the one in Theorem 2.3.4. The main point of the following discussion is due to [37], which we rephrase in our framework.

Let $[\mu : \mathfrak{m} \to \mathfrak{n}]$ be an equivalence class of a crossed module μ with kernel V_1 and cokernel $\mathfrak{g}$. By the above Corollary 2.5.4, there exists a short exact sequence of $\mathfrak{g}$-modules

$$0 \to V_1 \to V_2 \to V_3 \to 0$$

and a 2-cocycle $\alpha \in C^2(\mathfrak{g}; V_3)$ such that

$$0 \to V_1 \to V_2 \to V_3 \times_\alpha \mathfrak{g} \to \mathfrak{g} \to 0 \qquad (2.12)$$

is equivalent to μ. Now, Goldberg's theorem ([37], cf. also [36] Theorem 4) states that (2.12) has trivial class $[\partial\alpha] = [\gamma] \in H^3(\mathfrak{g}; V_1)$ if and only if there exists an abelian extension $\mathfrak{e}$ of $\mathfrak{g}$ by V_2 such that $\mathfrak{e}/V_1 \cong V_3 \times_\alpha \mathfrak{g}$ as Lie algebras.

Indeed, in case there exists such an abelian extension $\mathfrak{e} = V_2 \times_{\tilde{\alpha}} \mathfrak{g}$, we can take $\tilde{\alpha}$ as β in the construction of $\partial\alpha$ in the proof of Theorem 2.5.1. Then $d^{V_2}\beta = 0$ and as a consequence $\partial\alpha = 0$.

On the other hand, consider the case $[\partial\alpha] = 0$, that is, $\partial\alpha = d^{V_1}\lambda$ for some $\lambda \in C^2(\mathfrak{g}; V_1)$. Recall the exact sequence of complexes used for the construction of $\partial\alpha$:

$$\begin{array}{ccccc}
C^2(\mathfrak{g}, V_1) & \overset{i}{\rightarrowtail} & C^2(\mathfrak{g}, V_2) & \overset{\pi}{\twoheadrightarrow} & C^2(\mathfrak{g}, V_3) \\
\downarrow{\scriptstyle d^{V_1}} & & \downarrow{\scriptstyle d^{V_2}} & & \downarrow{\scriptstyle d^{V_3}} \\
C^3(\mathfrak{g}, V_1) & \overset{i}{\rightarrowtail} & C^3(\mathfrak{g}, V_2) & \overset{\pi}{\twoheadrightarrow} & C^3(\mathfrak{g}, V_3)
\end{array}$$

We have $d^{V_2}i(\lambda) = d^{V_2}\beta$ and $\pi(\beta) = \alpha$. We claim that $-i(\lambda)+\beta$ is a 2-cocycle defining an abelian extension $\mathfrak{e}$ of $\mathfrak{g}$ by V_2, which solves the problem. Indeed, $d^{V_2}(-i(\lambda) + \beta) = -d^{V_2}\beta + d^{V_2}\beta = 0$, thus $\mathfrak{e}$ exists, and $\pi(\beta) = \alpha$ implies that $\mathfrak{e}/V_1 \cong V_3 \times_\alpha \mathfrak{g}$. This shows Goldberg's characterization of the zero class.

Observe that this is the reasoning of item (a) in the injectivity proof in Section 2.5.2 in this special case.

2.6 Relation to Lie algebra kernels

In this section, we will rephrase the theory of the previous sections in terms of Lie algebra kernels, that is, homomorphisms of Lie algebras $\psi : \mathfrak{g} \to \mathfrak{out}(\mathfrak{l})$. Glimpses of this have already been visible in the former treatment, but here we will put emphasis entirely on Lie algebra kernels. In this context, we will recover the main features of the previous sections, in particular the relation to third Lie algebra cohomology.

2.6.1 Lie algebra kernels and third cohomology

Definition 2.6.1. Let $\mathfrak{g}$ and $\mathfrak{l}$ be arbitrary Lie algebras. A *Lie algebra kernel* is a Lie algebra homomorphism $\psi : \mathfrak{g} \to \mathfrak{out}(\mathfrak{l})$. Sometimes, it is also the Lie algebra $\mathfrak{l}$, which is called a $\mathfrak{g}$-kernel in this situation.

Remark 2.6.2. A given homomorphism $\psi : \mathfrak{g} \to \mathfrak{out}(\mathfrak{l})$ induces the structure of a $\mathfrak{g}$-module on the center $\mathfrak{z}(\mathfrak{l})$ of $\mathfrak{l}$. The center $\mathfrak{z}(\mathfrak{l})$, endowed with this $\mathfrak{g}$-module structure, is then called the *nucleus* of ψ.

Definition 2.6.3. Let $\mathfrak{l}_1$, $\mathfrak{l}_2$ be $\mathfrak{g}$-kernels with fixed nucleus $\mathfrak{n} := \mathfrak{z}(\mathfrak{l})$. Denote by ψ_1 and ψ_2 the corresponding homomorphisms. Let us construct the *sum of the Lie algebra kernels* $\mathfrak{l}_1$ and $\mathfrak{l}_2$, denoted $\mathfrak{l}_1 + \mathfrak{l}_2$.

In the direct sum $\mathfrak{l}_1 \oplus \mathfrak{l}_2$, let $\mathfrak{j}$ be the ideal generated by the pairs $(n, -n)$ with $n \in \mathfrak{n}$. Denote by $\mathfrak{l} := \mathfrak{l}_1 \oplus \mathfrak{l}_2/\mathfrak{j}$ and identify the center

$$\mathfrak{z}(\mathfrak{l}) = \mathfrak{n} \oplus \mathfrak{n}/\mathfrak{j}$$

with $\mathfrak{n}$ by means of the isomorphism induced by $(n_1, n_2) \mapsto n_1 + n_2$. Define a $\mathfrak{g}$-kernel $\psi : \mathfrak{g} \to \mathfrak{out}(\mathfrak{l})$ as follows: For $x \in \mathfrak{g}$, let $\zeta_i \in \mathfrak{der}(\mathfrak{l}_i)$ be a derivation representing $\psi_i(x) \in \mathfrak{out}(\mathfrak{l}_i)$ for $i = 1, 2$. There exists a unique derivation ζ on $\mathfrak{l}$, which coincides with ζ_i on $\mathfrak{l}_i$. The class of ζ modulo $\mathrm{ad}(\mathfrak{l})$ is then denoted $\psi(x)$. The nucleus of ψ is clearly the $\mathfrak{g}$-module $\mathfrak{n}$.

Definition 2.6.4. Call a $\mathfrak{g}$-kernel ψ *extendible* in case there exists a general extension of Lie algebras

$$0 \to \mathfrak{l} \to \mathfrak{e} \to \mathfrak{g} \to 0$$

such that its outer action is $\psi : \mathfrak{g} \to \mathfrak{out}(\mathfrak{l})$.

Let $\mathfrak{l}_1$, $\mathfrak{l}_2$ be $\mathfrak{g}$-kernels with fixed nucleus $\mathfrak{n} := \mathfrak{z}(\mathfrak{l})$. Denote by ψ_1 and ψ_2 the corresponding homomorphisms. The kernels $\mathfrak{l}_1$ and $\mathfrak{l}_2$ are called *equivalent* in case there exist extendible $\mathfrak{g}$-kernels $\mathfrak{u}$ and $\mathfrak{v}$ with nucleus $\mathfrak{n}$ such that

$$\mathfrak{l}_1 + \mathfrak{u} = \mathfrak{l}_2 + \mathfrak{v}.$$

Lemma 2.6.1. *Equivalence of $\mathfrak{g}$-kernels is an equivalence relation. Equipped with the above defined sum, the set of equivalence classes of $\mathfrak{g}$-kernels fixed nucleus $\mathfrak{n}$ becomes an abelian group.*

Proof. Concerning the equivalence relation, symmetry and reflexivity are clear. For transitivity, it suffices to show that the sum of extendible $\mathfrak{g}$-kernels is extendible.

For this, let $\phi_j : \mathfrak{e}_j \to \mathfrak{g}$ for $j = 1, 2$ be extensions of $\mathfrak{g}$ by $\mathfrak{l}_1$ and $\mathfrak{l}_2$, respectively, realizing the corresponding extendible $\mathfrak{g}$-kernels with common nucleus $\mathfrak{n}$. Let $\mathfrak{s} \subset \mathfrak{e}_1 \oplus \mathfrak{e}_2$ be the subalgebra consisting of pairs (e_1, e_2) such that $\phi_1(e_1) = \phi_2(e_2)$. The Lia algebra $\mathfrak{s}$ is the pullback of the pair of arrows (ϕ_1, ϕ_2). Divide $\mathfrak{s}$ by the ideal generated by the pairs $(n, -n)$ for $n \in \mathfrak{n}$ to obtain a Lie algebra $\mathfrak{e}$. Then $\mathfrak{e}$ contains $\mathfrak{l}_1 + \mathfrak{l}_2$ as an ideal, and the homomorphism ϕ_1 (or equally well ϕ_2) realizes $\mathfrak{e}$ as an extension of $\mathfrak{g}$ by $\mathfrak{l}_1 + \mathfrak{l}_2$ with nucleus $\mathfrak{n}$.

The associativity and commutativity relations of the sum are clear. It remains to construct an inverse to a $\mathfrak{g}$-kernel $\mathfrak{l}$ with nucleus $\mathfrak{n}$. For this, divide the direct sum $\mathfrak{l} \oplus \mathfrak{n}$ by the ideal $\mathfrak{j}$ generated by the diagonal (n, n) for $n \in \mathfrak{n}$. Its center $\mathfrak{n} \oplus \mathfrak{n}/\mathfrak{j}$ can be identified to $\mathfrak{n}$ using the homomorphism $(n_1, n_2) \mapsto -n_1 + n_2$. The $\mathfrak{g}$-kernel structure on $\mathfrak{l}$ induces a $\mathfrak{g}$-kernel structure on

$$-\mathfrak{l} := \mathfrak{l} \oplus \mathfrak{n}/\mathfrak{j}.$$

The zero class is the class of extendible $\mathfrak{g}$-kernels. Thus, it remains to show that $-\mathfrak{l} + \mathfrak{l}$ is an extendible kernel and that for an extendible $\mathfrak{g}$-kernel $\mathfrak{u}$ with nucleus $\mathfrak{n}$ the extendibility of $\mathfrak{u} + \mathfrak{l}$ implies the extendibility of $\mathfrak{l}$. This will follow from the extendibility criterion derived from the description by cocycles in the next lemma. The present proof will be completed in Remark 2.6.6. □

Let us recall how $\mathfrak{g}$-kernels are described in terms of cocycles. First, lift the map $\psi : \mathfrak{g} \to \mathfrak{out}(\mathfrak{g})$ to a linear map $S : \mathfrak{g} \to \mathfrak{der}(\mathfrak{g})$. As in the previous sections, the failure of S to be a homomorphism is denoted

$$R_S(x, y) := [S(x), S(y)] - S([x, y])$$

for all $x, y \in \mathfrak{g}$. Furthermore, for all $x, y \in \mathfrak{g}$, there exists $\omega(x, y) \in \mathfrak{l}$ such that $R_S(x, y) = \mathrm{ad} \circ \omega(x, y)$. The map ω may be taken bilinear and alternating. The characteristic class $[\chi] \in H^3(\mathfrak{g}, \mathfrak{z}(\mathfrak{l}))$ is then defined to be $\chi = d_S\omega$. Here, d_S denotes the formal coboundary operator of Lie algebra cohomology of $\mathfrak{g}$ with values in $\mathfrak{l}$, the "action" of $\mathfrak{g}$ on $\mathfrak{l}$ being defined by S (while this is not an action in general); see equation (B.1) in Appendix B.

One computes that $d_S\omega$ has values in the center $\mathfrak{z}(\mathfrak{l})$; cf. Lemma 2.2.2 and Definition 2.4.8, and that it is a 3-cocycle; cf. Lemma 2.2.3 and Definition 2.4.8. Still like in Definition 2.4.8, one shows that the cohomology class $[d_S\omega]$ does only depend on the $\mathfrak{g}$-kernel $\psi : \mathfrak{g} \to \mathfrak{out}(\mathfrak{l})$ and does not depend on the choice of S or ω.

Remark 2.6.5. Observe that in case $\mathfrak{l}$ is extendible, S and ω may be chosen such that $d_S\omega = 0$. Indeed, let $\phi : \mathfrak{e} \to \mathfrak{g}$ be an extension realizing the $\mathfrak{g}$-kernel ψ. Then we can find a linear map $\rho : \mathfrak{g} \to \mathfrak{e}$, $x \mapsto \rho(x)$ such that $\phi(\rho(x)) = x$ for all $x \in \mathfrak{g}$. We may then define $S(x)(l) := S_\rho(x)(l) := [\rho(x), l]$ for all $l \in \mathfrak{l}$. We obtain then $\omega(x,y) = [\rho(x), \rho(y)] - \rho([x,y])$. A direct computation shows $d_S\omega = 0$ (cf. Lemma 2.2.2) as ω defines the 2-cocycle corresponding to this extension.

Lemma 2.6.2. *Let $\mathfrak{l}$ be a $\mathfrak{g}$-kernel. Suppose that one may choose the corresponding S and ω such that $d_S\omega = 0$. Then $\mathfrak{l}$ is extendible.*

Proof. Since $d_S\omega = 0$, one can define an extension $\phi : \mathfrak{e} \to \mathfrak{g}$ by $\mathfrak{l}$ as follows: As a vector space, $\mathfrak{e}$ is the direct sum $\mathfrak{l} \oplus \mathfrak{g}$. The bracket on $\mathfrak{e}$ is defined by equation (2.7) and it has been shown to satisfy the Jacobi identity in Lemma 2.4.1. It is clear that the $\mathfrak{g}$-module structure on the center of $\mathfrak{l}$ is the given one. □

Remark 2.6.6. Observe that the maps S and ω can be carried along the construction of the sum of two $\mathfrak{g}$-kernels in a linear fashion. More precisely, for given $\mathfrak{g}$-kernels $\mathfrak{l}_1$ and $\mathfrak{l}_2$ with common nucleus $\mathfrak{n}$, the sum $S_1 + S_2$ of linear liftings S_1 and S_2 of the homomorphisms $\psi_j : \mathfrak{g} \to \mathfrak{out}(\mathfrak{l}_j)$ for $j = 1, 2$ is a lifting for $\psi : \mathfrak{g} \to \mathfrak{out}(\mathfrak{l}_1 + \mathfrak{l}_2)$, and $\omega_1 + \omega_2$ satisfies $R_{S_1+S_2} = \mathrm{ad} \circ (\omega_1 + \omega_2)$. Therefore, the class $[\chi]$ corresponding to $\mathfrak{l}_1 + \mathfrak{l}_2$ may be represented by $d_{S_1+S_2}(\omega_1 + \omega_2)$.

In the same way, if S lifts $\psi : \mathfrak{g} \to \mathfrak{out}(\mathfrak{l})$, then $-S$ lifts $-\psi : \mathfrak{g} \to \mathfrak{out}(-\mathfrak{l})$ and thus the class of $-\mathfrak{l}$ may be represented by $d_S(-\omega) = -d_S\omega$.

The next theorem is an analogue of Theorem 2.2.1 in the present context of $\mathfrak{g}$-kernels. We give an independent proof of it in order to compare the two proofs; here, the injectivity is rather formal, while in the proof of Theorem 2.2.1 in Section 2.5, it was the surjectivity which was formal.

Theorem 2.6.3. *The map $\xi : \psi \mapsto [d_S\omega]$ induces an isomorphism of abelian groups between the group of equivalence classes of $\mathfrak{g}$-kernels with fixed nucleus $\mathfrak{n} = \mathfrak{z}(\mathfrak{l})$ and the cohomology space $H^3(\mathfrak{g}, \mathfrak{z}(\mathfrak{l}))$.*

Proof. According to Remark 2.6.5, the map ξ sends an extendible $\mathfrak{g}$-kernel to the zero class. On the other hand, according to Remark 2.6.6, the additivity of the map ξ is clear. In total, ξ is a well-defined additive map.

The injectivity of ξ is shown as follows: In case $[d_S\omega] = 0$, we have $d_S\omega = d_{\mathfrak{z}(\mathfrak{l})}\tau$ for some cochain $\tau \in C^2(\mathfrak{g}, \mathfrak{z}(\mathfrak{l}))$. Now replace ω by $\omega - \tau$ to get $d_S\omega = 0$. Then Lemma 2.6.2 implies that the $\mathfrak{g}$-kernel is extendible.

The surjectivity is, as already stated, more involved. Hochschild's idea [43] is to construct an associative kernel for the universal enveloping algebra $U\mathfrak{g}$ corresponding to the given 3-cohomology class, which gives then rise to the wanted $\mathfrak{g}$-kernel. He uses for this an associative cocycle instead of a Lie algebra cocycle.

Denote by U and U' two copies of the vector space underlying $U\mathfrak{g}$. One may consider U' as being the image of U under a vector space isomorphism $U \to U'$. Denote by

I the sum of the homogeneous components of degree 1 and 2 of $T(U') = \bigoplus_{i=0}^{\infty}(U')^{\otimes i}$, that is, $I = U' \oplus (U' \otimes U')$. Let C be the associative (nonunital) algebra $C = \mathbb{K}e \oplus \mathbb{K}f$ of dimension two where the generators e and f satisfy the relations $e^2 = e, f^2 = f, ef = f$ and $fe = e$.

The direct sum $J = C \oplus I$ carries the structure of an associative algebra by imposing that C is a subalgebra, that $IJ = \{0\}$, and that $ev = v = fv$ for all $v \in I$. Let K denote the vector space direct sum $J \oplus (U \otimes U)$. There is a structure of an associative (nonunital) algebra on K such that J is an ideal and such that:

(a) $(U' \otimes U' + U \otimes U)K = \{0\}$.
(b) $e(u_1 \otimes u_2) = u_1' \otimes u_2', f(u_1 \otimes u_2) = 0$ and $u'(u_1 \otimes u_2) = (uu_1)' \otimes u_2' - u' \otimes (u_1u_2)'$ for all $u, u_1, u_2 \in U$.

The center of K is reduced to $\{0\}$. Indeed, denote by $c = j + u$ a central element with $j \in J$ and $u \in U \otimes U$. As $uj = 0$, we have that $ju = 0$ and, therefore, either $j = \alpha f + v$ for some $v \in U' \otimes U'$ and $\alpha \in \mathbb{K}$, or $u = 0$. In the first case, commutation with f shows $v = 0$. Commutation with e then shows $\alpha = 0$ and $u = 0$. In the second case, j must be a central element of J, and thus zero.

The $\mathfrak{g}$-module $\mathfrak{n}$ (i. e., the given nucleus) gives rise to an $U\mathfrak{g}$-module $\mathfrak{n}$. Define an associative product on $L := K \oplus \mathfrak{n}$ with $K\mathfrak{n} = \mathfrak{n}K = \{0\}$ (extending the given product on K). The center of L is then clearly $\mathfrak{n}$.

We will now define a $\mathfrak{g}$-module structure on L that takes into account the given 3-cocycle $\gamma \in Z^3(U\mathfrak{g}, \mathfrak{n})$. The cocycle property is here understood in the associative or Hochschild sense, that is, γ is a trilinear map from $U\mathfrak{g}$ to $\mathfrak{n}$ satisfying for all $x, y, z, t \in U\mathfrak{g}$ the identity $x \cdot \gamma(y, z, t) - \gamma(xy, z, t) + \gamma(x, yz, t) - \gamma(x, y, zt) = 0$. The passage from the given class $[\gamma'] \in H^3(\mathfrak{g}, \mathfrak{n})$ to the class $[\gamma] \in HH^3(U\mathfrak{g}, \mathfrak{n})$ is possible using the isomorphism $H^n(\mathfrak{g}, M^{\mathrm{ad}}) \cong HH^n(U\mathfrak{g}, M)$ for any $U\mathfrak{g}$-bimodule M. Here, M^{ad} is the adjoint $\mathfrak{g}$-module associated to a bimodule M, that is, $x \in \mathfrak{g}$ acts on $m \in M$ by $x \cdot_{\mathrm{ad}} m := x \cdot m - m \cdot x$. This isomorphism is shown in Theorem 3.3.2, page 98, in [61] (in the homological framework and for right $\mathfrak{g}$-modules).

We make L into a left $U := U\mathfrak{g}$-module, containing $\mathfrak{n}$ as a submodule, and such that $U \cdot J = \{0\}$ and $u \cdot (u_1 \otimes u_2) = uu_1 \otimes u_2 - u \otimes u_1u_2 + \gamma(u, u_1, u_2)$. We claim that $u \cdot (ab) = (u \cdot a)b$ for all $a, b \in L$ and all $u \in U$. Indeed, since J is an ideal in L which is annihilated by U, $u \cdot (ab)$ and $(u \cdot a)b$ are both 0 if $a \in J$. If $a \in U \otimes U + \mathfrak{n}$, then $ab = 0$ and $u \cdot a \in U \otimes U + \mathfrak{n}$, so that $(u \cdot a)b = 0$. This suffices in order to establish the claim.

Next, we define right U-operators in L such that $(U' \otimes U' + U \otimes U + \mathfrak{n}) \cdot U = \{0\}$, $e \cdot u = u', f \cdot u = 0$ and $u_1' \cdot u = (u_1u)' + u_1' \otimes u'$. These right U-operators commute with the left U-operators. On J, this follows at once from the fact that $J \cdot U \subset J$ and $U \cdot J = \{0\}$. On $U \otimes U + \mathfrak{n}$, this follows from the fact that $U \cdot (U \otimes U + \mathfrak{n}) \subset U \otimes U + \mathfrak{n}$ and $(U \otimes U + \mathfrak{n}) \cdot U = \{0\}$.

Furthermore, $(ab) \cdot u = a(b \cdot u)$ for all $a, b \in L$ and all $u \in U$. Indeed, if b lies in $U' \otimes U' + U \otimes U + \mathfrak{n}$, then so does ab, whence $(ab) \cdot u = 0 = a(b \cdot u)$. Hence we may assume $b \in C + U'$. Then, if $a \in I + U \otimes U + \mathfrak{n}$, $ab = 0$ and $b \cdot u \in I$, so that $a(b \cdot u) = 0$.

Hence we may assume that $a \in C$ and $b \in C + U'$, and in this case it is easy to verify directly that $(ab) \cdot u = a(b \cdot u)$.

Finally, we claim that $(a \cdot u)b = a(u \cdot b)$. Indeed, if $a \in U \otimes U + \mathfrak{n}$, we have $(a \cdot u)b = 0 = a(u \cdot b)$. Hence we may suppose that $a \in J$. Then $a \cdot u \in I$. Now if $b \in J + \mathfrak{n}$, we have $u \cdot b \in \mathfrak{n}$ and, therefore, $(a \cdot u)b = 0 = a(u \cdot b)$. Hence we may assume that $a \in J$ and $b \in U \otimes U$. Now if $a \in \mathbb{K}f + U' \otimes U'$, we have $a \cdot u = 0$ and $a(u \cdot b) = 0$. There remains the case $a \in \mathbb{K}e + U'$ and $b \in U \otimes U$. In this case, the relation $(a \cdot u)b = a(u \cdot b)$ is easily verified directly.

With our right U-operators, L is *not* a right U-module. Instead of $(a \cdot u_1) \cdot u_2 = a \cdot (u_1 u_2)$, we have $(a \cdot u_1) \cdot u_2 - a \cdot (u_1 u_2) = a(u_1 \otimes u_2)$. Indeed, each side is 0 whenever $a \in \mathbb{K}f + U' \otimes U' + U \otimes U + \mathfrak{n}$. The remaining case $a \in \mathbb{K}e + U'$ is easily checked directly.

Now regard L as a Lie algebra using the bracket $[a, b] := ab - ba$ for all $a, b \in L$. For $x \in \mathfrak{g}$, we define $S(x) \in \mathfrak{Der}(L)$ by $S(x)(a) = x \cdot a - a \cdot x$ for all $a \in L$. S is clearly a linear map $S : \mathfrak{g} \to \mathfrak{Der}(L)$. Its failure to be a homomorphism of Lie algebras is computed to be

$$\begin{aligned}(S(x)S(y) &- S(y)S(x) - S([x, y]))(a)\\ &= (a \cdot y) \cdot x - a \cdot (yx) - (a \cdot x) \cdot y + a \cdot (xy)\\ &= [x \otimes y - y \otimes x, a].\end{aligned}$$

This means that S is a homomorphism up to inner derivations, and thus the map $\psi : \mathfrak{g} \to \mathfrak{Der}(L)$ defined by $\psi(x) = [S(x)]$ is a $\mathfrak{g}$-kernel. It is easy to see that the induced action on the center $\mathfrak{n}$ is the action we started with.

The above equation for $S(x)S(y) - S(y)S(x) - S([x, y])$ implies at once $\omega(x, y) = x \otimes y - y \otimes x$. A direct computation shows that $d_S\omega$ gives

$$d_S\omega(x, y, z) = \gamma(x, y, z) - \gamma(x, z, y) + \gamma(z, x, y) - \gamma(z, y, x) + \gamma(y, z, x) - \gamma(y, x, z).$$

This is the antisymmetrization of the cocycle γ, which is an expression for the Lie algebra 3-cocycle corresponding to the associative cocycle γ; see [61] page 98.

In conclusion, L is the wanted $\mathfrak{g}$-kernel corresponding to the cocycle γ. □

Remark 2.6.7. This method of construction of $\mathfrak{g}$-kernels is not very explicit, because an explicit map sending Lie algebra cocycles to associative cocycles (inducing the isomorphism between the corresponding cohomology groups) is complicated. It can be constructed using the contraction s_* of the Chevalley–Eilenberg complex $(C^*(\mathfrak{g}, U\mathfrak{g}), d)$ appearing in Section 2.8.2; see [88].

Note that the construction in the proof above become clearer by studying Chapter 3 on crossed modules of associative algebras.

Remark 2.6.8. In fact, there is a more direct isomorphism between the group of equivalence classes of $\mathfrak{g}$-kernels with fixed nucleus $\mathfrak{z}(\mathfrak{l})$ and the group of equivalence classes of crossed modules with cokernel $\mathfrak{g}$ and kernel $\mathfrak{z}(\mathfrak{l})$. The isomorphism is induced by the map ξ, which pulls back the canonical crossed module

$$0 \to \mathfrak{z}(\mathfrak{l}) \to \mathfrak{l} \to \mathfrak{der}(\mathfrak{l}) \to \mathfrak{out}(\mathfrak{l}) \to 0$$

via the homomorphism $\psi : \mathfrak{g} \to \mathfrak{out}(\mathfrak{l})$ to a crossed module with kernel $\mathfrak{z}(\mathfrak{l})$ and cokernel $\mathfrak{g}$; cf. Remark 2.4.9.

One easily verifies that ξ is well-defined and additive. Indeed, in case ψ is extendible, the class of the pullback crossed module is zero thanks to Theorem 2.4.8. On the other hand, examining the constructions of the sum of crossed modules and of the sum of $\mathfrak{g}$-kernels one easily sees the additivity.

Now ξ is injective. Indeed, $\xi(\psi) = 0$ implies that the class of the crossed module $\mu : \mathfrak{l} \to \mathfrak{g}^S$ is zero for $S : \mathfrak{g} \to \mathfrak{der}(\mathfrak{l})$ a lift of $\psi : \mathfrak{g} \to \mathfrak{out}(\mathfrak{l})$; see Remark 2.4.9. Now Theorem 2.4.8 implies that there exists a general extension

$$0 \to \mathfrak{l} \to \mathfrak{e} \to \mathfrak{g} \to 0$$

realizing ψ. This means that ψ is extendible and represents therefore the zero class.

On the other hand, to each crossed module, we have an associated outer action, which shows the surjectivity of the map ξ.

2.6.2 Finite dimensional kernels

In this subsection, we will sketch the results of Hochschild on the possibility to represent cohomology classes by finite dimensional kernels [44].

First of all, Hochschild constructs a geometric model for the Chevalley–Eilenberg complex. Let us fix the base field $\mathbb{K}$ to be $\mathbb{C}$. Let $P := \mathbb{C}[[x_1, \dots, x_n]]$ be the ring of formal power series in n formal variables. Denote by D_i the partial differentiation with respect to x_i for $i = 1, \dots, n$, and by D_i^{-1} the formal integration with respect to x_i (the "constant of integration" is taken to be zero). For a subspace $U \subset P$, its *calculus closure* is the subspace of P spanned by all elements of P obtained from applying a finite number of operators D_i and D_j^{-1} to elements of U. Denote by $\Omega^1(P)$ the *module of de Rham differentials* over P, that is, the free P-module with basis $dx_1, \dots, dx_n$. Let $\Omega^*(P)$ be the *formal de Rham algebra*, that is, the exterior algebra generated by $\Omega^1(P)$. The de Rham differential $d : P \to \Omega^1(P)$ turns $\Omega^*(P)$ into a commutative differential graded algebra. The formal de Rham Lemma for $\mathbb{C}^n$ now reads the following.

Lemma 2.6.4. *Let $b \in \Omega^{\geq 1}(P)$ such that $db = 0$. Let U_b be the subspace of P, which is spanned by the coefficients of b when expressed as a P-linear combination of the ordered products of the dx_i. Then there exists an element $c \in \Omega^*(P)$ such that $dc = b$ and U_c lies in the calculus closure of U_b.*

Proof. We refer to Lemma 1 in [44]. □

Now let $\mathfrak{l}$ be a solvable Lie algebra of dimension n. Let $\mathfrak{k}$ be another Lie algebra which contains $\mathfrak{l}$ as an ideal. Then there is a filtration

$$\{0\} := \mathfrak{l}_n \subset \mathfrak{l}_{n-1} \subset \cdots \subset \mathfrak{l}_r := [\mathfrak{k}, \mathfrak{l}] \subset \mathfrak{l}_{r-1} \subset \cdots \subset \mathfrak{l}_0 := \mathfrak{l},$$

such that each $\mathfrak{l}_i$ is an ideal in $\mathfrak{l}$ and has dimension $n - i$. This is due to the fact that over $\mathbb{C}$, simple $\mathfrak{l}$-modules for the solvable Lie algebra $\mathfrak{l}$ must be 1-dimensional.

According to this filtration, there is an adapted basis $u_1, \dots, u_n$ of $\mathfrak{l}$ such that $\mathfrak{l}_i$ is generated by $u_{i+1}, \dots, u_n$. Therefore, the structure constants c_{ijk} defined by $[u_i, u_j] = \sum_{k=1}^n c_{ijk} u_k$ satisfy $c_{ijk} = 0$ unless $k \geq i$, $k \geq j$ and $k > r$. Furthermore, $c_{ijj} = 0$ if $i > r$ (this follows from Lie's theorem, see [44]).

Next, consider a finite dimensional $\mathfrak{k}$-module V. Viewing it as an $\mathfrak{l}$-module and with the same reasoning as above, there exists an appropriate basis $\{v_j\}_{j=1,\dots,n}$ of V such that for each k, $\mathfrak{l} \cdot v_k$ is contained in the subspace spanned by $v_{k+1}, \dots, v_m$. Writing $u_i \cdot v_j = \sum_{k=1}^n e_{ijk} v_k$, we still have for these "structure constants" $e_{ijk} = 0$ for $k \geq j$ and $e_{ijj} = 0$ for $i > r$.

Introduce the formal series $c_i := \exp(\sum_{p=1}^n c_{pii} x_p)$ and $e_j := \exp(\sum_{p=1}^n e_{pjj} x_p)$. The subalgebra of P generated by $\mathbb{C}$, the x_k for $k = 1, \dots, n$, the c_i and their inverses for $i = 1, \dots, n$ and the e_j and their inverses for $j = 1, \dots, n$ is denoted Q. Denote by $\Omega^*(Q)$ the subalgebra of $\Omega^*(P)$ generated by Q and the differentials dx_k for $k = 1, \dots, n$. Hochschild now constructs an embedding of the Chevalley–Eilenberg complex $(C^*(\mathfrak{l}), d)$ for a Lie algebra $\mathfrak{l}$ into the formal de Rham complex $(\Omega^*(Q), d)$. The explicit construction is done in *loc. cit.* One aspect of it is that it comes from viewing elements of the Lie algebra $\mathfrak{l}$ as left invariant vector fields on the (connected, 1-connected) Lie groups L associated to $\mathfrak{l}$ in a formal neighborhood of $1 \in L$. From this point of view, formal series correspond to formal Taylor expansions and the group elements are formal exponentials. The product in L can be expressed by successive bracketing in $\mathfrak{l}$ using exponentials (cf. Baker–Campbell–Hausdorff formula for the group product).

More to the point, the Chevalley–Eilenberg coboundary operator d is induced by the bracket and can therefore be expressed in terms of the structure constants $da_i = -\sum_{j<k} c_{jki} a_j a_k$ for the a_i being a dual basis in $C^1(\mathfrak{l})$, dual to the basis u_j of $\mathfrak{l}$. An embedding of the Chevalley–Eilenberg complex $C^*(\mathfrak{l})$ into $\Omega^*(Q)$ should send the a_i to some differential forms b_i which must then satisfy the same relations. Hochschild shows by induction that one can find elements b_i meeting these conditions using the restrictions on the structure constants established earlier and Lemma 2.6.4.

Hochschild then defines the usual operators i_X and L_X of Cartan calculus on $(C^*(\mathfrak{l}), d)$ by pulling them back from $(\Omega^*(Q), d)$ via the embedding. Using the operators L_X, Q becomes an $\mathfrak{l}$-module.

Lemma 2.6.5. *With this structure, Q becomes a locally finite $\mathfrak{l}$-module, that is, every finite dimensional subspace of Q is contained in a finite-dimensional $\mathfrak{l}$-submodule of Q. Furthermore, $[\mathfrak{k}, \mathfrak{l}]$ is nilpotent on every finite-dimensional $\mathfrak{l}$-submodule of Q.*

Proof. This is Lemma 2 in *loc. cit.* □

Hochschild now tensors $(\Omega^*(Q), d)$ with V in order to embed the tensor product complex $(C^*(\mathfrak{l}) \otimes V = C^*(\mathfrak{l}, V), d)$ into $(\Omega_V^*(Q) := \Omega^*(Q) \otimes V, d)$. Extending the differential to V this times uses the power series e_j in the same way we used the c_i in the first place. On the other hand, the complex $(\Omega_V^*(Q), d)$ is shown to be isomorphic to $(\Omega^*(Q)_m, d)$, where $\Omega^*(Q)_m$ is just the direct sum of m copies of $\Omega^*(Q)$ and the differential is applied coordinatewise (and m is the dimension of V). This leads Hochschild then to the following

Lemma 2.6.6. *Let $\mathfrak{k}$ be a Lie algebra containing $\mathfrak{l}$ as a finite dimensional solvable ideal. Let V be a finite dimensional $\mathfrak{l}$-module. Then there exists a finite dimensional $\mathfrak{l}$-module U on which $[\mathfrak{k}, \mathfrak{l}]$ is nilpotent, which contains V as a $\mathfrak{l}$-submodule, and which is such that the canonical homomorphism $H^*(\mathfrak{l}, V) \to H^*(\mathfrak{l}, U)$ annihilates every cohomology class of positive dimension.*

Proof. This is Lemma 3 of *loc. cit.*:

One uses the embedding of the Chevalley–Eilenberg complex $(C^*(\mathfrak{l}, V), d)$ into $(\Omega^*(Q)_m, d)$. The formal de Rham Lemma 2.6.4 is easily generalized to $(\Omega^*(Q)_m, d)$ and shows thus that every cohomology class of degree > 0 for $\mathfrak{l}$ with values in V can be represented in $(\Omega^*(Q)_m, d)$ by an element of the form dc. It therefore suffices to show that the coefficients of c are contained in a finite dimensional $\mathfrak{l}$-submodule in order to annihilate this given cohomology class.

The coefficients of c in $\Omega_V^0(Q)_m =: Q_m$ when expressed in terms of the b_i (i. e., the images of the a_i under the embedding), together with the elements in V (identified with a subspace of Q_m), span a finite dimensional subspace of Q_m, which by Lemma 2.6.5 is contained in a finite dimensional $\mathfrak{l}$-submodule of Q_m on which $[\mathfrak{k}, \mathfrak{l}]$ acts in a nilpotent way. This shows the annihilation for one class.

Since the vector space $H^*(\mathfrak{l}, V)$ is finite dimensional, it is now clear that the $\mathfrak{l}$-module U can be found as an $\mathfrak{l}$-submodule of Q_m. □

For the next result, Hochschild uses a lemma which refines Zassenhaus' module enlargement.

Lemma 2.6.7. *Let $\mathfrak{k}$ be a finite dimensional Lie algebra and suppose that $\mathfrak{k} = \mathfrak{s} + \mathfrak{l}$ where $\mathfrak{s}$ is a subalgebra, $\mathfrak{l}$ an ideal and $\mathfrak{s} \cap \mathfrak{l} = \{0\}$. Let N be a finite dimensional $\mathfrak{l}$-module and suppose that $[\mathfrak{s}, \mathfrak{l}]$ is contained in an ideal of $\mathfrak{l}$ which is nilpotent on N. Let M be a $\mathfrak{k}$-module which, as an $\mathfrak{l}$-module, is contained in N. Then there exists a finite dimensional $\mathfrak{k}$-module Z, which contains M as a $\mathfrak{k}$-submodule and N as an $\mathfrak{l}$-submodule.*

Proof. This is Lemma 4 in *loc. cit.* □

Definition 2.6.9. A cohomology class c in $H^*(\mathfrak{l}, V)$ is called *effaceable* in case there exists a finite dimensional $\mathfrak{l}$-module U containing V such that the canonical homomorphism $H^*(\mathfrak{l}, V) \to H^*(\mathfrak{l}, U)$ annihilates c.

Remark 2.6.10. Observe that the finite dimensionality condition of U is essential, because every cohomology class is effaced in some (possibly infinite dimensional) $\mathfrak{l}$-module. Indeed, each $\mathfrak{l}$-module may be embedded into an injective $\mathfrak{l}$-module, and for an injective $\mathfrak{l}$-module I, we have $H^*(\mathfrak{l}, I) = 0$.

Theorem 2.6.8. *Let $\mathfrak{k}$ be a finite dimensional Lie algebra and let M be a finite dimensional $\mathfrak{k}$-module. Let $\mathfrak{s}$ be a maximal semisimple subalgebra of $\mathfrak{k}$. Then a cohomology class for $\mathfrak{k}$ in M is effaceable if and only if it is annihilated by the restriction homomorphism $H^*(\mathfrak{k}, M) \to H^*(\mathfrak{s}, M)$.*

Remark 2.6.11. The Levi decomposition theorem affirms that every finite dimensional real or complex Lie algebra $\mathfrak{k}$ is the semidirect product of a semisimple subalgebra and its *radical*, that is, the maximal solvable ideal of $\mathfrak{k}$.

Proof. Since every finite dimensional $\mathfrak{s}$-module is *semisimple*, that is, each submodule splits, the canonical homomorphism $H^*(\mathfrak{k}, M) \to H^*(\mathfrak{s}, N)$ is an isomorphism whenever N is finite dimensional and $M \subset N$. Therefore, the condition of the theorem is necessary.

Now let $\mathfrak{l}$ denote the radical of $\mathfrak{k}$. By Levi's theorem, we have $\mathfrak{k} = \mathfrak{l} + \mathfrak{s}$ and $\mathfrak{s} \cap \mathfrak{l} = \{0\}$. Using the Hochschild–Serre spectral sequence, Hochschild and Serre [46] found that $H^*(\mathfrak{k}, M)$ is isomorphic to the tensor product

$$H^*(\mathfrak{s}, \mathbb{C}) \otimes H^*(\mathfrak{l}, M)^{\mathfrak{k}},$$

where $H^*(\mathfrak{l}, M)^{\mathfrak{k}}$ denotes the subspace of invariants (cf. Appendix B). They show that their isomorphism ϕ_M satisfies the following properties: If ρ denotes the restriction homomorphism $H^*(\mathfrak{k}, M) \to H^*(\mathfrak{s}, M)$, then $\rho \circ \phi_M$ is an isomorphism on $H^*(\mathfrak{s}, \mathbb{C}) \otimes H^0(\mathfrak{l}, M)^{\mathfrak{k}}$. Furthermore, if N is a finite dimensional $\mathfrak{k}$-module containing M, and if γ denotes the canonical homomorphisms $H^*(\mathfrak{k}, M) \to H^*(\mathfrak{k}, N)$ and $H^*(\mathfrak{l}, M) \to H^*(\mathfrak{l}, N)$, we have $\gamma \circ \phi_M(u \otimes v) = \phi_N(u \otimes \gamma(v))$ for all $u \in H^*(\mathfrak{s}, M)$ and all $v \in H^*(\mathfrak{l}, M)^{\mathfrak{k}}$.

Now suppose that $c \in H^n(\mathfrak{k}, M)$ and $\rho(c) = 0$. Write $c = \phi_M(\sum_{p+q=n} w_{p,q})$, with

$$w_{p,q} \in H^p(\mathfrak{s}, \mathbb{C}) \otimes H^q(\mathfrak{l}, M)^{\mathfrak{k}}.$$

If $q > 0$, we have $\rho \circ \phi_M(w_{p,q}) = 0$ because no $q > 0$-term appears in the explicit expression for the isomorphism $\rho \circ \phi_M$. Thus $\rho(c) = 0$ implies that $\rho \circ \phi_M(w_{n,0}) = 0$, whence we conclude that $w_{n,0} = 0$. Thus $c = \phi_M(w_{n-1,1} + \cdots + w_{0,n})$. Combining Lemmas 2.6.6 and 2.6.7, we see that there exists a finite dimensional $\mathfrak{k}$-module N containing M such that γ annihilates every $H^q(\mathfrak{l}, M)$ with $q > 0$. For this module N, we have $\gamma(c) = 0$, because c does only have components with $q > 0$. We can take the $\mathfrak{k}$-module N as U in the definition of effaceability. □

We introduce now a notion which is related to effaceability. For this, instead of reasoning directly with classes in the Chevalley–Eilenberg complex of a Lie algebra $\mathfrak{k}$, we consider their images in the Hochschild complex of $U\mathfrak{k}$; see Appendix C. Let M be a $\mathfrak{k}$-module. In case $\mathcal{Q}$ is an ideal of $U\mathfrak{k}$, which annihilates M, M may be regarded as an $U\mathfrak{k}/\mathcal{Q}$-module. We have then a homomorphism $HH^*(U\mathfrak{k}/\mathcal{Q}, M) \to HH^*(U\mathfrak{k}, M)$.

Definition 2.6.12. Call a cohomology class c of $\mathfrak{k}$ with values in M *of finite type*, if it belongs to the image of $HH^*(U\mathfrak{k}/\mathcal{Q}, M)$ in $HH^*(U\mathfrak{k}, M)$ for some finite codimensional ideal $\mathcal{Q}$ of $U\mathfrak{k}$, which annihilates M.

Theorem 2.6.9. *Every effaceable cohomology class of degree 3 is of finite type.*

Proof. Let c be a class in degree 2 with values in M and let N be a finite dimensional effacing $\mathfrak{k}$-module for c, which contains M. Let $\mathcal{Q}$ be the annihilator ideal in $U\mathfrak{k}$ of N, and let f be a cocycle representing c. There is a 1-cochain g on $U\mathfrak{k}$ with values in N such that $f(r,s) = r \cdot g(s) - g(rs)$ for all $r, s \in U\mathfrak{k}^+$. Taking $r \in \mathcal{Q}$, we see that $g(\mathcal{Q}U\mathfrak{k}^+) \subset M$. Take a 1-cochain h on $U\mathfrak{k}$ with values in M, which coincides with g on $\mathcal{Q}U\mathfrak{k}^+$. Then the cocycle $f_1 = f - dh$ still represents c. We have $f_1(\mathcal{Q}, U\mathfrak{k}^+) = \{0\}$, and a fortiori $f_1(\mathcal{Q}^2, U\mathfrak{k}^+) = \{0\}$. Using the cocycle identity for $df_1(r, q_1, q_2)$ with $r \in U\mathfrak{k}^+$ and $q_1, q_2 \in \mathcal{Q}$, we find that $f_1(U\mathfrak{k}^+, \mathcal{Q}^2) = \{0\}$. Hence c belongs to the canonical image of $HH^*(U\mathfrak{k}^+/\mathcal{Q}^2, M)$. Since $\mathcal{Q}$ is of finite codimension, so is $\mathcal{Q}^2$, and we have shown that c is of finite type.

Now suppose c of degree 3 is effaced in a finite dimensional $\mathfrak{k}$-module N containing M. Then, by the long exact sequence induced in cohomology by the short exact sequence of $\mathfrak{k}$-modules

$$0 \to M \to N \to N/M \to 0,$$

the class c is the image of some class $e \in H^2(\mathfrak{k}, N/M)$ under the connecting homomorphism. As we have shown above, e is of finite type. Let $\mathcal{P}$ be an ideal of finite codimension in $U\mathfrak{k}^+$, which reduces e. Then we can find a 2-cochain g on $U\mathfrak{k}^+$ with values in N which vanishes on $\mathcal{P} \otimes (U\mathfrak{k}^+) + (U\mathfrak{k}^+) \otimes \mathcal{P}$ and which, when taken modulo M, represents e. Its coboundary dg takes values in M and represents c. Let $\mathcal{Q}$ be the annihilator of N in $U\mathfrak{k}^+$. Then all three terms in the cocycle identity $dg(u,v,w) = 0$ are separately zero, whenever u, v or w lies in $\mathcal{P} \cap \mathcal{Q}$. Hence $\mathcal{P} \cap \mathcal{Q}$ reduces c and the theorem is proved. □

The reasoning of this last proof generalizes to give the following.

Theorem 2.6.10. *Let $\mathfrak{k}$ be a finite dimensional solvable Lie algebra and M a finite dimensional $\mathfrak{k}$-module. Then every cohomology class on $\mathfrak{k}$ with values in M is of finite type.*

Proof. First of all, every cohomology class in degree 0 is of finite type. Indeed, a cohomology class of degree 0 is simply an element of M and one may take as $\mathcal{Q}$ the annihilator of M in $U\mathfrak{k}^+$.

By Theorem 2.6.8, every cohomology class of positive degree is effaceable. One may then reason by induction, the inductive step being exactly like in the above proof: The class c is effaceable and effaced in a finite dimensional module N, the exact sequence gives then rise to a class e of degree one less to which applies the induction hypothesis. $\mathcal{P}$ is as before an ideal reducing e. $\mathcal{Q}$ is as before the annihilator of N. The cocycle identity for a cochain with values in N representing e modulo M is valid by the vanishing of all terms whenever one of the elements is in $\mathcal{P} \cap \mathcal{Q}$. □

Now come back to $\mathfrak{k}$-kernels. We have seen in Theorem 2.6.3 how to construct a $\mathfrak{k}$-kernel with center N to a given (associative) 3-cocycle of $U\mathfrak{k}^+$ with values in N such that the obstruction class of the kernel is represented by the given cocycle. Now, if the cocycle is the canonical image of a 3-cocycle on $U\mathfrak{k}^+/\mathcal{Q}$ with values in N, where $\mathcal{Q}$ is an ideal of finite codimension in $U\mathfrak{k}^+$, which annihilates N, we can carry out the standard construction with the algebra $U\mathfrak{k}^+/\mathcal{Q}$ instead of $U\mathfrak{k}^+$. For finite dimensional $\mathfrak{k}$ and N, this yields a finite dimensional $\mathfrak{k}$-kernel with the same obstruction class. Hence we have the following result.

Theorem 2.6.11. *Let $\mathfrak{k}$ be a finite dimensional Lie algebra and N be a finite dimensional $\mathfrak{k}$-module. Then every cohomology class in $H^3(\mathfrak{k}, N)$, which is of finite type, or, equivalently, which is effaceable, is the obstruction of a finite dimensional $\mathfrak{k}$-kernel with center N.*

In order to prove the converse (here in characteristic zero), Hochschild shows the following.

Theorem 2.6.12. *Let $\mathfrak{k}$ be a finite dimensional Lie algebra and N be a finite dimensional $\mathfrak{k}$-module. Let $\mathfrak{s}$ be a maximal semisimple subalgebra of $\mathfrak{k}$. Then an element of $H^3(\mathfrak{k}, N)$ is the obstruction of a finite dimensional $\mathfrak{k}$-kernel if and only if its restriction to $\mathfrak{s}$ is zero.*

Remark 2.6.13. Let us first remark why this theorem shows the converse of the previous theorem, namely, that every cohomology class c in $H^3(\mathfrak{k}, N)$, which is the obstruction of a finite dimensional $\mathfrak{k}$-kernel with center N is of finite type. Indeed, by the theorem, the restriction of c to $\mathfrak{s}$ is zero. But then Theorem 2.6.8 implies that c is effaceable, which in degree 3, is equivalent to being of finite type.

Proof. The previous remark implies that if the restriction of c to $\mathfrak{s}$ is zero, then c is of finite type. Theorem 2.6.11 now implies that c is the obstruction of a finite dimensional $\mathfrak{k}$-kernel with center N.

Conversely, suppose that c is the obstruction of a finite dimensional $\mathfrak{k}$-kernel $\psi : \mathfrak{k} \to \mathfrak{out}(\mathfrak{m})$ with center N. Then $\psi(\mathfrak{s})$ is a semisimple subalgebra of $\mathfrak{out}(\mathfrak{m}) = \mathfrak{der}(\mathfrak{m})/\mathrm{ad}(\mathfrak{m})$. There is a subalgebra $\mathfrak{e}$ of $\mathfrak{der}(\mathfrak{m})$, which contains $\mathrm{ad}(\mathfrak{m})$ and for which $\mathfrak{e}/\mathrm{ad}(\mathfrak{m}) = \psi(\mathfrak{s})$. Observe that by Theorem 2.4.8 together with the fact that $H^2(\mathfrak{s}, L) = 0$ for any (finite dimensional) semisimple Lie algebra and any finite dimensional $\mathfrak{s}$-module L, every extension by a semisimple Lie algebra is trivial.

Therefore, we can find an embedding ρ of $\psi(\mathfrak{s})$ into $\mathfrak{e}$ (and thus into $\mathfrak{der}(\mathfrak{m})$), which is inverse to the canonical homomorphism of the extension $\mathfrak{e}$ onto $\psi(\mathfrak{s})$. Now we may use a lift S of $\psi : \mathfrak{k} \to \mathfrak{out}(\mathfrak{m})$ which coincides with the map $\rho \circ \psi$ to compute the 3-cohomology class associated to the $\mathfrak{k}$-kernel. It follows that the failure ω of S to be a homomorphism of Lie algebras vanishes on $\mathfrak{s} \otimes \mathfrak{s}$. Thus $d_{\mathfrak{S}}\omega$ vanishes on $\mathfrak{s} \otimes \mathfrak{s} \otimes \mathfrak{s}$, which shows that the restriction of the 3-cocycle $d_{\mathfrak{S}}\omega$ to $\mathfrak{s}$ is zero. □

Remark 2.6.14. To put the previous results into perspective, recall that a (finite-dimensional) simple complex Lie algebra $\mathfrak{s}$ has $H^3(\mathfrak{s}, \mathbb{C}) \cong \mathbb{C}$, generated by the class represented by the Cartan cocycle

$$(x, y, z) \mapsto \kappa([x, y], z),$$

where κ is a nondegenerate symmetric bilinear form (e. g., the Killing form) on $\mathfrak{s}$. The previous results imply that this class cannot be represented by a crossed module $\mu : \mathfrak{m} \to \mathfrak{n}$ with *finite-dimensional* Lie algebras $\mathfrak{m}$ and $\mathfrak{n}$. We will see in the section of examples crossed modules representing the Cartan cohomology class. These are called *string Lie algebra*, because they are (in some sense) the Lie algebra of the string group introduced in Section 1.7.2. The fact that there is no crossed module with finite-dimensional components representing the Cartan cohomology class is at the origin of the difficulty for finding finite-dimensional models for the string group; cf. [90, 80].

2.7 Relation to the Hochschild–Serre spectral sequence

The goal of this section is to state the conditions on an ideal $\mathfrak{l} \subset \mathfrak{n}$ and a central extension $\hat{\mathfrak{l}}$ of $\mathfrak{l}$ to form a crossed module in terms of the Hochschild–Serre spectral sequence for $\mathfrak{l} \subset \mathfrak{n}$.

Recall the Hochschild–Serre spectral sequence from Appendix B. Suppose $\mathfrak{n}$ is a Lie algebra and $\mathfrak{l} \subset \mathfrak{n}$ is a subalgebra. Let furthermore V be an $\mathfrak{n}$-module, which is also regarded as an $\mathfrak{l}$-module by restriction.

In the usual Chevalley–Eilenberg complex $C^*(\mathfrak{n}, V)$, one introduces the following *filtration*:

$$F^p C^{p+q}(\mathfrak{n}, V) = \{c \in C^{p+q}(\mathfrak{n}, V) \mid c(x_1, \ldots, x_{p+q}) = 0 \text{ for } x_1, \ldots, x_{q+1} \in \mathfrak{l}\}. \tag{2.13}$$

The $F^p C^{p+q}(\mathfrak{n}, V)$ form a filtration of the Chevalley–Eilenberg complex and permit to compute its cohomology by a spectral sequence, the *Hochschild–Serre spectral sequence*; see [46].

We restate here the corresponding theorem of Hochschild–Serre.

Theorem 2.7.1. *The spectral sequence*

$$\{E_r^{p,q}, d_r^{p,q} : E_r^{p,q} \to E_r^{p+r,q-r+1}\}$$

corresponding to the above filtration has the following properties:

(a) $E_1^{p,q} = H^q(\mathfrak{l}, \mathrm{Hom}(\Lambda^p(\mathfrak{n}/\mathfrak{l}), V))$;

(b) $E_2^{p,q} = H^p(\mathfrak{n}, \mathfrak{l}, V)$;

(c) *if $\mathfrak{l}$ is an ideal, then* $E_2^{p,q} = H^p(\mathfrak{n}/\mathfrak{l}, H^q(\mathfrak{l}, V))$;
(d) *the spectral sequence converges to* $H^*(\mathfrak{n}, V)$; *the edge homomorphisms*

$$H^q(\mathfrak{n}, V) \to E_\infty^{0,q} \to E_1^{0,q} = H^q(\mathfrak{l}, V),$$

and

$$H^p(\mathfrak{n}, \mathfrak{l}, V) = E_2^{p,0} \to E_\infty^{p,0} \to H^p(\mathfrak{n}, V)$$

may be identified with the natural homomorphisms $H^q(\mathfrak{n}, V) \to H^q(\mathfrak{l}, V)$ *and* $H^p(\mathfrak{n}, \mathfrak{l}, V) \to H^p(\mathfrak{n}, V)$, *respectively;*
(e) *if* $V = \mathbb{K}$ *or if* V *is an associative commutative algebra on which* $\mathfrak{n}$ *acts by derivations, then the spectral sequence is multiplicative; in this case, the isomorphisms in (a)–(c) are also multiplicative, as well as the convergence in (d).*

Explanations about the statement are given in Appendix B.

Now let us specify to the situation, which is of interest for us in this section. We consider an ideal $\mathfrak{l}$ of a Lie algebra $\mathfrak{n}$, with corresponding quotient Lie algebra $\mathfrak{g}$. We have stated in Section 2.3 conditions on a central extension $\hat{\mathfrak{l}}$ of $\mathfrak{l}$ implying that $\hat{\mathfrak{l}} \to \mathfrak{n}$ forms a crossed module. We will now state these conditions in terms of the Hochschild–Serre spectral sequence.

Let us first list the given data for our problem. We consider given

- an ideal $\mathfrak{l}$ of a Lie algebra $\mathfrak{n}$, and thus an extension

$$0 \to \mathfrak{l} \to \mathfrak{n} \xrightarrow{\pi} \mathfrak{g} \to 0,$$

- a central extension $\hat{\mathfrak{l}}$ of $\mathfrak{l}$ by a center V,
- and an action of $\mathfrak{n}$ on V.

Associated to the extension

$$0 \to \mathfrak{l} \to \mathfrak{n} \xrightarrow{\pi} \mathfrak{g} \to 0,$$

there is a Hochschild–Serre spectral sequence. By Hochschild–Serre's theorem, we have for the E_2-term:

$$E_2^{p,q} = H^p(\mathfrak{n}/\mathfrak{l}, H^q(\mathfrak{l}, V)) = H^p(\mathfrak{g}, H^q(\mathfrak{l}, V)).$$

The central extension $\hat{\mathfrak{l}} \to \mathfrak{l}$ is described (up to equivalence) by a cohomology class $[\omega] \in H^2(\mathfrak{l}, V)$. Observe that the first condition to have a crossed module $\hat{\mathfrak{l}} \to \mathfrak{n}$ is that the $\mathfrak{l}$-action on $\hat{\mathfrak{l}}$ extends to an $\mathfrak{n}$-action on $\hat{\mathfrak{l}}$ by derivations. This is condition (a) in Lemma 2.3.1. We claim that this condition is equivalent to the condition that the class $[\omega] \in H^2(\mathfrak{l}, V)$ is $\mathfrak{n}$-invariant. Indeed, Lemma 2.3.1 states this condition (a) as for all

$x \in \mathfrak{n}$, $x \cdot \omega = d\theta_x$, which means that the class $[\omega] \in H^2(\mathfrak{l}, V)$ is invariant with respect to $\mathfrak{n}$. From the point of view of the Hochschild–Serre spectral sequence, this means that

$$[\omega] \in H^0(\mathfrak{g}, H^2(\mathfrak{l}, V)) = E_2^{0,2}.$$

By the very position of $E_2^{0,2}$ in the spectral sequence, no nonzero differential has its image in $E_2^{0,2}$. In the same line of thoughts, there are only two possibly nonzero differentials starting from $E_2^{0,2}$. It is interesting to consider the differentials $d_2^{0,2} : E_2^{0,2} \to E_2^{2,1}$ and $d_3^{0,2} : E_3^{0,2} \to E_3^{3,0}$. The situation may be depicted as follows:

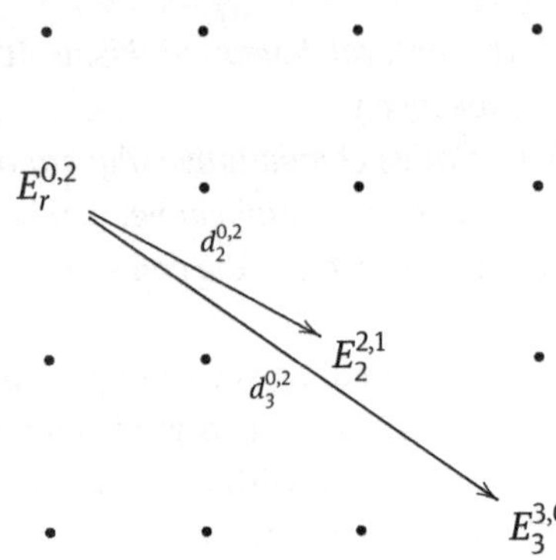

for $r = 2, 3$.

Definition 2.7.1. An element $e \in E_2^{0,2}$ is called *transgressive* in case $d_2^{0,2} e = 0$. Given a transgressive element $e \in E_2^{0,2}$, the *transgression* of e is the image $d_3^{0,2} e$.

Now we can state the main theorem of this section.

Theorem 2.7.2. *Given the data which we listed above, the conditions for $\hat{\mathfrak{l}} \to \mathfrak{n}$ to give a crossed module are equivalent to the conditions that the class $[\omega]$ of the central extension $\hat{\mathfrak{l}}$ is $\mathfrak{n}$-invariant and transgressive.*

Moreover, if the conditions hold true, the characteristic class is then the transgression of $[\omega]$.

Proof. We have already shown that the $\mathfrak{n}$-invariance is exactly condition (a) in Lemma 2.3.1. It remains to show that condition (b) in the same lemma is the transgressivity of $[\omega]$ and that $d_3^{0,2}[\omega]$ equals the class $[\chi] \in H^3(\mathfrak{g}, V)$.

Observe that the computation of $d_2^{0,2}$ implies that we have extended ω to a map $\theta : \mathfrak{n} \times \mathfrak{l} \to V$. We do this respecting the $\mathfrak{n}$-module structure of the central extension $\hat{\mathfrak{l}}$ and obtain a map θ like in equation (2.4). Now compute $d_2^{0,2}\theta(x_1, x_2, l)$ for $x_1, x_2 \in \mathfrak{n}$ and $l \in \mathfrak{l}$. Strictly speaking, x_1 and x_2 are only representatives of classes in $\mathfrak{g} = \mathfrak{n}/\mathfrak{l}$. Observe further that $d_2^{0,2}$ is just the ordinary Lie algebra coboundary operator, but with values in $C^2(\mathfrak{g}, C^1(\mathfrak{l}, V))$.

$$\begin{aligned} -d_2^{0,2}\theta(x_1, x_2, l) &= x_1 \cdot \theta(x_2, l) - x_2 \cdot \theta(x_1, l) + l \cdot \theta(x_1, x_2) \\ &\quad - \theta([x_1, x_2], l) + \theta([x_1, l], x_2) - \theta([x_2, l], x_1). \end{aligned}$$

Observe that the lower line are the last three terms of condition (b) in Lemma 2.3.1, and the first two terms are the first two terms in the condition (b) in Lemma 2.3.1. As $\mathfrak{l}$ acts trivially on V, the term $l \cdot \theta(x_1, x_2) = 0$, and this shows that the $d_2^{0,2}\theta(x_1, x_2, l) = 0$ for all $x_1, x_2 \in \mathfrak{n}$ and $l \in \mathfrak{l}$ is equivalent to the fact that equation (2.4) defines a *representation* of $\mathfrak{n}$ on $\widehat{\mathfrak{l}}$. In conclusion, we have shown the first part of the theorem, namely, that $\widehat{\mathfrak{l}} \to \mathfrak{n}$ is a crossed module if and only if $[\omega]$ is $\mathfrak{n}$-invariant and $d_2^{0,2}[\omega] = 0$.

Now for the computation of $d_3^{0,2}$ (in case the first two conditions are fulfilled), we need to extend θ further to a cochain in $C^2(\mathfrak{g}, V)$, because the coboundary of it should lie in $C^3(\mathfrak{g}, V)$. By taking representatives in $\mathfrak{n}$ for the arguments in $\mathfrak{g}$, this means extending θ to a cochain in $C^2(\mathfrak{n}, V)$ such that its coboundary vanishes as soon as one argument is in $\mathfrak{l}$.

This is exactly what $\tilde{\theta}$ in Section 2.3 does. Thus $d_3^{0,2}[\omega]$ is represented by the cochain in $C^3(\mathfrak{g}, V)$ corresponding to $d^{\mathfrak{n}}\tilde{\theta}$, and this shows the last claim. □

Remark 2.7.2. In conclusion, fixing $\mathfrak{m} = \widehat{\mathfrak{l}}$, $\mathfrak{l} \subset \mathfrak{n}$ (and not only $\mathfrak{g}$ and the $\mathfrak{g}$-module V), one obtains with this construction classes in $E_\infty^{3,0}$ in

$$H^3(\mathfrak{n}, V) = \bigoplus_{i+j=3} E_\infty^{i,j}.$$

2.8 Examples

We have already seen elementary examples in the first section. Let us construct in this section some more involved examples. We will construct crossed modules associated to some famous 3-cohomology classes.

The point with these examples is that although *all* 3-cohomology classes may be represented by crossed modules, it is not always obvious to find a *meaningful* crossed module corresponding to a given cohomology class. Meaningful means here geometrically inspired by the string gerbe in the first example and geometrically inspired by the relation between Gelfand–Fuks cocycle and Godbillon–Vey cocycle (i. e., by the process of fiber integration) in the second example.

2.8.1 The string Lie algebra

Let $\mathfrak{g}$ be a finite dimensional real Lie algebra and denote by I the closed interval $I = [0, 1]$. We consider the *smooth path algebra*

$$P(\mathfrak{g}) := \{\xi \in C^\infty(I, \mathfrak{g}) \mid \xi(0) = 0\}$$

of $\mathfrak{g}$. Then evaluation in $1 \in I$, denoted ev_1, leads to a short exact sequence

$$0 \to \mathfrak{l} \to P(\mathfrak{g}) \to \mathfrak{g} \to 0,$$

where $\mathfrak{l} := \mathrm{Ker}(\mathrm{ev}_1)$ is the ideal of closed paths in $P(\mathfrak{g})$ and a linear section $\rho : \mathfrak{g} \to P(\mathfrak{g})$ is given by $\rho(x)(t) := tx$. Note that $\mathfrak{l}$ is larger than the Lie algebra $C^\infty(S^1, \mathfrak{g})$, which corresponds to those elements of $\mathfrak{l}$ for which all derivatives have the same boundary value in 0 and 1.

Let $\kappa : \mathfrak{g} \times \mathfrak{g} \to \mathfrak{z}$ be an invariant bilinear form. The invariance condition means that for all $x, y, z \in \mathfrak{g}$, we have

$$\kappa([x, y], z) = \kappa(x, [y, z]).$$

We consider in the following $\mathfrak{z}$ as a trivial $P(\mathfrak{g})$-module. Then the Lie algebra $\mathfrak{l}$ has a central extension $\widehat{\mathfrak{l}} := \mathfrak{z} \times_\omega \mathfrak{l}$ where the cocycle ω is given by

$$\omega(\xi, \eta) := \int_0^1 \kappa(\xi, \eta') := \int_0^1 \kappa(\xi, \eta')(t)dt. \tag{2.14}$$

We define $\widetilde{\omega} \in C^2(P(\mathfrak{g}), \mathfrak{z})$ by

$$\widetilde{\omega}(\xi, \eta) := \frac{1}{2}\int_0^1 (\kappa(\xi, \eta') - \kappa(\eta, \xi')) = \frac{1}{2}\int_0^1 (2\kappa(\xi, \eta') - \kappa(\eta, \xi)')$$
$$= \int_0^1 \kappa(\xi, \eta') - \frac{1}{2}\kappa(\xi, \eta)(1).$$

We observe that for $(\xi, \eta) \in P(\mathfrak{g}) \times \mathfrak{l}$, we have $\widetilde{\omega}(\xi, \eta) = \theta(\xi, \eta) := \int_0^1 \kappa(\xi, \eta')$. This is the map θ which we used in Section 2.3 in order to associate a 3-class to a crossed module. It is therefore clear that $\widetilde{\omega}$ may serve as $\tilde{\theta}$, the skew symmetric extension of θ to $P(\mathfrak{g}) \times P(\mathfrak{g})$.

For the following construction of the crossed module related to the string Lie algebra, we compute

$$\sum_{\text{cycl.}} \int_0^1 \kappa([\xi, \eta], \zeta') = \int_0^1 \kappa([\xi, \eta], \zeta') + \kappa([\eta, \zeta], \xi') + \kappa([\zeta, \xi], \eta')$$
$$= \int_0^1 \kappa([\xi, \eta], \zeta') + \kappa([\xi', \eta], \zeta) + \kappa([\xi, \eta'], \zeta)$$
$$= \int_0^1 \kappa([\xi, \eta], \zeta)' = \kappa([\xi, \eta], \zeta)(1).$$

This implies

$$\sum_{\text{cycl.}} \int_0^1 \kappa([\xi, \eta]', \zeta) = \sum_{\text{cycl.}} \int_0^1 \kappa([\xi', \eta], \zeta) + \kappa([\xi, \eta'], \zeta)$$

$$= \sum_{\text{cycl.}} \int_0^1 \kappa(\xi', [\eta, \zeta]) + \kappa([\zeta, \xi], \eta')$$
$$= \kappa([\eta, \zeta], \xi)(1) + \kappa([\zeta, \xi], \eta)(1)$$
$$= 2\kappa([\eta, \zeta], \xi)(1) = 2\kappa([\xi, \eta], \zeta)(1).$$

Here, we used invariance of κ and skew symmetry of $[,]$ to rearrange the terms in order to use the above computation. We deduce from here

$$(d_{P(\mathfrak{g})}\tilde{\omega})(\xi, \eta, \zeta) = \frac{1}{2} \int_0^1 \sum_{\text{cycl.}} (\kappa(\xi, [\eta, \zeta]') - \kappa([\eta, \zeta], \xi'))$$
$$= \frac{1}{2}(2\kappa([\eta, \zeta], \xi)(1) - \kappa([\eta, \zeta], \xi)(1)) = \frac{1}{2}\kappa([\eta, \zeta], \xi)(1),$$

and the expression $d_{P(\mathfrak{g})}\tilde{\omega}$ thus vanishes on $(P(\mathfrak{g}))^2 \times \mathfrak{l}$. As in Section 2.3, the cochain $d_{P(\mathfrak{g})}\tilde{\omega}$ passes to the quotient $\mathfrak{g} = P(\mathfrak{g})/\mathfrak{l}$ to give rise to a 3-cocycle $\chi \in Z^3(\mathfrak{g}, \mathfrak{z})$ with $\mathrm{ev}_1^*\chi = d_{P(\mathfrak{g})}\tilde{\omega}$. By Lemma 2.3.1, the formula

$$x \cdot (z, l) = (\theta(x, l), [x, l])$$

defines a representation of $P(\mathfrak{g})$ on $\mathfrak{l}$. Moreover, the map $\mu : \hat{\mathfrak{l}} \to P(\mathfrak{g})$ defines a crossed module, which we call the *string Lie algebra crossed module*. By our computations, we have computed the 3-cocycle χ, which reads explicitly

$$\chi(x, y, z) = \frac{1}{2}\kappa([x, y], z),$$

and $\chi \in Z^3(\mathfrak{g}, \mathfrak{z})$. Up to the factor $\frac{1}{2}$, we get back the Cartan cocycle from Remark 2.6.14. The cocycle χ represents the class in $H^3(\mathfrak{g}, \mathfrak{z})$ of the crossed module $\mu : \hat{\mathfrak{l}} \to P(\mathfrak{g})$. In case $\kappa \neq 0$, the class $[\chi] \in H^3(\mathfrak{g}, \mathfrak{z})$ is nonzero. For a simple Lie algebra $\mathfrak{g}$, taking κ to be the *Killing form*, the 1-dimensional cohomology space $H^3(\mathfrak{g}, \mathbb{R})$ is generated by the class $[\chi]$.

Remark 2.8.1. This example should be compared to the example of the string group (after Baez–Crans–Schreiber–Stevenson) in [5]. The string Lie algebra here is the Lie algebra crossed module corresponding to the Lie group crossed module given by the string group in Section 1.7.

2.8.2 The string Lie algebra revisited

In this subsection, we give a different representative for the equivalence class of the crossed module representing the generator of $H^3(\mathfrak{g}, \mathbb{C})$ for a (finite-dimensional) simple complex Lie algebra $\mathfrak{g}$. Namely, we give a representative coming from the principal construction in Section 2.5.

We will develop here a procedure for lifting 3-cocycles on a Lia algebra $\mathfrak{g}$ with values in the trivial module $\mathbb{C}$ to 2-cocycles with values in $(U\mathfrak{g}^+)^\vee$, the linear dual of the augmentation ideal $U\mathfrak{g}^+$ of the enveloping algebra $U\mathfrak{g}$ of $\mathfrak{g}$. The main point is the explicit formula for the corresponding 2-cocycle. The procedure works for any complex finite-dimensional Lie algebra $\mathfrak{g}$ and any 3-cocycle with values in the trivial module $\mathbb{C}$.

We denote by $U\mathfrak{g}$ the universal enveloping algebra of a Lie algebra $\mathfrak{g}$, and by $U\mathfrak{g}^+$ the kernel of the augmentation map

$$\varepsilon : U\mathfrak{g} \to \mathbb{C}.$$

Identifying $\mathbb{C}$ with its dual, this gives rise to a short exact sequence of (left) $\mathfrak{g}$-modules:

$$0 \to \mathbb{C} \to U\mathfrak{g}^\vee \to (U\mathfrak{g}^+)^\vee \to 0. \tag{2.15}$$

Recall that $U\mathfrak{g}$ is a cocommutative Hopf algebra. In particular, its vector space of endomorphisms $\mathrm{End}_\mathbb{C}(U\mathfrak{g})$ is equipped with the *convolution product* defined for $f, g \in \mathrm{End}_\mathbb{C}(U\mathfrak{g})$ by

$$f \star g = \mathrm{mult} \circ (f \otimes g) \circ \Delta,$$

where mult is the associative product and Δ is the coproduct on $U\mathfrak{g}$. The coproduct is written in Sweedler notation

$$\Delta(x) = x_1 \otimes x_2,$$

that is, the two tensor components are distinguished by subscripts and the sum is not written. The composition $\eta\epsilon$ of unit and counit of $U\mathfrak{g}$ gives a unit for the convolution product.

The $(n-1)$-fold iterated coproduct of an element x of $U\mathfrak{g}$ will be written

$$\Delta^{(n-1)}x = x_1 \otimes \cdots \otimes x_n \in U\mathfrak{g}^{\otimes n}$$

in Sweedler's notation.

$\mathbb{C}[t]$ denotes the $\mathbb{C}$-algebra of polynomials in the variable t with complex coefficients. If λ is a complex number, and V is a vector space, we have an evaluation map $\mathrm{ev}_\lambda : V \otimes \mathbb{C}[t] \to V$, which sends $v \otimes t^n$ to $\lambda^n v$, for all v in V. In particular, every $\mathbb{C}[t]$-linear map $B : V \otimes \mathbb{C}[t] \to V \otimes \mathbb{C}[t]$ is uniquely determined by its $\mathbb{C}$-linear restriction to V: $B_t : V \to V \otimes \mathbb{C}[t]$. We will employ the notation B_λ to denote the composition $\mathrm{ev}_\lambda \circ B_t : V \to V$.

Let $\mathfrak{g}$ be a Lie algebra over $\mathbb{C}$ with Lie bracket $[-,-]$, and denote by $\eta : \mathbb{C} \to U\mathfrak{g}$ the unit map.

Definition 2.8.2. If x is an element of $U\mathfrak{g}$ and g is an element of $\mathfrak{g}$, then we set
(a) $\mathrm{pr}(x) := \sum_{k\geq 0} \frac{(-1)^k}{k+1} (\mathrm{Id} - \eta\epsilon)^{\star(k+1)}(x)$, and this defines an endomorphism pr of $U\mathfrak{g}$;

(b) $\phi_t(x) := \sum_{n\geq 0} \frac{t^n}{n!}\mathrm{pr}^{\star n}(x)$, which defines an element of $U\mathfrak{g} \otimes \mathbb{C}[t]$, and thus a $\mathbb{C}[t]$-linear endomorphism of $U\mathfrak{g} \otimes \mathbb{C}[t]$;
(c) $A_t(x, g) := \phi_{-t}(x_1)\phi_t(x_2 g)$.

The following proposition summarizes the different properties of the *Eulerian idempotent* pr and the operators ϕ_t and A_t, which we will need in the sequel.

Proposition 2.8.1. *Let $\triangle$ denote the unique $\mathbb{C}[t]$-linear coproduct of $U\mathfrak{g}\otimes\mathbb{C}[t]$ extending the comultiplication on $U\mathfrak{g}$. Then,*

1. *pr is idempotent* ($\mathrm{pr} \circ \mathrm{pr} = \mathrm{pr}$) *and takes its values in $\mathfrak{g} \subset U\mathfrak{g}$. More precisely, if $g_1, \dots, g_n$ are elements of $\mathfrak{g}$, then*

$$\mathrm{pr}(g_1 \cdots g_n) = \frac{1}{n^2} \sum_{\sigma\in\Sigma_n} (-1)^{d(\sigma)} \binom{n-1}{d(\sigma)}^{-1} [g_{\sigma(1)}, \dots, g_{\sigma(n)}], \tag{2.16}$$

where the notation $[h_1, \dots, h_n]$ stands for the iterated Lie bracket

$$[h_1, [h_2, [\dots, [h_{n-1}, h_n] \dots]]]$$

of elements $h_1, \dots, h_n$ of $\mathfrak{g}$ and $d(\sigma)$ denotes the number of descents of the permutation σ.

2. *ϕ_t is a morphism of coalgebras, that is,*

$$\triangle(\phi_t(x)) = (\phi_t \otimes \phi_t) \triangle (x) = \phi_t(x_1) \otimes \phi_t(x_2)$$

for all x in $U\mathfrak{g} \subset U\mathfrak{g} \otimes \mathbb{C}[t]$.

3. *$\phi_0 = \eta\epsilon$ and $\phi_1 = \mathrm{id}$.*
4. *$\phi_t \star \phi_{-t} = \eta\epsilon$ on $U\mathfrak{g} \subset U\mathfrak{g} \otimes \mathbb{C}[t]$.*
5. *For all x in $U\mathfrak{g}$ and g in $\mathfrak{g}$, $A_t(x, g)$ is an element of $\mathfrak{g}$ and*

$$A_t(x, g) = -\phi_{-t}(x_1 g)\phi_t(x_2).$$

Proof. Formula (2.16) for the Eulerian idempotent pr is due to Solomon in the form

$$\mathrm{pr}(g_1 \cdots g_n) = \frac{1}{n} \sum_{\sigma\in\Sigma_n} (-1)^{d(\sigma)} \binom{n-1}{d(\sigma)}^{-1} g_{\sigma(1)} \cdots g_{\sigma(n)};$$

see Proposition 4.2, page 174, in [63]. It is also closely related to Théorème (19), page 183, in [39], where it is explained that the combinatorial prefactors can be rearranged like in Proposition 4.2, page 174, in [63]. The image of $g_1 \cdots g_n \in U\mathfrak{g}$ according to the formula in Théorème (19) in [39] lies in the free Lie algebra $\mathcal{L}(\mathfrak{g}) \subset T\mathfrak{g}$. It is thus an eigenvector for the Dynkin idempotent, which introduces the additional factor n in the denominator (see the main theorem of [101]—Dynkin–Specht–Wever theorem). The resulting formula is then viewed in $\mathfrak{g}$ (as a quotient of the free Lie algebra).

All other computations are straightforward. □

Example 2.8.3. Let $\mathfrak{g}$ be a Lie algebra and $g_1, g_2, \dots, g_n$ be elements in $\mathfrak{g}$. The above formula for $\mathrm{pr} : U\mathfrak{g} \to \mathfrak{g}$ gives

$$\mathrm{pr}(g_1) = g_1,$$
$$\mathrm{pr}(g_1g_2) = \frac{1}{4}([g_1, g_2] - [g_2, g_1]) = \frac{1}{2}[g_1, g_2],$$

and, using Jacobi's identity to obtain the second equality and the skew symmetry of the bracket to obtain the third one,

$$\begin{aligned}\mathrm{pr}(g_1g_2g_3) &= \frac{1}{9}\Big([g_1, [g_2, g_3]] - \frac{1}{2}[g_1, [g_3, g_2]] + [g_3, [g_2, g_1]] - \frac{1}{2}[g_2, [g_1, g_3]] \\ &\quad - \frac{1}{2}[g_3, [g_1, g_2]] - \frac{1}{2}[g_2, [g_3, g_1]]\Big) \\ &= \frac{1}{9}\Big(\frac{3}{2}[g_1, [g_2, g_3]] + \frac{3}{2}[g_3, [g_2, g_1]]\Big) = \frac{1}{6}([g_1, [g_2, g_3]] + [[g_1, g_2], g_3]).\end{aligned}$$

Recall that by definition, $\mathrm{pr}(1) = 0$, which implies that $\mathrm{pr}^{\star n}$ vanishes on words of length strictly lower than n. Since $\phi_t = \sum_{n\geq 0} \frac{t^n}{n!}\mathrm{pr}^{\star n}$, we have thanks to the expressions of pr obtained above:

$$\phi_t(g_1) = t\,\mathrm{pr}(g_1) = t\,g_1,$$
$$\phi_t(g_1g_2) = t\,\mathrm{pr}(g_1g_2) + \frac{t^2}{2}\mathrm{pr}^{\star 2}(g_1g_2) = \frac{t}{2}[g_1, g_2] + \frac{t^2}{2}(g_1g_2 + g_2g_1),$$

and

$$\begin{aligned}\phi_t(g_1g_2g_3) &= t\,\mathrm{pr}(g_1g_2g_3) + \frac{t^2}{2}\mathrm{pr}^{\star 2}(g_1g_2g_3) + \frac{t^3}{6}\mathrm{pr}^{\star 3}(g_1g_2g_3) \\ &= \frac{t}{6}([g_1, [g_2, g_3]] + [[g_1, g_2], g_3]) + \frac{t^2}{4}([g_1, g_2]g_3 + [g_1, g_3]g_2 + [g_2, g_3]g_1 \\ &\quad + g_1[g_2, g_3] + g_2[g_1, g_3] + g_3[g_1, g_2]) + \frac{t^3}{6}(g_1g_2g_3 + g_1g_3g_2 + g_2g_1g_3 \\ &\quad + g_3g_2g_1 + g_3g_1g_2 + g_2g_3g_1).\end{aligned}$$

Setting $t = 1$ in the last equality and using the fact that the bracket coincides with the commutator in $U\mathfrak{g}$, gives

$$\begin{aligned}&\phi_1(g_1g_2g_3) \\ &= \frac{1}{12}(2g_1g_2g_3 - 2g_1g_3g_2 - 2g_2g_3g_1 + 2g_3g_2g_1 + 2g_1g_2g_3 - 2g_3g_1g_2 \\ &\quad - 2g_2g_1g_3 + 2g_3g_2g_1 + 3g_1g_2g_3 - 3g_2g_1g_3 + 3g_1g_3g_2 - 3g_3g_1g_2 \\ &\quad + 3g_2g_3g_1 - 3g_3g_2g_1 + 3g_1g_2g_3 - 3g_1g_3g_2 + 3g_2g_1g_3 - 3g_2g_3g_1 \\ &\quad + 3g_3g_1g_2 - 3g_3g_2g_1 + 2g_1g_2g_3 + 2g_1g_3g_2 + 2g_2g_1g_3 + 2g_3g_2g_1\end{aligned}$$

$$+ 2g_3g_1g_2 + 2g_2g_3g_1)$$
$$= g_1g_2g_3,$$

which is consistent with the fact that $\phi_1 = \mathrm{id}_{U\mathfrak{g}}$.

Finally, the first values of A_t on words of low lengths are given by

$$\begin{aligned} A_t(1, g_1) &= \phi_{-t}(1)\phi_t(g_1) = tg_1, \\ A_t(g_1, g_2) &= \phi_{-t}(g_1)\phi_t(g_2) + \phi_{-t}(1)\phi_t(g_1g_2) \\ &= -t^2g_1g_2 + \frac{t}{2}[g_1, g_2] + \frac{t^2}{2}(g_1g_2 + g_2g_1) = \frac{t - t^2}{2}[g_1, g_2], \end{aligned}$$

and

$$\begin{aligned} &A_t(g_1g_2, g_3) \\ &= \phi_{-t}(g_1g_2)\phi_t(g_3) + \phi_{-t}(g_1)\phi_t(g_2g_3) + \phi_{-t}(g_2)\phi_t(g_1g_3) + \phi_t(g_1g_2g_3) \\ &= \frac{-t^2}{2}[g_1, g_2]g_3 + \frac{t^3}{2}(g_1g_2g_3 + g_2g_1g_3) - \frac{t^2}{2}g_1[g_2, g_3] - \frac{t^3}{2}(g_1g_2g_3 + g_1g_3g_2) \\ &\quad - \frac{t^2}{2}g_2[g_1, g_3] - \frac{t^3}{2}(g_2g_1g_3 + g_2g_3g_1) + \frac{t}{6}([g_1, [g_2, g_3]] + [[g_1, g_2], g_3]) \\ &\quad + \frac{t^2}{4}([g_1, g_2]g_3 + [g_1, g_3]g_2 + [g_2, g_3]g_1 + g_1[g_2, g_3] + g_2[g_1, g_3] + g_3[g_1, g_2]) \\ &\quad + \frac{t^3}{6}(g_1g_2g_3 + g_1g_3g_2 + g_2g_1g_3 + g_3g_2g_1 + g_3g_1g_2 + g_2g_3g_1) \\ &= \frac{t}{6}([g_1, [g_2, g_3]] + [[g_1, g_2], g_3]) + \frac{t^2}{4}(-[g_1, g_2]g_3 + g_3[g_1, g_2] \\ &\quad + [g_1, g_3]g_2 - g_2[g_1, g_3] + [g_2, g_3]g_1 - g_1[g_2, g_3]) \\ &\quad + \frac{t^3}{6}(g_1g_2g_3 - 2g_1g_3g_2 + g_2g_1g_3 + g_3g_2g_1 + g_3g_1g_2 - 2g_2g_3g_1) \\ &= \frac{t}{6}([g_1, [g_2, g_3]] + [[g_1, g_2], g_3]) + \frac{t^2}{4}(-[[g_1, g_2]g_3] + [[g_1, g_3]g_2] - [g_1[g_2, g_3]]) \\ &\quad + \frac{t^3}{6}(g_1g_2g_3 - 2g_1g_3g_2 + g_2g_1g_3 + g_3g_2g_1 + g_3g_1g_2 - 2g_2g_3g_1) \\ &= \frac{t}{6}([g_1, [g_2, g_3]] + [[g_1, g_2], g_3]) - \frac{t^2}{2}[g_1[g_2, g_3]] + \frac{t^3}{6}([g_1, [g_2, g_3]] + [g_2, [g_1, g_3]]). \end{aligned}$$

Let us note some equivariance properties of the operators pr, ϕ_t and A_t with respect to the adjoint action.

Proposition 2.8.2. *Let* $\mathfrak{g}$ *be a Lie algebra and* $U\mathfrak{g}$ *its universal enveloping algebra with its usual Hopf algebra structure. Let* $h \in \mathfrak{g}$. *Then we have:*

(a) $\Delta \circ \mathrm{ad}_h = (\mathrm{ad}_h \otimes \mathrm{Id} + \mathrm{Id} \otimes \mathrm{ad}_h) \circ \Delta$;

(b) $\mathrm{pr}(\mathrm{ad}_h(g_1 \cdots g_n)) = \mathrm{ad}_h(\mathrm{pr}(g_1 \cdots g_n))$ *for all* $g_1, \ldots, g_n \in \mathfrak{g}$;

(c) $\mathrm{ad}_h \circ \phi_t = \phi_t \circ \mathrm{ad}_h$;

(d) *For all $x \in U\mathfrak{g}$ and all $g \in \mathfrak{g}$, we have*

$$\mathrm{ad}_h(A_t(x,g)) = A_t(\mathrm{ad}_h(x),g) + A_t(x,\mathrm{ad}_h(g)).$$

Proof. (a) Using that h is primitive, we compute for all $x \in U\mathfrak{g}$,

$$\begin{aligned}\triangle[h,x] &= \triangle(hx - xh)\\ &= \triangle(h) \triangle (x) - \triangle(x) \triangle (h)\\ &= (1\otimes h + h\otimes 1) \triangle (x) - \triangle(x)(1\otimes h + h\otimes 1)\\ &= (\mathrm{ad}_h \otimes \mathrm{id} + \mathrm{id}\otimes \mathrm{ad}_h)\circ \triangle(x).\end{aligned}$$

(b) Here, we use Solomon's formula from the proof of Proposition 2.8.1:

$$\begin{aligned}\mathrm{pr}(\mathrm{ad}_h(g_1\cdots g_n)) &= \sum_{i=1}^n \mathrm{pr}(g_1\dots[h,g_i]\dots g_n)\\ &= \sum_{i=1}^n \frac{1}{n}\sum_{\sigma\in S_n}\frac{(-1)^{d(\sigma)}}{\binom{cn-1}{d(\sigma)}} g_{\sigma(1)}\dots[h,g_{\sigma(i)}]\dots g_{\sigma(n)}\\ &= \frac{1}{n}\sum_{\sigma\in S_n}\frac{(-1)^{d(\sigma)}}{\binom{cn-1}{d(\sigma)}}\mathrm{ad}_h(g_{\sigma(1)}\cdots g_{\sigma(n)})\\ &= \mathrm{ad}_h(\mathrm{pr}(g_1\cdots g_n)).\end{aligned}$$

(c) Observe first that the equivariance of $\triangle$ and pr imply that

$$\mathrm{ad}_h \circ \mathrm{pr}^{\star n} = \mathrm{pr}^{\star n}\circ \mathrm{ad}_h.$$

Indeed, we have for all $x \in U\mathfrak{g}$:

$$\begin{aligned}\mathrm{ad}_h(\mathrm{pr}(x_1)\cdots \mathrm{pr}(x_n)) &= \sum_{i=1}^n \mathrm{pr}(x_1)\cdots \mathrm{ad}_h(\mathrm{pr}(x_i))\cdots \mathrm{pr}(x_n)\\ &= \sum_{i=1}^n \mathrm{pr}(x_1)\cdots \mathrm{pr}(\mathrm{ad}_h(x_i))\cdots \mathrm{pr}(x_n)\\ &= \sum_{i=1}^n \mathrm{mult}^{\otimes n}\circ \mathrm{pr}^{\otimes n}(x_1\otimes\cdots\otimes \mathrm{ad}_h(x_i)\otimes\cdots\otimes x_n)\\ &= (\mathrm{mult}^{\otimes n}\circ \mathrm{pr}^{\otimes n}\circ \triangle^{(n-1)}\circ \mathrm{ad}_h)(x).\end{aligned}$$

Now we compute

$$\begin{aligned}\mathrm{ad}_h\circ \phi_t = \mathrm{ad}_h\circ\left(\sum_{n\geq 0}\frac{t^n}{n!}\mathrm{pr}^{\star n}\right) &= \sum_{n\geq 0}\frac{t^n}{n!}\mathrm{ad}_h\circ \mathrm{pr}^{\star n}\\ &= \sum_{n\geq 0}\frac{t^n}{n!}\mathrm{pr}^{\star n}\circ \mathrm{ad}_h = \phi_t\circ \mathrm{ad}_h.\end{aligned}$$

(d) Using the preceding computations, we obtain

$$\begin{aligned}\mathrm{ad}_h(A_t(x,g)) &= \mathrm{ad}_h(\phi_{-t}(x_1)\phi_t(x_2g))\\ &= \mathrm{ad}_h(\phi_{-t}(x_1))\phi_t(x_2g) + \phi_{-t}(x_1)\mathrm{ad}_h(\phi_t(x_2g))\\ &= \phi_{-t}(\mathrm{ad}_h(x_1))\phi_t(x_2g) + \phi_{-t}(x_1)\phi_t(\mathrm{ad}_h(x_2g))\\ &= (\phi_{-t}(\mathrm{ad}_h(x_1))\phi_t(x_2g) + \phi_{-t}(x_1)\phi_t(\mathrm{ad}_h(x_2)g) + \phi_{-t}(x_1)\phi_t(x_2\mathrm{ad}_h(g)))\\ &= A_t(\mathrm{ad}_h(x),g) + A_t(x,\mathrm{ad}_h(g)).\end{aligned}$$ □

Now we come to the procedure of lifting trivial valued 3-cocycles to 2-cocycles with values in $(U\mathfrak{g}^+)^\vee$. Let $f:\Lambda^3\mathfrak{g}\to\mathbb{C}$ be a Chevalley–Eilenberg 3-cochain of $\mathfrak{g}$ with values in $\mathbb{C}$. Using the identification,

$$C^2(\mathfrak{g},U\mathfrak{g}^\vee)\cong C_2(\mathfrak{g},U\mathfrak{g})^\vee$$

we can use f to produce a 2-cochain $\tilde{\alpha}$ in $C^2(\mathfrak{g},U\mathfrak{g}^\vee)$ by setting

$$\tilde{\alpha}(x,g_1,g_2) := \int_0^1 dt\, f(\mathrm{pr}(x_1),A_t(x_2,g_1),A_t(x_3,g_2)) \tag{2.17}$$

for all x in $U\mathfrak{g}$, g_1 and g_2 in $\mathfrak{g}$. This 2-cochain $\tilde{\alpha}$ induces then the 2-cochain α in $C^2(\mathfrak{g},(U\mathfrak{g}^+)^\vee)$ by restriction to $U\mathfrak{g}^+\subset U\mathfrak{g}$.

Example 2.8.4. If $x=\lambda$ is an element of $\mathbb{C}\subset U\mathfrak{g}$, then formula (2.17) reads

$$\alpha(\lambda,g_1,g_2)=0.$$

If $x=g$ is an element of $\mathfrak{g}\subset U\mathfrak{g}$, then formula (2.17) reads

$$\alpha(g,g_1,g_2)=\int_0^1 dt\, f(g,tg_1,tg_2)=\frac{1}{3}f(g,g_1,g_2).$$

If $x=gh$ is a product of two elements of $\mathfrak{g}$, then formula (2.17) reads

$$\begin{aligned}\alpha(gh,g_1,g_2) = \int_0^1 dt\,\Big(&\frac{1}{2}f([g,h],tg_1,tg_2)+f\Big(g,\frac{(t-t^2)}{2}[h,g_1],tg_2\Big)\\ &+f\Big(g,tg_1,\frac{(t-t^2)}{2}[h,g_2]\Big)+f\Big(h,\frac{(t-t^2)}{2}[g,g_1],tg_2\Big)\\ &+f\Big(h,tg_1,\frac{(t-t^2)}{2}[g,g_2]\Big)\Big)\\ =\frac{1}{24}(&4f([g,h],g_1,g_2)+f(g,[h,g_1],g_2)+f(g,g_1,[h,g_2])\\ &+f(h,[g,g_1],g_2)+f(h,g_1,[g,g_2])).\end{aligned}$$

Theorem 2.8.3. *If f is a 3-cocycle with values in $\mathbb{C}$, then the restriction of $d\tilde{\alpha}$ to $\mathbb{C}\otimes\Lambda^3\mathfrak{g} \subset U\mathfrak{g}\otimes\Lambda^3\mathfrak{g}$ is f. In particular, α is a 2-cocycle in $C^2(\mathfrak{g},(U\mathfrak{g}^+)^\vee)$ and $[f] = \partial[\alpha]$ in $H^3(\mathfrak{g},\mathbb{C})$, where ∂ is the connecting homomorphism of the long exact sequence*

$$\cdots \to H^2(\mathfrak{g},\mathbb{C}) \to H^2(\mathfrak{g},U\mathfrak{g}^\vee) \to H^2(\mathfrak{g},(U\mathfrak{g}^+)^\vee) \xrightarrow{\partial} H^3(\mathfrak{g},\mathbb{C}) \to H^3(\mathfrak{g},U\mathfrak{g}^\vee) \to \cdots \quad (2.18)$$

associated to the short exact sequence of coefficients (2.15).

Proof. This assertion is a consequence of a more general statement asserting that the right-hand side of formula (2.17) is just the degree 3 component of the dual $s_*^\vee$ of a certain contracting homotopy s_* of $(C_*(\mathfrak{g};U\mathfrak{g}),d)$ applied to f. However, in order not to digress too far, we will prove here only the special case. For x in $U\mathfrak{g}$ and g_1, g_2, g_3 in $\mathfrak{g}$, we have

$$\begin{aligned} d\alpha(x,g_1,g_2,g_3) &= \alpha(xg_1,g_2,g_3) - \alpha(xg_2,g_1,g_3) + \alpha(xg_3,g_1,g_2) \\ &\quad + \alpha(x,[g_1,g_2],g_3) - \alpha(x,[g_1,g_3],g_2) - \alpha(x,g_1,[g_2,g_3]). \end{aligned}$$

Using the fact that the g_i are primitive, the first part of the preceding equality reads

$$\begin{aligned} &\alpha(xg_1,g_2,g_3) - \alpha(xg_2,g_1,g_3) + \alpha(xg_3,g_1,g_2) \\ &= \int_0^1 dt\, f(\mathrm{pr}(x_1g_1),A_t(x_2,g_2),A_t(x_3,g_3)) + f(\mathrm{pr}(x_1),A_t(x_2g_1,g_2),A_t(x_3,g_3)) \\ &\quad + f(\mathrm{pr}(x_1),A_t(x_2,g_2),A_t(x_3g_1,g_3)) - f(\mathrm{pr}(x_1g_2),A_t(x_2,g_1),A_t(x_3,g_3)) \\ &\quad - f(\mathrm{pr}(x_1),A_t(x_2g_2,g_1),A_t(x_3,g_3)) - f(\mathrm{pr}(x_1),A_t(x_2,g_1),A_t(x_3g_2,g_3)) \\ &\quad + f(\mathrm{pr}(x_1g_3),A_t(x_2,g_1),A_t(x_3,g_2)) + f(\mathrm{pr}(x_1),A_t(x_2g_3,g_1),A_t(x_3,g_2)) \\ &\quad + f(\mathrm{pr}(x_1),A_t(x_2,g_1),A_t(x_3g_3,g_2)). \end{aligned}$$

Denote the nine terms in this last sum by (1) to (9). For y in $U\mathfrak{g}$ and g, h in $\mathfrak{g}$, properties (d) and (e) of ϕ_t and A_t listed in Proposition 2.8.1 imply that

$$\begin{aligned} A_t(yg,h) - A_t(yh,g) &= \phi_{-t}(y_1g)\phi_t(y_2h) - \phi_{-t}(y_1h)\phi_t(y_2g) \\ &\quad + \phi_{-t}(y_1)\phi_t(y_2gh) - \phi_{-t}(y_1)\phi_t(y_2hg) \\ &= \phi_{-t}(y_1g)\phi_t(y_2h) - \phi_{-t}(y_1h)\phi_t(y_2g) + A_t(y,[g,h]) \\ &= \phi_{-t}(y_1g)\phi_t(y_2)\phi_{-t}(y_3)\phi_t(y_4h) \\ &\quad - \phi_{-t}(y_1h)\phi_t(y_2)\phi_{-t}(y_3)\phi_t(y_4g) + A_t(y,[g,h]) \\ &= A_t(y,[g,h]) - [A_t(y_1,g),A_t(y_2,h)]. \end{aligned}$$

Here, the third equality comes from property (d): $\phi_t \star \phi_{-t} = \eta\epsilon$.

Now, the terms (2)+(5), (3)+(8) and (6)+(9) of the above sum make appear the expressions above $A_t(yg,h) - A_t(yh,g)$, for which we use then

$$A_t(yg,h) - A_t(yh,g) = A_t(y,[g,h]) - [A_t(y_1,g),A_t(y_2,h)].$$

The resulting terms involving $A_t(y,[g,h])$ then cancel the corresponding term of the form $\alpha(\dots,[g_j,g_k],\dots)$ in $d\alpha(x,g_1,g_2,g_3)$. Hence we are left with (the 3 remaining terms involving $[A_t(y_1,g),A_t(y_2,h)]$ plus the terms (1), (4) and (7)),

$$\begin{aligned}
&d\alpha(x,g_1,g_2,g_3)\\
&\quad=\int_0^1 dt\, f(\mathrm{pr}(x_1g_1),A_t(x_2,g_2),A_t(x_3,g_3))\\
&\qquad-f(\mathrm{pr}(x_1),[A_t(x_2,g_1),A_t(x_3,g_2)],A_t(x_4,g_3))-f(\mathrm{pr}(x_1g_2),A_t(x_2,g_1),A_t(x_3,g_3))\\
&\qquad+f(\mathrm{pr}(x_1),A_t(x_2,g_1),[A_t(x_3,g_2)A_t(x_4,g_3)])+f(\mathrm{pr}(x_1g_3),A_t(x_2,g_1),A_t(x_3,g_2))\\
&\qquad+f(\mathrm{pr}(x_1),[A_t(x_2,g_1),A_t(x_3,g_3)],A_t(x_4,g_2)).
\end{aligned}$$

Since f is a cocycle, we have that

$$df(\mathrm{pr}(x_1),A_t(x_2,g_1),A_t(x_3,g_2),A_t(x_4,g_3))=0,$$

which enables to rewrite the three terms of $d\alpha(x,g_1,g_2,g_3)$ involving Lie brackets to obtain

$$\begin{aligned}
&d\alpha(x,g_1,g_2,g_3)\\
&\quad=\int_0^1 dt\, f(\mathrm{pr}(x_1g_1),A_t(x_2,g_2),A_t(x_3,g_3))\\
&\qquad-f([\mathrm{pr}(x_1),A_t(x_2,g_1)],A_t(x_3,g_2),A_t(x_4,g_3))-f(\mathrm{pr}(x_1g_2),A_t(x_2,g_1),A_t(x_3,g_3))\\
&\qquad+f([\mathrm{pr}(x_1),A_t(x_2,g_2)],A_t(x_3,g_1),A_t(x_4,g_3))+f(\mathrm{pr}(x_1g_3),A_t(x_2,g_1),A_t(x_3,g_2))\\
&\qquad-f([\mathrm{pr}(x_1),A_t(x_2,g_3)],A_t(x_3,g_1),A_t(x_4,g_2)).
\end{aligned}$$

But one easily checks that for all y in $U\mathfrak{g}$ and g in $\mathfrak{g}$,

$$\mathrm{pr}(yg)-[\mathrm{pr}(y_1),A_t(y_2,g)]=\frac{d}{dt}A_t(y,g)$$

so that, using point (c) of Proposition 2.8.1,

$$\begin{aligned}
d\alpha(x,g_1,g_2,g_3)&=\int_0^1 dt\, f\Big(\frac{d}{dt}A_t(x_1,g_1),A_t(x_2,g_2),A_t(x_3,g_3)\Big)\\
&\quad+f\Big(A_t(x_1,g_1),\frac{d}{dt}A_t(x_2,g_2),A_t(x_3,g_3)\Big)\\
&\quad+f\Big(A_t(x_1,g_1),A_t(x_2,g_2),\frac{d}{dt}A_t(x_3,g_3)\Big)\\
&=\int_0^1 dt\,\frac{d}{dt}f(A_t(x_1,g_1),A_t(x_2,g_2),A_t(x_3,g_3))
\end{aligned}$$

$$= f(A_1(x_1,g_1), A_1(x_2,g_2), A_1(x_3,g_3)) - f(A_0(x_1,g_1), A_0(x_2,g_2), A_0(x_3,g_3))$$
$$= \epsilon(x) f(g_1,g_2,g_3)$$

from which the theorem follows immediately. □

Together with Theorem 2.5.1, this shows the following.

Corollary 2.8.4. *If f is a 3-cocycle, then the Yoneda product*

$$0 \to \mathbb{C} \to U\mathfrak{g}^\vee \to (U\mathfrak{g}^+)^\vee \times_\alpha \mathfrak{g} \to \mathfrak{g} \to 0$$

is a crossed module of Lie algebras whose equivalence class represents $[f]$.

Remark 2.8.5. Observe that Theorem 2.5.1 shows that there is *some* cocycle α such that the above Yoneda product is a crossed module whose equivalence class represents $[f]$, but Theorem 2.8.3 implies that *the* cocycle α defined using formula (2.17) does the job.

In other words, we have constructed abelian representatives (of the equivalence class of crossed modules associated) to each 3-cohomology class given by some constant-valued 3-cocycle f. This applies in particular to the Cartan cocycle.

The abelian case

When the Lie algebra $\mathfrak{g}$ is abelian, the enveloping algebra $U\mathfrak{g}$ is nothing but the symmetric algebra $S\mathfrak{g}$ and the idempotent pr is simply the canonical projection $\mathrm{pr} : S\mathfrak{g} \to \mathfrak{g}$ on words of length one. Thus, denoting by $|x|$ the length of a word in $S\mathfrak{g}$, one easily checks that ϕ_t is given by

$$\phi_t(x) = t^{|x|}x$$

for all homogeneous elements x in $S\mathfrak{g}$. Moreover, ϕ_t is a morphism of algebras (which is no longer true in general) so that

$$A_t(x,\mathfrak{g}) = \phi_{-t}(x_1)\phi_t(x_2 g) = \phi_{-t}(x_1)\phi_t(x_2)\phi_t(g)$$
$$= t(\phi_{-t} \star \phi_t)(x)g = t\epsilon(x)g.$$

As a consequence, the 2-cocycle $\alpha : S\mathfrak{g} \otimes \Lambda^2\mathfrak{g} \to \mathbb{K}$ lifting any given Chevalley–Eilenberg 3-cochain $f : \Lambda^3\mathfrak{g} \to \mathbb{K}$ takes the form

$$\alpha(x,g_1,g_2) = \int_0^1 t^2 f(\mathrm{pr}(x),g_1,g_2)\,dt = \begin{cases} \frac{1}{3}f(x,g_1,g_2) & \text{if } |x| = 1, \\ 0 & \text{otherwise,} \end{cases} \tag{2.19}$$

for any homogeneous element $x \in S\mathfrak{g}$.

The example of the Cartan cocycle of $\mathfrak{sl}_2(\mathbb{C})$

Here, $\mathfrak{g} := \mathfrak{sl}_2(\mathbb{C})$ and $f : \Lambda^3\mathfrak{g} \to \mathbb{C}$ is the Cartan 3-cocycle defined by

$$f(g_1, g_2, g_3) := \kappa(g_1, [g_2, g_3])$$

for all g_1, g_2 and g_3 in $\mathfrak{g}$, where κ is the Killing form of $\mathfrak{g} = \mathfrak{sl}_2(\mathbb{C})$ Our goal is to study the 2-cocycle α lifting f in order to get a concrete description of the crossed module encoding the Cartan cocycle. More precisely, we want to determine the values of

$$\alpha(x, g_1, g_2) := \int_0^1 dt\, \kappa(\mathrm{pr}(x_1), [A_t(x_2, g_1), A_t(x_3, g_2)])$$

when g_1, g_2 run over a given basis of $\mathfrak{g}$ and x runs over the associated Poincaré–Birkhoff–Witt basis of $U\mathfrak{g}^+$. Denote by (X, Y, H) the standard ordered basis of $\mathfrak{sl}_2(\mathbb{C})$, which satisfies the relations

$$[X, Y] = H, \quad [H, X] = 2X, \quad [H, Y] = -2Y.$$

In this basis, the problem amounts to determining the values of the following complex numbers:

$$B_{XY}(a, b, c) := \alpha(X^a Y^b H^c, X, Y)$$

and

$$B_{XH}(a, b, c) := \alpha(X^a Y^b H^c, X, H)$$

and

$$B_{YH}(a, b, c) := \alpha(X^a Y^b H^c, Y, H)$$

for all natural numbers a, b and c.

While obtaining a general formula for the values of B_{XY}, B_{XH} and B_{YH} seems out of reach for the moment, in particular because the combinatorics appearing in the expression (2.16) of $\mathrm{pr}(x)$, for a word x of $U\mathfrak{g}$, get more and more complicated as the length of x increases, we can at least show that these values are "often" 0.

Proposition 2.8.5. *Let a, b and c be three natural numbers.*

(a) *If $a \neq b$, then*

$$B_{XY}(a, b, c) = 0.$$

(b) *If $a \neq b - 1$, then*

$$B_{XH}(a, b, c) = 0.$$

(c) *If* $a \neq b+1$, *then*

$$B_{YH}(a,b,c) = 0.$$

Proof. The proposition follows from the following invariance property of the Cartan cocycle f, together with the equivariance properties of Proposition 2.8.2. For all $x, y, z, h \in \mathfrak{sl}_2(\mathbb{C})$, we have

$$f(\mathrm{ad}_h(x), y, z) + f(x, \mathrm{ad}_h(y), z) + f(x, y, \mathrm{ad}_h(z)) = 0,$$

using the invariance of κ and the Jacobi identity in the Lie algebra $\mathfrak{sl}_2(\mathbb{C})$.

From this, we have directly that the cocycle α from equation (2.17) is also invariant, that is, for all $h, x, y \in \mathfrak{sl}_2(\mathbb{C})$ and all $u \in U\mathfrak{sl}_2(\mathbb{C})$, we have $\alpha(\mathrm{ad}_h(u,x,y)) = 0$.

Therefore, we get

$$\begin{aligned} 0 = \alpha(\mathrm{ad}_h(u,x,y)) &= \alpha(\mathrm{ad}_h(u), x, y) + \alpha(u, \mathrm{ad}_h(x), y) + \alpha(u, x, \mathrm{ad}_h(y)) \\ &= \int_0^1 dt\, f(\mathrm{ad}_h(\mathrm{pr}(u_1)), A_t(u_2, x), A_t(u_3, y)) \\ &\quad + \int_0^1 dt\, f(\mathrm{pr}(u_1), \mathrm{ad}_h(A_t(u_2, x)), A_t(u_3, y)) \\ &\quad + \int_0^1 dt\, f(\mathrm{pr}(u_1), A_t(u_2, x), \mathrm{ad}_h(A_t(u_3, y))). \end{aligned}$$

We apply now Proposition 2.8.2 to obtain

$$\begin{aligned} 0 = \alpha(\mathrm{ad}_h(u,x,y)) &= \int_0^1 dt\, f(\mathrm{pr}(\mathrm{ad}_h u^{(1)}), A_t(u^{(2)}, x), A_t(u^{(3)}, y)) \\ &\quad + \int_0^1 dt\, f(\mathrm{pr}(u_1), A_t(\mathrm{ad}_h(u_2), x), A_t(u_3, y)) \\ &\quad + \int_0^1 dt\, f(\mathrm{pr}(u_1), A_t(u_2, \mathrm{ad}_h(x)), A_t(u_3, y)) \\ &\quad + \int_0^1 dt\, f(\mathrm{pr}(u_1), A_t(u_2, x), A_t(\mathrm{ad}_h(u_3), y)) \\ &\quad + \int_0^1 dt\, f(\mathrm{pr}(u_1), A_t(u_2, x), A_t(u_3, \mathrm{ad}_h(y))). \end{aligned}$$

We specialize then to $h =: H \in \mathfrak{sl}_2(\mathbb{C})$ and use that X, Y, but also $u := X^a Y^b H^c$ are eigenvectors for the adjoint action of H. The outcome is

$$0 = \alpha(\mathrm{ad}_h(u, X, Y)) = 2(a-b)\alpha(u, X, Y).$$

The first assertion of the proposition follows. The two other assertions are proved similarly. □

Example 2.8.6. Here, we specialize the last formula of Example 2.8.4 for $\mathfrak{g} = \mathfrak{sl}_2(\mathbb{C})$.

In this case, since the 2-fold coproduct of the monomial XY reads

$$\begin{aligned}\triangle^{(2)}(XY) = {} & XY \otimes 1 \otimes 1 + 1 \otimes XY \otimes 1 + 1 \otimes 1 \otimes XY + X \otimes Y \otimes 1 \\ & + Y \otimes X \otimes 1 + 1 \otimes X \otimes Y + 1 \otimes Y \otimes X + X \otimes 1 \otimes Y + Y \otimes 1 \otimes X,\end{aligned}$$

$\mathrm{pr}(1) = 0$, $A_t(1, g) = tg$ for all g in $\mathfrak{sl}_2(\mathbb{C})$, we can express $\alpha(XY, X, Y)$ as

$$\begin{aligned}\alpha(XY, X, Y) = \int_0^1 & dt\, f(\mathrm{pr}(XY), tX, tY) + f(X, A_t(Y, X), tY) \\ & + f(Y, A_t(X, X), tY) + f(X, tX, A_t(Y, Y)) + f(Y, tX, A_t(X, Y)).\end{aligned}$$

The third and fourth terms on the RHS of the previous expression vanish due to $[X, X] = 0$ and $[Y, Y] = 0$ (see Example 2.8.3), and we can use formulas for A_t and pr established above and the skew symmetry of f to get

$$\begin{aligned}\alpha(XY, X, Y) &= \int_0^1 dt\, \frac{t^2}{2} f([X, Y], X, Y) + \frac{t^2 - t^3}{2} f(X, [Y, X], Y) + \frac{t^2 - t^3}{2} f(Y, X, [X, Y]) \\ &= \frac{1}{6} f(X, Y, H) + \frac{1}{24} f(X, Y, H) - \frac{1}{24} f(X, Y, H) \\ &= \frac{1}{6} f(X, Y, H) = \frac{1}{6} \kappa(X, [Y, H]) = \frac{1}{3} \kappa(X, Y) = \frac{4}{3}.\end{aligned}$$

An explicit formula for the cocycle α corresponding to the Cartan cocycle of a simple complex Lie algebra is still to be found.

2.8.3 A crossed module representing the Godbillon–Vey cocycle

We now consider the Godbillon–Vey cocycle θ whose class $[\theta] \in H^3(\mathrm{Vect}(S^1), \mathbb{C})$ generates the third (complex valued) cohomology of the Lie algebra of vector fields on the circle $\mathrm{Vect}(S^1)$.

In order to construct a crossed module representing θ, we first consider the Lie algebra W_1 of (complex valued) polynomial vector fields on the line, in one formal variable z. As a complex vector space,

$$W_1 = \bigoplus_{n \geq -1} \mathbb{C} z^{n+1} \frac{d}{dz},$$

and the bracket is given by

$$\left[z^{m+1}\frac{d}{dz}, z^{n+1}\frac{d}{dz}\right] = (n-m)z^{n+m+1}\frac{d}{dz}.$$

Sometimes, one writes these generators as $e_n := z^{n+1}\frac{d}{dz}$ and then the bracket takes the form $[e_m, e_n] = (n-m)e_{n+m}$. Usually, when one speaks about *formal* vector fields, the coefficient functions are supposed to be formal series and not polynomials, so this corresponds then to $\widehat{W}_1 = \Pi_{n\geq -1}\mathbb{C}z^{n+1}\frac{d}{dz}$ and the bracket is also written formally as

$$\left[f(z)\frac{d}{dz}, g(z)\frac{d}{dz}\right] = (fg' - gf')(z)\frac{d}{dz},$$

for formal series $f, g \in \mathbb{C}[[z]]$. Usually, one identifies a vector field (polynomial or formal) with its coefficient function and writes simply f for $f\frac{d}{dz}$.

The following defines a one parameter family of 3-cocycles on W_1 (and $\widehat{W}_1$):

$$\theta_z(f,g,h) := \begin{vmatrix} f & g & h \\ f' & g' & h' \\ f'' & g'' & h'' \end{vmatrix}(z).$$

Usually one takes as Godbillon–Vey cocycle the evaluation in 0, that is, $\theta := \theta_0$, but actually all θ_z are cohomologuous; see Remark 2.8.9.

In the same way as $\widehat{W}_1$ represents formal vector fields, there are formal functions and formal differential forms. More generally, let F_λ be the *space of (polynomial) λ-densities*, that is, as a complex vector space $F_\lambda = \bigoplus_{n\geq -1}\mathbb{C}z^{n+1}(dz)^\lambda$ and F_λ becomes a W_1-module by setting

$$f(z)\frac{d}{dz}\cdot g(z)(dz)^\lambda = (fg' + \lambda gf')(z)(dz)^\lambda.$$

In the same way, the space of formal λ-densities is denoted by $\widehat{F}_\lambda$. Observe that $\widehat{F}_0$ are formal functions on the complex line $\mathbb{C}$, $\widehat{F}_1$ are formal 1-forms, and $\widehat{F}_{-1} = \widehat{W}_1$ as a $\widehat{W}_1$-module (where $\widehat{W}_1$ is regarded as a $\widehat{W}_1$-module using the adjoint action).

It is easy to verify that the formal de Rham sequence

$$0 \to \mathbb{C} \to \widehat{F}_0 \overset{d_{\mathrm{dR}}}{\to} \widehat{F}_1 \to 0 \qquad (2.20)$$

is a short exact sequence of $\widehat{W}_1$-modules.

There is another cocycle which plays a role in the construction. Namely, let $\alpha \in Z^2(\widehat{W}_1, \widehat{F}_1)$ be defined by

$$\alpha(f,g) := \begin{vmatrix} f' & g' \\ f'' & g'' \end{vmatrix}(z)(dz)^1.$$

Remark 2.8.7. This cocycle is the integrand of the *Gelfand–Fuchs cocycle*

$$\omega(f,g) := \int_{S^1} \begin{vmatrix} f' & g' \\ f'' & g'' \end{vmatrix} (t)dt$$

whose class $[\omega]$ generates $H^2(\mathrm{Vect}(S^1), \mathbb{C})$. The Gelfand–Fuchs cocycle defines the (continuous) universal central extension of $\mathrm{Vect}(S^1)$ which is also known as the *Virasoro algebra* and which plays an prominent role in string theory; see, for example, [38].

The key relation between the two cocycles θ_z and α is described in the following lemma. In its statement, θ_z is viewed as a function in z to which one applies the de Rham differential and obtains a 1-form. On the other hand, to the cocycle $\alpha \in Z^2(\widehat{W}_1, \widehat{F}_1)$, one may apply the Chevalley–Eilenberg differential $d^{\mathbb{C}}$ (corresponding to coefficients in the trivial $\widehat{W}_1$-module $\mathbb{C}$), and the result is nontrivial.

Lemma 2.8.6. $d_{\mathrm{dR}}\theta_z = d^{\mathbb{C}}\alpha$.

Proof. Let $f, g, h \in \widehat{W}_1$ be formal vector fields. Compute the derivatives of the bracket:

$$[f,g]' = (fg' - g'f)' = fg'' - gf''.$$

In the same vein:

$$[f,g]'' = (fg'' - gf'')' = f'g'' - g'f'' + fg''' - gf'''.$$

Now compute

$$\begin{aligned} d^{\mathbb{C}}\alpha(f,g) &= \sum_{\text{cycl.}} \begin{vmatrix} [f,g]' & h' \\ [f,g]'' & h'' \end{vmatrix} (z)(dz)^1 \\ &= \sum_{\text{cycl.}} ((fg'' - gf'')h'' - (f'g'' - g'f'' + fg''' - gf''')h')(z)(dz)^1. \end{aligned}$$

Such a cyclic sum over (f, g, h) is obviously zero in case two of the three functions in the expression have the same number of derivatives. This is the case for $\sum_{\text{cycl.}}(fg'' - gf'')h''$ and $\sum_{\text{cycl.}}(f'g'' - g'f'')h'$. Therefore, the only terms which remain are

$$d^{\mathbb{C}}\alpha(f,g) = -\sum_{\text{cycl.}} (fg''' - gf''')h'.$$

On the other hand, we have

$$d_{\mathrm{dR}}\theta_z(f,g,h) = \begin{vmatrix} f & g & h \\ f' & g' & h' \\ f'' & g'' & h'' \end{vmatrix}' (z)$$

$$
\begin{aligned}
&= (f(g'h'' - h'g'') - f'(gh'' - hg'') + f'(gh' - hg'))'(z) \\
&= f'(g'h'' - h'g'') + f(g'h''' - h'g''') - f''(gh'' - hg'') \\
&\quad - f'(g'h'' - h'g'') - f'(gh''' - hg''') + f'''(gh' - hg') + f''(gh'' - hg'') \\
&= f(g'h''' - h'g''') - f'(gh''' - hg''') + f'''(gh' - hg').
\end{aligned}
$$

This shows our claim. □

Corollary 2.8.7. *The connecting homomorphism induced by the short exact sequence (2.20) sends α to (the negative of) θ_0, that is,*

$$\partial\alpha = -\theta_0.$$

Proof. Let $f, g \in \widehat{W}_1$ be formal vector fields. The expression $\alpha(f,g)$ is a formal series in z and admits a primitive denoted $A(f,g)$. Now compute

$$
\begin{aligned}
(d^{\widehat{F}_0}A)(f,g,h) &= \sum_{\text{cycl.}} A([f,g],h) - \sum_{\text{cycl.}} f\cdot A'(g,h) \\
&= \int_0^z (d^{\mathbb{C}}\alpha)(f,g,h)(t)dt - \sum_{\text{cycl.}} f\alpha(g,h) \\
&= \int_0^z d_{\mathrm{dR}}\theta_t(f,g,h)dt - \theta_z(f,g,h) \\
&= \theta_z(f,g,h) - \theta_0(f,g,h) - \theta_z(f,g,h) \\
&= -\theta_0(f,g,h).
\end{aligned}
$$

In the computation, we have used the lemma going from line 2 to line 3 and, furthermore, the following computation:

$$
\begin{aligned}
\sum_{\text{cycl.}} f\alpha(g,h) &= \sum_{\text{cycl.}} f(g'h'' - h'g'')(z) \\
&= (f(g'h'' - h'g'') - f'(gh'' - hg'') + f'(gh' - hg'))(z) \\
&= \theta_z(f,g,h).
\end{aligned}
$$

Now $A(f,g)$ may be taken as a lift of $\alpha(f,g)$ to a cochain with values in $\widehat{F}_0$, and we have just computed that its coboundary $d^{\widehat{F}_0}A$ is the image in $C^3(\widehat{W}_1, \widehat{F}_0)$ of the cochain $-\theta_0$. But this is just rephrasing $\partial\alpha = -\theta_0$. □

Corollary 2.8.8. *The de Rham sequence (2.20) and the abelian extension of $\widehat{W}_1$ by $\widehat{F}_1$ using the 2-cocycle α fit together to give a crossed module of Lie algebras*

$$0 \to \mathbb{C} \to \widehat{F}_0 \to \widehat{F}_1 \times_\alpha \widehat{W}_1 \to \widehat{W}_1 \to 0.$$

This crossed module represents (up to a non-zero multiple) the Godbillon–Vey class in $H^3(\widehat{W}_1, \mathbb{C})$.

Proof. The statement follows immediately from Theorem 2.5.1 and the above corollary. □

Remark 2.8.8. It is possible to construct similar crossed modules for the corresponding Godbillon–Vey classes in related Lie algebras like W_1, $\mathrm{Vect}(S^1)$ or even $\mathrm{Hol}(\Sigma_k)$, the Lie algebra of holomorphic vector fields on the open Riemann surface $\Sigma_k := \Sigma \setminus \{p_1, \dots, p_k\}$ where Σ is a compact connected Riemann surface, $p_1, \dots, p_k$ are pairwise distinct points of Σ and $k \geq 1$, and also $\mathrm{Vect}_{1,0}(\Sigma)$ or $\mathrm{Vect}_{0,1}(\Sigma)$, the Lie algebras of smooth vector fields of type $(1,0)$ (resp., of type $(0,1)$) on the compact connected Riemann surface Σ.

Let us here only comment on $\mathrm{Vect}(S^1)$, inviting the interested reader to consult [95] concerning the other constructions.

Indeed, the de Rham sequence for S^1 instead of the line is not suitable for the construction, as it reads

$$0 \to \mathbb{R} \to C^\infty(S^1) \to \Omega^1(S^1) \to H^1(S^1, \mathbb{R}) \to 0,$$

and thus has four terms instead of three. The way out is to lift vector fields on the circle to its universal covering which is the real line, and to make them in this way act on the de Rham sequence on the line. This idea leads to the suitable crossed module representing the Godbillon–Vey cocycle for $\mathrm{Vect}(S^1)$, and the same idea also let to the crossed module for the group of diffeomorphisms on the circle; see Section 1.7.3.

Remark 2.8.9. Let us comment on the construction of the Gelfand–Fuchs cocycle

$$\omega(f,g) = \int_{S^1} \begin{vmatrix} f' & g' \\ f'' & g'' \end{vmatrix} (t)dt$$

as a fiber integral of the Godbillon–Vey cocycle

$$\theta_z(f,g,h) = \begin{vmatrix} f & g & h \\ f' & g' & h' \\ f'' & g'' & h'' \end{vmatrix} (z).$$

The Godbillon–Vey cocycle θ_z is here seen as a formal function on the line with values in Lie algebra 1-cocycles of $\widehat{W}_1$. As such we may apply the de Rham differential to θ_z, and the above lemma shows that $d_{\mathrm{dR}}\theta_z = d^{\mathbb{C}}\alpha$.

In order to define now the *fiber integral* $\int_{S^1} \alpha(t)dt$, observe that vector fields on the circle may be Taylor expanded in some point $t \in S^1$ in order to give formal vector fields. As the circle S^1 has trivial tangent bundle, the expansions in the different points $t \in S^1$ form a smooth function in $t \in S^1$ with values in formal vector fields. The expression $\int_{S^1} \alpha(t)dt$ takes these expansions at t, inserts them into α and integrates the obtained function over S^1. One sees that this fiber integral gives here the Gelfand–Fuchs cocycle.

The cocycle identity for the fiber integral follows directly from the formula $d_{\mathrm{dR}}\theta_z = d^{\mathbb{C}}\alpha$. Indeed,

$$d^{\mathbb{C}}\int_{S^1}\alpha(t)dt = \int_{S^1} d^{\mathbb{C}}\alpha(t)dt = \int_{S^1} d_{\mathrm{dR}}\theta_z = 0$$

by Stokes' theorem.

This is an example of a quite general construction procedure for $\mathrm{Vect}(M)$ cocycles from cocycles for the Lie algebra in n formal variables by fiber integral, provided that M is an n-dimensional closed manifold with trivialized tangent bundle. Namely, the Taylor expansion map (at $x \in M$) induces a Lie algebra morphism $\phi_x : \mathrm{Vect}(M) \to \widehat{W}_n$, and thus a morphism of complexes $\phi_x^* : C^*(\widehat{W}_n, \mathbb{C}) \to C^*(\mathrm{Vect}(M), \mathbb{C})$. For a given formal cocycle $\theta \in Z^q(\widehat{W}_n, \mathbb{C})$, we denote by $\theta_x := \phi_x^*\theta$ the corresponding function on M with values in cocycles. An equation $d_{\mathrm{dR}}\theta_x = d^{\mathbb{C}}\alpha$ implies as before that $\int_M \alpha$ is a cocycle. Inductively, one can get in this way a family of cochain valued differential forms

$$\theta_i \in \Omega^i(M) \otimes C^{q-i}(\mathrm{Vect}(M)),$$

linked by the equations $d_{\mathrm{dR}}\theta_i = d^{\mathbb{C}}\theta_{i+1}$. By Stokes' Theorem, once again the integral $\int_\sigma \theta_i$ of θ_i over some singular i-cycle σ in M gives a fiber integral cocycle.

Remark 2.8.10. The two examples above (the string Lie algebra and the crossed module representing the Godbillon-Vey cocycle) give rise to the possibility to compare different crossed modules representing the same cohomology class.

Indeed, take the string crossed module for the Lie algebra $\mathfrak{sl}_2(\mathbb{C})$. We worked in Subsection 2.8.1 in the real setting, but everything carries over to the complex framework to give a crossed module

$$0 \to \mathbb{C} \to \Omega(\mathfrak{sl}_2(\mathbb{C})) \to P(\mathfrak{sl}_2(\mathbb{C})) \to \mathfrak{sl}_2(\mathbb{C}) \to 0.$$

Now observe that the formal vector fields $e_{-1} = \frac{d}{dz}$, $e_0 = z\frac{d}{dz}$ and $e_1 = z^2\frac{d}{dz}$ generate a subalgebra in $\widehat{W}_1$ isomorphic to $\mathfrak{sl}_2(\mathbb{C})$. This makes it possible to pullback the crossed module of $\widehat{W}_1$ to a crossed module of $\mathfrak{sl}_2(\mathbb{C})$. Up to a factor $\frac{1}{2}$ and a minus sign, this crossed module represents the same cohomology class (see Lemma 4 in [96]).

The crossed module coming from the Godbillon–Vey cocycle may be put into a slightly more comprehensible form. For this, it is also shown in [96] that the de Rham sequence, seen as a sequence of $\mathfrak{sl}_2(\mathbb{C})$-modules, is isomorphic to

$$0 \to L(0)^\vee \to M(0)^\vee \to N(0)^\vee \to 0,$$

where $M(\lambda)$ is the standard Verma module for $\mathfrak{sl}_2(\mathbb{C})$ of highest weight λ, $N(\lambda)$ is the maximal submodule and $L(\lambda)$ is the simple quotient. As before, $\vee$ means the corresponding dual modules (which correspond to the *formal* functions and *formal* 1-forms). The crossed module reads then

$$0 \to \mathbb{C} = L(0)^\vee \to M(0)^\vee \to N(0)^\vee \times_\alpha \mathfrak{sl}_2(\mathbb{C}) \to \mathfrak{sl}_2(\mathbb{C}) \to 0,$$

where $N(0)^\vee \times_\alpha \mathfrak{sl}_2(\mathbb{C})$ is the abelian extension of $\mathfrak{sl}_2(\mathbb{C})$ by $N(0)^\vee$ by means of the cocycle α.

It is amazing to see that these two totally different crossed modules represent the same cohomology class. Hochschild's Theorem 2.6.12 shows that in any case, we have to encounter infinite dimensional Lie algebras here. But while the first crossed module is geometric in flavor, the second one is more representation theoretic. It seems hopeless to try to construct the zig-zag of morphisms of crossed modules linking the two.

2.9 Relation to relative cohomology

In this section, we report on a different approach to the relation between crossed modules and cohomology. Namely, when one looks for the data to fix in a crossed module $\mu : \mathfrak{m} \to \mathfrak{n}$ in order to make it correspond to a *relative* cohomology class, it comes out that this is not only the kernel $\mathrm{Ker}(\mu) = V$ and the cokernel $\mathrm{Coker}(\mu) = \mathfrak{g}$, but one has to fix the whole quotient morphism $\pi : \mathfrak{n} \to \mathfrak{g}$. Then the theory becomes much simpler, because fixing π, the vector space $\mathfrak{m} \cong V \oplus \mathrm{Im}(\mu)$ is already fixed. This approach is due to Kassel and Loday in [53].

For the sake of this section only, we thus introduce a new notion of equivalence of crossed modules. Fix an epimorphism of Lie algebras $\pi : \mathfrak{n} \to \mathfrak{g}$ and a $\mathfrak{g}$-module V.

Definition 2.9.1. Two crossed modules $\mu_i : \mathfrak{m}_i \to \mathfrak{n}$ for $i = 1, 2$ are called *relatively equivalent* if there exists a homomorphism of Lie algebras $\varphi : \mathfrak{n}_1 \to \mathfrak{n}_2$ such that the diagram

$$\begin{array}{ccccccccc}
0 & \longrightarrow & V & \xrightarrow{i_1} & \mathfrak{m}_1 & \xrightarrow{\mu_1} & \mathfrak{n} & \xrightarrow{\pi} & \mathfrak{g} & \longrightarrow & 0 \\
 & & \downarrow{\scriptstyle \mathrm{id}_V} & & \downarrow{\scriptstyle \varphi} & & \downarrow{\scriptstyle \mathrm{id}_{\mathfrak{n}}} & & \downarrow{\scriptstyle \mathrm{id}_{\mathfrak{g}}} & & \\
0 & \longrightarrow & V & \xrightarrow{i_2} & \mathfrak{m}_2 & \xrightarrow{\mu_2} & \mathfrak{n} & \xrightarrow{\pi} & \mathfrak{g} & \longrightarrow & 0
\end{array}$$

is commutative and such that φ is equivariant for the actions η_i of $\mathfrak{n}$ on $\mathfrak{m}_i$ for $i = 1, 2$.

Remark 2.9.2. By the Five Lemma, the homomorphism φ is necessarily an isomorphism. Therefore, it is clear that relative equivalence is an equivalence relation.

The main theorem of this section reads then the following.

Theorem 2.9.1. *The map associating to a relative crossed module the relative 3-cocycle $\kappa^* f$ (to be defined below) induces a natural isomorphism*

$$\mathrm{crmod}(\mathfrak{g},\mathfrak{n},V)\cong H^3(\mathfrak{g},\mathfrak{n},V)$$

between the abelian group of relative equivalence classes of crossed modules $\mu:\mathfrak{m}\to\mathfrak{n}$ with fixed quotient morphism $\pi:\mathfrak{n}\to\mathfrak{g}$ and fixed $\mathfrak{g}$-module $V=\mathrm{Ker}(\mu)$, and the third relative cohomology group $H^3(\mathfrak{g},\mathfrak{n},V)$.

Proof. Let $\mu:\mathfrak{m}\to\mathfrak{n}$ be a crossed module with quotient morphism $\pi:\mathfrak{n}\to\mathfrak{g}$ and with $\mathrm{Ker}(\mu)$ identified with V as a $\mathfrak{g}$-module. Kassel and Loday associate to the crossed module μ a 2-cochain f with values in V associated to a section $s:\mathfrak{g}\to\mathfrak{n}$ of π and a section $\sigma:\mathrm{Im}(\mu)=\mathrm{Ker}(\pi)\to\mathfrak{m}$ of μ. For all $x,y\in\mathfrak{g}$, they set

$$g(x,y)=\sigma([s(x),s(y)]-s[x,y])\in\mathfrak{m}.$$

This corresponds to our $\beta(x,y)$ in equation (2.1). Furthermore, they set for all $n\in\mathfrak{n}$,

$$\Psi(n)=\sigma(n-s\circ\pi(n))\in\mathfrak{m}.$$

With these notation, they define

$$f(n,n')=g(\pi(n),\pi(n'))-n'\cdot\Psi(n)+n\cdot\Psi(n')-[\Psi(n),\Psi(n')]-\Psi[n,n'].$$

Observe that with respect to Kassel–Loday's article, the sign of the last term is changed. This 2-cochain in defined on $\mathfrak{n}$. Let us show that f has values in V:

$$\begin{aligned}\mu f(n,n')&=\mu g(\pi n,\pi n')-\mu(n'\cdot\Psi(n))+\mu(n\cdot\Psi(n'))-\mu[\Psi(n),\Psi(n')]-\mu\Psi[n,n']\\&=\mu\sigma([s\pi(n),s\pi(n')]-s[\pi n,\pi n'])-[n',\mu\Psi(n')]+[n,\mu\Psi(n')]\\&\quad-[\mu\Psi(n),\mu\Psi(n')]-\mu\Psi[n,n']\\&=[s\pi(n),s\pi(n')]-s[\pi n,\pi n']-[n',n-s\pi(n)]+[n,n'-s\pi(n')]\\&\quad-[n-s\pi(n),n'-s\pi(n')]-[n,n']+s\pi[n,n']\\&=[s\pi(n),s\pi(n')]-s[\pi n,\pi n']-[n',n]+[n',s\pi(n)]+[n,n']\\&\quad-[n,s\pi(n')]-[n,n']+[s\pi(n),n']+[n,s\pi(n')]-[s\pi(n),s\pi(n')]\\&\quad-[n,n']+s\pi[n,n']\\&=0.\end{aligned}$$

Kassel and Loday need a relative cocycle. The complex of relative Lie algebra cohomology is by definition the following quotient complex:

$$0\to C^*(\mathfrak{g},V)\xrightarrow{\pi^*}C^*(\mathfrak{n},V)\xrightarrow{\kappa^*}C^*(\mathfrak{g},\mathfrak{n},V)\to 0.$$

The relative cocycle they associate to the crossed module μ is defined to be $\kappa^* f \in C^2(\mathfrak{g}, \mathfrak{n}, V)$. The cohomology in $C^2(\mathfrak{g}, \mathfrak{n}, V)$ is denoted $H^3(\mathfrak{g}, \mathfrak{n}, V)$. In order to show that $\kappa^* f$ is a cocycle, Kassel and Loday introduce a cochain k defined by

$$k(x,y,z) = \sum_{\text{cycl.}} g(x,[y,z]) + \sum_{\text{cycl.}} s(x)\cdot g(y,z) \in V.$$

Now the situation is the following:

$$\begin{array}{ccccc} C^2(\mathfrak{g},V) & \xrightarrow{\pi^*} & C^2(\mathfrak{n},V) & \xrightarrow{\kappa^*} & C^2(\mathfrak{g},\mathfrak{n},V) \\ \downarrow d & & \downarrow d & & \downarrow d \\ C^3(\mathfrak{g},V) & \xrightarrow{\pi^*} & C^3(\mathfrak{n},V) & \xrightarrow{\kappa^*} & C^3(\mathfrak{g},\mathfrak{n},V) \end{array}$$

Now let us show that $df = \pi^* k$:

$$\begin{aligned} df(n,n',n'') = {} & g(\pi n, \pi[n',n'']) + g(\pi n', \pi[n'',n]) + g(\pi n'', \pi[n,n']) \\ & + n\cdot g(\pi n', \pi n'') + n'\cdot g(\pi n'', \pi n) + n''\cdot g(\pi n, \pi n') \\ & - [n',n'']\cdot\Psi(n) - [n'',n]\cdot\Psi(n') - [n,n']\cdot\Psi(n'') \\ & - n\cdot(n''\cdot\Psi(n')) - n'\cdot(n\cdot\Psi(n'')) - n''\cdot(n'\cdot\Psi(n)) \\ & + n\cdot(n'\cdot\Psi(n'')) + n'\cdot(n''\cdot\Psi(n)) + n''\cdot(n\cdot\Psi(n')) \\ & + n\cdot\Psi([n',n'']) + n'\cdot\Psi([n'',n]) + n''\cdot\Psi([n,n']) \\ & - [\Psi(n),\Psi([n',n''])] - [\Psi(n'),\Psi([n'',n])] - [\Psi(n''),\Psi([n,n'])] \\ & - n\cdot[\Psi(n'),\Psi(n'')] - n'\cdot[\Psi(n''),\Psi(n)] - n''\cdot[\Psi(n),\Psi(n')] \\ & - \Psi([n,[n',n'']]) - \Psi([n',[n'',n]]) - \Psi([n'',[n,n']]) \\ & - n\cdot\Psi([n',n'']) - n'\cdot\Psi([n'',n]) - n''\cdot\Psi([n,n']). \end{aligned}$$

Observe that third, fourth and fifth line together cancel because of the property of the action. Further, sixth and last line are opposite to each other. The last but first line cancels because of the Jacobi identity. We are thus left with

$$\begin{aligned} df(n,n',n'') = {} & g(\pi n, \pi[n',n'']) + n\cdot g(\pi n', \pi n'') - [\Psi(n),\Psi([n',n''])] \\ & - n\cdot[\Psi(n'),\Psi(n'')] + \text{cycl.} \end{aligned} \tag{2.21}$$

What we need is

$$\pi^* k(n,n',n'') = g(\pi n, \pi[n',n'']) + (s\pi(n))\cdot g(\pi n', \pi n'') + \text{cycl.}$$

In order to make appear $\pi^* k(n,n',n'')$ in (2.21), compute the difference of the first three terms of (2.21) with $\pi^* k(n,n',n'')$:

$$(n - s\pi(n))\cdot g(\pi n', \pi n'') - [\Psi(n),\Psi([n',n''])]$$

$$
\begin{aligned}
&= \mu\Psi(n) \cdot g(\pi n', \pi n'') - [\Psi(n), \Psi([n', n''])] \\
&= [\Psi(n), \sigma([s\pi n', s\pi n''] - s\pi[n', n'']) - \sigma([n', n''] - s\pi[n', n''])] \\
&= -([s\pi n', s\pi n''] - [n', n'']) \cdot \Psi(n).
\end{aligned} \tag{2.22}
$$

The last term of Expression (2.21), up to cyclic permutations, is

$$
\begin{aligned}
-n' \cdot [\Psi(n''), \Psi(n)] + \text{cycl.} &= -[n' \cdot \Psi(n''), \Psi(n)] - [\Psi(n''), n' \cdot \Psi(n)] + \text{cycl.} \\
&= -[n' \cdot \Psi(n'') - n'' \cdot \Psi(n'), \Psi(n)] + \text{cycl.} \\
&= -\mu(n' \cdot \Psi(n'') - n'' \cdot \Psi(n')) \cdot \Psi(n) + \text{cycl.} \\
&= -([n', n'' - s\pi n''] - [n'', n' - s\pi n']) \cdot \Psi(n) + \text{cycl.} \\
&= -(2[n', n''] - [n', s\pi n''] + [n'', s\pi n']) \cdot \Psi(n) + \text{cycl.}
\end{aligned}
$$

Joining this expression and equation (2.22) gives for the difference of equation (2.21) with $\pi^* k(n, n', n'')$:

$$
\begin{aligned}
&([n', s\pi n''] - [n'', s\pi n'] - [n', n''] - [s\pi n', s\pi n'']) \cdot \Psi(n) + \text{cycl.} \\
&= -[n' - s\pi(n'), n'' - s\pi n''] \cdot \Psi(n) + \text{cycl.} \\
&= -[\mu\Psi(n'), \mu\Psi(n'')] \cdot \Psi(n) + \text{cycl.} \\
&= -\mu([\Psi(n'), \Psi(n'')]) \cdot \Psi(n) + \text{cycl.} \\
&= -[[\Psi(n'), \Psi(n'')], \Psi(n)] + \text{cycl.} = 0.
\end{aligned}
$$

This shows the identity $df = \pi^* k$. The identity then implies that $d\kappa^* f = \kappa^* df = \kappa^* \pi^* f = 0$ and, therefore, $\kappa^* f$ is a cocycle. What we have done so far (cf. Remark 2.9.3) can be resumed in the existence of a well-defined map

$$
\mathrm{crmod}(\mathfrak{g}, \mathfrak{n}, V) \to H^3(\mathfrak{g}, \mathfrak{n}, V), \quad [\mu : \mathfrak{m} \to \mathfrak{n}] \mapsto [\kappa^* f].
$$

Conversely, suppose given a cocycle in $C^2(\mathfrak{g}, \mathfrak{n}, V)$, which we lift to a cochain $f \in C^2(\mathfrak{n}, V)$. As $\kappa^* f$ is a cocycle, we have a cochain $k \in C^3(\mathfrak{g}, V)$ such that $df = \pi^* k$. In particular, the restriction of f to $\mathrm{Ker}(\pi) =: \mathfrak{l}$ gives a cocycle in $C^2(\mathfrak{l}, V)$. We get thus a Lie algebra structure on the direct sum $\mathfrak{m} = V \oplus \mathfrak{l}$, which makes it a central extension using the bracket

$$
[(z, l), (z', l')] = (f(l_1, l_2), [l_1, l_2]).
$$

Restriction onto $\mathfrak{n} \times \mathfrak{l}$, we obtain from f an action of $\mathfrak{n}$ on $\mathfrak{m}$ by the formula

$$
n \cdot (z, l) = (\pi(n) \cdot z + f(n, l), [n, l]);
$$

cf. equation (2.4). One checks with Lemma 2.3.1 that with these data, the map $\mu : \mathfrak{m} \to \mathfrak{n}$, given by $(z, l) \mapsto l$, is a crossed module. Indeed, both conditions (a) and (b) of Lemma 2.3.1 follow at once from the cocycle condition of f, written for the different elements of $\mathfrak{n} \cong \mathfrak{l} \oplus \mathfrak{g}$. The addition of a coboundary to f does not affect the (relative) equivalence class of this crossed module.

We thus get a well-defined map

$$H^3(\mathfrak{g}, \mathfrak{n}, V) \to \text{crmod}(\mathfrak{g}, \mathfrak{n}, V), \quad [\kappa^* f] \mapsto [\mu : \mathfrak{m} \to \mathfrak{n}].$$

By construction, we obtain as associated cohomology class to this crossed module the class of f. In the other direction, the two maps also compose to the identity. This is shown by constructing explicitly a morphism of crossed modules from the given crossed module to the one which we have constructed from its cohomology class. □

Remark 2.9.3. In this remark, we link elements of the above proof to the proof of Gerstenhaber's Theorem 2.2.1.

Our first claim is that f is an extension to $\mathfrak{n}$ of the cochain θ in equation (2.4). Indeed, equation (2.4) reads for $n \in \mathfrak{n}$ and $(0, n') \in \text{Im}(\mu) \subset \mathfrak{n}$ simply

$$n \cdot (0, n') = (\theta(n, n'), [n, n']).$$

Therefore, all we have to show is that $f(n, n') + \sigma[n, n'] = n \cdot \sigma(n')$ for all $n \in \mathfrak{n}$ and all $n' \in \text{Im}(\mu) = \text{Ker}(\pi)$.

To show this, note that for $n' \in \text{Im}(\mu) = \text{Ker}(\pi)$, we have $\Psi(n') = \sigma(n')$. Recall that $\text{Im}(\mu) = \text{Ker}(\pi)$ is an ideal, so that $\Psi[n, n'] = \sigma[n, n']$, and we have $n' = \mu(\sigma(n'))$, which implies

$$n' \cdot \Psi(n) = \mu(\sigma(n')) \cdot \Psi(n) = [\sigma(n'), \Psi(n)]$$

by the axioms of a crossed module. Observe also that for $n' \in \text{Im}(\mu) = \text{Ker}(\pi)$, $g(\pi(n), \pi(n')) = 0$. In total, we get

$$\begin{aligned} f(n, n') &= -[\sigma(n'), \Psi(n)] + n \cdot \sigma(n') - [\Psi(n), \sigma(n')] - \sigma([n, n']) \\ &= n \cdot \sigma(n') - \sigma([n, n']). \end{aligned}$$

This shows our claim. We conclude that the cochain f corresponds to the extension $\tilde{\theta}$ (defined in equation (2.5)) of θ to a cochain in $C^2(\mathfrak{n}, V)$.

Obviously, k is (up to a sign) close to our expression $d^{\mathfrak{m}}\beta$ appearing in Lemma 2.2.2. But as Kassel and Loday consider k to have values in V, one should identify it with the 3-cocycle γ.

The identity $df = \nu^* k$ is just $d^{\mathfrak{n}}\tilde{\theta} = \pi^* \gamma$ which was shown in Proposition 2.3.3.

The explicit modifications of the cocycles γ and $\chi = d_S\tilde{\theta}$ due to the choice of different sections which appear in Lemmas 2.2.4 and in 2.3.2 are easily used to show that the class of $\kappa^* f$ does not depend on the choice of the sections s and σ.

Lemma 2.2.5 shows in the same way that relatively equivalent crossed modules give rise to the same cohomology class.

Remark 2.9.4. The relation of the relative class $[\kappa^* f] \in H^3(\mathfrak{g}, \mathfrak{n}, V)$ to the absolute class $[d_S\tilde{\theta}] \in H^3(\mathfrak{g}, V)$ is given by the connecting homomorphism in the long exact sequence in cohomology associated to the short exact sequence of complexes

$$0 \to C^*(\mathfrak{g}, V) \xrightarrow{\pi^*} C^*(\mathfrak{n}, V) \xrightarrow{\kappa^*} C^*(\mathfrak{g}, \mathfrak{n}, V) \to 0.$$

Indeed, by definition of the connecting homomorphism ∂, the image $\partial(\kappa^* f)$ is obtained by first lifting $\kappa^* f$ to a cochain in $C^2(\mathfrak{n}, V)$, for which we may take f, then by taking its coboundary df and finally by identifying df with the image $\pi^* k$ of some element $k \in C^3(\mathfrak{g}, V)$. By definition, $[\partial(\kappa^* f)]$ is then set to be $[\partial(\kappa^* f)] = [k]$. We see that the connecting homomorphism ∂ sends Kassel–Loday's relative class to the absolute class.

In order to state this once again more neatly, introduce the forgetful map

$$D : \mathrm{crmod}(\mathfrak{g}, \mathfrak{n}, V) \to \mathrm{crmod}(\mathfrak{g}, V),$$

which forgets the fixed quotient morphism $\pi : \mathfrak{n} \to \mathfrak{g}$. It is well-defined. Then we have a commutative diagram

$$\begin{array}{ccc} H^3(\mathfrak{g}, \mathfrak{n}, V) & \xrightarrow{\cong} & \mathrm{crmod}(\mathfrak{g}, \mathfrak{n}, V) \\ \downarrow \partial & & \downarrow D \\ H^3(\mathfrak{g}, V) & \xrightarrow{\cong} & \mathrm{crmod}(\mathfrak{g}, V) \end{array}$$

Remark 2.9.5. We have seen in the preceding two remarks that the reasoning in the proof of the isomorphy between relative crossed modules and cohomology is largely equivalent to the reasoning in the absolute statement. In fact, Theorem 2.9.1 even implies Theorem 2.2.1.

Indeed, given an epimorphism $\pi : \mathfrak{n} \to \mathfrak{g}$ and a $\mathfrak{g}$-module V, consider the long exact sequence in cohomology induced by the short exact sequence of complexes

$$0 \to C^*(\mathfrak{g}, V) \to C^*(\mathfrak{n}, V) \to C^*(\mathfrak{g}, \mathfrak{n}, V) \to 0.$$

There is furthermore an exact sequence

$$\mathrm{Ext}(\mathfrak{n}, V) \to \mathrm{crmod}(\mathfrak{g}, \mathfrak{n}, V) \to \mathrm{crmod}(\mathfrak{g}, V) \to \mathrm{crmod}(\mathfrak{n}, V),$$

where V is viewed as an $\mathfrak{n}$-module via $\pi : \mathfrak{n} \to \mathfrak{g}$. Together, we have an exact ladder

$$\begin{array}{ccccccccc} \cdots & \longrightarrow & H^2(\mathfrak{n}, V) & \longrightarrow & H^3(\mathfrak{g}, \mathfrak{n}, V) & \longrightarrow & H^3(\mathfrak{g}, V) & \longrightarrow & \cdots \\ & & \uparrow & & \uparrow & & \uparrow & & \\ \cdots & \longrightarrow & \mathrm{Ext}(\mathfrak{n}, V) & \longrightarrow & \mathrm{crmod}(\mathfrak{g}, \mathfrak{n}, V) & \longrightarrow & \mathrm{crmod}(\mathfrak{g}, V) & \longrightarrow & \cdots \end{array}$$

Such an exact ladder exists for each choice of V and $\pi : \mathfrak{n} \to \mathfrak{g}$. Now suppose that V is an injective $\mathfrak{g}$-module. Then $H^3(\mathfrak{g}, V) = 0$. We claim that the isomorphism between relative crossed modules and relative 3-cohomology implies that in this case $\mathrm{crmod}(\mathfrak{g}, V) = 0$. Indeed, this follows from diagram chasing: Given a crossed module $\mu : \mathfrak{m} \to \mathfrak{n}$, there is a quotient morphism $\pi : \mathfrak{n} \to \mathfrak{g} := \mathrm{Coker}(\mu)$. We consider the exact ladder corresponding to this map π. Here, the class $[a]$ of $\mu : \mathfrak{m} \to \mathfrak{n}$ lifts to the class $[b]$ of a relative crossed module by construction. The class $[a]$ is sent to $0 \in H^3(\mathfrak{g}, V)$, thus $[b]$ is sent isomorphically (by the hypothesis that the relative isomorphism statement is true!) to $[c] \in H^3(\mathfrak{g}, \mathfrak{n}, V)$ which is sent to zero. By exactness, there exists $[d] \in H^2(\mathfrak{n}, V)$ which is sent to $[c]$. The class $[d]$ corresponds isomorphically to a class $[e] \in \mathrm{Ext}(\mathfrak{n}, V)$. But it is clear that $[e]$ must be sent to $[b]$, thus by exactness $[a] = 0$.

For the general case, embed V into an injective $\mathfrak{g}$-module I with quotient Q:

$$0 \to V \to I \to Q \to 0.$$

This short exact sequence of coefficients induces long exact sequences both in cohomology and gives an exact ladder

$$\begin{array}{ccccccc}
\ldots H^2(\mathfrak{g}, V) & \longrightarrow & H^2(\mathfrak{g}, Q) & \longrightarrow & H^3(\mathfrak{g}, V) & \longrightarrow & H^3(\mathfrak{g}, I) \ldots \\
\uparrow & & \uparrow & & \uparrow & & \uparrow \\
\ldots \mathrm{Ext}(\mathfrak{g}, V) & \longrightarrow & \mathrm{Ext}(\mathfrak{g}, Q) & \longrightarrow & \mathrm{crmod}(\mathfrak{g}, V) & \longrightarrow & \mathrm{crmod}(\mathfrak{g}, I) \ldots
\end{array}$$

Here, we have $H^3(\mathfrak{g}, I) = 0$ and $\mathrm{crmod}(\mathfrak{g}, I) = 0$ by the preceding claim, and the maps $\mathrm{Ext}(\mathfrak{g}, Q) \to H^2(\mathfrak{g}, Q)$ and $\mathrm{Ext}(\mathfrak{g}, V) \to H^2(\mathfrak{g}, V)$ are isomorphisms by the standard interpretation of H^2 in terms of abelian extensions; see Appendix B. Therefore, we can conclude by the Five Lemma that $H^3(\mathfrak{g}, V) \cong \mathrm{crmod}(\mathfrak{g}, V)$, and this shows how Theorem 2.9.1 implies Theorem 2.2.1.

2.10 Strict Lie 2-algebras

In this section, we describe the relation of crossed modules of Lie algebras to a categorified version of Lie algebras, namely *Lie 2-algebras* and cat^1-Lie algebras. *Categorification* means the creation of a new algebraic structure by replacing sets and maps in the definition of some standard algebraic structure by categories and functors. This opens up the possibility for relaxing some axioms of this algebraic structure, that is, rather than imposing them as equations, one imposes the axioms only up to a natural isomorphism. This describes *weak* categorification. In order to be able to work with such a weak categorification, one imposes then coherence conditions on these natural isomorphisms. On the other hand, when one still imposes the old axioms as equations, one speaks about *strict* categorification.

The main point of this section is that crossed modules of Lie algebras are in one-to-one correspondence to strict Lie 2-algebras; see Theorem 2.10.2, and to cat^1-Lie algebras. In order to avoid introducing too much of the heavy categorical machinery, let us concentrate on the object side and neglect the morphism side of almost all statements.

2.10.1 Crossed modules and cat^1-Lie algebras

As a warm-up in the categorical structures, we will consider in this section cat^1-Lie algebras and show that the notion of a cat^1-Lie algebra is equivalent to the notion of a crossed module.

Definition 2.10.1. A cat^1-*Lie algebra* $(\mathfrak{g}_1, \mathfrak{g}_0, s, t)$ consists of a Lie algebra $\mathfrak{g}_1$ together with a Lie subalgebra $\mathfrak{g}_0$ and two homomorphisms of Lie algebras $s, t : \mathfrak{g}_1 \to \mathfrak{g}_0$ such that

$$s|_{\mathfrak{g}_0} = t|_{\mathfrak{g}_0} = \mathrm{id}_{\mathfrak{g}_0}, \quad \text{and} \quad [\mathrm{Ker}(s), \mathrm{Ker}(t)] = 0.$$

Theorem 2.10.1. cat^1-*Lie algebras are in one-to-one correspondence with crossed modules of Lie algebras.*

Proof. The proof of the theorem is in fact a half-way journey through the proof of Theorem 2.10.2. Indeed, given a crossed module of Lie algebras $\mu : \mathfrak{m} \to \mathfrak{n}$, we define $\mathfrak{g}_1 := \mathfrak{m} \rtimes \mathfrak{n}$, the semidirect product of $\mathfrak{n}$ and $\mathfrak{m}$, using the action of $\mathfrak{n}$ on $\mathfrak{m}$ given in the crossed module. Furthermore, let $\mathfrak{g}_0 := \mathfrak{n}$. The map $s : \mathfrak{g}_1 \to \mathfrak{g}_0$ is defined to be the projection onto $\mathfrak{n}$, while the map $t : \mathfrak{g}_1 \to \mathfrak{g}_0$ is defined by $(m, n) \mapsto \mu(m) + n$.

Let us compute $[\mathrm{Ker}(s), \mathrm{Ker}(t)]$. It is obvious that $\mathrm{Ker}(s) = \{(m, 0) \in \mathfrak{m} \rtimes \mathfrak{n}\}$ and that

$$\mathrm{Ker}(t) = \{(m', n') \in \mathfrak{m} \rtimes \mathfrak{n} \mid \mu(m') = -n'\}.$$

But the bracket of two such elements gives

$$\begin{aligned}[(m, 0), (m', n')] &= ([m, m'] - n' \cdot m, 0)\\ &= ([m, m'] + \mu(m') \cdot m, 0)\\ &= ([m, m'] + [m', m], 0) = (0, 0).\end{aligned}$$

Conversely, given a cat^1-Lie algebra $(\mathfrak{g}_1, \mathfrak{g}_0, s, t)$, we define $\mathfrak{n} := \mathfrak{g}_0$, $\mathfrak{m} := \mathrm{Ker}(s)$ and $\mu := t|_{\mathrm{Ker}(s)}$. The subalgebra $\mathfrak{n}$ acts via the bracket on $\mathfrak{g}_1$ by derivations. This action preserves $\mathrm{Ker}(s)$, as $s([x, y]) = [s(x), s(y)] = [s(x), 0] = 0$ for all $x \in \mathfrak{n}$ and $y \in \mathrm{Ker}(s)$. Condition (a) for a crossed module reads for all $n \in \mathfrak{n}$ and all $m \in \mathfrak{m}$,

$$\mu(n \cdot m) = t([n, m]) = [t(n), t(m)] = [n, t(m)] = [n, \mu(m)].$$

Condition (b) reads for all $m, m' \in \mathfrak{m}$,

$$\mu(m) \cdot m' = [t(m), m'] = [m + (t(m) - m), m'] = [m, m'],$$

where we have used $(t(m) - m) \in \mathrm{Ker}(t)$, $m' \in \mathrm{Ker}(s)$ and $[\mathrm{Ker}(t), \mathrm{Ker}(s)] = 0$. □

2.10.2 Strict 2-vector spaces and 2-term complexes

Fix a field $\mathbb{K}$ of characteristic 0. There exist competing definitions for a 2-vector space, but for us here, a *2-vector space* V over $\mathbb{K}$ is simply a category object in Vect, the category of vector spaces (cf. Definition 5 in [4]). This means that V consists of a vector space of *arrows* V_1, a vector space of *objects* V_0, linear maps $V_1 \underset{t}{\overset{s}{\rightrightarrows}} V_0$ called *source* and *target*, a linear map $i : V_0 \to V_1$, called *object inclusion*, and a linear map

$$m : V_1 \times_{V_0} V_1 \to V_1,$$

which is called the *categorical composition*. These data are supposed to satisfy the usual axioms of a category.

An equivalent point of view is to view a 2-vector space as a 2-term complex of vector spaces $d : C_1 \to C_0$. The equivalence is spelt out in Section 3 of [4]: One passes from a category object in Vect (given by $V_1 \underset{t}{\overset{s}{\rightrightarrows}} V_0$, $i : V_0 \to V_1$, etc.) to a 2-term complex $d : C_1 \to C_0$ by taking $C_1 := \mathrm{Ker}(s)$, $d := t|_{\mathrm{Ker}(s)}$ and $C_0 = V_0$. In the reverse direction, to a given 2-term complex $d : C_1 \to C_0$, one associates $V_1 = C_0 \oplus C_1$, $V_0 = C_0$, $s(c_0, c_1) = c_0$, $t(c_0, c_1) = c_0 + d(c_1)$, and $i(c_0) = (c_0, 0)$. The only subtle point is here that the categorical composition m is already determined by $V_1 \underset{t}{\overset{s}{\rightrightarrows}} V_0$ and $i : V_0 \to V_1$ (see Lemma 6 in [4]). Namely, writing an arrow $c_1 =: f$ with $s(f) = x$, $t(f) = y$, that is, $f : x \mapsto y$, one denotes the *arrow part* of f by $\vec{f} := f - i(s(f))$, and for two composable arrows $f, g \in V_1$, the composition m is then defined by

$$f \circ g := m(f, g) := i(x) + \vec{f} + \vec{g}.$$

Observe that we use here Baez–Crans convention on the composition, that is, we compose from left to right (the source of $f \circ g$ is the source of f and not the source of g like in usual composition).

2.10.3 Strict Lie 2-algebras and crossed modules

Definition 2.10.2. A *strict Lie 2-algebra* is a category object in the category Lie of Lie algebras over $\mathbb{K}$.

This means that it is the data of two Lie algebras, $\mathfrak{g}_0$, the *Lie algebra of objects*, and $\mathfrak{g}_1$, the *Lie algebra of arrows*, together with morphisms of Lie algebras $s, t : \mathfrak{g}_1 \to \mathfrak{g}_0$, *source* and *target*, a morphism $i : \mathfrak{g}_0 \to \mathfrak{g}_1$, the *object inclusion*, and a morphism $m : \mathfrak{g}_1 \times_{\mathfrak{g}_0} \mathfrak{g}_1 \to \mathfrak{g}_1$, the *composition of arrows*, such that the usual axioms of a category are satisfied.

Theorem 2.10.2. *Strict Lie 2-algebras are in one-to-one correspondence with crossed modules of Lie algebras.*

Proof. Given a Lie 2-algebra $\mathfrak{g}_1 \underset{t}{\overset{s}{\rightrightarrows}} \mathfrak{g}_0$, $i : \mathfrak{g}_0 \to \mathfrak{g}_1$, the corresponding crossed module is defined by

$$\mu := t|_{\mathrm{Ker}(s)} : \mathfrak{m} := \mathrm{Ker}(s) \to \mathfrak{n} := \mathfrak{g}_0.$$

The action of $\mathfrak{n}$ on $\mathfrak{m}$ is given by

$$n \cdot m := [i(n), m],$$

for $n \in \mathfrak{n}$ and $m \in \mathfrak{m}$ (where the bracket is taken in $\mathfrak{g}_1$). This is well-defined and an action by derivations. Axiom (a) follows from

$$\mu(n \cdot m) = \mu([i(n), m]) = [\mu \circ i(n), \mu(m)] = [n, \mu(m)].$$

Axiom (b) follows from

$$\mu(m) \cdot m' = [i \circ \mu(m), m'] = [i \circ t(m), m'] = [m + r, m'] = [m, m']$$

by writing $i \circ t(m) = m + r$ with $r \in \mathrm{Ker}(t)$ and by using that $\mathrm{Ker}(t)$ and $\mathrm{Ker}(s)$ in a Lie 2-algebra commute. This is shown in Lemma 2.10.3 after the proof.

On the other hand, given a crossed module of Lie algebras $\mu : \mathfrak{m} \to \mathfrak{n}$, associate to it

$$\mathfrak{m} \rtimes \mathfrak{n} \underset{t}{\overset{s}{\rightrightarrows}} \mathfrak{n}, \quad i : \mathfrak{n} \to \mathfrak{m} \rtimes \mathfrak{n}$$

with $s(m, n) = n$, $t(m, n) = \mu(m) + n$, $i(n) = (0, n)$, where the semidirect product Lie algebra $\mathfrak{m} \rtimes \mathfrak{n}$ is built from the given action of $\mathfrak{n}$ on $\mathfrak{m}$. Let us emphasize that $\mathfrak{m} \rtimes \mathfrak{n}$ is built from the Lie algebra $\mathfrak{n}$ and the $\mathfrak{n}$-module $\mathfrak{m}$; the bracket of $\mathfrak{m}$ does not intervene here. The composition of arrows is already encoded in the underlying structure of 2-vector space, as remarked in the previous subsection. In the second lemma below, we show that the composition is a morphism of Lie algebras. □

Lemma 2.10.3. $[\mathrm{Ker}(s), \mathrm{Ker}(t)] = 0$ *in a Lie 2-algebra.*

Proof. The fact that the composition of arrows is a homomorphism of Lie algebras gives the following "middle four exchange" (or functoriality) property

$$[g_1, g_2] \circ [f_1, f_2] = [g_1 \circ f_1, g_2 \circ f_2]$$

for composable arrows $f_1, f_2, g_1, g_2 \in \mathfrak{g}_1$. Now suppose that $g_1 \in \mathrm{Ker}(s)$ and $f_2 \in \mathrm{Ker}(t)$. Then denote by f_1 and by g_2 the identity (with respect to the composition) in $0 \in \mathfrak{g}_0$. As these are identities, we have $g_1 = g_1 \circ f_1$ and $f_2 = g_2 \circ f_2$. On the other hand, i is a morphism of Lie algebras and sends $0 \in \mathfrak{g}_0$ to the $0 \in \mathfrak{g}_1$. Therefore, we may conclude

$$[g_1, f_2] = [g_1 \circ f_1, g_2 \circ f_2] = [g_1, g_2] \circ [f_1, f_2] = 0.$$ □

Lemma 2.10.4. *Given a crossed module of Lie algebras $\mu : \mathfrak{m} \to \mathfrak{n}$, the composition of the underlying 2-vector space of the precategory* $\mathfrak{m} \rtimes \mathfrak{n} \overset{s}{\underset{t}{\rightrightarrows}} \mathfrak{n}$ *is a morphism of Lie algebras.*

Proof. Every morphism $f = (m, n) \in \mathfrak{m} \rtimes \mathfrak{n}$ is described in the underlying 2-vector space by its starting point $s(f) = s(m, n) = n$ and its arrow part $\vec{f} = f - i(s(f)) = (m, 0)$. We have to show the double-four-exchange law, that is, for composable arrows f_1, f_2, g_1, g_2 we need to show

$$[g_1 \circ f_1, g_2 \circ f_2] = [g_1, g_2] \circ [f_1, f_2].$$

Here, it is understood that $g_i : x_i \mapsto y_i$ and $f_i : y_i \mapsto z_i$ for $i = 1, 2$. The fact that the two are composable means that $t(g_i) = y_i = s(f_i)$ for $i = 1, 2$. Now translate all this into elements of the semidirect product. We call $g_i = (m_i, n_i)$ and $f_i = (m_i', n_i')$ for $i = 1, 2$ and the composability means now $t(m_i, n_i) = \mu(m_i) + n_i = n_i' = s(f_i)$ for $i = 1, 2$.

We compute

$$\begin{aligned}
[g_1 \circ f_1, g_2 \circ f_2] &= [i(s(g_1)) + \vec{g_1} + \vec{f_1}, i(s(g_2)) + \vec{g_2} + \vec{f_2}] \\
&= [(0, n_1) + (m_1, 0) + (m_1', 0), (0, n_2) + (m_2, 0) + (m_2', 0)] \\
&= [(m_1 + m_1', n_1), (m_2 + m_2', n_2)] \\
&= ([m_1 + m_1', m_2 + m_2'] + n_1 \cdot (m_2 + m_2') - n_2 \cdot (m_1 + m_1'), [n_1, n_2]).
\end{aligned}$$

This is to compare with

$$\begin{aligned}
[g_1, g_2] \circ [f_1, f_2] &= i(s([g_1, g_2])) + \overrightarrow{[g_1, g_2]} + \overrightarrow{[f_1, f_2]} \\
&= ([m_1, m_2] + n_1 \cdot m_2 - n_2 \cdot m_1 + [m_1', m_2'] + n_1' \cdot m_2' - n_2' \cdot m_1', [n_1, n_2])
\end{aligned}$$

because

$$\begin{aligned}
[g_1, g_2] &= [(m_1, n_1), (m_2, n_2)] \\
&= ([m_1, m_2] + n_1 \cdot m_2 - n_2 \cdot m_1, [n_1, n_2])
\end{aligned}$$

and

$$\overrightarrow{[g_1, g_2]} = ([m_1, m_2] + n_1 \cdot m_2 - n_2 \cdot m_1, 0)$$

and

$$\overrightarrow{[f_1, f_2]} = ([m_1', m_2'] + n_1' \cdot m_2' - n_2' \cdot m_1', 0).$$

The equality of these two expressions now follows from the following computation, where we use the $\mu(m_i) + n_i = n_i'$ for $i = 1, 2$ and property (b) (i. e., $\mu(m) \cdot m' = [m, m']$):

$$\begin{aligned}
&[m_1', m_2] + [m_1, m_2'] + n_1 \cdot m_2' - n_2 \cdot m_1' - n_1' \cdot m_2 + n_2' \cdot m_1' \\
&= [m_1', m_2] + [m_1, m_2'] + n_1 \cdot m_2' - n_2 \cdot m_1' - (n_1 + \mu(m_1)) \cdot m_2' + (n_2 + \mu(m_2)) \cdot m_1' \\
&= [m_1', m_2] + [m_1, m_2'] - \mu(m_1) \cdot m_2 + \mu(m_2) \cdot m_1' = 0.
\end{aligned}$$

□

Remark 2.10.3. (1) It is implicit in the previous proof that starting from a crossed module $\mu : \mathfrak{m} \to \mathfrak{n}$, passing to the Lie 2-algebra $\mathfrak{g}_1 \overset{s}{\underset{t}{\rightrightarrows}} \mathfrak{g}_0$, $i : \mathfrak{g}_0 \to \mathfrak{g}_1$ (and thus forgetting the bracket on $\mathfrak{m}$!), one may finally reconstruct the bracket on $\mathfrak{m}$. This is due to the fact that the bracket on $\mathfrak{m}$ is encoded in the action by the property (b) of a crossed module

$$[m, m'] = \mu(m) \cdot m'.$$

(2) It is shown in Section 2.5 that for a given third cohomology class (or, equivalently, for a given equivalence class of crossed modules), there is a crossed module $\mu : \mathfrak{m} \to \mathfrak{n}$ representing this class such that the bracket on $\mathfrak{m}$ is trivial.

2.10.4 Semistrict Lie 2-algebras and 2-term L_∞-algebras

An equivalent point of view is to view a strict Lie 2-algebra as a Lie algebra object in the category Cat of (small) categories. From this second point of view, we have a functorial Lie bracket which is supposed to be antisymmetric and must fulfill the Jacobi identity. Weakening the antisymmetry axiom and the Jacobi identity up to coherent isomorphisms leads then to semistrict Lie 2-algebras (here antisymmetry holds strictly, but Jacobi is weakened), hemistrict Lie 2-algebras (here Jacobi holds strictly, but antisymmetry is weakened) or even to (general) Lie 2-algebras (both axioms are weakened). Let us record the definition of a semistrict Lie 2-algebra (see [4] Definition 22).

Definition 2.10.4. A *semistrict Lie 2-algebra* consists a 2-vector space L together with a skew symmetric, bilinear and functorial bracket $[,] : L \times L \to L$ and a completely antisymmetric trilinear natural isomorphism

$$J_{x,y,z} : [[x, y], z] \to [x, [y, z]] + [[x, z], y],$$

called the *Jacobiator*. The Jacobiator is required to satisfy the Jacobiator identity (see [4] Definition 22).

Semistrict Lie 2-algebras together with morphisms of semistrict Lie 2-algebras (see Definition 23 in [4]) form a strict 2-category (see Proposition 25 in [4]). Strict Lie algebras form a full sub-2-category of this 2-category; see Proposition 42 in [4]. In order to view a strict Lie 2-algebra $\mathfrak{g}_1 \to \mathfrak{g}_0$ as a semistrict Lie 2-algebra, the functorial bracket is constructed for $f : x \mapsto y$ and $g : a \mapsto b$, $f, g \in \mathfrak{g}_1$ and $x, y, a, b \in \mathfrak{g}_0$ by defining its source $s([f, g])$ and its arrow part $\overrightarrow{[f, g]}$ to be $s([f, g]) := [x, a]$ and $\overrightarrow{[f, g]} := [x, \vec{g}] + [\vec{f}, b]$ (see the proof of Theorem 36 in [4]). By construction, it is compatible with the composition, that is, functorial.

Remark 2.10.5. One observes that the functorial bracket on a strict Lie 2-algebra $\mathfrak{g}_1 \to \mathfrak{g}_0$ is constructed from the bracket in $\mathfrak{g}_0$, and the bracket between $\mathfrak{g}_1$ and $\mathfrak{g}_0$, but does not involve the bracket on $\mathfrak{g}_1$ itself.

There is a 2-vector space underlying every semi-strict Lie 2-algebras, thus one may ask which structure is inherited from a semistrict Lie 2-algebra by the corresponding 2-term complex of vector spaces. This leads us to 2-term L_∞-algebras; see [4] Theorem 36.

Definition 2.10.6. An L_∞*-algebra* is a graded vector space L together with a sequence $l_k(x_1,\dots,x_k)$, $k>0$, of graded antisymmetric operations of degree $k-2$ such that the following identity is satisfied:

$$\sum_{k=1}^{n}(-1)^k \sum_{\substack{i_1<\cdots<i_k;\, j_1<\cdots<j_{n-k}\\ \{i_1,\dots,i_k\}\cup\{j_1,\dots,j_{n-k}\}=\{1,\dots,n\}}} (-1)^\epsilon l_n(l_k(x_{i_1},\dots,x_{i_k}),x_{j_1},\dots,x_{j_{n-k}})=0.$$

Here, the sign $(-1)^\epsilon$ equals the product of the sign of the shuffle permutation and the Koszul sign.

We will be mainly concerned with 2-term L_∞-algebras. These are L_∞-algebras L such that the graded vector space L consists only of two components L_0 and L_1. An L_∞-algebra $L = L_0 \oplus L_1$ has at most l_1, l_2 and l_3 as its nontrivial "brackets." l_1 is a differential (i. e., here just a linear map $L_0 \to L_1$), l_2 is a bracket with components $[,] : L_0 \otimes L_0 \to L_0$, $[,] : L_1 \otimes L_0 \to L_1$ and $[,] : L_0 \times L_1 \to L_1$, and l_3 is some kind of 3-cocycle $l_3 : L_0 \otimes L_0 \otimes L_0 \to L_1$. More precisely, in case $l_1 = 0$, L_0 is a Lie algebra, L_1 is an L_0-module and l_3 is then an actual 3-cocycle. This kind of 2-term L_∞-algebra is called *skeletal*; see Section 6 in [4]. The complete list of axioms satisfied by l_1, l_2 and l_3 in a 2-term L_∞-algebra is given in Lemma 33 of [4].

As said before, the passage from a 2-vector space to its associated 2-term complex induces a passage from semistrict Lie 2-algebras to 2-term L_∞-algebras, which turns out to be an equivalence of 2-categories (see [4] Theorem 36).

Theorem 2.10.5. *The 2-categories of semi-strict Lie 2-algebras and of 2-term L_∞-algebras are equivalent.*

Remark 2.10.7. In particular, restricting to the sub-2-category of strict Lie 2-algebras, there is an equivalence between crossed modules of Lie algebras and 2-term L_∞-algebras with trivial l_3. In other words, there is an equivalence between crossed modules and differential graded Lie algebras (with two terms).

2.10.5 Classification of semistrict Lie 2-algebras

Baez and Crans show in [4] that every semistrict Lie 2-algebra is equivalent to a skeletal Lie 2-algebra (i. e., one where the differential d of the underlying complex of vector spaces is zero). Then they go on showing that skeletal Lie 2-algebras are classified by triples consisting of an honest Lie algebra $\mathfrak{g}$, a $\mathfrak{g}$-module V and a class $[\gamma] \in H^3(\mathfrak{g}, V)$.

This is achieved using the homotopy equivalence of the underlying complex of vector spaces with its cohomology. In total, they get in this way a classification, up to equivalence, of semistrict Lie 2-algebras in terms of triples $(\mathfrak{g}, V, [\gamma])$.

On the other hand, strict Lie 2-algebras are in one-to-one correspondence with crossed modules of Lie algebras, as we have seen in a previous subsection. In conclusion, there are two ways to classify strict Lie 2-algebras: by the associated crossed module or, regarding them as special semistrict Lie 2-algebras, by Baez–Crans classification. Let us show here that these two classifications are compatible, that is, that they lead to the same triple $(\mathfrak{g}, V, [\gamma])$.

For this, let us denote by `sLie2` the 2-category of strict Lie 2-algebras, by `ssLie2` the 2-category of semistrict Lie 2-algebras, by `sssLie2` the 2-category of skeletal semistrict Lie 2-algebras, by `triples` the (trivial) 2-category of triples of the above form $(\mathfrak{g}, V, [\gamma])$, and by `crmod` the (trivial) 2-category of crossed modules of Lie algebras.

Theorem 2.10.6. *The following diagram is commutative:*

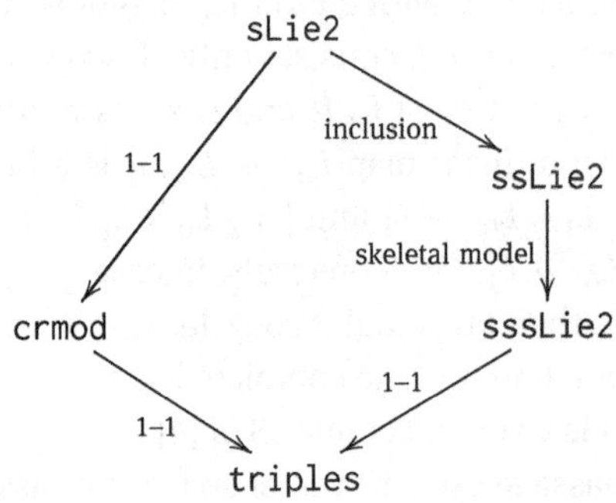

Proof. Let us first describe the arrows. The arrow `sLie2` $\to$ `crmod` has been investigated in Theorem 2.10.2. The arrow `crmod` $\to$ `triples` sends a crossed module $\mu : \mathfrak{m} \to \mathfrak{n}$ to the triple $(\mathrm{Coker}(\mu), \mathrm{Ker}(\mu), [\gamma])$, where the cohomology class $[\gamma] \in H^3(\mathrm{Coker}(\mu), \mathrm{Ker}(\mu))$ is defined choosing sections—the procedure is described in detail in Section 2.2. The arrow `ssLie2` $\to$ `sssLie2` is the choice of a skeletal model for a given semistrict Lie 2-algebra—it is given by the homotopy equivalence of the underlying 2-term complex with its cohomology displayed in the extremal lines of the following diagram:

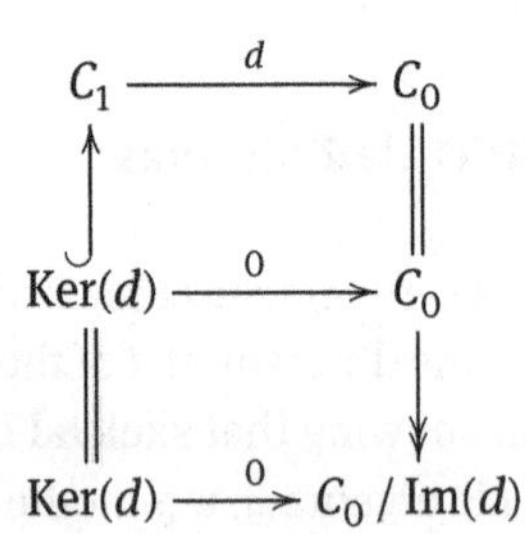

Now let $d : C_1 \to C_0$ with some bracket $[,]$ and $l_3 = 0$ be a 2-term L_∞-algebra corresponding to regarding a strict Lie 2-algebra as a semistrict Lie 2-algebra, and build its skeletal model. The model comes together with a morphism of semistrict Lie 2-algebras (ϕ_2, ϕ_1, ϕ_0) given by

$$\begin{array}{ccc} C_1 & \xrightarrow{d} & C_0 \\ {\scriptstyle\phi_1}\uparrow & & \uparrow{\scriptstyle\phi_0} \\ \mathrm{Ker}(d) & \xrightarrow{0} & C_0 \,/\, \mathrm{Im}(d) \end{array}$$

Here, $\phi_0 =: \sigma$ is a linear section of the quotient map. The structure of a semi-strict Lie 2-algebra is transferred to the lower line in order to make (ϕ_2, ϕ_1, ϕ_0) a morphism of semistrict Lie 2-algebras. In order to compute now the l_3 term of the lower semistrict Lie 2-algebra, one first finds that (first equation in Definition 34 of [4]) $\phi_2 : C_0 / \mathrm{Im}(d) \times C_0 / \mathrm{Im}(d) \to C_1$ is such that

$$d\phi_2(x, y) = \sigma[x, y] - [\sigma(x), \sigma(y)],$$

the failure of the section σ to be a homomorphism of Lie algebras. Then l_3 is related to ϕ_2 by the second formula in Definition 34 of [4]. This gives here

$$l_3(x, y, z) = d\phi_2(x, y, z)$$

for $x, y, z \in C_0 / \mathrm{Im}(d)$. As usual, d is the formal Chevalley–Eilenberg differential of the cochain $\phi_2 : C_0 / \mathrm{Im}(d) \times C_0 / \mathrm{Im}(d) \to C_1$ with values in C_1 as if C_1 was a $C_0 / \mathrm{Im}(d)$-module (which is usually not the case). This is exactly the expression of the cocycle γ associated to the crossed module of Lie algebras $d : C_1 \to C_0$ obtained using the section σ; see Section 2.2. □

Corollary 2.10.7. *Every semistrict Lie 2-algebra is equivalent (as an object of the 2-category* `ssLie2`*) to a strict Lie 2-algebra.*

Remark 2.10.8. This corollary is known to hold for abstract reasons (MacLane's coherence theorem). Here, we have proved a result which is somewhat more refined: The procedure to strictify a semistrict Lie 2-algebra is rather easy to perform. First, the Jacobiator gives a 3-cohomology class. Then one has to construct a crossed module corresponding to this cohomology class. This can be done in several way, using free Lie algebras, using injective modules, etc. and one may adapt the construction method to the problem at hand.

Remark 2.10.9. Let us describe the string Lie 2-algebra as an example of semistrict Lie 2-algebras. A skeletal model of the string Lie 2-algebra may be described as $0 : \mathfrak{g}_1 \to \mathfrak{g}_0$, where $\mathfrak{g}_1 = \mathbb{R}$ and $\mathfrak{g}_0$ some fixed real simple Lie algebra. $\mathbb{R}$ is seen as a trivial

$\mathfrak{g}_0$-module, and $l_3(x,y,z) = \langle [x,y],z\rangle$ for all $x,y,z \in \mathfrak{g}_0$ is the Cartan 3-cocycle (defined starting from the Killing form $\langle,\rangle$ and the bracket $[,]$ of $\mathfrak{g}_0$). On the other hand, as a semistrict Lie 2-algebra, the string Lie algebra has the Cartan cocycle as its Jacobiator. On still another hand, as a strict Lie 2-algebra, the string Lie algebra corresponds to one of the crossed modules in Section 2.8.1 and its following subsection.

2.11 Exercises

Exercise 2.11.1. Show that $\mathrm{crmod}(\mathfrak{g},-)$ is an additive functor and in particular that we have for an equivalence class $[\mu : \mathfrak{m} \to \mathfrak{n}] \in \mathrm{crmod}(\mathfrak{g},V)$ and for $\mathfrak{g}$-module morphisms $\alpha_1,\alpha_2 : V \to V'$

$$(\alpha_1+\alpha_2)[\mu : \mathfrak{m} \to \mathfrak{n}] = \alpha_1[\mu : \mathfrak{m} \to \mathfrak{n}] + \alpha_2[\mu : \mathfrak{m} \to \mathfrak{n}].$$

Exercise 2.11.2. State and prove the *abstract Bianchi identity* from the proof of Proposition 2.3.3. Use it to prove $d_{S_1}R_S = 0$ in Definition 2.4.8.

Exercise 2.11.3. Complete the proof of Lemma 2.6.1 using Remark 2.6.6.

Exercise 2.11.4. Let $\mathfrak{g}$ be a Lie algebra and $U\mathfrak{g}$ be its universal enveloping algebra. Let M be a $U\mathfrak{g}$-bimodule and let M^{ad} be the corresponding *adjoint* $\mathfrak{g}$-module, that is, $x \in \mathfrak{g}$ acts on $m \in M$ by

$$x \cdot m := xm - mx.$$

Define using antisymmetrization a map of complexes

$$(C^*(U\mathfrak{g},M),d) \to (C^*(\mathfrak{g},M^{\mathrm{ad}}),d)$$

from the Hochschild complex of $U\mathfrak{g}$ to the Chevalley–Eilenberg complex of $\mathfrak{g}$; cf. proof of Theorem 2.6.3.

Exercise 2.11.5. As an explicit example of the crossed module

$$0 \to \mathfrak{z}(\mathfrak{g}) \to \mathfrak{g} \to \mathfrak{aut}(\mathfrak{g}) \to \mathfrak{out}(\mathfrak{g}) \to 0,$$

consider as Lie algebra $\mathfrak{g}$ the *loop algebra* $L\mathfrak{k} := C^\infty(S^1,\mathfrak{k})$ of some finite-dimensional Lie algebra $\mathfrak{k}$. Writing its elements $f \otimes x$ with $f \in C^\infty(S^1)$ and $x \in \mathfrak{k}$, the *current bracket* reads

$$[f\otimes x, g\otimes y] := fg \otimes [x,y].$$

For a simple complex Lie algebra $\mathfrak{k}$, $L\mathfrak{k}$ has a universal central extension $\widetilde{L\mathfrak{k}}$ given by the *Kac–Moody cocycle*

$$\omega(f\otimes x, g\otimes y) := \int_0^1 \kappa(x,y) f'(t) g(t)\, dt,$$

where κ is the Killing form on $\mathfrak{k}$; cf. equation (2.14). Compute $\mathfrak{aut}(\widetilde{L\mathfrak{k}})$, $\mathfrak{out}(\widetilde{L\mathfrak{k}})$ and write down the corresponding crossed module. Is its 3-class nontrivial?

Exercise 2.11.6. Show the commutativity of the diagrams at the end of Section 2.9, in the spirit of Exercise 1.11.9 for the corresponding objects for groups.

Exercise 2.11.7. This exercise retraces the lines of the main theorem from Gerstenhaber's article [35].

Associate to a Lie algebra $\mathfrak{g}$ and a $\mathfrak{g}$-module V a cohomological δ-functor in the following way. To $(\mathfrak{g}, V)$, we associate the class of exact sequences $\mathrm{Seq}^n(\mathfrak{g}, V)$ with n terms which are obtained from splicing together an $n-1$ term exact sequence of $\mathfrak{g}$-modules

$$0 \to V_1 = V \to \cdots \to V_{n-1} \to 0$$

with an abelian extension

$$0 \to V_{n-1} \to V_{n-1} \times_\alpha \mathfrak{g} \to \mathfrak{g} \to 0$$

of $\mathfrak{g}$ by V_{n-1} by means of the 2-cocycle $\alpha \in Z^2(\mathfrak{g}, V_{n-1})$. Pass to classes with respect to the natural equivalence relation on $\mathrm{Seq}^n(\mathfrak{g}, V)$ to obtain a set $\widetilde{H}^n(\mathfrak{g}, V)$.

(a) Show that the functor $(\mathfrak{g}, V) \mapsto \widetilde{H}^{n+1}(\mathfrak{g}, V)$ satisfies the axioms of the Ext-functor $\mathrm{Ext}^n_{U\mathfrak{g}}(k, V)$ and deduce that

$$\widetilde{H}^{n+1}(\mathfrak{g}, V) \cong H^n(\mathfrak{g}, V).$$

(b) Deduce that the subfunctor $(\mathfrak{g}, V) \mapsto \widetilde{H}^4(\mathfrak{g}, V)$ given by equivalence classes of exact sequences spliced together from a short exact coefficients sequence and an abelian extension is isomorphic to $H^3(\mathfrak{g}, V)$. How to obtain from this result Gerstenhaber's Theorem 2.2.1?

As a further question in this direction, one may show that the splicing together of an exact sequence of $\mathfrak{g}$-modules (starting with V) with a crossed module with cokernel $\mathfrak{g}$ also gives a cohomological δ-functor which is (up to a degree shift) isomorphic to $H^*(\mathfrak{g}, V)$. From the Yoneda Ext point of view, this represents Ext classes as products with classes in Ext^3 (crossed modules). Does this also work with objects representing 1-cohomology classes or 0-classes?

2.12 Bibliographical notes

The definition of a crossed module of Lie algebras seems to be due to Gerstenhaber [36]. The procedure of associating a 3-cocycle to a crossed module is an adaptation of the case of groups. Theorem 2.2.1 has been stated for the first time by Gerstenhaber in [36], according to the historical note by Mac Lane [67]. For the proof, Gerstenhaber refers to [35] where he proves in fact that splicing together long exact sequences of R-modules with abelian extensions yields a cohomology theory which is again $\mathrm{Ext}^*_R(M, N)$. The proof is by stating that both theories satisfy the axioms and vanish on injectives (without further details).

The lemmas in Section 2.2 are either from [96] or taken from Neeb's article [77]. Sections 2.3 and 2.4 are entirely due to Neeb [79], where he adapts the corresponding theory in the case of groups to Lie algebras. It depicts crossed modules as obstructions to the existence of extensions.

Section 2.5 is my approach in [96] to the subject. It is completely analogous to the case of groups. Section 2.6 is due to Hochschild [43, 44]. It exposes the theory of Lie algebra kernels (analogous to the theory of kernels in the framework of groups) and culminates in Hochschild's theorem in Section 2.6.2 for representing cohomology classes by finite dimensional kernels.

In Section 2.7, we link crossed modules to the Hochschild–Serre spectral sequence, as in the case of groups. In Section 2.8, the string Lie algebra version in 2.8.1 is due to Neeb [77], while the version in 2.8.2 is joint work with Rivière [96, 88, 89]. Section 2.8.3 is from [96], following discussion with B. Shoikhet.

Section 2.9 is due to Kassel and Loday [53]. Again, I understood only from [16] and private communication with M. Ladra that the relative framework implies the absolute framework and is thus completely equivalent; see Remark 2.9.5.

The theory of strict Lie 2-algebras and their relation to crossed modules of Lie algebras is due to Baez and Crans [4]; only Theorem 2.10.6 is from [1].

3 Crossed modules of associative algebras

In this chapter, we discuss crossed modules of associative algebras. Once again, in order to emphasize the similarities with the previous chapters, we proceed in the same order, namely after general definitions in Section 3.1, we discuss crossed modules in relation to third cohomology (Sections 3.2 to 3.5). Section 3.6 is dedicated to associative algebra kernels (emphasizing the resemblance with Section 2.6). Associative 2-algebras appear in Section 3.7 as a categorized version of associative algebras. Section 3.7 reports on the passage from crossed modules of associative algebras to/from crossed modules of Lie algebras and groups. In Section 3.9, we list some examples of crossed modules of associative algebras. The chapter closes with a collection of exercises in Section 3.10.

Associative algebras are usually considered to be nonunital (i. e., do not necessarily possess a unit) and are taken over a field $\mathbb{K}$, that is, multiplication is $\mathbb{K}$-linear. There are unital algebras occurring in this chapter and we try to emphasize when we need them.

3.1 Definitions

We will say that an associative algebra A *acts on* an associative algebra B if B possesses an A-bimodule structure such that for all $a \in A$ and all $b, b' \in B$,

$$a \cdot (bb') = (a \cdot b)b', \quad (b \cdot a)b' = b(a \cdot b'), \quad (bb') \cdot a = b(b' \cdot a).$$

More precisely, this is already an action of A on B *compatible with the product on* B. We will denote the left action by $\eta_l : A \to \mathrm{End}(B)$ or sometimes simply by $b \mapsto a \cdot b$, and the right action by $\eta_r : A \to \mathrm{End}(B)$ or sometimes simply by $b \mapsto b \cdot a$ for all $a \in A$ and all $b \in B$.

Definition 3.1.1. A *crossed module of associative algebras* is the data of a morphism of associative algebras $\mu : M \to N$ together with an action η of N on M (compatible with the product on M, i. e., in the above sense) such that:
(a) $\mu(n \cdot m) = n\mu(m)$ and $\mu(m \cdot n) = \mu(m)n$ for all $m \in M$ and all $n \in N$;
(b) $\mu(m) \cdot m' = mm' = m \cdot \mu(m')$ for all $m, m' \in M$.

Remark 3.1.2. Property (a) means that μ is a homomorphism of N-bimodules with respect to the N-action on M given by η_l and η_r and left and right multiplication on N itself. Property (b) is called Peiffer's identity.

Remark 3.1.3. To each crossed module $\mu : M \to N$, one associates a four-term exact sequence

$$0 \to V \xrightarrow{i} M \xrightarrow{\mu} N \xrightarrow{\pi} A \to 0,$$

https://doi.org/10.1515/9783110750959-003

where $\mathrm{Ker}(\mu) =: V$ and $\mathrm{Coker}(\mu) =: A$. This is in fact a short exact sequence of associative algebras, as is explained in the following remark.

Remark 3.1.4.

(a) By property (a), $\mathrm{Im}(\mu)$ is a two-sided ideal of N. Therefore, the quotient $A = N/\mathrm{Im}(\mu)$ is an associative algebra. Observe that this implies that as soon as M is unital and μ respects the unit, the ideal $\mathrm{Im}(\mu)$ must be equal to N, and the quotient A is trivial. This is why we will never suppose M to be unital. On the other hand, we will suppose in some situations N to be unital, because for unital algebras, the bar complex has better properties (has a contraction). One can always achieve this by adjoining a unit to N.

(b) By property (b), the multiplication on $\mathrm{Ker}(\mu) = V$ is trivial, thus V is a square zero ideal of M. Furthermore, $\mathrm{Im}(\mu)$ acts trivially on V on both sides and the multiplication by elements of M on both sides of V is trivial.

(c) The homomorphism π is surjective, so we can lift elements from A to N and then act with these on M via η. This prescription defines an *outer (bimodule) action* of A on M. More precisely, let $\rho : A \to M$ and $\rho' : A \to M$ be linear sections of π and $a \in A$, then $\eta(\rho(a))$ and $\eta(\rho'(a))$ differ by the action of an element of $\mathrm{Im}(\mu)$. Indeed, $\pi(\rho(a) - \rho'(a)) = 0$, thus there exists $m \in M$ such that $\mu(m) = \rho(a) - \rho'(a)$. We deduce for all $m' \in M$,

$$\eta_l(\rho(a) - \rho'(a))(m') = \eta_l(\mu(m))(m') = mm'.$$

Therefore, the action on V is independent of the section, because $\mathrm{Im}(\mu)$ acts trivially on V. This is true on both sides.

Moreover, $\pi(\rho(aa') - \rho(a)\rho(a')) = 0$, and there exists $m \in M$ such that $\mu(m) = \rho(aa') - \rho(a)\rho(a')$. This implies that up to multiplication by the element $m \in M$, the composition $\eta_l \circ \rho$ is a left action. In the same vein, $\eta_r \circ \rho$ is a right action up to multiplication by the same element $m \in M$ (but this time from the right).

On the other hand, it is clear that $\eta_l \circ \rho$ and $\eta_r \circ \rho$ form together a bimodule action as soon as they are actions, because the compatibility between the two actions is immediate from the compatibility of η_l and η_r. In this sense, $\eta_l \circ \rho$ and $\eta_r \circ \rho$ form together an outer bimodule action, that is, a bimodule action up to the multiplication by an element of M.

Observe that in any case V becomes a genuine A-bimodule using $\eta_l \circ \rho$ and $\eta_r \circ \rho$, because the multiplication action of M on V is trivial by property (b).

(d) Note that by this lifting procedure, $\eta_l \circ \rho$ and $\eta_r \circ \rho$ do not render M an A-bimodule in general.

Remark 3.1.5. Let us explain the link between the notion of a crossed module of algebras as a homomorphism of algebras $\mu : M \to N$ and the notion of a crossed bimodule as in [6]. A *crossed bimodule* $\partial : V \to A$ in [6] is the data of a $\mathbb{K}$-algebra A, an A-bimodule V and a map of A-bimodules $\partial : V \to A$ such that $(\partial v) \cdot w = v \cdot (\partial w)$

for all $v, w \in V$. Observe that in the case of a crossed bimodule $\partial : V \to A$, V acquires the structure of an algebra by setting $vw := v \cdot (\partial w)$ (or $= (\partial v) \cdot w$, which is the same). Translating in this way crossed bimodules into crossed modules, the notions are equivalent.

Remark 3.1.6. Let us list some elementary examples of crossed modules:
(a) Each A-bimodule V gives rise to a crossed module $\mu : V \to A$, where μ sends all elements of V to zero.
(b) Each square-zero extension with trivial bimodule action is a crossed module. Indeed, such extensions correspond exactly to the case where $\mu : M \to N$ is surjective (and then M is a square-zero extension of N by V where N acts trivially on V).
(c) Each (inclusion of a) two-sided ideal constitutes a crossed module. Indeed, the inclusion of an ideal corresponds exactly to the case where the map $\mu : M \to N$ is injective.
(d) For each associative algebra L, there is a crossed module $L \to M(L)$, where $M(L)$ is the *multiplication algebra* of L. It is by definition the set of all two-sided multiplications in L, that is, all pairs of endomorphisms satisfying the above requirements of an action (see the definition in Appendix C). As explained in Appendix C, $M(L)$ becomes an associative algebra such that left and right multiplication by a fixed element $l \in L$ gives an element of $M(L)$, called an *inner multiplication* of L. The map μ is then simply the map sending an element $l \in L$ to the inner multiplication it defines. Its kernel, that is, the elements $l \in L$ such that $lL = Ll = 0$ is called the *nucleus* of L. In fact, the set of inner multiplications $\mathrm{Inn}(L)$ forms an ideal in $M(L)$, and the corresponding quotient $M(L)/\mathrm{Inn}(L) =: \mathrm{Out}(L)$ is called the *algebra of outer multiplications* of L.

Definition 3.1.7. Two crossed modules $\mu : M \to N$ (with action η) and $\mu' : M' \to N'$ (with action η') such that $\mathrm{Ker}(\mu) = \mathrm{Ker}(\mu') =: V$ and $\mathrm{Coker}(\mu) = \mathrm{Coker}(\mu') =: A$ are called *elementary equivalent* if there are algebra homomorphisms $\varphi : M \to M'$ and $\psi : N \to N'$, which are compatible with the actions, that is,

$$\varphi(\eta(n)(m)) = \eta'(\psi(n))(\varphi(m))$$

for all $n \in N$ and all $m \in M$, and such that the following diagram is commutative:

$$\begin{array}{ccccccccccc}
0 & \longrightarrow & V & \xrightarrow{i} & M & \xrightarrow{\mu} & N & \xrightarrow{\pi} & A & \longrightarrow & 0 \\
 & & \downarrow{\scriptstyle \mathrm{id}_V} & & \downarrow{\scriptstyle \varphi} & & \downarrow{\scriptstyle \psi} & & \downarrow{\scriptstyle \mathrm{id}_A} & & \\
0 & \longrightarrow & V & \xrightarrow{i'} & M' & \xrightarrow{\mu'} & N' & \xrightarrow{\pi'} & A & \longrightarrow & 0
\end{array}$$

We call *equivalence of crossed modules* the equivalence relation generated by elementary equivalence. One easily sees that two crossed modules are equivalent in case there exists a zig-zag of elementary equivalences going from one to the other (where the arrows do not necessarily all go into the same direction).

Let us denote by $\mathrm{crmod}(A, V)$ the set of equivalence classes of crossed modules with a fixed kernel V and a fixed cokernel A.

Remark 3.1.8. Compare this equivalence relation to the equivalence of square-zero extensions; cf. Appendix C. In the framework of extensions, equivalence imposes that the underlying $\mathbb{K}$-vector space of the extension is the direct sum, up to bijection. For crossed modules, the exactness of the sequence does not prescribe the $\mathbb{K}$-vector space of the two middle terms, the end terms being fixed. This leads to much more different representatives of the same equivalence class of a crossed module.

Remark 3.1.9. The set of equivalence classes of crossed modules of associative algebras carries the *Baer sum* and becomes thus an abelian group. The definitions are completely analogous to the corresponding definitions given in Chapter 2 for Lie algebras.

3.1.1 Crossed modules with free algebras

Lemma 3.1.1. *Suppose $\mu : M \to N$ is a crossed module and $f : N' \to N$ a morphism of associative algebras. Let M' (together with $\mu' : M' \to N'$ and $\varphi : M' \to M$) be the pullback of the diagram*

$$\begin{array}{ccc} & & N' \\ & & \downarrow f \\ M & \xrightarrow{\mu} & N \end{array}$$

Then there exists a unique structure of a crossed module on $\mu' : M' \to N'$ such that the diagram

$$\begin{array}{ccccccccccc} 0 & \longrightarrow & V & \longrightarrow & M' & \xrightarrow{\mu'} & N' & \longrightarrow & A & \longrightarrow & 0 \\ & & \| & & \downarrow \varphi & & \downarrow f & & \| & & \\ 0 & \longrightarrow & V & \longrightarrow & M & \xrightarrow{\mu} & N & \longrightarrow & A & \longrightarrow & 0 \end{array}$$

defines a morphism of crossed modules.

Proof. By definition, the pullback M' is $\{(m, n') \in M \times N' \mid f(n') = \mu(m)\}$. In this description, $\varphi : M' \to M$ and $\mu' : M' \to N'$ are induced by the projections onto the first and second component, respectively. We make elements $n', n'' \in N'$ act on M' by setting $(m, n') \cdot n'' := (m \cdot f(n''), n'n'')$ and $n' \cdot (m, n'') := (f(n') \cdot m, n'n'')$. These left and right actions respect the subspace M'. This bimodule action is compatible with the product on M' by associativity of N' and by the crossed module properties for $\mu : M \to N$. It is clear that $\mu' : M' \to N'$ is a morphism of algebras and a crossed module. Finally, it is also easy to see that (φ,f) is a morphism of crossed modules.

The unicity stems from the fact that the morphisms $\varphi : M' \to M$ and $f : N' \to N$ impose the N'-bimodule structure on the pullback $M' = \{(m,n') \in M \times N' \mid f(n') = \mu(m)\}$. □

The equivalence classes in $\mathrm{crmod}(A,V)$ are also called *connected components*, because the zig-zag of morphisms linking two equivalent crossed modules may be viewed as a path leading from one to the other.

Corollary 3.1.2. *In each connected component of* $\mathrm{crmod}(A,V)$, *there exists a crossed module*

$$0 \to V \to M' \xrightarrow{\mu'} N' \to A \to 0$$

such that $N' = \mathcal{F}(S)$ *is a free algebra on the set* S. *For any other object* $\mu : M \to N$, *there is a morphism* $(\varphi,f) : (\mu' : M' \to \mathcal{F}(S)) \to (\mu : M \to N)$. *Thus two crossed modules* $\mu : M \to N$, $\mu : M'' \to N''$ *are in the same component in* $\mathrm{crmod}(A,V)$ *if and only if there exists a diagram of the form*

$$(\mu : M \to N) \leftarrow (\mu' : M' \to \mathcal{F}(S)) \to (\mu'' : M'' \to N'').$$

3.2 Relation to third cohomology I

Crossed modules of associative algebras serve once again as explicit representatives of third cohomology classes. This is the essence of the following theorem, which is still in the realm of Gerstenhaber's theorem; see [37] and for the proof therein [36].

Theorem 3.2.1. *The map associating to a crossed module the* 3*-cocycle* γ *(defined below) induces an isomorphism of abelian groups:*

$$b : \mathrm{crmod}(A,V) \cong HH^3(A,V).$$

The proof of Theorem 3.2.1 will already be completed in this section, breaking with the structure of the sections on group and Lie algebra crossed modules.

3.2.1 Constructing the map *b*

In a first step, we will discuss in this subsection how to associate a Hochschild cohomology class in $HH^3(A,V)$ to a given crossed module.

We shall use the notation $b([\mu : M \to N])$ or simply $b([\mu])$ for b applied to a crossed module $\mu : M \to N$ representing an equivalence class $[\mu : M \to N]$.

Let us show how to associate to a crossed module a 3-cocycle of A with values in V. For this, recall the exact sequence from Remark 3.1.3:

$$0 \to V \xrightarrow{i} M \xrightarrow{\mu} N \xrightarrow{\pi} A \to 0.$$

The first step is to take a linear section ρ of π and to compute the failure of ρ to be a homomorphism of associative algebras, that is,

$$\alpha(x_1, x_2) = \rho(x_1)\rho(x_2) - \rho(x_1 x_2),$$

for all $x_1, x_2 \in A$. The expression $\alpha(x_1, x_2)$ is bilinear in (x_1, x_2). We have obviously $\pi(\alpha(x_1, x_2)) = 0$, because π is an algebra homomorphism, so $\alpha(x_1, x_2) \in \mathrm{Im}(\mu) = \mathrm{Ker}(\pi)$. This means by exactness that there exists $\beta(x_1, x_2) \in M$ such that

$$\mu(\beta(x_1, x_2)) = \alpha(x_1, x_2).$$

Choosing a linear section σ on $\mathrm{Im}(\mu)$, one can choose β as

$$\beta(x_1, x_2) = \sigma(\alpha(x_1, x_2)) \tag{3.1}$$

showing that we can also suppose β bilinear in x_1, x_2. Now $\mu(d^M\beta(x_1, x_2, x_3))$, that is, the μ-image of the Hochschild coboundary $d^M\beta(x_1, x_2, x_3)$ (see Formula (C.1) in Appendix C), is zero for all $x_1, x_2, x_3 \in A$ by the following lemma.

Lemma 3.2.2. *$\mu(d^M\beta(x_1, x_2, x_3)) = 0$ where d^M is the formal expression of the Hochschild coboundary operator (see Formula (C.1) in Appendix C) corresponding to cohomology of A with values in M, and $x_1, x_2, x_3 \in A$.*

Remark 3.2.1. For general information about Hochschild cohomology, please refer to Appendix C.

By Hochschild cohomology of A with values in M, we mean that A "acts" on M by $\eta \circ \rho$. This gives just formally the expression of a Hochschild coboundary operator, while the map $\eta \circ \rho$ is not an algebra action on M in general.

Proof.

$$\begin{aligned}
&\mu(d^M\beta(x_1, x_2, x_3)) \\
&\quad = \mu(\rho(x_1) \cdot \beta(x_2, x_3) - \beta(x_1 x_2, x_3) + \beta(x_1, x_2 x_3) - \beta(x_1, x_2) \cdot \rho(x_3)) \\
&\quad = \rho(x_1)\alpha(x_2, x_3) - \alpha(x_1 x_2, x_3) + \alpha(x_1, x_2 x_3) - \alpha(x_1, x_2)\rho(x_3) \\
&\quad = \rho(x_1)(\rho(x_2)\rho(x_3) - \rho(x_2 x_3)) - (\rho(x_1 x_2)\rho(x_3) - \rho(x_1 x_2 x_3)) \\
&\qquad + (\rho(x_1)\rho(x_2 x_3) - \rho(x_1 x_2 x_3)) - (\rho(x_1)\rho(x_2) - \rho(x_1 x_2))\rho(x_3) \\
&\quad = 0.
\end{aligned}$$

□

This means that $d^M\beta(x_1,x_2,x_3) \in \mathrm{Ker}(\mu) = \mathrm{Im}(i) = i(V)$, that is, there exists $\gamma(x_1,x_2,x_3) \in V$ such that $d^M\beta(x_1,x_2,x_3) = i(\gamma(x_1,x_2,x_3))$. The explicit formula of $d^M\beta(x_1,x_2,x_3)$ shows trilinearity in the three variables x_1, x_2 and x_3. Choosing a linear section τ on $i(V) = \mathrm{Ker}(\mu)$, one can choose γ to be $\tau \circ d^M\beta$ (in the obvious sense) gaining that γ is also trilinear in x_1, x_2 and x_3. By the following lemma, γ is a Hochschild 3-cocycle of A with values in V.

Lemma 3.2.3. *The trilinear map γ is a Hochschild 3-cocycle of A with values in V.*

Proof. It is sufficient to show that $i(d^V\gamma) = 0$, because i is injective. Therefore, we compute

$$\begin{aligned}
&i(d^V\gamma(x_1,x_2,x_3,x_4))\\
&\quad = i(\rho(x_1)\cdot\gamma(x_2,x_3,x_4) - \gamma(x_1x_2,x_3,x_4) + \gamma(x_1,x_2x_3,x_4)\\
&\qquad - \gamma(x_1,x_2,x_3x_4) + \gamma(x_1,x_2,x_3)\cdot\rho(x_4))\\
&\quad = \rho(x_1)\cdot d^M\beta(x_2,x_3,x_4) - d^M\beta(x_1x_2,x_3,x_4) + d^M\beta(x_1,x_2x_3,x_4)\\
&\qquad - d^M\beta(x_1,x_2,x_3x_4) + d^M\beta(x_1,x_2,x_3)\cdot\rho(x_4)\\
&\quad = \rho(x_1)\cdot(\rho(x_2)\cdot\beta(x_3,x_4) - \beta(x_2x_3,x_4) + \beta(x_2,x_3x_4)\\
&\qquad - \beta(x_2,x_3)\cdot\rho(x_4)) - (\rho(x_1x_2)\cdot\beta(x_3,x_4) - \beta(x_1x_2x_3,x_4)\\
&\qquad + \beta(x_1x_2,x_3x_4) - \beta(x_1x_2,x_3)\cdot\rho(x_4)) + (\rho(x_1)\cdot\beta(x_2x_3,x_4)\\
&\qquad - \beta(x_1x_2x_3,x_4) + \beta(x_1,x_2x_3x_4) - \beta(x_1,x_2x_3)\cdot\rho(x_4))\\
&\qquad - (\rho(x_1)\cdot\beta(x_2,x_3x_4) - \beta(x_1x_2,x_3x_4) + \beta(x_1,x_2x_3x_4)\\
&\qquad - \beta(x_1,x_2)\cdot\rho(x_3x_4)) + (\rho(x_1)\cdot\beta(x_2,x_3) - \beta(x_1x_2,x_3)\\
&\qquad + \beta(x_1,x_2x_3) - \beta(x_1,x_2)\cdot\rho(x_3))\cdot\rho(x_4).
\end{aligned}$$

After simplifying this expression, we obtain as remaining terms:

$$\begin{aligned}
&i(d^V\gamma(x_1,x_2,x_3,x_4))\\
&\quad = (\rho(x_1)\rho(x_2) - \rho(x_1x_2))\cdot\beta(x_3,x_4) + \beta(x_1,x_2)\cdot(\rho(x_3x_4) - \rho(x_3)\rho(x_4)).
\end{aligned}$$

Now use that $\rho(x_1)\rho(x_2) - \rho(x_1x_2) = \mu(\beta(x_1,x_2))$ to obtain with property (b) of a crossed module:

$$\begin{aligned}
i(d^V\gamma(x_1,x_2,x_3,x_4)) &= \mu(\beta(x_1,x_2))\cdot\beta(x_3,x_4) - \beta(x_1,x_2)\cdot\mu(\beta(x_3,x_4))\\
&= \beta(x_1,x_2)\beta(x_3,x_4) - \beta(x_1,x_2)\beta(x_3,x_4)\\
&= 0.
\end{aligned}$$

□

Lemma 3.2.4. *The class of the Hochschild 3-cocycle γ does not depend on the choice of the sections ρ, σ and τ.*

Proof. First, we show that for a fixed section ρ, the class of γ does not depend on the sections σ and τ. In the following, x, y, z are elements of A.

(1) Suppose that we chose two different linear sections σ and σ' of μ on $\mathrm{Im}(\mu)$, leading to different lifts $\beta = \sigma\alpha$ and $\beta' = \sigma'\alpha$. We have then $\beta - \beta' \in \mathrm{Ker}(\mu) = \mathrm{Im}(i)$ and, therefore, there exists an element $\zeta(x,y)$ in V such that

$$\beta'(x,y) = \beta(x,y) + i(\zeta)(x,y).$$

When applying in the next step d^M to all terms, the ζ will give a coboundary, because $d^M(i(\zeta)) = i(d^V\zeta)$. This shows that the corresponding γ and γ' differ by a coboundary.

(2) Two "different" sections τ and τ' of i have to be the same; they are both inverses of the isomorphism i onto its image.

It remains thus to show that the class of γ does not depend on the section ρ.

(3) Let ρ and ρ' be two sections of π. Denote by $\alpha(x,y)$, respectively, $\alpha'(x,y)$, $\beta(x,y)$, respectively, $\beta'(x,y)$, $\gamma(x,y,z)$, respectively, $\gamma'(x,y,z)$, the elements of N, M and V constructed above with respect to ρ and ρ'. Here, $x,y,z \in A$. By construction, $\rho - \rho'$ takes values in $\mathrm{Ker}(\pi) \subset N$. Define $h := \sigma \circ (\rho - \rho')$, and further $\beta''(x,y) \in M$ by

$$\beta''(x,y) := \rho(x)\cdot h(y) - h(xy) + h(x)\cdot\rho'(y).$$

Observe that $\mu(h(x)) = \rho(x) - \rho'(x)$, and thus

$$\begin{aligned}\mu(\beta''(x,y)) &= \mu(\rho(x)\cdot h(y) - h(xy) + h(x)\cdot\rho'(y))\\ &= \rho(x)\rho(y) - \rho(x)\rho'(y) - \rho(xy) + \rho'(xy) + \rho(x)\rho'(y) - \rho'(x)\rho'(y)\\ &= \rho(x)\rho(y) - \rho(xy) + \rho'(xy) - \rho'(x)\rho'(y)\\ &= \mu(\beta(x,y) - \beta'(x,y)).\end{aligned}$$

Therefore, there exists $\zeta(x,y) \in V$ such that

$$\beta(x,y) - \beta'(x,y) = \beta''(x,y) + i(\zeta(x,y)).$$

Now compute

$$\begin{aligned}i\circ(\gamma - \gamma')(x,y,z) &= d^M\beta(x,y,z) - d^M\beta'(x,y,z)\\ &= \rho(x)\cdot\beta(y,z) - \beta(xy,z) + \beta(x,yz) - \beta(x,y)\cdot\rho(z)\\ &\quad - \rho'(x)\cdot\beta'(y,z) - \beta'(xy,z) + \beta'(x,yz) - \beta'(x,y)\cdot\rho'(z).\end{aligned}$$

We introduce the mixed coboundary d^{mix} as follows:

$$d^{\mathrm{mix}}\omega(x,y,z) := \rho(x)\cdot\omega(y,z) - \omega(xy,z) + \omega(x,yz) - \omega(x,y)\cdot\rho'(z).$$

With this mixed coboundary operator, we can express the preceding as

$$i\circ(\gamma - \gamma')(x,y,z) = d^{\mathrm{mix}}(\beta - \beta')(x,y,z) + (\rho(x) - \rho'(x))\cdot\beta'(y,z)$$

$$
\begin{aligned}
&- \beta(x,y)\cdot(\rho(z)-\rho'(z))\\
&= d^{\mathrm{mix}} i(\zeta)(x,y,z) + (\rho(x)-\rho'(x))\cdot\beta'(y,z)\\
&\quad - \beta(x,y)\cdot(\rho(z)-\rho'(z)) + d^{\mathrm{mix}}\beta''(x,y,z).
\end{aligned}
$$

Observe that $d^{\mathrm{mix}} \circ i = i \circ d^V$, because the action of A on V does not depend on the choice of the section, as shown before. We therefore obtain

$$
\begin{aligned}
i\circ(\gamma-\gamma')(x,y,z) &= i(d^V(\zeta))(x,y,z) + (\rho(x)-\rho'(x))\cdot\beta'(y,z)\\
&\quad - \beta(x,y)\cdot(\rho(z)-\rho'(z)) + \rho(x)\cdot(\rho(y)\cdot h(z)\\
&\quad - h(yz) + h(y)\cdot\rho'(z)) - \rho(xy)\cdot h(z)\\
&\quad + h(xyz) - h(xy)\cdot\rho'(z) + \rho(x)\cdot h(yz)\\
&\quad - h(xyz) + h(x)\cdot\rho'(yz) - (\rho(x)\cdot h(y)\\
&\quad - h(xy) + h(x)\cdot\rho'(y))\cdot\rho'(z)\\
&= i(d^V(\zeta))(x,y,z) + (\rho(x)-\rho'(x))\cdot\beta'(y,z)\\
&\quad - \beta(x,y)\cdot(\rho(z)-\rho'(z))\\
&\quad + (\rho(x)\rho(y)-\rho(xy))\cdot h(z) + h(x)\cdot(\rho'(y)\rho'(z)-\rho'(yz)).
\end{aligned}
$$

We use again the fact that $\rho(x)\rho(y) - \rho(xy) = \mu(\beta(x,y))$ and $\rho'(x)\rho'(y) - \rho'(xy) = \mu(\beta'(x,y))$ to arrive at the expression

$$
\begin{aligned}
i\circ(\gamma-\gamma')(x,y,z) &= i(d^V(\zeta))(x,y,z) + (\rho(x)-\rho'(x))\cdot\beta'(y,z)\\
&\quad - \beta(x,y)\cdot(\rho(z)-\rho'(z)) + \mu(\beta(x,y))\cdot h(z) + h(x)\cdot\mu(\beta'(y,z))\\
&= i(d^V(\zeta))(x,y,z) + (\rho(x)-\rho'(x))\cdot\beta'(y,z)\\
&\quad - \beta(x,y)\cdot(\rho(z)-\rho'(z)) + \beta(x,y)\cdot\mu(h(z)) + \mu(h(x))\cdot\beta'(y,z)\\
&= i(d^V(\zeta))(x,y,z).
\end{aligned}
$$

This shows finally that γ and γ' represent the same class in $HH^3(A,V)$. □

Lemma 3.2.5. *Let $\mu : M \to N$ (with action η) and $\mu' : M' \to N'$ (with action η') such that* $\mathrm{Ker}(\mu) = \mathrm{Ker}(\mu') =: V$ *and* $\mathrm{Coker}(\mu) = \mathrm{Coker}(\mu') =: A$ *be two elementary equivalent crossed modules.*

Then the corresponding cohomology classes $b([\mu]) = [\gamma]$ and $b([\mu']) = [\gamma']$ coincide in $H^3(\mathfrak{g}, V)$.

Proof. Denote by (φ, ψ) the morphism rendering the two crossed modules elementary equivalent; see Definition 3.1.7. Denote further by γ (resp., γ') the cocycle associated to $\mu : M \to N$ (resp., to $\mu' : M' \to N'$) by choosing sections ρ, σ and τ (resp., ρ', σ' and τ') as in the preceding construction of the 3-cocycle.

As ρ is a section of π, $\tilde{\rho}' := \psi \circ \rho$ is a section of π'. Lemma 3.2.4 shows that α' (defined by the section ρ') and $\tilde{\alpha}' := \psi \circ \alpha$ (defined by the section $\tilde{\rho}'$) give rise to the same cohomology class $[\gamma']$ represented by γ'.

Now let us compute

$$(\tilde{d}^{M'}(\sigma' \circ \psi \circ \alpha) - \tilde{d}^{M'}(\varphi \circ \beta))(x,y,z)$$

for $x, y, z \in A$ (where $\tilde{d}^{M'}$ is the formal Hochschild coboundary operator with values in M' and with the formal action $\eta' \circ \psi \circ \rho$). Let us call $\tilde{\beta}' := \sigma' \circ \psi \circ \alpha$.

First, we remark that $(\tilde{\beta}' - \varphi \circ \beta)(x,y) \in \mathrm{Ker}(\mu')$, thus we introduce $v(x,y) \in V$ such that $(\tilde{\beta}' - \varphi \circ \beta)(x,y) = i'v(x,y)$ for all $x, y \in A$. Now compute $\tilde{d}^{M'}(\varphi \circ \beta)$ using that (φ, ψ) satisfies the conditions of Definition 3.1.7:

$$\begin{aligned}\tilde{d}^{M'}(\varphi \circ \beta)(x,y,z) &= \eta'_l(\psi(\rho(x)))(\varphi(\beta(y,z))) - \varphi(\beta(xy,z)) + \varphi(\beta(x,yz)) \\ &\quad - \eta'_r(\psi(\rho(z)))(\varphi(\beta(x,y))) \\ &= \varphi(\eta_l(\rho(x)))(\beta(y,z)) - \varphi(\beta(xy,z)) + \varphi(\beta(x,yz)) \\ &\quad - \varphi(\eta_r(\rho(z)))(\beta(x,y)) \\ &= \varphi(d^M \beta(x,y,z)).\end{aligned}$$

On the other hand, compute $(\tilde{d}^{M'} i'v)$

$$\begin{aligned}(\tilde{d}^{M'} i'v)(x,y,z) &= \eta'_l(\psi(\rho(x)))(i'v(y,z)) - i'v(xy,z) + i'v(x,yz) \\ &\quad - \eta'_r(\psi(\rho(z)))(i'v(x,y)) \\ &= \eta'_l(\psi(\rho(x)))(\varphi iv(y,z)) - \varphi(iv(xy,z)) + \varphi(iv(x,yz)) \\ &\quad - \eta'_r(\psi(\rho(z)))(\varphi iv(x,y)) \\ &= \varphi(\eta_l(\rho(x))(iv(y,z))) - \varphi(iv(xy,z)) + \varphi(iv(x,yz)) \\ &\quad - \varphi(\eta_r(\rho(z))(iv(x,y))) \\ &= i'(d^V v(x,y,z)).\end{aligned}$$

We thus obtain

$$(\gamma' - \gamma)(x,y,z) = d^V v(x,y,z),$$

showing that the two elementary equivalent crossed modules have the same cohomology class. □

In conclusion, we have constructed a well-defined map b associating to an equivalence class of crossed modules with kernel V and cokernel A a cohomology class in $HH^3(A, V)$. Theorem 3.2.1 states that this map is an isomorphism of abelian groups.

3.2.2 Bijectivity of the map *b*

We will now suppose that the associative algebra A is unital (i. e., possesses a unit). This simplifies things when using the bar-complex (e. g., it has a contraction, see Appendix C). In case one starts without a unit, one can always adjoin a unit in order to proceed.

For the surjectivity of the map b, we will use the following lemma.

Lemma 3.2.6. *Let B be a unital associative algebra, V be a $\mathbb{K}$-module and $d : V \to B$ be a $\mathbb{K}$-linear map. Consider the maps*

$$D_2 : B \otimes V \otimes B \otimes V \otimes B \to B \otimes V \otimes B$$

and

$$D_1 : B \otimes V \otimes B \to B$$

defined by

$$D_2(b \otimes v \otimes b' \otimes v' \otimes b'') := bd(v)b' \otimes v' \otimes b'' - b \otimes v \otimes b'd(v')b''$$

and

$$D_1(b \otimes v \otimes b') := bd(v)b'$$

for all $b, b', b'' \in B$ and all $v, v' \in V$. Then we have $D_1 \circ D_2 = 0$ and D_1 induces a map

$$\mu : C := (B \otimes V \otimes B)/\mathrm{Im}(D_2) \to B,$$

which is a crossed module.

Proof. First, compute

$$\begin{aligned} D_1(D_2(b \otimes v \otimes b' \otimes v' \otimes b'')) &= D_1(bd(v)b' \otimes v' \otimes b'') - D_1(b \otimes v \otimes b'd(v')b'') \\ &= bd(v)b'd(v')b'' - bd(v)b'd(v')b'' = 0, \end{aligned}$$

thus we have as expected $D_1 \circ D_2 = 0$, which implies that μ is well-defined. The maps D_1 and D_2 are clearly bimodule morphisms. Furthermore, it is clear that the image of a morphism of bimodules is a subbimodule of the codomain, thus $\mathrm{Im}(D_2) \subset B \otimes V \otimes B$ is a subbimodule and the quotient C is a B-bimodule.

Let us show that $\mu : C \to B$ is a crossed bimodule in the sense of Remark 3.1.5. This will show that it is a crossed module. For this, it is enough to show that

$$\mu(b \otimes v \otimes b') \cdot b'' \otimes w \otimes b''' = b \otimes v \otimes b' \cdot \mu(b'' \otimes w \otimes b'').$$

This boils down to showing that

$$\begin{aligned} &\mu(b \otimes v \otimes b') \cdot b'' \otimes w \otimes b''' - b \otimes v \otimes b' \cdot \mu(b'' \otimes w \otimes b'') \\ &\quad = D_1(b \otimes v \otimes b') \cdot b'' \otimes w \otimes b''' - b \otimes v \otimes b' \cdot D_1(b'' \otimes w \otimes b'') \\ &\quad = bd(v)b'b'' \otimes w \otimes b''' - b \otimes v \otimes b'b''d(w)b''' \\ &\quad = D_2(b \otimes v \otimes b'b'' \otimes w \otimes b''') \in \mathrm{Im}(D_2). \end{aligned}$$

□

In order to show the surjectivity of the map b, we will apply the lemma to the tensor algebra on A:

$$B := T(A) = \bigoplus_{n\geq 0} A^{\otimes n},$$

and to the $\mathbb{K}$-module $V := A \otimes A$ together with the map

$$d(a \otimes a') = a \otimes a' - aa'$$

for all $a, a' \in A$. The preceding lemma gives us a crossed module

$$\mu : C := T(A) \otimes (A \otimes A) \otimes T(A)/\mathrm{Im}(D_2) \to B = T(A).$$

Let us show that the cokernel of this map is exactly A. For this, we will use the multiplication map $\pi : T(A) \to A$ defined by $\pi(a_1 \otimes \cdots \otimes a_n) = a_1 \dots a_n$. We want to show that $\mathrm{Im}(\mu) = \mathrm{Ker}(\pi)$. An arbitrary element $b \in T(A)$ is written

$$b = \sum_{j=1}^{s} a_1^j \otimes \cdots \otimes a_{n_j}^j$$

with some integers $s, n_1, \dots, n_s$ and a_i^j elements in A for all $1 \leq j \leq s$ and all $1 \leq i \leq n_j$. For this element b, we compute

$$\begin{aligned}
b &= \sum_{j=1}^{s} a_1^j \otimes \cdots \otimes a_{n_j}^j \\
&= \sum_{j=1}^{s} \left\{ \sum_{k=1}^{n_j-2} (a_1^j \dots a_k^j \otimes a_{k+1}^j \otimes \cdots \otimes a_{n_j}^j - a_1^j \dots a_{k+1}^j \otimes a_{k+2}^j \otimes \cdots \otimes a_{n_j}^j) + a_1^j \dots a_{n_j}^j \right\} \\
&= \sum_{j=1}^{s} \sum_{k=1}^{n_j-2} \mu([1 \otimes (a_1^j \dots a_k^j \otimes a_{k+1}^j) \otimes a_{k+2}^j \dots \otimes a_{n_j}^j \otimes 1]) + \sum_{j=1}^{s} a_1^j \dots a_{n_j}^j \\
&= \mu\left(\sum_{j=1}^{s} \sum_{k=1}^{n_j-2} [1 \otimes (a_1^j \dots a_k^j \otimes a_{k+1}^j) \otimes a_{k+2}^j \dots \otimes a_{n_j}^j \otimes 1] \right) + \pi(b).
\end{aligned}$$

This shows that $b \in \mathrm{Ker}(\pi)$ if and only if $b \in \mathrm{Im}(\mu)$ and, therefore, $\mathrm{Ker}(\pi) = \mathrm{Im}(\mu)$. Observe that by the axioms of a crossed module, $\mathrm{Im}(\mu)$ is an ideal, thus the quotient is an algebra. Moreover, the quotient is isomorphic to A *as an algebra*, because the product in the quotient $T(A)/\mathrm{Ker}(\pi)$ is $(a, a') \mapsto a \otimes a'$, which is equal to aa' modulo $\mathrm{Ker}(\pi)$.

Denote by $K := \mathrm{Ker}(\mu)$ the kernel of μ, then we obtain an exact sequence

$$0 \to K \to C \xrightarrow{\mu} T(A) \xrightarrow{\pi} A \to 0.$$

Remark 3.2.2. Observe that $T(A)$ is the free associative (unital) algebra on the vector space A and the crossed module constructed above is thus is a *free crossed module* in the category of associative algebras.

Now we construct a morphism to a piece of the bar resolution

$$(B(A) := A^{\otimes(n+2)}, d_n)_{n\in\mathbb{N}}$$

of A, where the augmentation is given by the multiplication map of A, namely mult : $A \otimes A \to A$ defined by $a \otimes a' \mapsto aa'$. As the bar resolution is acyclic (see Appendix C), it induces an exact sequence

$$0 \to A^{\otimes 5}/\mathrm{Im}(d_4) \overset{d_3}{\to} A^{\otimes 4} \overset{d_2}{\to} A^{\otimes 3} \overset{d_1}{\to} A^{\otimes 2} \overset{\mathrm{mult}}{\to} A.$$

Observe that the multiplication map is surjective, because we supposed that A possesses a unit, but we will not use this. We construct a morphism of exact sequences in the following way:

$$\begin{array}{ccccccccc}
0 & \longrightarrow & K & \overset{i}{\longrightarrow} & C & \overset{\mu}{\longrightarrow} & T(A) & \overset{\pi}{\longrightarrow} & A & \longrightarrow & 0 \\
 & & \downarrow h & & \downarrow h_1 & & \downarrow h_0 & & & & \\
0 & \longrightarrow & A^{\otimes 5}/\mathrm{Im}(d_4) & \overset{d_3}{\longrightarrow} & A^{\otimes 4} & \overset{d_2}{\longrightarrow} & A^{\otimes 3} & \overset{d_1}{\longrightarrow} & A^{\otimes 2} & \overset{\mathrm{mult}}{\longrightarrow} & A
\end{array}$$

Indeed, the maps h_0, h_1, h are defined as follows:

- $h_0(a) := 1\otimes a\otimes 1$ for all $a \in A$ and then extended by $h_0(xy) = h_0(x)\cdot\pi(y)+\pi(x)\cdot h_0(y)$ for all $x, y \in T(A)$;
- $h_1([x \otimes (a \otimes a') \otimes y]) := \pi(x) \otimes a \otimes a' \otimes \pi(y)$ for all $a, a' \in A$ and all $x, y \in T(A)$;
- $h(k) := [1 \otimes h_1(i(k))]$ for all $k \in K$.

Let us show that the squares are commutative: Indeed, for all $x, y \in T(A)$ and all $a, a' \in A$, we compute

$$\begin{aligned}
&h_0(\mu([x \otimes (a \otimes a') \otimes y])) \\
&\quad = h_0(x \otimes a \otimes a' \otimes y) - h_0(x \otimes aa' \otimes y) \\
&\quad = h_0(x \otimes a)\pi(a' \otimes y) + \pi(x \otimes a)h_0(a' \otimes y) \\
&\qquad - h_0(x \otimes aa')\pi(y) - \pi(x \otimes aa')h_0(y) \\
&\quad = h_0(x)aa'\pi(y) + \pi(x)(1 \otimes a \otimes 1)a'\pi(y) \\
&\qquad + \pi(x)a(1 \otimes a' \otimes 1)\pi(y) + \pi(x)aa'h_0(y) \\
&\qquad - h_0(x)aa'\pi(y) - \pi(x)(1 \otimes aa' \otimes 1)\pi(y) - \pi(x)aa'h_0(y) \\
&\quad = \pi(x) \otimes a \otimes (a'\pi(y)) - \pi(x) \otimes aa' \otimes \pi(y) + (\pi(x)a) \otimes a' \otimes \pi(y) \\
&\quad = d_2(\pi(x) \otimes a \otimes a' \otimes \pi(y))
\end{aligned}$$

$$= d_2(h_1(x \otimes a \otimes a' \otimes y)).$$

This shows $h_0 \circ \mu = d_2 \circ h_1$ as it should. Moreover, for $i(k)$ in $\mathrm{Ker}(\mu)$, we have by definition of d_3:

$$\begin{aligned} d_3(h(k)) &= d_3([1 \otimes h_1(i(k))]) \\ &= h_1(i(k)) + 1 \otimes d_2(h_1(i(k))) \\ &= h_1(i(k)) + 1 \otimes h_0(\mu(i(k))) \\ &= h_1(i(k)). \end{aligned}$$

This shows that the diagram is commutative.

Now consider given a Hochschild 3-cocycle c on A with values in V. We have seen in Appendix C that a 3-cocycle on A with values in V can be viewed as a map of A-bimodules $c : A^{\otimes 5}/\mathrm{Im}(d_4) \to V$. Now associate to c the composition $\tilde{c} = c \circ h : K \to V$. Viewing V as a $T(A)$-bimodule via π, we have the following commutative diagram of $T(A)$-bimodules:

$$\begin{array}{ccc} K & \xrightarrow{i} & C \\ \Big\downarrow{\scriptstyle \tilde{c}} & & \Big\downarrow{\scriptstyle \mu} \\ V & \xrightarrow{0} & T(A) \end{array}$$

On the other hand, we have the pushout of $T(A)$-bimodules:

$$\begin{array}{ccc} K & \xrightarrow{i} & C \\ \Big\downarrow{\scriptstyle \tilde{c}} & & \Big\downarrow{\scriptstyle i_2} \\ V & \xrightarrow{i_1} & \tilde{C} \end{array}$$

Combining these two diagrams, we obtain by the universal property of the pushout the $T(A)$-bimodule map $\tilde{\mu} : \tilde{C} \to T(A)$ in the following diagram:

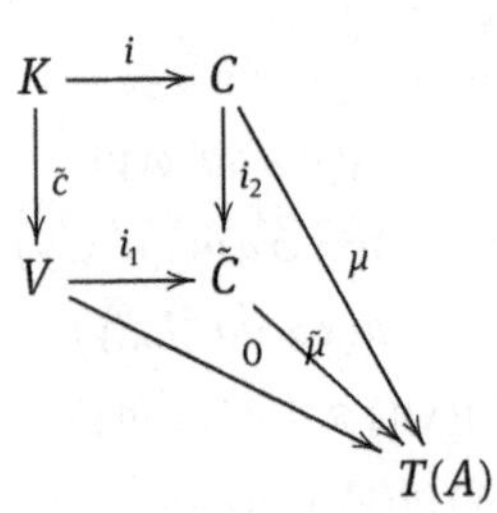

We claim that the pushout crossed module

$$0 \to K \xrightarrow{i_1} \tilde{C} \xrightarrow{\tilde{\mu}} T(A) \xrightarrow{\pi} A \to 0$$

is a preimage of the 3-cocycle c with respect to the map b. First of all, as remarked before, one can view the $T(A)$-bimodule as an associative algebra by defining $xy := x \cdot \tilde{\mu}(y)$ for all $x, y \in \tilde{C}$. Therefore, we obtain a crossed module.

In order to compute its 3-class, let $\rho : A \to T(A)$ be the linear section of $\pi : T(A) \to A$ given by $\rho(a) = a \in T(A)$, and define the section $\sigma : \mathrm{Im}(\mu) \to C$ by

$$\sigma(a \otimes a' - aa') = [1 \otimes (a \otimes a') \otimes 1]$$

for all $a \otimes a' - aa' \in \mathrm{Im}(\mu)$. It is clear that the map $i_2 \circ \sigma : \mathrm{Im}(\tilde{\mu}) \to \tilde{C}$ is a section of $\tilde{\mu} : \tilde{C} \to \mathrm{Im}(\tilde{\mu}) = \mathrm{Ker}(\pi)$. Let us compute a cocycle representing the class $b([\tilde{\gamma}])$, associated to the crossed module $\tilde{\mu} : \tilde{C} \to T(A)$. We will denote by β resp. $\tilde{\beta}$ the 2-cochains constructed from the sections ρ and $\tilde{\rho}$ resp. in the process of the definition of the 3-class. For all a, a', a'' in A, we obtain

$$\begin{aligned} i_1(\tilde{\gamma})(a,a',a'') &= d\tilde{\beta}(a \otimes a' \otimes a'') \\ &= [(0, \beta(a \otimes a' \otimes a''))] \\ &= [(0, [a \otimes a' \otimes a'' \otimes 1 - 1 \otimes aa' \otimes a'' \otimes 1 \\ &\quad + 1 \otimes a \otimes a'a'' \otimes 1 - 1 \otimes a \otimes a' \otimes a''])] \\ &= [(c \circ h(a \otimes a' \otimes a'' \otimes 1 - 1 \otimes aa' \otimes a'' \otimes 1 \\ &\quad + 1 \otimes a \otimes a'a'' \otimes 1 - 1 \otimes a \otimes a' \otimes a''), 0)] \\ &= [(c([1 \otimes a \otimes a' \otimes a'' \otimes 1 - 1 \otimes 1 \otimes aa' \otimes a'' \otimes 1 \\ &\quad + 1 \otimes 1 \otimes a \otimes a'a'' \otimes 1 - 1 \otimes 1 \otimes a \otimes a' \otimes a'']), 0)]. \end{aligned}$$

But we have

$$\begin{aligned} &d_4(1 \otimes 1 \otimes a \otimes a' \otimes a'' \otimes 1) \\ &\quad = 1 \otimes a \otimes a' \otimes a'' \otimes 1 - 1 \otimes a \otimes a' \otimes a'' \otimes 1 \\ &\qquad + 1 \otimes 1 \otimes aa' \otimes a'' \otimes 1 - 1 \otimes 1 \otimes a \otimes a'a'' \otimes 1 + 1 \otimes 1 \otimes a \otimes a' \otimes a'' \\ &\quad = 1 \otimes 1 \otimes aa' \otimes a'' \otimes 1 - 1 \otimes 1 \otimes a \otimes a'a'' \otimes 1 + 1 \otimes 1 \otimes a \otimes a' \otimes a''. \end{aligned}$$

Thus we have

$$[1 \otimes 1 \otimes aa' \otimes a'' \otimes 1 - 1 \otimes 1 \otimes a \otimes a'a'' \otimes 1 + 1 \otimes 1 \otimes a \otimes a' \otimes a''] = 0$$

in $A^{\otimes 5}/\mathrm{Im}(d_4)$. Our expression reduces thus to

$$i_1(\tilde{\gamma})(a,a',a'') = [(c([1 \otimes a \otimes a' \otimes a'' \otimes 1]), 0)] = i_1(c(a \otimes a' \otimes a'')).$$

This shows that the constructed crossed module is a preimage of the class $[c]$ in $HH^3(A, V)$, and thus the surjectivity of the map b.

We close this subsection by showing the injectivity of the map b. We will use the above constructed crossed module $\mu : C \to T(A)$ and its class $b([\mu : C \to T(A)]) = [c]$ again. Let

$$0 \to V \xrightarrow{i'} M \xrightarrow{\mu'} N \to A \to 0$$

be another crossed module with the same class $[c]$. We will show that $\mu' : M \to N$ is elementary equivalent to $\mu : C \to T(A)$. This implies injectivity of the map b. The assumption that $\mu' : M \to N$ has class $[c]$ means that taking sections $\rho' : A \to N$ and $\sigma' : \mathrm{Im}(\mu') \to M$, the corresponding 3-cocycle γ' (constructed via α', and then β') differs from c by a 3-coboundary, that is, there exists $\lambda \in C^2(A, V)$ such that $c = \gamma' + d\lambda$. We have by construction a commutative diagram:

$$\begin{array}{ccc} A^{\otimes 2} & \xrightarrow{\beta' + i'\lambda} & M \\ & \searrow^{\alpha'} & \downarrow{\scriptstyle \mu'} \\ & & \mathrm{Im}(\mu') \end{array}$$

As $\mathrm{Im}(\alpha') \subset \mathrm{Im}(\mu')$, let us consider a basis $\{e_i\}_{i \in I}$ of $\mathrm{Im}(\alpha')$, which one completes to a basis of $\mathrm{Im}(\mu')$, denoted $\{e_i\}_{i \in J}$ with $I \subset J$. For all $i \in I$, choose $y_i \in (\alpha')^{-1}(\{e_i\})$ and set $t(e_i) := (\beta' + i' \circ \lambda)(y_i)$. For $j \in J \setminus I$, set $t'(e_j) := c_j$ where c_j is any preimage of e_j with respect to μ'. In this way, we have constructed a linear map $t' : \mathrm{Im}(\mu') \to M$ such that $\mu' \circ t' = \mathrm{id}_{\mathrm{Im}(\mu')}$ and

$$t' \circ \alpha' = \beta' + i \circ \lambda.$$

By construction, this section t' (together with ρ') leads to the 3-cocycle $c = \gamma' + d\lambda$. Now consider the diagram

$$\begin{array}{ccccccccc} 0 & \longrightarrow & K & \xrightarrow{i} & C & \xrightarrow{\mu} & T(A) & \xrightarrow{\pi} & A & \longrightarrow & 0 \\ & & \downarrow{\scriptstyle \bar{c}} & & \downarrow{\scriptstyle \bar{g}} & & \downarrow{\scriptstyle \bar{\rho}} & & \| & & \\ 0 & \longrightarrow & V & \longrightarrow & M & \xrightarrow{\mu'} & N & \longrightarrow & A & \longrightarrow & 0 \end{array}$$

where $\bar{\rho} : T(A) \to N$ is the algebra morphism induced by ρ' and $\bar{g} : C \to M$ is the linear map induced by $\beta' : A^{\otimes 2} \to M$. The square most to the right is clearly commutative, as ρ is a section of $N \to A$ and $\bar{\rho}$ extends it. Let us verify the commutativity of the other two squares: Let $x, y \in T(A)$ and $a, a' \in A$. Then

$$\begin{aligned} \mu'(\bar{g}([x \otimes (a \otimes a') \otimes y])) &= \mu'(\bar{\rho}(x) \cdot \beta'(a \otimes a') \cdot \bar{\rho}(y)) \\ &= \bar{\rho}(x)\mu'(\beta'(a \otimes a'))\bar{\rho}(y) \\ &= \bar{\rho}(x)(\bar{\rho}(a)\bar{\rho}(a') - \bar{\rho}(aa'))\bar{\rho}(y) \end{aligned}$$

$$
\begin{aligned}
&= \bar{\rho}(x \otimes a \otimes a' \otimes y - x \otimes (aa') \otimes y) \\
&= \bar{\rho}(\mu([x \otimes (a \otimes a') \otimes y])).
\end{aligned}
$$

The left most square commutes because the section t' (together with ρ') leads to the 3-cocycle c (inducing $\tilde{c}$) via β' (which induces $\bar{g}$). Now we have the following situation:

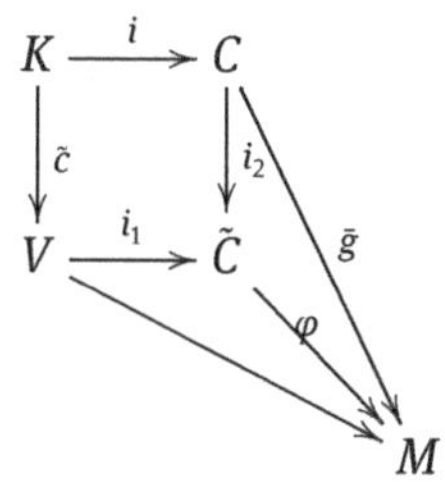

The map of $T(A)$-bimodules $\varphi : \tilde{C} \to M$ exists uniquely by the universal property of the pushout. Therefore, the diagram

$$
\begin{array}{ccccccccccc}
0 & \longrightarrow & V & \xrightarrow{i_1} & \tilde{C} & \xrightarrow{\tilde{\mu}} & T(A) & \xrightarrow{\pi} & A & \longrightarrow & 0 \\
 & & \| & & \downarrow{\scriptstyle \varphi} & & \downarrow{\scriptstyle \bar{\rho}} & & \| & & \\
0 & \longrightarrow & V & \longrightarrow & M & \xrightarrow{\mu'} & N & \longrightarrow & A & \longrightarrow & 0
\end{array}
$$

is an elementary equivalence. This shows that the map b is injective and completes the proof of Theorem 3.2.1.

3.3 Relation to third cohomology II

Exactly as with groups and Lie algebras, we sketch in this section a different point of view on crossed modules of associative algebras. Instead of considering the four-term exact sequence

$$0 \to V \xrightarrow{i} M \xrightarrow{\mu} N \xrightarrow{\pi} A \to 0,$$

we rather think of $\mu : M \to N$ as being spliced together from a square-zero extension with nucleus Z

$$0 \to Z \to \widehat{L} \xrightarrow{p} L \to 0,$$

of the ideal L in a general extension of associative algebras

$$0 \to L \to \widehat{A} \to A \to 0.$$

Recall square-zero extensions from Appendix C. As Z is in the nucleus of $\widehat{L}$ (meaning that L acts trivially on Z), L acts on $\widehat{L}$ in a well-defined way by $p(l)\cdot l' := ll'$ and $l\cdot p(l') := ll'$. The induced bimodule action of L on Z is then trivial. As before, in order for these two to form a crossed module, this bimodule action of L on $\widehat{L}$ should extend to a bimodule action of $\widehat{A}$ on $\widehat{L}$. We have seen in Appendix C that the product in a square-zero extension takes the following form in product coordinates:

$$(z,l)(z',l') = (l\cdot z' + z\cdot l' + \omega(l,l'), ll') = (\omega(l,l'), ll'),$$

where $\omega \in Z^2(L,Z)$ is a Hochschild 2-cocycle determining the square-zero extension up to equivalence and where the action of L on Z is trivial. We denote the extension then in product coordinates $\widehat{L} = Z \times_\omega L$. Observe that the cocycle identity for ω reads for all $l,l',l'' \in L$ simply $\omega(ll',l'') = \omega(l,l'l'')$, because the action terms are zero.

The extension of the action of L on $\widehat{L}$ to an action of $\widehat{A}$ on $\widehat{L}$ must also extend the bimodule action of A on Z, denoted $(a,z,a') \mapsto a\cdot z\cdot a'$, and is then of the form

$$a\cdot(z,l)\cdot a' = (a\cdot z\cdot a' + \theta(a,l,a'), ala'), \tag{3.2}$$

with some map $\theta : \widehat{A}\times L\times\widehat{A} \to Z$. Observe that $ala' \in L$, because L is a two-sided ideal of $\widehat{A}$. Instead of the map θ, one may also think of the left and right actions θ_l and θ_r defined by $\theta(a,l,0) =: \theta_l(a,l) =: \theta_{l,a}(l)$ and $\theta(0,l,a) =: \theta_r(a,l) =: \theta_{r,a}(l)$. The maps $\theta_l : \widehat{A}\oplus L \to Z$ and $\theta_r : L\oplus\widehat{A} \to Z$ are linear. Property (b) of a crossed module implies for $l,l'' \in L \subset \widehat{A}$,

$$l\cdot(z',l') = (z,l)(z,l') = (\omega(l,l'), ll') \quad\text{and}\quad (z',l')\cdot l'' = (z,l')(z'',l'') = (\omega(l',l''), l'l'').$$

This means that $\theta_r|_{L\times L} = \theta_l|_{L\times L} = \omega$.

Lemma 3.3.1. *Fix an element $a \in \widehat{A}$.*

(a) *The linear maps $\theta_{l,a}, \theta_{r,a} \in \mathrm{Hom}(L,Z)$ define maps*

$$\widehat{L}\to\widehat{L},\quad (z,l)\mapsto (a\cdot z + \theta_{l,a}(l), al) \quad\text{and}\quad (z,l)\mapsto (z\cdot a + \theta_{r,a}(l), la),$$

which are compatible with the algebra product if and only if:
 (i) $\theta_{l,a}(ll') + a\cdot\omega(l,l') = \omega(al,l')$;
 (ii) $\omega(la,l') = \omega(l,al')$;
 (iii) $\theta_{r,a}(ll') + \omega(l,l')\cdot a = \omega(l,l'a)$;
 for all $l,l' \in L$ and all $a \in \widehat{A}$.

(b) *Suppose that the maps $\theta_l,\theta_r : \widehat{A} \to \mathrm{Hom}(L,Z)$ satisfy the conditions (i), (ii) and (iii) stated in (a). Then (θ_l,θ_r) define a bimodule action of $\widehat{A}$ on $\widehat{L}$ by homomorphisms via the formula*

$$a\cdot(z,l)\cdot a' = (a\cdot z\cdot a' + \theta(a,l,a'), ala')$$

if and only if:

(i) $a \cdot \theta_{l,a'}(l) + \theta_{l,a}(a'l) = \theta_{l,aa'}(l)$;
(ii) $\theta_{l,a}(l) \cdot a' + \theta_{r,a'}(al) = a \cdot \theta_{r,a'}(l) + \theta_{l,a}(la')$;
(iii) $\theta_{r,a}(l) \cdot a' + \theta_{r,a'}(la) = \theta_{r,aa'}(l)$;
where $a, a' \in \widehat{A}$ and $l \in L$.

Proof.

(a) The compatibility of the action of $\widehat{A}$ on $\widehat{L}$ with the algebra product means $a \cdot (bb') = (a \cdot b)b'$, $(b \cdot a)b' = b(a \cdot b')$ and $(bb') \cdot a = b(b' \cdot a)$ for all $a \in \widehat{A}$ and all $b, b' \in \widehat{L}$. Writing $b = (z, l)$ and $b' = (z', l')$, we obtain for the first condition

$$a \cdot (z, l)(z', l') = a \cdot (\omega(l, l'), ll') = (a \cdot \omega(l, l') + \theta_{l,a}(ll'), all')$$

on the one hand, and on the other hand,

$$(a \cdot b)b' = (a \cdot z + \theta_{l,a}(l), al)(z', l') = (\omega(al, l'), (al)l').$$

Comparing the first components, we get the first condition.
The second condition gives on the one hand,

$$((z, l) \cdot a)(z', l') = (z \cdot a + \theta_{r,a}(l), la)(z', l') = (\omega(la, l'), (la)l'),$$

and on the other hand,

$$(z, l)(a \cdot (z', l')) = (z, l)(a \cdot z' + \theta_{l,a}(l'), al') = (\omega(l, al'), l(al')).$$

Comparing the first components, we get the second condition.
The third condition gives one the one hand,

$$(z, l)(z', l') \cdot a = (\omega(l, l'), ll') \cdot a = (\omega(l, l') \cdot a + \theta_{r,a}(ll'), (ll')a),$$

and on the other hand,

$$(z, l)((z', l') \cdot a) = (z, l)(z' \cdot a + \theta_{r,a}(l'), l'a) = (\omega(l, l' \cdot a), l(l'a)).$$

Comparing the first components, we get the third condition.

(b) Here, the conditions are for all $a, a' \in \widehat{A}$ and all $b = (z, l) \in \widehat{L}$ that $a \cdot (a' \cdot b) = aa' \cdot b$, $a \cdot (b \cdot a') = (a \cdot b) \cdot a'$ and $b \cdot (aa') = (b \cdot a) \cdot a'$.
The first condition gives on the one hand,

$$a \cdot (a' \cdot (z, l)) = a \cdot (a' \cdot z + \theta_{l,a'}(l), a'l) = (a \cdot (a' \cdot z + \theta_{l,a'}(l)) + \theta_{l,a}(a'l), a(a'l)),$$

and on the other hand,

$$aa' \cdot (z, l) = (aa' \cdot z + \theta_{l,aa'}(l), aa'l).$$

Comparing the first components, we get the first condition.
The second condition gives on the one hand,

$$a \cdot (b \cdot a') = a \cdot (z \cdot a' + \theta_{r,a'}(l), la') = (a \cdot (z \cdot a' + \theta_{r,a'}(l)) + \theta_{l,a}(la'), a(la')),$$

and on the other hand,

$$(a \cdot b) \cdot a' = (a \cdot z + \theta_{l,a}(l), al) \cdot a' = ((a \cdot z + \theta_{l,a}(l)) \cdot a' + \theta_{r,a'}(al), (al)a').$$

Comparing the first components, we get the second condition.
The third condition gives on the one hand,

$$b \cdot (aa') = (z, l) \cdot aa' = (z \cdot aa' + \theta_{r,aa'}(l), l(aa')),$$

and on the other hand,

$$\begin{aligned}(b \cdot a) \cdot a' &= ((z, l) \cdot a) \cdot a' = (z \cdot a + \theta_{r,a}(l), la) \cdot a' \\ &= ((z \cdot a + \theta_{r,a}(l)) \cdot a' + \theta_{r,a'}(la), (la)a').\end{aligned}$$

Comparing the first components, we get the third condition. □

The action of $\widehat{A}$ on $\widehat{L}$ is determined by the map θ. The previous lemma states conditions on θ to be an algebra action of $\widehat{A}$ by algebra automorphisms on $\widehat{L}$. In case we are given a crossed module, these conditions are fulfilled and θ extends. As before for groups and Lie algebras, we can use this extension to define the 3-class associated to the crossed module. Extend θ_l and θ_r to $\tilde{\theta} : \widehat{A} \times \widehat{A} \to Z$ such that $\tilde{\theta}|_{\widehat{A}\times L} = \theta_l$ and $\tilde{\theta}|_{L\times\widehat{A}} = \theta_r$. Note that this is possible. Indeed, on $L \times L$, both coincide and give ω. One extends from $\widehat{A} \times L$ and from $L \times \widehat{A}$ to $\widehat{A} \times \widehat{A}$ using a section ρ of the quotient map $\pi : \widehat{A} \to A$.

We obtain thus a 3-cocycle $d\tilde{\theta}$ which vanishes by construction on $L \times \widehat{A}^2$, $\widehat{A}^2 \times L$ and $\widehat{A} \times L\widehat{A}$ because of the cocycle identities in part (b) of the lemma. Therefore, $d\tilde{\theta}$ induces a 3-cocycle in $Z^3(A, Z)$ using $\widehat{A}/L \cong A$.

We leave it to the interested reader to investigate the independence of the class of $d\tilde{\theta}$ from the choices of the extension $\tilde{\theta}$ and the choice of the cocycle ω; cf. Lemma 2.3.2, and to show that the corresponding cohomology class coincides with the previously defined class; cf. Proposition 2.3.3.

3.4 Relation to general extensions

In this section, we suppose that in a crossed module

$$0 \to V \to M \to N \xrightarrow{\pi} A \to 0,$$

the algebras A and N are unital and that the morphism π respects the unit.

We call *general extension* of A by M an exact sequence of associative algebras

$$0 \to L \xrightarrow{i} E \xrightarrow{\pi} A \to 0,$$

where the multiplication on L is not necessarily trivial (observe that L is in general not unital). A general extension gives rise to an *outer A-bimodule structure* on L, namely

$$a \cdot l := \rho(a)i(l), \quad l \cdot a := i(l)\rho(a),$$

where $\rho : A \to E$ is a linear section of π. As there need not exist a ρ, which is a morphism of associative algebras, this defines a bimodule action only up to multiplication by an element; cf Remark 3.1.4(c). Thus to any general extension we can associate a morphism of algebras $\varphi : A \to M(L)$, the algebra of multiplications of L; cf. Remark 3.1.4(c).

We have seen in Remark 3.1.4 that we can also associate to a crossed module an outer bimodule action. The question which we address in this section is whether in return every outer bimodule action comes from a general extension. Once again, the answer will be that to an outer bimodule action, one can associate a crossed module, called the *obstruction crossed module*, such that its class is zero if and only if there exists a general extension whose associated outer bimodule action is the action we started with. While we show this here concentrating on the extension side, we will focus on the *associative algebra kernel* side in the next section.

Suppose given a crossed module

$$0 \to V \to M \xrightarrow{\mu} N \to A \to 0.$$

The following definition expresses the fact that the outer bimodule action associated to it is realized by a general extension.

Definition 3.4.1. A *μ-extension* of A by M is a commutative diagram with exact rows

$$\begin{array}{ccccccccccc} & & 0 & \longrightarrow & M & \xrightarrow{i} & S & \xrightarrow{\pi} & A & \longrightarrow & 0 \\ & & & & \| & & \downarrow{\scriptstyle \xi} & & \| & & \\ 0 & \longrightarrow & V & \xrightarrow{\text{incl}} & M & \xrightarrow{\mu} & N & \xrightarrow{p} & A & \longrightarrow & 0 \end{array}$$

where S is a unital associative algebra and $\pi : S \to A$ a morphism of unital associative algebras. Furthermore, we assume that $i(x)s = i(x \cdot \xi(s))$ and $si(x) = i(\xi(s) \cdot x)$ holds for all $x \in M$ and all $s \in S$.

Observe that $i(x)s = i(x \cdot \xi(s))$ can be interpreted as $i(i(x)s) = i(x \cdot \xi(s))$ (the product of an element of the ideal $i(M)$ and an element of S is in $i(M)$!), and thus implies $xs = i(x)s = x \cdot \xi(s)$ by injectivity of i, viewing xs in $i(M)$. The same holds for $si(x) = i(\xi(s) \cdot x)$.

Lemma 3.4.1. *Given a μ-extension of A by M, the associative product in M coincides with*

$$xy = \mu(x) \cdot y = x \cdot \mu(y)$$

for all $x, y \in M$. Moreover, one has a square zero extension

$$0 \to V \xrightarrow{i|_V} S \xrightarrow{\xi} N \to 0, \tag{3.3}$$

where we consider $V \subset M$ as a subspace.

Proof. The relations $xy = \mu(x) \cdot y = x \cdot \mu(y)$ for all $x, y \in M$ hold by definition in the crossed module $\mu : M \to N$. The product in V is trivial, thus it remains to show that the sequence is exact. $\xi \circ i|_V = \mu \circ \mathrm{incl} = 0$ by exactness of the four-term exact sequence of the crossed module. Moreover, $i|_V$ is injective. Suppose that $s \in S$ is in $\mathrm{Ker}(\xi)$. Then $\pi(s) = p(\xi(s)) = 0$, thus there exists $m \in M$ such that $i(m) = s$. Then $\mu(m) = \xi(i(m)) = 0$ and, therefore, there exists $v \in V = \mathrm{Ker}(\mu)$ with $v = \mathrm{incl}(v) = m$. Finally, $i|_V(v) = s$.

For the surjectivity of ξ, let $n \in N$ be given. For $p(n) \in A$, there exists a preimage $s \in S$ such that $\pi(s) = p(n)$. Thus $p(\xi(s) - n) = \pi(s) - p(n) = 0$, and thus there exists $m \in M$ with $\mu(m) = \xi(s) - n$. Now $\xi(s - i(m)) = \xi(s) - \mu(m) = n$. □

Observe that the outer bimodule action in the sequence (3.3) is the natural bimodule action of N on V associated to the crossed module. The equivalence relation on crossed modules induces a relation on μ-extensions. Let us denote by $\mathrm{Ext}_\mu(A, M)$ the equivalence classes of μ-extensions.

Theorem 3.4.2. *The equivalence class of a crossed module*

$$0 \to V \to M \xrightarrow{\mu} N \xrightarrow{p} A \to 0$$

is the zero class in $HH^3(A, V)$ if and only if $\mathrm{Ext}_\mu(A, V)$ is nonempty. If this is the case, the abelian group $HH^2(A, V)$ acts transitively and effectively on $\mathrm{Ext}_\mu(A, V)$.

Proof. Consider sections $\rho : A \to N$ of p and $\sigma : \mathrm{Im}(\mu) \to M$ of μ, which we need in order to define the 3-class associated to the crossed module $\mu : M \to N$. We will assume $\rho(1) = 1$. The 3-class associated to $\mu : M \to N$ is represented by the cocycle

$$d\beta(x, y, z) = \rho(x) \cdot \beta(y, z) - \beta(xy, z) + \beta(x, yz) - \beta(x, y) \cdot z,$$

where $\beta(x, y) = \sigma\alpha(x, y)$ and $\alpha(x, y) = \rho(x)\rho(y) - \rho(xy)$ for all $x, y, z \in A$.

Given a μ-extension of A by M

$$\begin{array}{ccccccccccc} & & 0 & \longrightarrow & M & \xrightarrow{i} & S & \xrightarrow{\pi} & A & \longrightarrow & 0 \\ & & & & \| & & \downarrow{\scriptstyle \xi} & & \| & & \\ 0 & \longrightarrow & V & \xrightarrow{\mathrm{incl}} & M & \xrightarrow{\mu} & N & \xrightarrow{p} & A & \longrightarrow & 0 \end{array}$$

choose a linear section $\tau : N \to S$ of ξ with $\tau(1) = 1$. This is possible by the surjectivity of ξ, which we have already seen in the proof of Lemma 3.4.1.

By composition, we obtain a section $\zeta := \tau \circ \rho : A \to S$ of the quotient map $\pi : S \to A$. One defines its associated 2-cochain $\alpha'(x,y) := \zeta(x)\zeta(y) - \zeta(xy)$ for all $x, y \in A$. Observe that α' is a 2-cochain on A with values in $\mathrm{Im}(i) = M$, thus there exists $\beta'(x,y) \in M$ with $i(\beta'(x,y)) = \alpha'(x,y)$. We claim that

$$\rho(x) \cdot \beta'(y,z) - \beta'(xy,z) + \beta'(x,yz) - \beta'(x,y) \cdot \rho(z) = 0. \tag{3.4}$$

Indeed, recall that by definition in a μ-extension

$$\xi(\zeta(x)) \cdot \beta'(y,z) = \rho(x) \cdot \beta'(y,z),$$

because for all $s \in S$ and all $c \in M$, we have $sc = \xi(s) \cdot c$ (here sc is the product in S, the result lies in the ideal M). We thus obtain

$$\rho(x) \cdot \beta'(y,z) = \zeta(x)\zeta(y)\zeta(z) - \zeta(x)\zeta(yz) \in S.$$

In the same way, we obtain

$$\beta'(x,y) \cdot \rho(z) = \zeta(x)\zeta(y)\zeta(z) - \zeta(xy)\zeta(z) \in S.$$

Therefore, as an element in S,

$$\begin{aligned}
&\rho(x) \cdot \beta'(y,z) - \beta'(xy,z) + \beta'(x,yz) - \beta'(x,y) \cdot \rho(z) \\
&\quad = \zeta(x)\zeta(y)\zeta(z) - \zeta(x)\zeta(yz) \\
&\qquad - \zeta(xy)\zeta(z) + \zeta(xyz) + \zeta(x)\zeta(yz) - \zeta(xyz) - \zeta(x)\zeta(y)\zeta(z) + \zeta(xy)\zeta(z) \\
&\quad = 0.
\end{aligned}$$

Since $\beta(x,y) = \sigma\mu\beta'(x,y)$ (because $\mu\beta'(x,y) = \xi(i(\beta'(x,y))) = \xi(\alpha'(x,y)) = \alpha(x,y)$), it follows that $\beta''(x,y) := \beta(x,y) - \beta'(x,y)$ lies in V. We obtain a well-defined map $\beta'' : A \otimes A \to V$. By equation (3.4), we have

$$d\beta(x,y,z) = x \cdot \beta''(y,z) - \beta''(xy,z) + \beta''(x,yz) - \beta''(x,y) \cdot z,$$

and this means that the class of $\mu : M \to N$ in $HH^3(A,V)$ is zero (because it is the coboundary of the cochain $\beta'' : A \otimes A \to V$).

Given any normalized 2-cocycle $\omega : A \otimes A \to V$, we can define a new μ-extension S_ω of A by M by putting on $S_\omega := M \oplus A$ the product

$$(m,a)(m',a') := (mm' + \rho(a) \cdot m' + m \cdot \rho(a') + \omega(a,a'), aa'),$$

where $mm' = m \cdot \mu(m')$ is the product from Lemma 3.4.1. This construction yields a transitive and effective action of $HH^2(A,V)$ on $\mathrm{Ext}_\mu(A,V)$.

Conversely, assume that the class $[\gamma]$ of $\mu : M \to N$ is the zero class in $HH^3(A, V)$. As usual, we suppose that γ is constructed using a section ρ and forming then α and β via $d\beta = \gamma$.

Then there exists a linear map $\beta' : A \otimes A \to V$ such that

$$\gamma(x,y,z) = x \cdot \beta'(y,z) - \beta'(xy,z) + \beta'(x,yz) - \beta'(x,y) \cdot z.$$

Now define $\beta'' : A \otimes A \to M$ by $\beta''(x,y) := \beta(x,y) - \beta'(x,y)$. This new cochain β'' is then clearly a cocycle and may serve to define an extension $S := M \oplus A$ with the product

$$(m,a)(m',a') := (mm' + \rho(a) \cdot m' + m \cdot \rho(a') + \beta''(a,a'), aa').$$

Together with the obvious inclusion $M \to S$ and projection $S \to A$ and $\xi : S \to N$, $\xi(m,a) := \mu(m) + \rho(a)$, this gives then a μ-extension. □

3.5 Construction of crossed modules

Here, we transpose the methods of proof of Gerstenhaber's theorem using the construction of crossed modules to the framework of associative algebras. Exactly as in the framework of groups and Lie algebras, the point of view is here to glue a crossed module

$$0 \to V \to M \xrightarrow{\mu} N \to A \to 0$$

from short exact sequences. More precisely, given an associative algebra A, a short exact sequence of A-bimodules

$$0 \to V_1 \to V_2 \to V_3 \to 0 \tag{3.5}$$

and a square zero extension of A by V_3 (viewed as an associative algebra with zero product)

$$0 \to V_3 \to E \to A \to 0, \tag{3.6}$$

the splicing together of both sequences will give a crossed module.

Theorem 3.5.1. *The Yoneda product of the sequences (3.5) and (3.6) is in a natural way a crossed module whose associated 3-class is the image under the connecting homomorphism in the long exact sequence in cohomology corresponding to the short exact sequence of coefficients (3.5).*

Proof. The map $\mu : V_2 \to E$, as part of the data of the crossed module, is obtained by the composition of $V_2 \to V_3$ and $V_3 \to E$. In product coordinates, $E = V_3 \oplus A$ as a vector space. The algebra A acts on V_2 by the given bimodule action, and we set V_3 to act on

V_2 trivially. This defines an action on E on the (trivial) associative algebra V_2, which is compatible with the product. It remains to show conditions (a) and (b) of a crossed module. Condition (b) is trivially true, because the elements coming from $V_3 \subset E$ act trivially on V_2. Condition (a) is true, because the product in the square zero extension E reduces to the bimodule action for elements of the form $(v, 0) \in V_3 \oplus A = E$ and $(0, a) \in V_3 \oplus A = E$.

The second part of the proof is formally exactly the same as in the case of groups or Lie algebras and left to the reader. □

Once again, this construction method may serve for showing surjectivity in Gerstenhaber's theorem. More precisely, we show that every equivalence class of crossed modules admits a representative which is of the above form, that is, spliced together from a square zero extension and a short exact sequence of coefficients. We call such a representative an *abelian representative* of the given class.

Theorem 3.5.2. *Let $[\mu : M \to N]$ be an equivalence class of crossed modules. Then there exists an abelian representative of the class $[\mu : M \to N]$.*

Proof. We have seen that the equivalence class $[\mu : M \to N]$ is equally well described by the 3-cohomology class $[\theta] \in HH^3(A, V)$, obtained in Theorem 3.2.1, where $V = \mathrm{Ker}(\mu)$ and $A = \mathrm{Coker}(\mu)$. As already described in detail in the beginning of this chapter, V is an A-bimodule in a natural way. Now, the category of A-bimodules has enough injectives, because it is simply the category of modules over the *enveloping algebra* $A \otimes A^{\mathrm{op}}$ of A (and the category of modules over an algebra has enough injective objects by a standard argument; see, for example, [103] Exercise 2.3.5, p. 42). Therefore, V may be embedded into an injective A-bimodule I, with quotient A-bimodule Q, that is, we have a short exact sequence of A-bimodules,

$$0 \to V \to I \to Q \to 0.$$

This short exact sequence induces a long exact sequence in cohomology, part of which looks like

$$\cdots \to HH^2(A, I) \to HH^2(A, Q) \xrightarrow{\partial} HH^3(A, V) \to HH^3(A, I) \to \cdots.$$

By injectivity of I, $HH^2(A, I) = HH^3(A, I) = 0$, thus the connecting homomorphism $\partial : HH^2(A, Q) \to HH^3(A, V)$ is an isomorphism. The cohomology class $[\theta] \in HH^3(A, V)$ corresponds thus under ∂^{-1} to a cohomology class $[\alpha] \in HH^2(A, Q)$. The cocycle α gives then rise to a square zero extension

$$0 \to Q \to E \to A \to 0.$$

The splicing together of this square zero extension with the above injective presentation of V gives by Theorem 3.5.1 a crossed module whose cohomology class is $\partial[\alpha] = [\theta]$. Thus the constructed crossed module is equivalent to $[\mu : M \to N]$ by Theorem 3.2.1. □

Remark 3.5.1.
(a) Observe that we obtain as a corollary once again the surjectivity in Theorem 3.2.1. There are thus (at least) two methods of constructing crossed modules. Either by free algebras or by injective bimodules, that is, one resolves either the algebra or the bimodule.
(b) Observe that, exactly as in the case of groups or Lie algebras, we do not need an *injective* bimodule in the above proof. It would be enough to have a bimodule I in which V embeds with quotient Q such that the induced map $\partial : HH^2(A, Q) \to HH^3(A, V)$ is surjective.
(c) In fact, again exactly as in the case of groups and Lie algebras, this theorem generalizes to show that the classes in $HH^n(A, V)$ may be represented by exact sequences obtained from splicing together a square zero extension with an exact sequence of A-bimodules; see [36] and Exercise 2.11.7.

3.6 Associative algebra kernels

3.6.1 Definition of an associative algebra kernel

Recall the algebra of multiplications $M(L)$ of an associative algebra L. Its elements are pairs (u, v) of endomorphisms $u, v \in \mathrm{End}(L)$ such that u and v constitute left and right halves of a bimodule action on L. The elements of the algebra L itself give rise to such multiplications, called *inner multiplications*, and the quotient $\mathrm{Out}(L) := M(L)/L$ is called the *algebra of outer multiplications* of L. Observe that the left (resp., right) component of every inner multiplication commutes with the right (resp., left) component of every multiplication due to associativity in L. Indeed, let l^l be the left component of the inner multiplication by $l \in L$ and (u, v) be any multiplication. Then the property $v(l_1 l_2) = l_1 v(l_2)$ for all $l_1, l_2 \in L$ of the right component of the multiplication (u, v) implies $l^l \circ v = v \circ l^l$, because applied to any $\bar{l} \in L$, both sides give

$$l^l \circ v(\bar{l}) = l v(\bar{l}) = v(l\bar{l}) = v \circ l^l(\bar{l}).$$

In the following, we will be interested only in subsets of $M(L)$ consisting of multiplications for which every right component commutes with every left component.

Consider associative algebra homomorphisms $\varphi : A \to \mathrm{Out}(L)$. Any such homomorphism arises as composition of a linear map $A \to M(L)$ and the canonical quotient map $\pi : M(L) \to \mathrm{Out}(L)$. A linear map $\rho : A \to M(L)$ will be called *regular* if the right component of every element of $\rho(A)$ commutes with the left component of every element of $\rho(A)$. A homomorphism $\varphi : A \to \mathrm{Out}(L)$ will be called *regular* if there exists a regular map $\rho : A \to M(L)$ such that $\varphi = \pi \circ \rho$.

Definition 3.6.1. Let A and L be associative algebras over $\mathbb{K}$. A regular homomorphism $\varphi : A \to \mathrm{Out}(L)$ of A to $\mathrm{Out}(L)$ is called an *associative algebra kernel*.

For a given kernel $\varphi : A \to \mathrm{Out}(L)$, any regular map $\rho : A \to M(L)$ such that $\varphi = \pi \circ \rho$ will be called *compatible* with φ. We will feel free to write for $\rho(a) = (u, v)$ simply $a \cdot l$ for $u(l)$ and $l \cdot a$ instead of $v(l)$ for all $l \in L$.

The kernel of the canonical map $L \to M(L)$, which sends each element $l \in L$ to the corresponding left and right multiplication by l is called the *nucleus* of L and denoted $Z := Z(L)$ (do not confuse the nucleus with the center of L!). Any associative algebra kernel $\varphi : A \to \mathrm{Out}(L)$ induces the structure of an A-bimodule on $Z(L)$ (cf. Remark 3.1.4). In particular, $Z(L)$ carries a natural structure of an L-bimodule.

3.6.2 Cohomology class associated to a kernel

We now associate a 3-class to every associative algebra kernel $\varphi : A \to \mathrm{Out}(L)$ using the usual scheme (cf. Section 3.2.1): Let $\rho : A \to M(L)$ be a linear map compatible with the regular homomorphism φ. The multiplication $\rho(a_1)\rho(a_2) - \rho(a_1a_2)$ is an inner multiplication (because $\pi(\rho(a_1)\rho(a_2) - \rho(a_1a_2)) = \varphi(a_1)\varphi(a_2) - \varphi(a_1a_2) = 0$ as φ is a homomorphism, and $\mathrm{Ker}(\pi) = \mathrm{Im}(L \to M(L))$). Hence we can find a bilinear map $h : A \times A \to L$ such that for all $a_1, a_2 \in A$, $h(a_1, a_2) = \rho(a_1)\rho(a_2) - \rho(a_1a_2)$. The map h will be called a *hindrance* for the associative algebra kernel φ.

Lemma 3.6.1. *The hindrance $h : A \otimes A \to L$ gives rise to a 3-cocycle $dh \in Z^3(A, Z)$.*

Proof. We will write simply $a \cdot l$ for $\rho(a)l$. For all $a_1, a_2, a_3 \in A$ and all $l \in L$, we have

$$\begin{aligned} a_1 \cdot (a_2 \cdot (a_3 \cdot l)) &= a_1 \cdot (a_2a_3 \cdot l + h(a_2, a_3)l) \\ &= a_1a_2a_3 \cdot l + h(a_1, a_2a_3)l + (a_1 \cdot h(a_2, a_3)l), \end{aligned}$$

and on the other hand,

$$\begin{aligned} a_1 \cdot (a_2 \cdot (a_3 \cdot l)) &= a_1a_2 \cdot (a_3 \cdot l) + h(a_1, a_2)(a_3 \cdot l) \\ &= a_1a_2a_3 \cdot l + h(a_1a_2, a_3)l + (h(a_1, a_2) \cdot a_3)l. \end{aligned}$$

Hence we have $dh(a_1, a_2, a_3)l = 0$ for all $l \in L$ and for the formal Hochschild coboundary operator d (as if ρ was a homomorphism); see equation (C.1) in Appendix C. In a similar way, one finds $ldh(a_1, a_2, a_3) = 0$ for all $l \in L$. Therefore, $dh(a_1, a_2, a_3) \in Z$, the nucleus of L. We continue to show that the cochain $dh \in C^3(A, Z)$ is indeed a cocycle. For this, we cannot use the fact that $d^2 = 0$ (because ρ is not a homomorphism in general), but we have to show it.

Let now $a_1, a_2, a_3, a_4 \in A$ and compute

$$\begin{aligned} &ddh(a_1, a_2, a_3, a_4) \\ &\quad = a_1 \cdot dh(a_2, a_3, a_3) - dh(a_1a_2, a_3, a_4) + dh(a_1, a_2a_3, a_4) \\ &\qquad - dh(a_1, a_2, a_3a_4) + dh(a_1, a_2, a_3) \cdot a_4 \end{aligned}$$

$$
\begin{aligned}
&= a_1 \cdot (a_2 \cdot h(a_3, a_4) - h(a_2a_3, a_4) + h(a_2, a_3a_4) - h(a_2, a_3) \cdot a_4) \\
&\quad - a_1a_2 \cdot h(a_3, a_4) + h(a_1a_2a_3, a_4) - h(a_1a_2, a_3a_4) + h(a_1a_2, a_3) \cdot a_4 \\
&\quad + a_1 \cdot h(a_2a_3, a_4) - h(a_1a_2a_3, a_4) + h(a_1, a_2a_3a_4) - h(a_1, a_2a_3) \cdot a_4 \\
&\quad - a_1 \cdot h(a_2, a_3a_4) + h(a_1a_2, a_3a_4) - h(a_1, a_2a_3a_4) + h(a_1, a_2) \cdot a_3a_4 \\
&\quad + (a_1 \cdot h(a_2, a_3) - h(a_1a_2, a_3) + h(a_1, a_2a_3) - h(a_1, a_2) \cdot a_3) \cdot a_4 \\
&= a_1 \cdot (a_2 \cdot h(a_3, a_4)) - a_1a_2 \cdot h(a_3, a_4) + h(a_1, a_2) \cdot a_3a_4 - (h(a_1, a_2) \cdot a_3) \cdot a_4 \\
&= h(a_1, a_2)h(a_3, a_4) - h(a_1, a_2)h(a_3, a_4) = 0.
\end{aligned}
$$

Observe that this is exactly the same computation as in Lemma 3.2.3. □

Remark 3.6.2. There is another way of associating a 3-cohomology class to an associative algebra kernel $\varphi : A \to \mathrm{Out}(L)$. Namely, one can use the homomorphism φ to form the pullback crossed module

$$
\begin{array}{ccccccccc}
0 & \longrightarrow & Z & \longrightarrow & L & \longrightarrow & M(L) & \stackrel{p}{\longrightarrow} & \mathrm{Out}(L) & \longrightarrow & 0 \\
 & & \| & & \uparrow & & \uparrow & & \uparrow{\scriptstyle \varphi} & & \\
0 & \longrightarrow & Z & \stackrel{i}{\longrightarrow} & M & \stackrel{\mu}{\longrightarrow} & N & \stackrel{\pi}{\longrightarrow} & A & \longrightarrow & 0
\end{array}
$$

and then take the 3-class associated to the crossed module $\mu : M \to N$. Observe that this is exactly the same procedure as above, and thus leads to the same 3-cocycle and class.

Indeed, as $N = \{(m, a) \in M(L) \oplus A \mid \varphi(a) = p(m)\}$, the map $\rho : A \to M(L)$ (viewed as a map $\rho : A \to N, \rho(m, a) := (\rho(a), a))$ may serve as a section of π. The hindrance h is then the failure of ρ to be a homomorphism of associative algebras, and the 3-cocycle dh is exactly the cocycle we would associate to the crossed module. Note that this gives crossed modules where the left most term is necessarily the nucleus $Z \subset L$.

Lemma 3.6.2. *The cohomology class of dh in $HH^3(A, Z)$ does not depend on the choice of the compatible map ρ, but only on the L-kernel $\varphi : A \to \mathrm{Out}(L)$.*

Proof. Let ρ_1 and ρ_2 be linear maps from A to $M(L)$, which are both compatible with φ. Then we can find a linear map $g : A \to L$ such that for all $a \in A$, $\rho_1(a) - \rho_2(a)$ is the inner multiplication by the element $g(a) \in L$. Let h_1 and h_2 be the hindrances corresponding to ρ_1 and ρ_2. We indicate the A-operations in L defined by ρ_1 and ρ_2 using $\cdot$ and $*$, respectively. Associated to ρ_1 and ρ_2, we also have formal Hochschild coboundary operators d_1 and d_2. Recall that the A-operations via $\cdot$ and $*$ coincide on the nucleus Z of φ, and thus the coboundary operators d_1 and d_2 coincide on a cochain with values in Z. We have to show that there exists a cochain $f \in C^2(A, Z)$ such that $d_1h_1 - d_2h_2 = df$.

We compute for all $a_1, a_2 \in A$ and all $l \in L$,

$$
(h_1(a_1, a_2) - h_2(a_1, a_2))l = a_1 \cdot (a_2 \cdot l) - a_1a_2 \cdot l - a_1 * (a_2 * l) + a_1a_2 * l
$$

$$\begin{aligned}
&= -g(a_1a_2)l + a_1 \cdot (a_2 * l + g(a_2)l) - a_1 * (a_2 * l) \\
&= -g(a_1a_2)l + a_1 \cdot g(a_2)l + (a_1 \cdot - a_1 *)(a_2 * l) \\
&= -g(a_1a_2)l + a_1 \cdot g(a_2)l + g(a_1)(a_2 * l) \\
&= -g(a_1a_2)l + a_1 \cdot g(a_2)l + (g(a_1) * a_2)l.
\end{aligned}$$

A similar computation shows

$$l(h_1(a_1,a_2) - h_2(a_1,a_2)) = l(a_1 \cdot g(a_2) - g(a_1a_2) + g(a_1) * a_2).$$

Therefore, we get that for all $a_1, a_2 \in A$, the inner multiplication by $h_1(a_1,a_2)-h_2(a_1,a_2)$ is equal to the inner multiplication by $p(a_1,a_2) := a_1 \cdot g(a_2) - g(a_1a_2) + g(a_1) * a_2$.

Now we define $f \in C^2(A,Z)$ for all $a_1, a_2 \in A$ by

$$f(a_1,a_2) := h_1(a_1,a_2) - h_2(a_1,a_2) - p(a_1,a_2).$$

Observe that f has its values in Z, because the inner multiplications by $h_1(a_1,a_2) - h_2(a_1,a_2)$ and $p(a_1,a_2)$ coincide, thus the inner multiplication by their difference must be multiplication by an element of the nucleus.

Applying the coboundary operator d_1, we obtain

$$df = d_1f = d_1h_1 - d_1h_2 - d_1p.$$

Compute for all $a_1, a_2, a_3 \in A$

$$\begin{aligned}
d_1h_2(a_1,a_2,a_3) &= a_1 \cdot h_2(a_2,a_3) - h_2(a_1a_2,a_3) + h_2(a_1,a_2a_3) - h_2(a_1,a_2) \cdot a_3 \\
&= (a_1 * + g(a_1))h_2(a_2,a_3) - h_2(a_1a_2,a_3) + h_2(a_1,a_2a_3) \\
&\quad - h_2(a_1,a_2)(g(a_3) + * a_3) \\
&= d_2h_2(a_1,a_2,a_3) + g(a_1)h(a_2,a_3) + h(a_1,a_2)g(a_3),
\end{aligned}$$

and also

$$\begin{aligned}
d_1p(a_1,a_2,a_3) = a_1 \cdot (&a_2 \cdot g(a_3) - g(a_2a_3) + g(a_2) * a_3) \\
&- a_1a_2 \cdot g(a_3) + g(a_1a_2a_3) - g(a_1a_2) * a_3 \\
&+ a_1 \cdot g(a_2a_3) - g(a_1a_2a_3) + g(a_1) * a_2a_3 \\
&- (a_1 \cdot g(a_2) - g(a_1a_2) + g(a_1) * a_2) * a_3 \\
= h_2(&a_1,a_2)g(a_3) - g(a_1)h_2(a_2,a_3).
\end{aligned}$$

Hence we have $d_1h_1 = df + d_2h_2$. Observe that this proof is very similar to the proof of Lemma 3.2.4. □

3.6.3 Relation to general extensions

Now let B be a general extension of A, that is, B and A are associative algebras over the same base field $\mathbb{K}$ and there is a surjective homomorphism $\beta : B \to A$. Let us denote by L the ideal which is the kernel of β. Let $\zeta : A \to B$ be a linear section of β. Then considering for all $a \in A$ the left and right multiplication by $\zeta(a)$ defines the structure of an associative algebra kernel $\varphi : A \to \mathrm{Out}(L)$ on L. This kernel does not depend on the choice of ζ, because two different sections ζ_1, ζ_2 lead for all $a \in A$ to a multiplication by $\zeta_1(a) - \zeta_2(a)$, which is in the kernel of β, thus is an inner multiplication by an element of L. Obstruction theory now asks conversely whether each associative algebra kernel $\psi : A \to \mathrm{Out}(L)$ comes from a general extension $\beta : B \to A$, that is, whether there exists a general extension $\beta : B \to A$ such that the induced map $\varphi : A \to \mathrm{Out}(L)$ is equal to ψ. We call a kernel $\psi : A \to \mathrm{Out}(L)$ for which there exists such an extension $\beta : B \to A$ a *special kernel*.

Theorem 3.6.3. *A kernel $\psi : A \to \mathrm{Out}(L)$ is special if and only if its associated 3-class $[dh] \in HH^3(A, Z)$ is zero.*

Proof. Let $\psi : A \to \mathrm{Out}(L)$ be a kernel induced from an extension $\beta : B \to A$. Then for a section ζ of β, consider the map $\rho : A \to M(L)$, which associates to each $a \in A$ the multiplication by $\zeta(a)$. This ρ is compatible with the kernel ψ as we remarked before. Therefore, the bilinear map $h : A \times A \to L$ defined for all $a_1, a_2 \in A$ by

$$h(a_1, a_2) := \zeta(a_1)\zeta(a_2) - \zeta(a_1 a_2)$$

is a hindrance for the associative algebra kernel ψ. Now compute the Hochschild coboundary using ρ:

$$\begin{aligned} dh(a_1, a_2, a_3) &= \zeta(a_1)(\zeta(a_2)\zeta(a_2) - \zeta(a_1 a_2)) - \zeta(a_1 a_2)\zeta(a_3) + \zeta(a_1 a_2 a_3) \\ &\quad + \zeta(a_1)\zeta(a_2 a_3) - \zeta(a_1 a_2 a_3) - (\zeta(a_1)\zeta(a_2) - \zeta(a_1 a_2))\zeta(a_3) \\ &= 0 \end{aligned}$$

by the associativity in the algebra L. Therefore, the 3-cohomology class of a special kernel is zero.

Conversely, suppose that $\psi : A \to \mathrm{Out}(L)$ is a kernel whose associated 3-class is zero. Then there exists a compatible map $\rho : A \to M(L)$ and a corresponding hindrance h with $dh = 0$. Indeed, a priori the fact that the class of dh is zero implies only that dh is a coboundary df for a cochain $f \in C^2(A, Z)$, but it is clear in the definition of the hindrance that one may alter h by any 2-cochain with values in Z, because Z is the nucleus.

Now we can construct an extension $\beta : B \to A$ as follows: The vector space B is the direct sum $A \oplus L$. On $A \oplus L$, we define the associative product

$$(a_1, l_1)(a_2, l_2) := (a_1 a_2, a_1 \cdot l_2 + l_1 \cdot a_2 + h(a_1, a_2) + l_1 l_2).$$

We have seen in Appendix C that the fact that $h \in Z^2(A,L)$ is a 2-cocycle is equivalent to the associativity of this product. Hence B is an associative algebra. The morphism β is defined by $\beta : B \to A$, $(a,l) \mapsto a$. Then $\beta : B \to A$ is an extension with kernel L and the multiplication of A on L defined by this extension is obviously $l \mapsto a \cdot l = \rho(a)(l)$ on the left and similarly on the right. This completes the proof. Observe that this proof is very similar to the proof of Theorem 3.4.2. □

3.6.4 The different extensions corresponding to a special kernel

We determine in this subsection the structure of the set of extensions, which determine the same (special) associative algebra kernel $\varphi : A \to \mathrm{Out}(L)$. Let $\beta : B \to A$ and $\gamma : C \to A$ be two extensions of A with the same kernel $\mathrm{Ker}(\beta) = \mathrm{Ker}(\gamma) = L$ and such that the induced A-bimodule structure on L coincides. Therefore, there exist sections ζ of β and τ of γ such that the A-operations in L derived from ζ and τ coincide.

It follows that for all $a_1, a_2 \in A$, there exists an element $f(a_1,a_2) \in Z$, with Z the nucleus of the kernel $\varphi : A \to \mathrm{Out}(L)$, such that

$$\tau(a_1)\tau(a_2) - \tau(a_1a_2) = \zeta(a_1)\zeta(a_2) - \zeta(a_1a_2) + f(a_1,a_2),$$

because the hindrances corresponding to τ and ζ must coincide up to an element in the nucleus. Moreover, we have $df = 0$, that is, $f \in Z^2(A,Z)$, again because the hindrances corresponding to τ and ζ coincide up to f. Hence we may define an associative algebra $B(f,\beta)$ on the vector space B corresponding to f and $\beta : B \to A$ by defining for all $b_1, b_2 \in B$,

$$b_1 \circ b_2 := b_1b_2 + f(\beta(b_1),\beta(b_2)).$$

The associativity of this new multiplication on B reduces to $df = 0$. Furthermore, $\beta : B(f,\beta) \to A$, $b \mapsto \beta(b)$, is still a homomorphism onto A.

Lemma 3.6.4. *The extension $\gamma : C \to A$ is equivalent to $\beta : B(f,\beta) \to A$.*

Proof. For $c \in C$, we set

$$\sigma(c) := \zeta(\gamma(c)) + (c - \tau(\gamma(c))) \in B(f,\beta).$$

Observe first of all that $\beta(\sigma(c)) = \gamma(c) \in A$, because $c - \tau(\gamma(c)) \in \mathrm{Ker}(\gamma) = L = \mathrm{Ker}(\beta)$. Using the fact that ζ and γ define the same A-operations in L, we can verify that

$$\sigma(c_1)\sigma(c_2) = \zeta(\gamma(c_1))\zeta(\gamma(c_2)) + (c_1c_2 - \tau(\gamma(c_1))\tau(\gamma(c_2))).$$

Indeed,

$$\sigma(c_1)\sigma(c_2) = (\zeta(\gamma(c_1)) + (c_1 - \tau(\gamma(c_1))))(\zeta(\gamma(c_2)) + (c_2 - \tau(\gamma(c_2)))).$$

In the mixed terms, we have, using fact that ζ and γ define the same A-operations in L,

$$\begin{aligned}\zeta(\gamma(c_1))(c_2 - \tau(\gamma(c_2))) &= \tau(\gamma(c_1))(c_2 - \tau(\gamma(c_2))) \\ &= \tau(\gamma(c_1))c_2 - \tau(\gamma(c_1))\tau(\gamma(c_2))\end{aligned}$$

and in the same way

$$(c_1 - \tau(\gamma(c_1)))\zeta(\gamma(c_2)) = c_1\tau(\gamma(c_2)) - \tau(\gamma(c_1))\tau(\gamma(c_2)).$$

On the other hand, we have in C

$$(c_1 - \tau(\gamma(c_1)))(c_2 - \tau(\gamma(c_2))) = c_1c_2 - (\tau(\gamma(c_1))c_2 + c_1\tau(\gamma(c_2))) + \tau(\gamma(c_1))\tau(\gamma(c_2)).$$

It follows that we have in B

$$\sigma(c_1)\sigma(c_2) = \zeta(\gamma(c_1))\zeta(\gamma(c_2)) + (c_1c_2 - \tau(\gamma(c_1))\tau(\gamma(c_2))).$$

We deduce from here that

$$\sigma(c_1) \circ \sigma(c_2) = \sigma(c_1c_2).$$

Indeed, compute

$$\begin{aligned}\sigma(c_1) \circ \sigma(c_2) &= \sigma(c_1)\sigma(c_2) + f(\beta(\sigma(c_1)), \beta(\sigma(c_2))) \\ &= \zeta(\gamma(c_1))\zeta(\gamma(c_2)) + (c_1c_2 - \tau(\gamma(c_1))\tau(\gamma(c_2))) + f(\gamma(c_1), \gamma(c_2)) \\ &= \zeta(\gamma(c_1))\zeta(\gamma(c_2)) + (c_1c_2 - \tau(\gamma(c_1))\tau(\gamma(c_2))) \\ &\quad + \tau(\gamma(c_1))\tau(\gamma(c_2)) - \tau(\gamma(c_1c_2)) - \zeta(\gamma(c_1))\zeta(\gamma(c_2)) + \zeta(\gamma(c_1c_2)) \\ &= \zeta(\gamma(c_1c_2)) + (c_1c_2 - \tau(\gamma(c_1c_2))) = \sigma(c_1c_2).\end{aligned}$$

Thus $\sigma : C \to B(f, \beta)$ is a homomorphism of associative algebras.

Now, for a given $b \in B(f, \beta)$, we set

$$\theta(b) := \tau(\beta(b)) + (b - \zeta(\beta(b))).$$

Exactly as $\beta(\sigma(c)) = \gamma(c)$ for all $c \in C$, we have here $\gamma(\theta(b)) = \beta(b)$ for all $b \in B$. Observe that σ and θ are the identity maps on $L = \mathrm{Ker}(\beta) = \mathrm{Ker}(\gamma)$. Then $\beta(\sigma(\theta(b))) = \beta(b)$ and, therefore, $\sigma \circ \theta : B \to B$ restricts to L as the identity and to A as well, thus $\sigma \circ \theta$ must be the identity. Finally, both $\theta \circ \sigma$ and $\sigma \circ \theta$ are identity maps. Thus $\sigma : C \to B(f, \beta)$ is an isomorphism of extensions of A by L. □

Theorem 3.6.5. *Let $\beta : B \to A$ be an extension of A with kernel L and with nucleus Z, and let $f_1, f_2 \in Z^2(A, Z)$. Then the extensions $\beta : B(f_1, \beta) \to A$ and $\beta : B(f_2, \beta) \to A$ are equivalent if and only if f_1 and f_2 are cohomologuous.*

Proof. Suppose that $f_1 - f_2 = dg$ for some cochain $g \in C^2(A, Z)$. For $b \in B(f_1, \beta)$, define $\sigma(b) := b + g(\beta(b)) \in B(f_2, \beta)$. We denote by $\circ$ the multiplication in $B(f_1, \beta)$ and by $*$ the multiplication in $B(f_2, \beta)$. Then we compute for all $b_1, b_2 \in B$,

$$\begin{aligned}\sigma(b_1) * \sigma(b_2) &= b_1 b_2 + f_2(\beta(b_1), \beta(b_2)) + \beta(b_1) \cdot g(\beta(b_2)) + g(\beta(b_1)) \cdot \beta(b_2)\\ &= b_1 b_2 + f_1(\beta(b_1), \beta(b_2)) + g(\beta(b_1 b_2))\\ &= \sigma(b_1 \circ b_2).\end{aligned}$$

This means that σ is an isomorphism from $B(f_1, \beta)$ onto $B(f_2, \beta)$. It is easy to see that σ is even an equivalence of extensions.

Conversely, suppose σ is an equivalence of extensions $B(f_1, \beta) \to B(f_2, \beta)$. Then in particular for all $b_1, b_2 \in B$,

$$\sigma(b_1 \circ b_2) = \sigma(b_1) * \sigma(b_2),$$

meaning

$$\sigma(b_1 b_2) + f_1(\beta(b_1), \beta(b_2)) = \sigma(b_1)\sigma(b_2) + f_2(\beta(b_1), \beta(b_2)).$$

For $b_1 = l \in L = \mathrm{Ker}(\beta)$, this gives $lb_2 = l\sigma(b_2)$, because σ restricts to the identity on L and L is an ideal. Similarly, $b_1 l = \sigma(b_1) l$, thus $b - \sigma(b) \in Z$. Also $b - \sigma(b)$ only depends on the coset of b modulo the ideal L, whence we may write $b - \sigma(b) = g(\beta(b))$ for some cochain $g \in C^1(A, Z)$. Compute now $dg(\beta(b_1), \beta(b_2))$. For this, we need to lift the elements $\beta(b_i)$ for $i = 1, 2$ to B in order to make them act on L; we may lift them as b_i or $\sigma(b_i)$, because they act on elements of Z where the difference does not matter:

$$\begin{aligned}dg(\beta(b_1), \beta(b_2)) &= b_1 \cdot g(\beta(b_2)) - g(\beta(b_1 b_2)) + g(\beta(b_1)) \cdot \sigma(b_2)\\ &= b_1(b_2 - \sigma(b_2)) - (b_1 b_2 - \sigma(b_1 b_2)) + (b_1 - \sigma(b_1))\sigma(b_2)\\ &= \sigma(b_1 b_2) - \sigma(b_1)\sigma(b_2).\end{aligned}$$

But

$$\sigma(b_1 b_2) - \sigma(b_1)\sigma(b_2) = dg(\beta(b_1), \beta(b_2))$$

for all $b_1, b_2 \in B$ implies with the above relation $f_2 - f_1 = dg$. This completes the proof. □

Now let E be the equivalence class of an extension $\beta : B \to A$ with nucleus Z and let $[f] \in HH^2(A, Z)$, that is, the 2-cocycle f represents the class $[f] \in HH^2(A, Z)$. Then the map $(\beta : B \to A) \mapsto (\beta : B(f, \beta) \to A)$ sends E to a unique equivalence class $[f] \cdot E$ with the same kernel L as E. We therefore arrive at the following.

Theorem 3.6.6. *Let $\varphi : A \to \mathrm{Out}(L)$ be a special kernel and denote its nucleus by Z. Then the abelian group $HH^2(A, Z)$ acts faithfully and transitively by $E \mapsto [f] \cdot E$ on the set of equivalence classes of extensions of A, which correspond to the associative algebra kernel $\varphi : A \mapsto \mathrm{Out}(L)$.*

Proof. The proof uses mainly Theorem 3.6.5 and we leave it as an exercise. The interested reader may consult Theorem 6.2, page 934, in [45]. □

Remark 3.6.3. Compare this theorem to the corresponding theorem for crossed modules instead of associative algebra kernels, that is, Theorem 3.4.2.

3.6.5 Gerstenhaber's theorem for kernels

In this subsection, we will come back to Gerstenhaber's theorem in the framework of associative algebra kernels in order to compare the methods used in the proof to those we have used in the proof of Theorem 3.2.1 and those we have used in Section 3.5.

Let $\varphi_i : A \to \mathrm{Out}(L_i)$ for $i = 1, 2$, be two associative algebra kernels with common nucleus $Z \subset L_1, L_2$. We call φ_1 and φ_2 *equivalent* in case there exist compatible maps $\rho_i : A \to M(L_i)$, $i = 1, 2$, together with an isomorphism $\sigma : L_1 \to L_2$, which leaves the elements of Z fixed and which is an isomorphism compatible with the A-operations with respect to the A-operations induced by ρ_1 and ρ_2 (recall that L_1 and L_2 are not necessarily A-bimodules with respect to these operations, only Z is canonically an A-bimodule). This relation is different from the equivalence of their associated crossed modules (see Remark 3.6.2). It turns out that this equivalence relation is not the correct one, which identifies the set of equivalence classes with $HH^3(A, Z)$ (cf. the case of Lie algebra kernels). In fact, one needs to quotient out the kernels coming from extensions, that is, from special associative algebra kernels.

Let us briefly introduce the sum of two associative algebra kernels. We leave the scalar multiplication as an exercise. Let $\varphi_i : A \to \mathrm{Out}(L_i)$ for $i = 1, 2$ be two associative algebra kernels with common nucleus Z. Consider the ideal Z^* of elements of the form $(z, -z)$ with $z \in Z$ in the direct product algebra $L_1 \times L_2$. The nucleus of $(L_1 \times L_2)/Z^*$ is $Z \oplus Z/Z^*$, which is isomorphic to Z using the addition map. The resulting associative algebra kernel $\varphi : A \to (L_1 \times L_2)/Z^*$ is called the *sum* of L_1 and L_2 and denoted $L_1 + L_2$. Choosing maps $\rho_i : A \to M(L_i)$ for $i = 1, 2$ compatible with φ_i for $i = 1, 2$, one obtains left and right multiplications $a \mapsto (a \cdot l_1, a \cdot l_2)$ and $a \mapsto (l_1 \cdot a, l_2 \cdot a)$, which constitute a regular map $A \to M(L_1 + L_2)$, and thus a compatible map for $\varphi : A \mapsto L_1 + L_2$. The nucleus of this kernel is by construction Z. Observe also that the obstruction cocycle/class corresponding to $L_1 + L_2$ is the sum of the obstruction cocycles/classes of L_1 and L_2. Observe further that the sum of two special kernels is again special, and the sum of two extensions, viewed as kernels, is the sum of the two extensions.

Definition 3.6.4. Two associative algebra kernels $\varphi_i : A \to \mathrm{Out}(L_i)$ for $i = 1, 2$ with common nucleus Z are called *similar* in case there exist special kernels $\sigma : S \to A$ and $\tau : T \to A$ with nucleus Z such that $S + L_1$ is equivalent to $T + L_2$.

By the additivity of the obstructions and the compatibility with equivalence, similar kernels have the same obstruction classes. This shows the injectivity part in the following theorem.

Theorem 3.6.7. *The set of similarity classes of associative algebra kernels of A with fixed nucleus Z has the structure of a vector space. The map which assigns to each similarity class its obstruction class is an isomorphism of this vector space to $HH^3(A, Z)$.*

Proof. The vector space structure on the set of similarity classes of associative algebra kernels with fixed nucleus is left to the reader as an exercise.

The proof consists in constructing an associative algebra kernel for a given 3-class. It turns out that this kernel may be constructed on the direct sum of Z with a complementary ideal L such that the algebra structure will only depend on A. This is what Hochschild [45] calls the *generic representation algebra*. The generic modules which Hochschild introduces in Section 4 of [44] are part of the bar resolution, and he needs a unit in order to define them. The cohomology class will then be used to define the A-operation on this algebra. We begin by constructing L.

Let E and F be 1-dimensional vector spaces over our ground field $\mathbb{K}$, with bases e and f of E and F, respectively. We first construct an auxiliary algebra C on the direct sum $E \oplus F$ by defining $e^2 = e, f^2 = f$, $ef = f$ and $fe = e$. C will become a subalgebra of L. Its function is merely that of ensuring that the nucleus of L be zero.

Next, we construct an ideal I of L: The product of any two elements of I will be defined to be zero; as a vector space, I is the direct sum of the following direct products: $E \times A$, $E \times A \times A$, $E \times A \times A \times A$. Furthermore, we define $IC = 0$ and $ev = v = fv$ for all $v \in I$.

In the course of the construction, we will need to adjoin a unit element to the algebra A. The corresponding algebra $A \oplus \mathbb{K}1$ will be denoted A^*. With this notation, we define L to be the direct sum of C, I and $A \times A \times A^*$ as a $\mathbb{K}$-vector space. Concerning the product, we define $(A \times A \times A^*)L = 0$, $F(A \times A \times A^*) = 0$, $(E \times A \times A \times A)(A \times A \times A^*) = 0$. Now $A \times A \times A^*$ has a basis consisting of elements of the form $a_1 \times a_2 \times 1$ and $a_1 \times a_2 \times a_3$. We complete the definition of the product in L by putting with respect to these basis elements:

$$\begin{aligned}
e(a_1 \times a_2 \times 1) &= e \times a_1 \times a_2, \\
e(a_1 \times a_2 \times a_3) &= e \times a_1 \times a_2a_3 - e \times a_1a_2 \times a_3 + e \times a_1 \times a_2 \times a_3, \\
(e \times a_1)(a_2 \times a_3 \times 1) &= e \times a_1 \times a_2 \times a_3 \\
(e \times a_1)(a_2 \times a_3 \times a_4) &= e \times a_1 \times a_2 \times a_3a_4 - e \times a_1 \times a_2a_3 \times a_4 + e \times a_1a_2 \times a_3 \times a_4.
\end{aligned}$$

We will skip the verification that the so-defined product on L is associative.

Let us show that the nucleus of L is zero. In fact, every element of L may be written in the form $\alpha e + \beta f + u + v + w + x$ where $\alpha, \beta \in \mathbb{K}$, $u \in A \times A \times A^*$, $v \in E \times A$, $w \in E \times A \times A$ and $x \in E \times A \times A \times A$. Now suppose that such an element is in the nucleus, that is, left and right multiplication by any element of L is zero. Multiplying on the right by $a_1 \times a_2 \times 1$, we find $\alpha e \times a_1 \times a_2 + v(a_1 \times a_2 \times 1) = 0$, whence $\alpha = v = 0$. Next, we multiply the element on the left by f and obtain $\beta f + w + x = 0$, whence $\beta = 0$, $w = 0$ and $x = 0$. Hence our element reduces to u. Now write $u = u_1 + u_2$ with $u_1 \in A \times A \times 1$ and $u_2 \in A \times A \times A$. If we multiply on the left by e, we see that $u_2 = 0$ (because at first, the component of $E \times A \times A \times A$ in u_2 must be zero, but this implies $u_2 = 0$). Therefore, finally also $u_1 = 0$, and the nucleus of L is trivial.

Now let K be the direct product of the algebras L and Z. By construction, Z is then the nucleus of K. We now proceed to define the A-operations in K. On the subspace $(A \times A \times A^*) \oplus Z$ of K, the operations will be given by

$$a \cdot (a_1 \times a_2 \times a_3^*) = aa_1 \times a_2 \times a_3^* - a \times a_1 a_2 \times a_3^* + a \times a_1 \times a_2 a_3^* + g(a, a_1, a_2) \cdot a_3^*,$$

where $g \in Z^3(A, Z)$ is a 3-cocycle representing the 3-class on which the construction depends, and by

$$(a_1 \times a_2 \times a_3^*) \cdot a = a_1 \times a_2 \times a_3.$$

The definition of the left A-multiplication in K is completed by setting $a \cdot u = 0$ for all $u \in C \oplus I$. The remaining right A-multiplications are determined by the following prescriptions:

$$\begin{aligned} e \cdot a &= e \times a, \quad f \cdot a = 0, \\ (e \times a_1) \cdot a &= e \times a_1 a + e \times a_1 \times a, \\ (e \times a_1 \times a_2) \cdot a &= e \times a_1 \times a_2 a - e \times a_1 a_2 \times a + e \times a_1 \times a_2 \times a_3, \\ (e \times a_1 \times a_2 \times a_3) \cdot a &= e \times a_1 \times a_2 \times a_3 a - e \times a_1 \times a_2 a_3 \times a + e \times a_1 a_2 \times a_3 \times a. \end{aligned}$$

Obviously, we have $(a_1 \cdot k) \cdot a_2 = a_1 \cdot (k \cdot a_2)$ for all $a_1, a_2 \in A$ and all $k \in K$. Similarly, the condition $(a \cdot k_1)k_2 = a \cdot k_1 k_2$ is satisfied almost trivially. In general, the only nontrivial identities are those in which factors in $A \times A \times A^*$ intervene. We shall exhibit only two of such relations in order to illustrate how the operators tie up with the multiplication in K. For example, for all $a, a_1, a_2 \in A$, we have

$$(e \cdot a)(a_1 \times a_2 \times 1) = (e \times a)(a_1 \times a_2 \times 1) = e \times a \times a_1 \times a_2,$$

and

$$\begin{aligned} e(a \cdot (a_1 \times a_2 \times 1)) &= e(aa_1 \times a_2 \times 1 - a \times a_1 a_2 \times 1 + a \times a_1 \times a_2 + g(a, a_1, a_2)) \\ &= e \times aa_1 \times a_2 - e \times a \times a_1 a_2 \\ &\quad + (e \times a \times a_1 a_2 - e \times aa_1 \times a_2 + e \times a \times a_1 \times a_2) \\ &= e \times a \times a_1 \times a_2 = (e \cdot a)(a_1 \times a_2 \times 1). \end{aligned}$$

Similarly,

$$((e \times a_1) \cdot a)(a_2 \times a_3 \times 1) = (e \times a_1 a + e \times a_1 \times a)(a_2 \times a_3 \times 1)$$
$$= e \times a_1 a \times a_2 \times a_3,$$

and

$$\begin{aligned}
&(e \times a_1)(a \cdot (a_2 \times a_3 \times 1)) \\
&\quad = (e \times a_1)(aa_2 \times a_3 \times 1 - a \times a_2 a_3 \times 1 + a \times a_2 \times a_3 + g(a, a_1, a_2)1) \\
&\quad = e \times a_1 \times aa_2 \times a_3 - e \times a_1 \times a \times a_2 a_3 \\
&\qquad + (e \times a_1 \times a \times a_2 a_3 - e \times a_1 \times aa_2 \times a_3 + e \times a_1 a \times a_2 \times a_3) \\
&\quad = e \times a_1 a \times a_2 \times a_3 = ((e \times a_1) \cdot a)(a_2 \times a_3 \times 1).
\end{aligned}$$

In fact, the relations $k_1(k_2 \cdot a) = (k_1 k_2) \cdot a$ for $k_1, k_2 \in K$ and $a \in A$ involve fewer complications. We shall omit the verification that this gives indeed the structure of an associative algebra kernel of A on K.

Finally, we claim that

$$a_1 \cdot (a_2 \cdot k) - a_1 a_2 \cdot k = (a_1 \times a_2 \times 1)k,$$

and $(k \cdot a_1) \cdot a_2 - k \cdot a_1 a_2 = k(a_1 \times a_2 \times 1)$ for all $k \in K$ and all $a_1, a_2 \in A$. The relations of the first type are all trivial since each side of the identity is always zero. This means that K is actually a left A-module. The relations of the second type are satisfied for similar reasons whenever k is an element of $Z \oplus F \oplus A \times A \times A^* \oplus E \times A \times A \times A$. Of the remaining nontrivial cases, we give two examples only:

$$\begin{aligned}
(e \cdot a_1) \cdot a_2 - e \cdot a_1 a_2 &= (e \times a_1) \cdot a_2 - e \times a_1 a_2 \\
&= e \times a_1 \times a_2 \\
&= e(a_1 \times a_2 \times 1).
\end{aligned}$$

Similarly,

$$\begin{aligned}
((e \times a) \cdot a_1) \cdot a_2 - (e \times a) \cdot a_1 a_2 &= (e \times aa_1 + e \times a \times a_1) \cdot a_2 - e \times aa_1 a_2 - e \times a \times a_1 a_2 \\
&= e \times aa_1 \times a_2 - e \times aa_1 \times a_2 + e \times a \times a_1 \times a_2 \\
&= e \times a \times a_1 \times a_2 \\
&= (e \times a)(a_1 \times a_2 \times 1).
\end{aligned}$$

Thus we have an associative algebra kernel $\varphi : A \to \mathrm{Out}(K)$ of A with hindrance h given by $h(a_1, a_2) = a_1 \times a_2 \times 1$ for all $a_1, a_2 \in A$. But then

$$dh(a_1, a_2, a_3) = a_1 \cdot h(a_2, a_3) - h(a_1 a_2, a_3) + h(a_1, a_2 a_3) - h(a_1, a_2) \cdot a_3$$

$$\begin{aligned}
&= a_1 \cdot a_2 \times a_3 \times 1 - a_1 a_2 \times a_3 \times 1 + a_1 \times a_2 a_3 \times 1 - a_1 \times a_2 \times 1 \cdot a_3 \\
&= (a_1 a_2 \times a_3 \times 1 - a_1 \times a_2 a_3 \times 1 + a_1 \times a_2 \times a_3 + g(a_1, a_2, a_3)1) \\
&\quad - a_1 a_2 \times a_3 \times 1 + a_1 \times a_2 a_3 \times 1 - a_1 \times a_2 \times a_3 \\
&= g(a_1, a_2, a_3).
\end{aligned}$$

Thus we have constructed an associative algebra kernel with prescribed hindrance. This ends the proof of the theorem. □

Remark 3.6.5. This proof has then been adapted by Hochschild to fit to the Lie algebra setting (see the proof of Theorem 2.6.3), using the relation between Lie algebra cohomology and Hochschild cohomology of the corresponding universal enveloping algebra.

3.7 Strict associative 2-algebras

In this section, we will discuss the easiest categorical version of associative algebras, namely strict associative 2-algebras. As before for groups or Lie algebras, these are category objects in the category of associative algebras. Before discussing them in detail, let us warm up with cat^1-associative algebras.

3.7.1 cat^1-associative algebras

Definition 3.7.1. A cat^1-*associative algebra* (A_1, A_0, s, t) consists of an associative algebra A_1 together with a subalgebra $A_0 \subset A_1$ and two homomorphisms $s, t : A_1 \to A_0$ such that

$$s|_{A_0} = t|_{A_0} = \mathrm{id}_{A_0}, \quad \text{and} \quad \mathrm{Ker}(s)\,\mathrm{Ker}(t) = \mathrm{Ker}(t)\,\mathrm{Ker}(s) = 0.$$

We will denote a cat^1-associative algebra also by $A_1 \underset{t}{\overset{s}{\rightrightarrows}} A_0$.

Remark 3.7.2. Some authors use the seemingly weaker condition $\mathrm{Ker}(s)\,\mathrm{Ker}(t) + \mathrm{Ker}(t)\,\mathrm{Ker}(s) = 0$ instead of the condition $\mathrm{Ker}(s)\,\mathrm{Ker}(t) = \mathrm{Ker}(t)\,\mathrm{Ker}(s) = 0$. Note that the sum $I + J$ of two ideals I and J contains I and J as subgroups. From here, it is not hard to show that both conditions are equivalent.

Theorem 3.7.1. *Crossed modules of associative algebras and* cat^1-*associative algebras are in one-to-one correspondence.*

Proof. Let a crossed module of associative algebras $\mu : M \to N$ be given. Form the semidirect product $M \rtimes N$ of M and N with respect to the bimodule action of N on M given in the crossed module. Then there are two algebra homomorphisms $M \rtimes N \underset{t}{\overset{s}{\rightrightarrows}} N$ defined by $s(m,n) = n$ and $t(m,n) = \mu(m) + n$. These two homomorphisms clearly satisfy $s|_N = t|_N = \mathrm{id}_N$, where N is regarded as a subalgebra of the semidirect product $M \rtimes N$ via $n \mapsto (0,n)$ for all $n \in N$. It remains to show that $\mathrm{Ker}(s)\,\mathrm{Ker}(t) + \mathrm{Ker}(t)\,\mathrm{Ker}(s) = 0$. Observe that $\mathrm{Ker}(s) = \{(m,0) \in M \rtimes N\}$ and $\mathrm{Ker}(t) = \{(m,n) \in M \rtimes N \mid \mu(m) = -n\}$. We compute in the semidirect product

$$\begin{aligned}(m,0)(m',-\mu(m')) &= (mm' + 0 \cdot m' - m \cdot \mu(m'), 0)\\ &= (mm' - mm', 0)\\ &= (0,0)\\ &= (m'm - \mu(m')m + m' \cdot 0, 0)\\ &= (m', -\mu(m'))(m,0).\end{aligned}$$

Thus $\mathrm{Ker}(s)\,\mathrm{Ker}(t) = \mathrm{Ker}(t)\,\mathrm{Ker}(s) = 0$.

Conversely, let $A_1 \underset{t}{\overset{s}{\rightrightarrows}} A_0$ be a cat^1-associative algebra. We put $N := A_0$, $M := \mathrm{Ker}(s)$ and $\mu := t|_{\mathrm{Ker}(s)}$. Then $\mu : M \to N$ is certainly a homomorphism of associative algebras. The subalgebra $N \subset A_1$ acts on the ideal $M = \mathrm{Ker}(s)$ by left and right multiplication in A_1. This gives a bimodule action compatible with the product on M. We have to show properties (a) and (b) of a crossed module.

Property (a) is readily computed:

$$\begin{aligned}\mu(n \cdot m) &= t(nm)\\ &= t(n)t(m)\\ &= nt(m)\\ &= n\mu(m).\end{aligned}$$

Here, for $n \in N = A_0$ and $m \in M = \mathrm{Ker}(s)$, we have used that t is a homomorphism, that the bimodule action is given by the product in A_1 and that t is the identity on A_0. A similar computation works for the right action.

Property (b) contains $\mu(m)\cdot m' = mm'$, which translates to the A_1-equality $t(m)m' = mm'$. This is true, because the element $t(m) - m \in \mathrm{Ker}(t)$ acts trivially on an element in $\mathrm{Ker}(s)$. The same argument works on the right. □

Remark 3.7.3. This one-to-one correspondence can be extended to an equivalence of categories (even of 2-categories) between the category of crossed modules and that of cat^1-associative algebras. We tend to omit the morphism part of the correspondence (and the parts about higher morphisms) in order to keep category technical constructions to a minimum.

3.7.2 Crossed modules and category objects in Alg

Definition 3.7.4. A *category object* $A_1 \underset{t}{\overset{s}{\rightrightarrows}} A_0$ in the category of associative algebras Alg is the data of an algebra of objects A_0, an algebra of morphisms (or arrows) A_1, source and target maps $s, t : A_1 \to A_0$, supposed to be homomorphisms of algebras, an object inclusion $i : A_0 \to A_1$, homomorphism of algebras, and a composition $\mathrm{comp} : A_1 \times_{A_0} A_1 \to A_1$, $(f, g) \mapsto g \circ f$, supposed to be a homomorphism of algebras, which satisfy the usual axioms of a category. Here, $A_1 \times_{A_0} A_1$ is the algebra of composable arrows, that is, pairs $(f, g) \in A_1 \times A_1$ such that $t(f) = s(g)$.

We call such a category object $A_1 \underset{t}{\overset{s}{\rightrightarrows}} A_0$ also a *strict associative 2-algebra*.

Theorem 3.7.2. *There is a one-to-one correspondence between strict associative 2-algebras and crossed modules of associative algebras.*

Proof. We shall actually prove a one-to-one correspondence between strict associative 2-algebras and cat^1-associative algebras. This correspondence may then be composed with the one from Theorem 3.7.1 in order to prove the theorem.

Comparing strict associative 2-algebras and cat^1-associative algebras, we see that on the one hand, strict associative 2-algebras have more data (the composition of arrows), while cat^1-associative algebras have more properties, namely

$$\mathrm{Ker}(s)\,\mathrm{Ker}(t) = \mathrm{Ker}(t)\,\mathrm{Ker}(s) = 0.$$

Given a category object $A_1 \underset{t}{\overset{s}{\rightrightarrows}} A_0$ in Alg, observe that we may use the object inclusion $i : A_0 \to A_1$ in order to regard A_0 as a subalgebra of A_1. Source and target map are then the identity on this subalgebra, because of the relations $s \circ i = t \circ i = \mathrm{id}_{A_0}$, which hold in a category.

In order to have a cat^1-associative algebra, we need to show that $\mathrm{Ker}(s)\,\mathrm{Ker}(t) = \mathrm{Ker}(t)\,\mathrm{Ker}(s) = 0$. This follows from the requirement that the composition of arrows should be a homomorphism of associative algebras, namely for composable arrows

$$(g_1 g_2) \circ (f_1 f_2) = (g_1 \circ f_1)(g_2 \circ f_2).$$

Once again, this is sometimes called the "middle four exchange law", because in the operation, one has exchanged the two arrows in the middle. Now suppose $g_1 \in \mathrm{Ker}(s)$ and $f_2 \in \mathrm{Ker}(t)$. Then denote by f_1 and by g_2 the identities (with respect to the composition) in $0 \in A_0$. We have with this choice $g_1 \circ f_1 = g_1$ and $g_2 \circ f_2 = f_2$. On the other hand, i is a morphism of associative algebras and, therefore, sends $0 \in A_0$ to $0 \in A_1$. The middle four exchange property therefore gives

$$g_1 f_2 = (g_1 \circ f_1)(g_2 \circ f_2) = (g_1 g_2) \circ (f_1 f_2) = (g_1 0) \circ (0 f_2) = 0.$$

This shows $\mathrm{Ker}(s)\,\mathrm{Ker}(t) = 0$, and a similar computation shows $\mathrm{Ker}(t)\,\mathrm{Ker}(s) = 0$.

Conversely, let $A_1 \underset{t}{\overset{s}{\rightrightarrows}} A_0$ be a cat^1-associative algebra. We define a category object as follows. The algebra A_1 becomes the algebra of arrows, the algebra A_0 becomes the algebra of objects. Source and target maps are the given maps s and t. The object inclusion i arises from the inclusion of the subalgebra A_0 into A_1, the requirement $s|_{A_0} = t|_{A_0} = \mathrm{id}_{A_0}$ shows then $s \circ i = t \circ i = \mathrm{id}_{A_0}$.

It remains to construct the composition map $\mathrm{comp} : A_1 \times_{A_0} A_1 \to A_1$, $(f,g) \mapsto g \circ f$. Once again, the composition does not constitute new data. It is encoded in the underlying strict 2-vector space, as we had already remarked in the framework of Lie 2-algebras. The proof of the double-four-exchange law, that is, of the fact that the composition is a morphism of associative algebras, is very similar to the proof of Lemma 2.10.4 and relegated to the exercises. □

3.8 Relation between crossed modules of associative and Lie algebras respective groups

Up to now, we have considered crossed modules of groups, Lie algebras and associative algebras. The categories Grp of groups, the category Lie of Lie algebras over a fixed field $\mathbb{K}$, and the category Alg of associative unital $\mathbb{K}$-algebras are linked by some canonical functors. We will discuss in this section how these functors extend to crossed modules over Grp, Lie and Alg.

3.8.1 Definition of the functors

First, let us introduce the functors. The functor $\mathbb{K}[-] : \mathtt{Grp} \to \mathtt{Alg}$ sends a group G to its *group algebra* $\mathbb{K}[G]$ over $\mathbb{K}$. $\mathbb{K}[G]$ is the $\mathbb{K}$-vector space with basis e_g indexed by the elements $g \in G$. We define an associative product on basis elements of $\mathbb{K}[G]$ by

$$e_g\, e_h := e_{gh}$$

for all elements $g, h \in G$ and extend it by $\mathbb{K}$-linearity to all elements of $\mathbb{K}[G]$.

In the other direction, there is the functor of unit group $-^\times : \mathtt{Alg} \to \mathtt{Grp}$, which sends an algebra A to the group $A^\times$ of its invertible elements. Note that we need a unit in the algebra A in order to have a group $A^\times$.

Proposition 3.8.1. *The functor* $\mathbb{K}[-] : \mathtt{Grp} \to \mathtt{Alg}$ *is left adjoint to the functor* $-^\times : \mathtt{Alg} \to \mathtt{Grp}$*, that is, we have for all groups G and all (unital) algebras A a natural isomorphism*

$$\mathrm{Hom}_{\mathtt{Alg}}(\mathbb{K}[G], A) \cong \mathrm{Hom}_{\mathtt{Grp}}(G, A^\times).$$

Proof. A homomorphism of algebras $f : \mathbb{K}[G] \to A$ is uniquely specified on the basis elements e_g for g ranging in G. The homomorphism property together with $e_g e_h = e_{gh}$ imply that we obtain a group homomorphism $\tilde{f} : G \to A^\times$ by defining $\tilde{f}(g) := f(e_g)$. All elements in the image of $\tilde{f}$ are invertible, thus $\tilde{f}$ has its image included in $A^\times$. Conversely, a group homomorphism $\tilde{f} : G \to A^\times$ extends to an algebra homomorphism $f : \mathbb{K}[G] \to A$ again because of the homomorphism property together with $e_g e_h = e_{gh}$. We leave the proof of the naturality to the interested reader. □

Now let us discuss the relation between the categories $\mathtt{Alg}$ and $\mathtt{Lie}$. There is the functor $_{\mathrm{Lie}} : \mathtt{Alg} \to \mathtt{Lie}$ associating to a unital associative algebra A its underlying Lie algebra A_{Lie} given by the same vector space as A, but with the bracket

$$[a, b] := ab - ba,$$

for all elements $a, b \in A$. In the reverse direction, there is the functor of the *universal enveloping algebra* $U : \mathtt{Lie} \to \mathtt{Alg}$ sending a Lie algebra $\mathfrak{g}$ to its universal enveloping algebra $U\mathfrak{g}$. As before, we have

Proposition 3.8.2. *The functor* $U : \mathtt{Lie} \to \mathtt{Alg}$ *is left adjoint to the functor* $_{\mathrm{Lie}} : \mathtt{Alg} \to \mathtt{Lie}$, *that is, we have for all algebras A and all Lie algebras $\mathfrak{g}$ a natural isomorphism,*

$$\mathrm{Hom}_{\mathtt{Alg}}(U\mathfrak{g}, A) \cong \mathrm{Hom}_{\mathtt{Lie}}(\mathfrak{g}, A_{\mathrm{Lie}}).$$

Proof. The adjointness follows right away from the universal property of the universal enveloping algebra, which reads as follows: Given any Lie algebra homomorphism $\tilde{f} : \mathfrak{g} \to A_{\mathrm{Lie}}$, there exists a unique homomorphism of associative algebras $f : U\mathfrak{g} \to A$ such that f restricts to $\tilde{f}$ on $\mathfrak{g} \subset U\mathfrak{g}$. Again, the proof of the naturality is left to the reader. □

Combining these functors, we obtain also functors between the categories $\mathtt{Lie}$ and $\mathtt{Grp}$, sending a Lie algebra $\mathfrak{g}$ to the unit group $(U\mathfrak{g})^\times$ of its universal enveloping algebra, and sending a group G to the Lie algebra $\mathbb{K}[G]_{\mathrm{Lie}}$ underlying its group algebra. Note that the group $(U\mathfrak{g})^\times$ is just $\mathbb{K}^\times$, thus not very interesting.

We will consider different functors between the categories $\mathtt{Lie}$ and $\mathtt{Grp}$ (or some subcategories) in later chapters, namely those related to integration of Lie algebras into Lie groups, when we consider crossed modules of Lie groups.

3.8.2 Crossed modules of groups and of associative algebras

The subject of this subsection is to extend the functors $\mathbb{K}[-] : \mathtt{Grp} \to \mathtt{Alg}$ and $-^\times : \mathtt{Alg} \to \mathtt{Grp}$ to crossed modules of groups, respectively, crossed modules of associative algebras.

Let $\mu : M \to N$ be a crossed module of groups. The functor $\mathbb{K}[-] : \mathtt{Grp} \to \mathtt{Alg}$ induces a morphism of associative algebras $\mu : \mathbb{K}[M] \to \mathbb{K}[N]$. Let us see that $\mathbb{K}[M]$ becomes a $\mathbb{K}[N]$-bimodule. First of all, the action of N on M extends by linearity to an action of N on $\mathbb{K}[M]$. This action corresponds to a group morphism $N \to \mathrm{Gl}(\mathbb{K}[M])$, which extends to a morphism of associative algebras $\mathbb{K}[N] \to \mathrm{End}(\mathbb{K}[M])$. Explicitly, this action is given for all $\sum \lambda_i n_i \in \mathbb{K}[N]$ and all $\sum \rho_j m_j \in \mathbb{K}[M]$ by

$$\sum \lambda_i n_i \cdot \sum \rho_j m_j = \sum_{i,j} \lambda_i \rho_j n_i \cdot m_j.$$

In order to obtain also a right action on $\mathbb{K}[M]$, we convert the left action into a right action by

$$m \cdot n := n^{-1} \cdot m$$

for all $n \in N$ and all $m \in M$. Unfortunately, these actions do not give an action by algebra homomorphisms as required. The problem is that while N acts by group automorphisms, that is, $n \cdot (m_1 m_2) = (n \cdot m_1)(n \cdot m_2)$ for all $n \in N$ and all $m_1, m_2 \in M$, we want $\mathbb{K}[N]$ to act on $\mathbb{K}[M]$ in a way compatible with the product on $\mathbb{K}[M]$, that is, $n \cdot (m_1 m_2) = (n \cdot m_1) m_2$. This cannot be true for the elements $n \in N \subset \mathbb{K}[N]$ and $m_1, m_2 \in M \subset \mathbb{K}[M]$, at least for nontrivial actions.

Therefore, we have to undertake a different approach in order to extend the functor $\mathbb{K}[-] : \mathtt{Grp} \to \mathtt{Alg}$ to crossed modules. One approach which works well is to use the passage through cat^1-groups and cat^1-algebras. The upshot is then the following theorem.

Theorem 3.8.3. *The functor $\mathbb{K}[-] : \mathtt{Grp} \to \mathtt{Alg}$ associating to a group G its group algebra $\mathbb{K}[G]$ extends to crossed modules sending a crossed module of groups $\mu : M \to N$ to a crossed modules of associative algebras $\bar{t}_{\mathrm{Ker}(\bar{s})} : \mathrm{Ker}(\bar{s}) \to \mathbb{K}[N]$, where $\bar{s}, \bar{t}$ are defined on some quotient $\mathbb{K}[M \rtimes N]/X$ of the group algebra on the semidirect product $M \rtimes N$.*

Proof. As already stated, the main idea is to use the passage to cat^1-objects. Let $\mu : M \to N$ be a crossed module of groups. We have seen that to $\mu : M \to N$ corresponds a cat^1-group $M \rtimes N \underset{t}{\overset{s}{\rightrightarrows}} N$, where the semidirect product $M \rtimes N$ is built from the action of N on M by automorphisms given in the crossed module $\mu : M \to N$. Applying the functor $\mathbb{K}[-]$, we obtain a diagram of associative algebras

$$\mathbb{K}[M \rtimes N] \underset{t}{\overset{s}{\rightrightarrows}} \mathbb{K}[N].$$

Here, we denoted the maps again s and t in order to simplify notation.

As $s|_N = t|_N = \mathrm{id}_N$, we obtain also $s|_{\mathbb{K}[N]} = t|_{\mathbb{K}[N]} = \mathrm{id}_{\mathbb{K}[N]}$. We need to impose the other condition of a cat^1-algebra, thus we consider the ideal

$$X := \mathrm{Ker}(s)\,\mathrm{Ker}(t) + \mathrm{Ker}(t)\,\mathrm{Ker}(s).$$

Quotienting out X, we obtain a diagram of associative algebras

$$\mathbb{K}[M \rtimes N]/X \underset{\bar{t}}{\overset{\bar{s}}{\rightrightarrows}} \mathbb{K}[N],$$

where now $\bar{s}$, $\bar{t}$ are the algebra morphisms induces by s and t respectively by passing to the quotient with respect to the ideal X. By construction, the result is a cat^1-algebra, and its corresponding crossed module of associative algebras is

$$\bar{t}_{\mathrm{Ker}(\bar{s})} : \mathrm{Ker}(\bar{s}) \to \mathbb{K}[N].$$

We leave the functoriality check to the reader. □

In order to better understand this extension of the functor $\mathbb{K}[-]$ to crossed modules, let us also show two properties this extension satisfies. In order to state these properties, let us introduce the following functors. The two ways of regarding an associative algebra A as a crossed module, namely as zero map $\mu_0 = 0 : 0 \to A$ or as the identity map $\mu_1 = \mathrm{id}_A : A \to A$, give rise to two functors

$$0 : \mathtt{Alg} \to \mathrm{crmod}_{\mathtt{Alg}},\ A \mapsto (0 : 0 \to A) \quad \text{and} \quad 1 : \mathtt{Alg} \to \mathrm{crmod}_{\mathtt{Alg}},\ A \mapsto (\mathrm{id}_A : A \to A).$$

In the same way, there are the corresponding functors for groups (which we denote in the same way in order to avoid to introduce awkward notation)

$$0 : \mathtt{Grp} \to \mathrm{crmod}_{\mathtt{Grp}},\ G \mapsto (0 : \{1\} \to G) \quad \text{and} \quad 1 : \mathtt{Grp} \to \mathrm{crmod}_{\mathtt{Grp}},\ G \mapsto (\mathrm{id}_G : G \to G).$$

Note that while in $\mathrm{id}_A : A \to A$ acts on itself by left and right multiplication, G acts on G in the crossed module $\mathrm{id}_G : G \to G$ by conjugation. These functors are full embeddings of the category $\mathtt{Alg}$ (resp., $\mathtt{Grp}$) as a subcategory of $\mathrm{crmod}_{\mathtt{Alg}}$ (resp., $\mathrm{crmod}_{\mathtt{Grp}}$).

One easily verifies that there is a commutative diagram:

$$\begin{array}{ccc} \mathtt{Grp} & \xrightarrow{\ 0\ } & \mathrm{crmod}_{\mathtt{Grp}} \\ \downarrow{\scriptstyle \mathbb{K}[-]} & & \downarrow{\scriptstyle \mathbb{K}[-]} \\ \mathtt{Alg} & \xrightarrow{\ 0\ } & \mathrm{crmod}_{\mathtt{Alg}} \end{array}$$

In the same way, we have the following.

Theorem 3.8.4. *There is a natural isomorphism of functors*

$$\mathbb{K}[-] \circ 1 = 1 \circ \mathbb{K}[-].$$

Proof. On the one hand, we have $(1 \circ \mathbb{K}[-])(G) = (\mathrm{id}_{\mathbb{K}[G]} : \mathbb{K}[G] \to \mathbb{K}[G])$. On the other hand, $(\mathbb{K}[-] \circ 1)(G)$ is the crossed module of associative algebras we have constructed in the previous theorem via the passage through cat^1-objects, here built on the crossed module of groups $\mathrm{id}_G : G \to G$. In the process of constructing $\mathbb{K}[-](\mathrm{id}_G : G \to G)$, one builds first the cat^1-group,

$$G \rtimes G \underset{\bar{t}}{\overset{\bar{s}}{\rightrightarrows}} G,$$

with source $s(g,h) = h$ and target map $t(g,h) = gh$. Note that there is a group homomorphism $\epsilon : G \to G \rtimes G, g \mapsto (g,1)$. The next step is then to construct the cat^1-algebra,

$$\mathbb{K}[G \rtimes G]/X \underset{\bar{t}}{\overset{\bar{s}}{\rightrightarrows}} \mathbb{K}[G],$$

where X is the ideal $\mathrm{Ker}(s)\,\mathrm{Ker}(t) + \mathrm{Ker}(t)\,\mathrm{Ker}(s)$. Denote by $\pi : \mathbb{K}[G \rtimes G] \to \mathbb{K}[G \rtimes G]/X$ the quotient map. Consider the composition $s \circ \pi \circ \epsilon : \mathbb{K}[G] \to \mathbb{K}[G]$ where we denoted still by $\epsilon : \mathbb{K}[G] \to \mathbb{K}[G \rtimes G]$ the map induced by $\epsilon : G \to G \rtimes G$. The composition gives clearly

$$s \circ \pi \circ \epsilon = s \circ \epsilon = 0,$$

the map sending $g \mapsto 1$ for all $g \in G$. In the same way, the composition with the target map gives $t \circ \pi \circ \epsilon = t \circ \epsilon = \mathrm{id}_G$. It follows that the homomorphism $\pi \circ \epsilon$ takes values in $\mathrm{Ker}(s)$ and is a right inverse for $t|_{\mathrm{Ker}(s)} : \mathrm{Ker}(s) \to \mathbb{K}[G]$. But $\mathrm{Ker}(s)$ is clearly generated by all elements of the form $(g,1)$ for $g \in G$ and, therefore, we have also $\pi \circ \epsilon \circ (t|_{\mathrm{Ker}(s)}) = \mathrm{id}_{\mathrm{Ker}(s)}$. Now it is easy to verify that the pair $(t|_{\mathrm{Ker}(s)}, \mathrm{id}_G)$ is an isomorphism of crossed modules of associative algebras from $t|_{\mathrm{Ker}(s)} : \mathrm{Ker}(s) \to \mathbb{K}[G]$ to $\mathrm{id}_{\mathbb{K}[G]} : \mathbb{K}[G] \to \mathbb{K}[G]$. The naturality check is left to the reader. □

The same procedure applies to the functor $-^\times : \mathtt{Alg} \to \mathtt{Grp}$.

Theorem 3.8.5. *The functor $-^\times : \mathtt{Alg} \to \mathtt{Grp}$ associating to a unital associative algebra A its group of units $A^\times$ extends to crossed modules to a functor associating to a crossed module of unital associative algebras $\mu : M \to N$ a crossed module of groups $\mu : M^\times \to N^\times$.*

Proof. Given a crossed module of unital associative algebras $\mu : M \to N$, we have seen how to associate to it its cat^1-algebra $M \rtimes N \underset{t}{\overset{s}{\rightrightarrows}} N$. Applying the unit group functor, we obtain a diagram of groups,

$$(M \rtimes N)^\times \underset{t}{\overset{s}{\rightrightarrows}} N^\times,$$

where we have denoted the maps restricted to the unit groups again s and t in order to simplify notation. As $s|_N = t|_N = \mathrm{id}_N$, we obtain also $s|_{N^\times} = t|_{N^\times} = \mathrm{id}_{N^\times}$.

It remains to show $[\mathrm{Ker}(s), \mathrm{Ker}(t)] = 1$ for the induced maps s and t on the level of unit groups. We claim that the elements of $\mathrm{Ker}(s)$ are of the form $(m, 1)$ such that there exists $m' \in M$ with $mm' + m + m' = m'm + m + m' = 0$. Indeed, recall the expression of the product in $M \rtimes N$:

$$(m, n)(m', n') = (mm' + n \cdot m' + m \cdot n', nn').$$

As we are considering groups of invertible elements, the elements of $\mathrm{Ker}(s)$ are those sent by $s : M \rtimes N \to N$, $s(m, n) = n$ to $1 \in N$. Thus they are of the form $(m, 1)$. Its inverse must be of the same form $(m', 1)$, and the condition to be a two-sided inverse reads

$$(m, 1)(m', 1) = (mm' + m + m', 1) = (0, 1) = (m', 1)(m, 1) = (m'm + m + m', 1),$$

whence the claim.

In order to obtain a description for $\mathrm{Ker}(t)$, recall that $t : M \rtimes N \to N$ is given by $t(m, n) = \mu(m) + n$. By the expression of t, the elements in $\mathrm{Ker}(t)$ have necessarily the form $(m, 1 - \mu(m))$ for some $m \in M$.

Now compute

$$\begin{aligned}(m, 1)(m', 1 - \mu(m')) &= (mm' + m' + (1 - \mu(m')) \cdot m, 1 - \mu(m')) \\ &= (mm' + m' + m - \mu(m') \cdot m, 1 - \mu(m')) \\ &= (-\mu(m') \cdot m, 1 - \mu(m')) \\ &= (m'm + m + m' - \mu(m') \cdot m, 1 - \mu(m')) \\ &= (m', 1 - \mu(m'))(m, 1).\end{aligned}$$

Therefore,

$$(M \rtimes N)^\times \underset{t}{\overset{s}{\rightrightarrows}} N^\times$$

is indeed a cat^1-group. We associate to it its crossed modules of groups

$$t|_{\mathrm{Ker}(s)} : \mathrm{Ker}(s) \to N^\times.$$

We leave the check of the functoriality to the reader. □

Again, the extension of the unit-algebra functor to crossed modules satisfies a similar property, namely the compatibility with the functor $1 : \mathtt{Alg} \to \mathrm{crmod}_{\mathtt{Alg}}$ sending a unital associative algebra A to the crossed module $\mathrm{id}_A : A \to A$, and its group counterpart $1 : \mathtt{Grp} \to \mathrm{crmod}_{\mathtt{Grp}}$.

Theorem 3.8.6. *There is a natural isomorphism of functors*

$$-^\times \circ 1 \cong 1 \circ -^\times.$$

Proof. On the one hand, we have for a unital associative algebra A that $(1 \circ -^{\times})(A)$ is the crossed module of groups $\mathrm{id}_{A^{\times}} : A^{\times} \to A^{\times}$. On the other hand, the crossed module $(-^{\times} \circ 1)(A)$ is the crossed module constructed in the theorem above, that is, the crossed module corresponding to the cat^1-group $(A \rtimes A)^{\times} \underset{t}{\overset{s}{\rightrightarrows}} A^{\times}$. Here, source and target map are defined as $s(a,b) = b$ and $t(a,b) = a + b$. We claim that $t|_{\mathrm{Ker}(s)} : \mathrm{Ker}(s) \to A^{\times}$ for the induced source and target maps on the corresponding unit groups, is an isomorphism of groups. Indeed, the map $u \mapsto (u - 1, 1) \in \mathrm{Ker}(s)$ with $u \in A^{\times}$ defines a two-sided inverse to $t|_{\mathrm{Ker}(s)}$. It is then easy to see that the pair $(t|_{\mathrm{Ker}(s)}, \mathrm{id}_{A^{\times}})$ is an isomorphism of crossed modules of groups from $t|_{\mathrm{Ker}(s)} : \mathrm{Ker}(s) \to A^{\times}$ to $\mathrm{id}_{A^{\times}} : A^{\times} \to A^{\times}$. The naturality check is again left to the reader. □

Theorem 3.8.7. *The functor* $\mathbb{K}[-]$ *between crossed modules is left adjoint to the unit group functor* $(-)^{\times}$ *between crossed modules.*

Proof. We have to construct a natural bijection

$$\mathrm{Hom}_{\mathrm{crmod}_{\mathrm{Grp}}}((\nu : H \to G), (\mu : M \to N)^{\times}) \cong \mathrm{Hom}_{\mathrm{crmod}_{\mathrm{Alg}}}(\mathbb{K}[\nu : H \to G], (\mu : M \to N))$$

for any crossed modules $\nu : H \to G$ of groups and $\mu : M \to N$ of unital associative algebras. Recall that this means that N is a unital associative algebra with 1 acting as identity.

For a given morphism $(\alpha, \beta) : (\nu : H \to G) \to (\mu : M \to N)^{\times}$, there is a corresponding morphism of cat^1-groups

$$\begin{array}{ccc} H \rtimes G & \underset{t}{\overset{s}{\rightrightarrows}} & G \\ \downarrow{\scriptstyle \alpha'} & & \downarrow{\scriptstyle \beta} \\ (M \rtimes N)^{\times} & \underset{t}{\overset{s}{\rightrightarrows}} & N^{\times}. \end{array}$$

Since the functor $\mathbb{K}[-]$ is left adjoint to the functor $(-)^{\times}$, we deduce a commutative diagram of associative algebras:

$$\begin{array}{ccc} \mathbb{K}[H \rtimes G] & \xrightarrow{s} & \mathbb{K}[G] \\ \downarrow{\scriptstyle \alpha'^{*}} & & \downarrow{\scriptstyle \beta^{*}} \\ M \rtimes N & \xrightarrow{s} & N. \end{array}$$

We obtain a similar diagram for the source maps t. As in the cat^1-associative algebra $M \rtimes N \underset{t}{\overset{s}{\rightrightarrows}} N$, we have $\mathrm{Ker}(s)\,\mathrm{Ker}(t) + \mathrm{Ker}(t)\,\mathrm{Ker}(s) = 0$, the above morphisms pass to the quotient and we obtain a morphism of cat^1-associative algebras

$$\begin{array}{ccc} \mathbb{K}[H \rtimes G]/X & \overset{s}{\underset{t}{\rightrightarrows}} & \mathbb{K}[G] \\ \downarrow{\scriptstyle \alpha'} & & \downarrow{\scriptstyle \beta} \\ M \rtimes N & \overset{s}{\underset{t}{\rightrightarrows}} & N. \end{array}$$

The corresponding morphism of crossed modules of associative algebras is the wanted morphism. The inverse assignment is similar. The naturality proof is left to the reader. □

3.8.3 Crossed modules of Lie algebras and of associative algebras

The subject of this subsection is to investigate how the functors $U : \mathtt{Lie} \to \mathtt{Alg}$ and $_{\mathrm{Lie}} : \mathtt{Alg} \to \mathtt{Lie}$ extend to crossed modules. Concerning the functor $U : \mathtt{Lie} \to \mathtt{Alg}$, we will present here one solution of the problem (due to Casas, Casado, Khamaladze and Ladra), while we have a different solution to the problem in the section on crossed modules of Hopf algebras.

Theorem 3.8.8. *The Lieization functor* $_{\mathrm{Lie}} : \mathtt{Alg} \to \mathtt{Lie}$ *passes to crossed modules and sends then a crossed module of associative algebras to a crossed module of Lie algebras.*

Proof. Consider the Lieization functor $_{\mathrm{Lie}} : \mathtt{Alg} \to \mathtt{Lie}$. Let $\mu : M \to N$ be a crossed module of associative algebras. The functor $_{\mathrm{Lie}}$ gives rise to morphism of Lie algebras (still denoted μ) $\mu : M_{\mathrm{Lie}} \to N_{\mathrm{Lie}}$. Now the bimodule action of N on M gives rise to a Lie algebra action of N_{Lie} on M_{Lie} in the following way. Put for all $n \in N$ and all $m \in M$,

$$n * m := n \cdot m - m \cdot n.$$

We have

$$\begin{aligned} n_1 * (n_2 * m) - n_2 * (n_1 * m) &= n_1 * (n_2 \cdot m - m \cdot n_2) - n_2 * (n_1 \cdot m - m \cdot n_1) \\ &= n_1 \cdot (n_2 \cdot m - m \cdot n_2) - (n_2 \cdot m - m \cdot n_2) \cdot n_1 \\ &\quad - n_2 \cdot (n_1 \cdot m - m \cdot n_1) + (n_1 \cdot m - m \cdot n_1) \cdot n_2 \\ &= n_1 n_2 \cdot m - n_2 n_1 \cdot m - m \cdot n_2 n_1 + m \cdot n_1 n_2 \\ &= [n_1, n_2] * m. \end{aligned}$$

On the other hand, N_{Lie} acts on M_{Lie} by derivations:

$$\begin{aligned} n * [m_1, m_2] &= n \cdot (m_1 m_2 - m_2 m_1) - (m_1 m_2 - m_2 m_1) \cdot n \\ &= (n \cdot m_1) m_2 - (n \cdot m_2) m_1 - m_1 (m_2 \cdot n) + m_2 (m_1 \cdot n) \end{aligned}$$

$$\begin{aligned}&-(m_1\cdot n)m_2+(m_2\cdot n)m_1+m_1(n\cdot m_2)-m_2(n\cdot m_1)\\&=(n*m_1)m_2-(n*m_2)m_1+m_1(n*m_2)-m_2(n*m_1)\\&=[n*m_1,m_2]+[m_1,n*m_2].\end{aligned}$$

Observe that the third line in this computation is zero! It remains to prove conditions (a) and (b) for a crossed module of Lie algebras. We have for all $n\in N$ and all $m\in M$,

$$\mu(n*m)=\mu(n\cdot m-m\cdot n)=n\mu(m)-\mu(m)n=[n,\mu(m)].$$

This shows condition (a). On the other hand, we have for all $m,m'\in M$,

$$\mu(m)*m'=\mu(m)\cdot m'-m'\cdot\mu(m)=mm'-m'm=[m,m'].$$

This shows condition (b). □

Again, let us show properties of this extension of the Lieization functor to crossed modules. Exactly as there two ways of regarding an associative algebra A as a crossed module (namely as the zero crossed module $0:0\to A$ or as the identity crossed module $\mathrm{id}_A:A\to A$), there are also two ways of regarding a Lie algebra $\mathfrak{g}$ as a crossed module, namely as $0:0\to\mathfrak{g}$ or as $\mathrm{id}_{\mathfrak{g}}:\mathfrak{g}\to\mathfrak{g}$. Once again, these assignments are functorial and embed the category of Lie algebras $\mathtt{Lie}$ (via functors denoted 0 and 1) as a full subcategory of the category $\mathrm{crmod}_{\mathtt{Lie}}$ of crossed modules of Lie algebras.

We have two commutative diagrams:

$$\begin{array}{ccc}\mathtt{Alg} & \xrightarrow{0,1} & \mathrm{crmod}_{\mathtt{Alg}}\\ \downarrow{\scriptstyle \mathrm{Lie}} & & \downarrow{\scriptstyle \mathrm{Lie}}\\ \mathtt{Lie} & \xrightarrow{0,1} & \mathrm{crmod}_{\mathtt{Lie}}\end{array}$$

Theorem 3.8.9. *The universal enveloping algebra functor $U:\mathtt{Lie}\to\mathtt{Alg}$ passes to crossed modules and sends then a crossed module of Lie algebras to a crossed module of associative algebras.*

Proof. The idea of the proof is here again to pass through cat^1-algebras, which are easier to handle. Namely, we have seen the passages from Lie algebra crossed modules to cat^1-Lie algebras in Theorem 2.10.1, and from cat^1-associative algebras to crossed modules of associative algebras. Thus it remains to send a cat^1-Lie algebra to a cat^1-associative algebra.

Let $\mathfrak{m}\rtimes\mathfrak{n}\underset{t}{\overset{s}{\rightrightarrows}}\mathfrak{n}$ be a cat^1-Lie algebra. Applying the enveloping functor U term-by-term, we obtain a diagram

$$U(\mathfrak{m}\rtimes\mathfrak{n})\underset{t}{\overset{s}{\rightrightarrows}}U\mathfrak{n}.$$

We still have for the induced maps (which we still denote by s and t) $s|_{U\mathfrak{n}} = t|_{U\mathfrak{n}} = \mathrm{id}_{U\mathfrak{n}}$, but the kernel condition for a cat^1-associative algebra need not be satisfied. We will put it in by hand by considering the diagram

$$U(\mathfrak{m} \rtimes \mathfrak{n})/X \underset{t}{\overset{s}{\rightrightarrows}} U\mathfrak{n},$$

where X is the ideal $X = \mathrm{Ker}(s)\,\mathrm{Ker}(t) + \mathrm{Ker}(t)\,\mathrm{Ker}(s)$. This is then the cat^1-associative algebra enveloping the given cat^1-Lie algebra. The functoriality check is left to the reader. □

We have a commutative diagram as before

$$\begin{array}{ccc} \mathtt{Lie} & \xrightarrow{0} & \mathrm{crmod}_{\mathtt{Lie}} \\ \Big\downarrow U & & \Big\downarrow U \\ \mathtt{Alg} & \xrightarrow{0} & \mathrm{crmod}_{\mathtt{Alg}} \end{array}$$

and for the full embedding as identity crossed modules:

Theorem 3.8.10. *There is a natural isomorphism of functors*

$$1 \circ U = U \circ 1.$$

Proof. On the one hand, $(1 \circ U)(\mathfrak{g})$ is the crossed module of associative algebras $\mathrm{id}_{U\mathfrak{g}} : U\mathfrak{g} \to U\mathfrak{g}$. We have to compare it to the crossed module corresponding to the above constructed cat^1-associative algebra,

$$U(\mathfrak{g} \rtimes \mathfrak{g})/X \underset{t}{\overset{s}{\rightrightarrows}} U\mathfrak{g}.$$

Going one step back, in the cat^1-Lie algebra $\mathfrak{g} \rtimes \mathfrak{g} \underset{t}{\overset{s}{\rightrightarrows}} \mathfrak{g}$, source and target map are defined by $s(x,y) = y$ and $t(x,y) = x + y$ for all $x, y \in \mathfrak{g}$. There is furthermore a homomorphism of Lie algebras $\epsilon : \mathfrak{g} \to \mathfrak{g} \rtimes \mathfrak{g}$ defined by $x \mapsto (x, 0)$ for all $x \in \mathfrak{g}$, such that $s \circ \epsilon = 0$ and $t \circ \epsilon = \mathrm{id}_{\mathfrak{g}}$. Let $\pi : U(\mathfrak{g} \rtimes \mathfrak{g}) \to U(\mathfrak{g} \rtimes \mathfrak{g})/X$ be the quotient morphism. Then we deduce $s \circ \pi \circ \epsilon = s \circ \epsilon = 0$ and $t \circ \pi \circ \epsilon = t \circ \epsilon = \mathrm{id}_{\mathfrak{g}}$ for the induced maps $s, t : U(\mathfrak{g} \rtimes \mathfrak{g})/X \to U\mathfrak{g}$, respectively, $s, t : U(\mathfrak{g} \rtimes \mathfrak{g}) \to U\mathfrak{g}$. Thus the homomorphism $\pi \circ \epsilon$ takes values in $\mathrm{Ker}(s)$ and is a right inverse to $t|_{\mathrm{Ker}(s)} : \mathrm{Ker}(s) \to U\mathfrak{g}$, the crossed module corresponding to the cat^1-associative algebra.

Now we show that $\pi \circ \epsilon \circ t|_{\mathrm{Ker}(s)} = \mathrm{id}_{\mathrm{Ker}(s)}$. We have that $\mathrm{Ker}(s) \subset U(\mathfrak{g} \rtimes \mathfrak{g})$ is generated as a $\mathbb{K}$-vector space by all elements of the form

$$(x_1, x_1') \dots (x_{i-1}, x_{i-1}')(x_i, 0)(x_{i+1}, x_{i+1}') \dots (x_n, x_n'),$$

with $(x_j, x_j') \in \mathfrak{g} \rtimes \mathfrak{g}$ for all $1 \leq j \leq n$. By definition of t and ϵ, the value of $\epsilon \circ t$ on such an element is

$$(x_1 + x_1', 0) \dots (x_{i-1} + x_{i-1}', 0)(x_i, 0)(x_{i+1} + x_{i+1}', 0) \dots (x_n + x_n', 0).$$

But in the quotient with respect to the ideal X, we have then

$$\begin{aligned}
&(x_1 + x_1', 0) \dots (x_{i-1} + x_{i-1}', 0)(x_i, 0)(x_{i+1} + x_{i+1}', 0) \dots (x_n + x_n', 0) \\
&\quad = (x_1, x_1')(x_2 + x_2', 0) \dots (x_{i-1} + x_{i-1}', 0)(x_i, 0)(x_{i+1} + x_{i+1}', 0) \dots (x_n + x_n', 0) \\
&\quad = \cdots \\
&\quad = (x_1, x_1') \dots (x_{i-1}, x_{i-1}')(x_i, 0)(x_{i+1}, x_{i+1}') \dots (x_n, x_n').
\end{aligned}$$

Here, we have used, for example, from the first to the second line that $(x_1 + x_1', 0) - (x_1, x_1') = (x_1', -x_1') \in \mathrm{Ker}(t)$ and so on. This means therefore that these two elements are equal in the quotient with respect to the ideal X. Hence $\pi \circ \epsilon \circ t|_{\mathrm{Ker}(s)} = \mathrm{id}_{\mathrm{Ker}(s)}$.

Now it is easy to see that the pair $(t|_{\mathrm{Ker}(s)}, \mathrm{id}_{\mathfrak{g}})$ is an isomorphism of the crossed modules of algebras $t|_{\mathrm{Ker}(s)} : \mathrm{Ker}(s) \to U\mathfrak{g}$ and $\mathrm{id}_{U\mathfrak{g}} : U\mathfrak{g} \to \mathfrak{g}$.

We leave the functoriality claim for this isomorphism to the reader. □

Theorem 3.8.11. *The functor U between crossed modules is left adjoint to the Lieization functor $_{\mathrm{Lie}}$ between crossed modules.*

Proof. We have to construct a natural bijection

$$\mathrm{Hom}_{\mathrm{crmod}_{\mathrm{Lie}}}((\nu : \mathfrak{h} \to \mathfrak{g}), (\mu : M \to N)_{\mathrm{Lie}}) \cong \mathrm{Hom}_{\mathrm{crmod}_{\mathrm{Alg}}}(U(\nu : \mathfrak{h} \to \mathfrak{g}), (\mu : M \to N))$$

for any crossed modules $\nu : \mathfrak{h} \to \mathfrak{g}$ of Lie algebras and $\mu : M \to N$ of associative algebras.

For a given morphism $(\alpha, \beta) : (\nu : \mathfrak{h} \to \mathfrak{g}) \to (\mu : M \to N)_{\mathrm{Lie}}$, there is a corresponding morphism of cat^1-Lie algebras

$$\begin{array}{ccc}
\mathfrak{h} \rtimes \mathfrak{g} & \overset{s}{\underset{t}{\rightrightarrows}} & \mathfrak{g} \\
\downarrow{\scriptstyle \alpha'} & & \downarrow{\scriptstyle \beta} \\
(M \rtimes N)_{\mathrm{Lie}} & \overset{s}{\underset{t}{\rightrightarrows}} & N_{\mathrm{Lie}}
\end{array}$$

with $\alpha'(x, y) = (\alpha(x), \beta(y))$ for all $x \in \mathfrak{h}$ and all $y \in \mathfrak{g}$. Since the enveloping functor U is left adjoint to the Lieization functor, we deduce a commutative diagram of algebras:

$$\begin{array}{ccc}
U(\mathfrak{h} \rtimes \mathfrak{g}) & \xrightarrow{s} & U\mathfrak{g} \\
\downarrow{\scriptstyle \alpha'^*} & & \downarrow{\scriptstyle \beta^*} \\
M \rtimes N & \xrightarrow{s} & N.
\end{array}$$

We obtain a similar diagram for the source maps t. As in the cat^1-associative algebra $M \rtimes N \underset{t}{\overset{s}{\rightrightarrows}} N$, we have $\mathrm{Ker}(s)\,\mathrm{Ker}(t) + \mathrm{Ker}(t)\,\mathrm{Ker}(s) = 0$, the above morphisms pass to the quotient and we obtain a morphism of cat^1-associative algebras

$$\begin{array}{ccc} U(\mathfrak{h} \rtimes \mathfrak{g})/X & \underset{t}{\overset{s}{\rightrightarrows}} & \mathfrak{g} \\ \downarrow{\scriptstyle \alpha'} & & \downarrow{\scriptstyle \beta} \\ M \rtimes N & \underset{t}{\overset{s}{\rightrightarrows}} & N. \end{array}$$

The corresponding morphism of crossed modules of associative algebras is the wanted morphism. The inverse assignment is similar. The naturality proof is left to the reader. □

3.9 Examples

3.9.1 The crossed module associated to a cochain algebra

Let C be a differential graded algebra over the ground field $\mathbb{K}$ with differential of degree $+1$. The data is thus a grading $C = \bigoplus_{i\geq 0} C^i$ compatible with the product $C^iC^j \subset C^{i+j}$ and a differential $d : C \to C$ of degree $+1$ which is a derivation with respect to the product $d(xy) = d(x)y + (-1)^{|x|}xd(y)$ and satisfies $d^2 = 0$. Consider the graded $\mathbb{K}$-vector space $M = \mathrm{Coker}(d)[1]$ and $N = \mathrm{Ker}(d)$. Here, we define for a graded vector space W the *shifted* graded vector space $W[1]$ by

$$(W[1])^{n+1} = W^n.$$

The elements in $(W[1])^{n+1}$ are denoted by $s(w)$ for $w \in W^n$. The notation s stands for *suspension*. As the cokernel $\mathrm{Coker}(d)$ is a graded vector space (because d is homogeneous of degree $+1$), this applies to $W = \mathrm{Coker}(d)$ and we obtain the shifted object $M = \mathrm{Coker}(d)[1]$. We denote by $s(\bar{x}) \in \mathrm{Coker}(d)[1]$ the element corresponding to $x \in C$ via the projection $C \twoheadrightarrow \mathrm{Coker}(d)$. Then d induces a map of graded vector spaces

$$\mu : M = \mathrm{Coker}(d)[1] \to N = \mathrm{Ker}(d), \quad s(\bar{x}) \mapsto dx.$$

The multiplication in C induces the structure of a $\mathbb{K}$-algebra on N thanks to the derivation property. Moreover, we obtain the structure of an N-bimodule on M by setting for all $x \in C$ and all $n \in N$,

$$\begin{aligned} n * s(\bar{x}) &:= (-1)^{|n|} s(\overline{nx}), \\ s(\bar{x}) * n &:= s(\overline{xn}). \end{aligned}$$

In fact, for $y = dz$ and $n \in N$, we have $(-1)^{|n|}ny = d(nz)$ by the derivation property and, therefore, the bimodule structure is well-defined. As explained before, in case we want to see M as an associative algebra, we consider the associative product on M to be defined such that property (b) of a crossed module is satisfied, that is,

$$s(\bar{x})s(\bar{y}) := \mu(s(\bar{x})) * s(\bar{y}) = s(\bar{x}) * \mu(s(\bar{y})).$$

Observe that we have indeed $\mu(s(\bar{x}))*s(\bar{y}) = s(\bar{x})*\mu(s(\bar{y}))$ due to the derivation property, the shift in degrees for M and the fact that $d(xy) = 0 \in M$:

$$\begin{aligned}\mu(s(\bar{x})) * s(\bar{y}) &= (dx) * s(\bar{y})\\ &= (-1)^{|x|+1}s(\overline{(dx)y})\\ &= s(\overline{x(dy)})\\ &= s(\bar{x}) * (dy)\\ &= s(\bar{x}) * \mu(s(\bar{y})).\end{aligned}$$

We now check that $\mu : M \to N$ is a crossed module. Given $n \in N$ and $s(\bar{x}) \in M$, we have

$$\begin{aligned}\mu(n * s(\bar{x})) &= (-1)^{|n|}\mu(s(\overline{nx}))\\ &= (-1)^{|n|}d(nx)\\ &= n\,d(x)\\ &= n\,\mu(s(\bar{x})).\end{aligned}$$

Thus the differential graded algebra C induces naturally a crossed module $\mu : M \to N$. The cokernel of μ is the cohomology $H^*(C)$ of C with respect to the differential d, and the kernel is the $H^*(C)$-bimodule whose underlying $\mathbb{K}$-vector space is $H^*(C)[1]$ and where the left multiplication is twisted, that is, $x*y = (-1)^{|x|}xy$, and the right multiplication is the ordinary one. This $H^*(C)$-bimodule is denoted by $\overline{H^*(C)[1]}$. The crossed module $\mu : M \to N$ gives rise by Theorem 3.2.1 to a class $[\theta] \in HH^3(H^*(C), \overline{H^*(C)[1]})$ which is called the *characteristic class* of the cochain algebra C.

3.9.2 The crossed module associated to a chain complex of vector spaces

Let $C = (\dots \xrightarrow{d} V_n \xrightarrow{d} V_{n-1} \xrightarrow{d} \dots)$ be a chain complex of vector spaces over a fixed field $\mathbb{K}$. The range of spaces V_n is in general not bounded, that is, there may exist nonzero V_n for all $n \in \mathbb{Z}$. Let $\mathfrak{gl}_0(C)$ be the unital associative algebra of chain maps $f : C \to C$ with the usual composition of chain maps as a product. The identity id_C is obviously a unit for this product.

Let $\mathrm{End}_i(C)$ denote the vector space of maps $f : C \to C$, which are homogeneous of degree i, but do not in general satisfy a compatibility relation with respect to the differential d. Obviously, $\mathfrak{gl}_0(C) \subset \mathrm{End}_0(C)$ consists exactly of the maps commuting with d. Composition yields a bilinear map $\mathrm{End}_i(C) \times \mathrm{End}_j(C) \to \mathrm{End}_{i+j}(C)$. A degree 1 map $s : C \to C$ will be called a *homotopy*, although we do not demand a compatibility with respect to the differential d. It consists thus of a family $\{s_i : V_i \to V_{i+1}\}_{i\in\mathbb{Z}}$ of $\mathbb{K}$-linear maps s_i. There is a map $\partial : \mathrm{End}_1(C) \to \mathfrak{gl}_0(C) \subset \mathrm{End}_0(C)$ defined by $\partial(s) = d \circ s + s \circ d$ for $s \in \mathrm{End}_1(C)$.

For $f \in \mathfrak{gl}_0(C)$ and $s \in \mathrm{End}_1(C)$, we have

$$\partial(f \circ s) = f \circ \partial(s) \quad \text{and} \quad \partial(s \circ f) = \partial(s) \circ f.$$

A degree 2 map $h : C \to C$ is a family $\{h_i : V_i \to V_{i+2}\}_{i\in\mathbb{Z}}$ of $\mathbb{K}$-linear maps $h_i : V_i \to V_{i+2}$. There is a map $\partial : \mathrm{End}_2(C) \to \mathrm{End}_1(C)$ defined by $\partial(h) = d \circ h + h \circ d$ for $h \in \mathrm{End}_2(C)$. Again, we have for $f \in \mathfrak{gl}_0(C)$ and $h \in \mathrm{End}_2(C)$, we have

$$\partial(f \circ h) = f \circ \partial(h) \quad \text{and} \quad \partial(h \circ f) = \partial(h) \circ f.$$

Define $\mathfrak{gl}_1(C) = \mathrm{End}_1(C)/\partial(\mathrm{End}_2(C))$. Then clearly $\partial : \mathrm{End}_1(C) \to \mathfrak{gl}_0(C)$ descends to a linear map $\mu : \mathfrak{gl}_1(C) \to \mathfrak{gl}_0(C)$. Moreover, the composition maps

$$\mathrm{End}_0(C) \times \mathrm{End}_1(C) \ni (f, s) \mapsto f \circ s \in \mathrm{End}_1(C)$$

and

$$\mathrm{End}_1(C) \times \mathrm{End}_0(C) \ni (s, f) \mapsto s \circ f \in \mathrm{End}_1(C)$$

descend to linear maps

$$\mathfrak{gl}_0(C) \times \mathfrak{gl}_1(C) \ni (f, s) \mapsto f \circ s \in \mathfrak{gl}_1(C)$$

and

$$\mathfrak{gl}_1(C) \times \mathfrak{gl}_0(C) \ni (s, f) \mapsto s \circ f \in \mathfrak{gl}_1(C).$$

Lemma 3.9.1. *For all $s, t \in \mathrm{End}_1(C)$, we have the following identity:*

$$\partial(s \circ t) = \partial(s) \circ t - s \circ \partial(t).$$

In particular, for $s, t \in \mathfrak{gl}_1(C)$, we obtain

$$\mu(s) \circ t = s \circ \mu(t) \in \mathfrak{gl}_1(C).$$

Proof. Indeed,

$$\begin{aligned}\partial(s) \circ t - s \circ \partial(t) &= (s \circ d + d \circ s) \circ t - s \circ (t \circ d + d \circ t)\\ &= d \circ s \circ t - s \circ t \circ d\\ &= \partial(s \circ t).\end{aligned}$$

□

Theorem 3.9.2. *For any chain complex* $C = (\ldots \xrightarrow{d} V_n \xrightarrow{d} V_{n-1} \xrightarrow{d} \ldots)$, *the map* $\mu : \mathfrak{gl}_1(C) \to \mathfrak{gl}_0(C)$ *defined by* $s \mapsto s \circ d + d \circ s$ *for* $s \in \mathfrak{gl}_1(C)$ *is a homomorphism of algebras. Moreover, with the left and right actions defined by*

$$\mathfrak{gl}_0(C) \times \mathfrak{gl}_1(C) \ni (f, s) \mapsto f \circ s \in \mathfrak{gl}_1(C)$$

and

$$\mathfrak{gl}_1(C) \times \mathfrak{gl}_0(C) \ni (s, f) \mapsto s \circ f \in \mathfrak{gl}_1(C),$$

and the product in $\mathfrak{gl}_1(C)$ *by* $s * t := s \circ \partial(t) = \partial(s) \circ t$, *the map* $\mu : \mathfrak{gl}_1(C) \to \mathfrak{gl}_0(C)$ *becomes a crossed module of associative algebras.*

Proof. We have

$$\begin{aligned}\mu(s * t) &= \mu(s \circ \partial(t))\\ &= s \circ (t \circ d + d \circ t) \circ d + d \circ s \circ (t \circ d + d \circ t)\\ &= (d \circ s + s \circ d) \circ (d \circ t + t \circ d)\\ &= \mu(s) \circ \mu(t),\end{aligned}$$

where we have used $d^2 = 0$ from the second to the third line.

The requirements for an action which is compatible with the product are clearly satisfied, because actions and product come from the associative composition of maps.

Property (b) of a crossed module is satisfied by definition of the product in $\mathfrak{gl}_1(C)$. The previous lemma shows that it is well-defined.

Property (a) follows from the above identities:

$$\partial(f \circ s) = f \circ \partial(s) \quad \text{and} \quad \partial(s \circ f) = \partial(s) \circ f$$

for $f \in \mathfrak{gl}_0(C)$ and $s \in \mathrm{End}_1(C)$. □

Remark 3.9.1.

(a) Note that while $\mathfrak{gl}_0(C)$ is a unital algebra and id_C acts as the identity on $\mathfrak{gl}_1(C)$, the associative algebra $(\mathfrak{gl}_1(C), *)$ is not unital. These are nevertheless the requirements to form a *unital crossed module* of associative algebras.

(b) Note that the above crossed module satisfies the additional property

$$(s \cdot f)(g \cdot t) = s((fg) \cdot t) = (s \cdot (fg))t$$

for all $s, t \in \mathfrak{gl}_1(C)$ and all $f, g \in \mathfrak{gl}_0(C)$. Once again, this follows from the fact that all actions and products involved are given by the composition of maps. This property is referred to as *left and right actions are fully interchangeable* in [27].

(c) Note that the construction in this subsection is a special case of the (dual to the) construction in the previous subsection. The graded algebra in question is just the algebra of endomorphisms, graded by the degree of an endomorphism. Note that the sign in the formulae is essential, while it can be completely avoided in this subsection (more precisely, it only appears in the statement of the lemma).

3.10 Exercises

Exercise 3.10.1. Find an example of an associative algebra L with nontrivial outer multiplication algebra $\mathrm{Out}(L)$. (Hint: Get, e. g., inspired by the loop Lie algebra.)

Exercise 3.10.2. Complete the construction of the vector space structure on the set of associative algebra kernels by defining the scalar multiplication; see Section 3.6.5 and especially the proof of Theorem 3.6.7.

Exercise 3.10.3. Complete Section 3.3 by showing the independence of the class of $d\tilde{\theta}$ from the choices of the extension $\tilde{\theta}$ and the choice of the cocycle ω (cf. Lemma 2.3.2), and that the corresponding cohomology class coincides with the previously defined class; cf Proposition 2.3.3.

Exercise 3.10.4. Complete the proof of Theorem 3.5.1.

Exercise 3.10.5. Provide a proof for Theorem 3.6.6.

Exercise 3.10.6. Complete the proof of Theorem 3.7.2, namely show the middle four exchange law for the categorical composition in the 2-vector space built from a cat^1-algebra. (Hint: Proceed as in Lemma 2.10.4.)

Exercise 3.10.7. Do the naturality and functoriality checks in Section 3.8, especially in the proofs of Theorem 3.8.7 and in Theorem 3.8.11.

Exercise 3.10.8. Show that there is (strictly speaking) in general no commutative diagram of the form (cf. Section 3.8.2)

$$\begin{array}{ccc} \mathtt{Alg} & \xrightarrow{0} & \mathrm{crmod}_{\mathtt{Alg}} \\ \downarrow{\scriptstyle (-)^{\times}} & & \downarrow{\scriptstyle (-)^{\times}} \\ \mathtt{Grp} & \xrightarrow{0} & \mathrm{crmod}_{\mathtt{Grp}}. \end{array}$$

Show that there is such a diagram in case $\mathbb{K} = \mathbb{F}_2$, the field with two elements. (In fact, in order to make sense of the diagram, first of all, the algebra A has to be unital. This gives a unital crossed module $0 : 0 \to A$. Then one has to adjoin a unit to the zero algebra 0, transforming $0 : 0 \to A$ into the unit of the algebra A, namely $\eta : \mathbb{K} \to A$. The corresponding crossed module of groups one obtains is then $\mathbb{K}^{\times} \to A^{\times}$.)

Exercise 3.10.9. The goal of this exercise is to show how to obtain explicit formulae for the cocycle α in the construction mechanism of crossed modules in Section 3.5. We address this question in some more generality at first, letting X be an arbitrary algebraic structure.

(a) Let $C^*(X, V)$ be the standard complex of the algebraic structure X and suppose that there is a short exact sequence of X-modules

$$0 \to V = V_1 \to V_2 \to V_3 \to 0$$

such that the complex $C^*(X, V_2)$ admits a chain homotopical contraction h. For a given 3-cocycle $\gamma \in Z^3(X, V)$, we would like to construct an explicit cocycle $\alpha \in Z^2(X, V_3)$ such that the connecting homomorphism sends $[\alpha]$ to $[\gamma]$. Show that one may construct α as follows:

$$\begin{array}{ccccc}
 & & h(i_*(\gamma)) & \xmapsto{\pi_*} & \alpha := \pi_*(h(i_*(\gamma))) \\
 & & \uparrow & & \downarrow \\
C^2(X,V) & \xrightarrow{i_*} & C^2(X,V_2) & \xrightarrow{\pi_*} & C^2(X,V_3) \\
\downarrow d^V & & \uparrow h & & \downarrow d^{V_3} \\
C^3(X,V) & \xrightarrow{i_*} & C^3(X,V_2) & \xrightarrow{\pi_*} & C^3(X,V_3) \\
 & & \uparrow & & \downarrow \\
\gamma & \xmapsto{i_*} & i_*(\gamma) & \xmapsto{\pi_*} & 0 = d^{V_3}\alpha
\end{array}$$

The cochain α is by definition $(\pi_* \circ h \circ i_*)(\gamma)$. (The main point to notice is here that by commutativity of the diagram, $d^{V_2} i_*(\gamma) = 0$ and, therefore,

$$(d^{V_2}h)(i_*\gamma) = i_*\gamma - (hd^{V_2})(i_*\gamma) = i_*\gamma$$

by the chain homotopy property of h.)

(b) Construct a contraction in the case of cohomology of groups (with values in the injective $\mathbb{K}[G]$-module $\mathbb{K}[G]^*$) and for associative algebras (with values in the injective A-bimodule $(A \otimes_{\mathbb{K}} A^{\mathrm{op}})^*$).

3.11 Bibliographical notes

In the theory of crossed modules of associative algebras, things are a bit different with respect to crossed modules of groups or Lie algebras. Nevertheless, Theorem 3.2.1 has been stated for the first time by Gerstenhaber in [37], according to the historical note by Mac Lane [71]. For the proof, Gerstenhaber refers to [36] where he proves in fact that splicing together long exact sequences of R-modules with abelian extensions yields a cohomology theory, which is again $\mathrm{Ext}^*_R(M, N)$. The proof is by stating that both theories satisfy the axioms and vanish on injectives (without further details).

Section 3.1.1 is due to Baues–Pirashvili [7]. In Section 3.2.1, we adapt the framework of groups and Lie algebras here to the case of associative algebras. Section 3.2.2 is due to Baues–Minian [6] with computations by Salim Rivière. Section 3.3 is new and an adaptation of the corresponding relation for Lie algebras here to associative algebras.

Section 3.4 is taken from Baues–Pirashvili [7]. It depicts as for the case of groups and Lie algebras crossed modules of associative algebras as obstruction for the existence of abelian extensions.

In Section 3.5, we adapt again our construction method of crossed modules to the case of associative algebras. In Section 3.6, we expose Hochschild's theory of associative algebra kernels [45], culminating in a version of Gerstenhaber's theorem for associative algebra kernels Theorem 3.6.7. The difference with Theorem 3.2.1 is that here the coefficients are necessarily in the nucleus Z of the associative algebra kernel.

In Section 3.7, we adapt the relation between strict 2-objects and crossed modules to associative algebras. Section 3.8 reports on the relation between crossed modules of associative and of Lie algebras and groups. The relation to crossed modules of groups is taken from Casas–Inassaridze–Khamaladze–Ladra [18]. The relation to crossed modules of Lie algebras is taken from Casas–Casado–Khamaladze–Ladra [17]. The examples in Section 3.9 are taken from Baues–Minian [6].

4 Crossed modules of Hopf algebras

In this section, the cohomological framework breaks down. We do not have cohomology classes corresponding to crossed modules of Hopf algebras. We study the relation between crossed modules of Lie and Hopf algebras in Section 4.2, and the relation between crossed modules of groups and Hopf algebras in Section 4.3. Because of the natural appearance of duality in this context, we also consider crossed *comodules* of Hopf algebras in Section 4.4. Section 4.5 is devoted to a first discussion (a second discussion follows in Section 5.1.6) of the integration of crossed modules of Lie algebra into crossed modules of groups; here via Hopf algebras. The categorical versions of Hopf algebras and their relation to crossed modules are reported on in Section 4.6, following [30]. Here, the work of Radford [89] on the semidirect product of Hopf algebras plays a fundamental role. We close the chapter with exercises in Section 4.8, after a section on strict quantum 2-groups taken from [75].

4.1 Preliminaries

Here, we collect standard notation and definitions for studying Hopf algebras coming from groups and Lie algebras, and their crossed modules. We introduce in Definition 4.1.3 the notion of a *crossed module of Hopf algebras*.

Let us denote by τ the twist in the symmetric monoidal category of vector spaces over a commutative field $\mathbb{K}$ of characteristic 0. Most considerations in this chapter can be done over a general symmetric monoidal category, but in order to keep abstraction to a minimum we refrain from this generality. In concrete terms, $\tau = \tau_{V,W} : V \otimes W \to W \otimes V$ is the isomorphism of vector spaces given on elementary tensors by $\tau(v \otimes w) = w \otimes v$. In geometric situations and some other situations, we will suppose $\mathbb{K} = \mathbb{C}$ in this chapter, but most of the time $\mathrm{car}(\mathbb{K}) \neq 2$ is sufficient. All Lie and Hopf algebras are supposed to be algebras over $\mathbb{K}$. All tensor products are over $\mathbb{K}$ unless otherwise indicated.

Definition 4.1.1. A *Hopf algebra H* over $\mathbb{K}$ is given by $(H, \mu_H, \eta_H, \triangle_H, \epsilon_H, S_H)$, where μ_H is the associative product, η_H its unit, $\triangle_H$ the coassociative coproduct, ϵ_H its counit, and S_H is the antipode.

We refer to [58] for an explanation of the terms occurring in this definition and for the explicit diagrams which are to be satisfied. We will use Sweedler notation in the following. Thus, for example, the coproduct is written

$$\triangle(h) = h_1 \otimes h_2,$$

where summation is understood. In case the Hopf algebra we work with is clear from the context, we feel free to drop the corresponding index (and write, e. g., μ instead

https://doi.org/10.1515/9783110750959-004

of μ_H). Morphisms of Hopf algebras are defined in the obvious way; in case you have doubts about this definition, see [58].

Let H be a Hopf algebra. Then H becomes a left H-module by the *adjoint action* which is defined for $h, k \in H$ as follows:

$$\begin{aligned}\mathrm{ad}_H(h)(k) &= \mu \circ (\mu \otimes S) \circ (\mathrm{id}_H \otimes \tau_{H,H}) \circ (\triangle \otimes \mathrm{id}_H)(h \otimes k)\\ &= h_1 k S(h_2),\end{aligned}$$

in Sweedler notation.

Recall the notions of an H-module algebra and an H-module coalgebra.

Definition 4.1.2. Let H be a Hopf algebra and (A, η, μ) be a unital associative algebra. A is called a *left H-module algebra* in case the vector space underlying A is a left H-module and the action is compatible with the algebra structure in the sense that μ (i. e., μ_A) and η (i. e., η_A) are morphisms of H-modules (where H acts on $A \otimes A$ and $\mathbb{K}$ using $\triangle_H$ and ϵ_H, resp.).

Similarly, a coalgebra $(C, \epsilon, \triangle)$ is a *left H-comodule coalgebra* in case the vector space C carries a left coaction of H and ϵ and $\triangle$ (i. e., ϵ_C and $\triangle_C$) are morphisms of H-comodules. In an analogous way, one defines a *left H-comodule algebra* and a *left H-module coalgebra.*

To be more explicit, let us write down the diagrams for an H-module algebra (A, μ, η). The vector space A is an H-module via a map $\phi_A : H \otimes A \to A$. As H is a bialgebra, $A \otimes A$ becomes an H-module using the coproduct, namely

$$\phi_{A\otimes A} = (\phi_A \otimes \phi_A) \circ (\mathrm{id}_H \otimes \tau \otimes \mathrm{id}_A) \circ (\triangle_H \otimes \mathrm{id}_A \otimes \mathrm{id}_A),$$

or with elements $h \in H$ and $a, b \in A$ in Sweedler notation

$$h \cdot a \otimes b := (h_1 \cdot a) \otimes (h_2 \cdot b).$$

Then the property that μ should be a morphism of H-modules is the commutativity of the diagram:

$$\begin{CD} H \otimes A \otimes A @>{\mathrm{id}_H \otimes \mu}>> H \otimes A \\ @V{\phi_{A\otimes A}}VV @VV{\phi_A}V \\ A \otimes A @>{\mu}>> A \end{CD}$$

The commutativity of this diagram means for all $h \in H$ and all $a, b \in A$,

$$h \cdot (ab) = (h_1 \cdot a)(h_2 \cdot b). \tag{4.1}$$

In the same way, $\mathbb{K}$ becomes an H-module using the counit $\epsilon : H \to \mathbb{K}$, that is, $\phi_{\mathbb{K}} = \epsilon : H \otimes \mathbb{K} \to \mathbb{K}$ (with the obvious identifications). Then the fact that η is a morphism of H-modules is the commutativity of the diagram

$$\begin{array}{ccc} H\otimes\mathbb{K} & \xrightarrow{\mathrm{id}_H\otimes\eta} & H\otimes A \\ \downarrow{\scriptstyle \phi_{\mathbb{K}}} & & \downarrow{\scriptstyle \phi_A} \\ \mathbb{K} & \xrightarrow{\eta} & A \end{array}$$

and this commutativity means for all $h \in H$,

$$\epsilon(h)1 = h\cdot 1. \tag{4.2}$$

Similarly, a coalgebra $(C, \triangle_C, \epsilon_C)$ becomes an H-module coalgebra in case the coproduct $\triangle_C$ and the counit ϵ_C are morphisms of H-modules, where $C\otimes C$ carries the above H-module structure $\phi_{C\otimes C}$ and $\mathbb{K}$ the above H-module structure using ϵ_H. This means then that the diagrams

$$\begin{array}{ccc} H\otimes C & \xrightarrow{\phi_C} & C \\ \downarrow{\scriptstyle \mathrm{id}_H\otimes\triangle_C} & & \downarrow{\scriptstyle \triangle_C} \\ H\otimes C\otimes C & \xrightarrow{\phi_{C\otimes C}} & C\otimes C \end{array}$$

and

$$\begin{array}{ccc} H\otimes C & \xrightarrow{\mathrm{id}_H\otimes\epsilon_C} & H\otimes\mathbb{K} \\ \downarrow{\scriptstyle \phi_C} & & \downarrow{\scriptstyle \epsilon_H} \\ C & \xrightarrow{\epsilon_C} & \mathbb{K} \end{array}$$

are commutative, or, with elements $c \in C$ and $h \in H$:

$$(h\cdot c)_1\otimes(h\cdot c)_2 = (h_1\cdot c_1)\otimes(h_2\cdot c_2), \quad \text{and} \quad \epsilon_C(h\cdot c) = \epsilon_H(h)\epsilon_C(c). \tag{4.3}$$

We leave to the reader the spelling out of the two diagrams characterizing an H-comodule algebra, and the two diagrams characterizing an H-comodule coalgebra; see Exercise 4.8.1. In Exercise 4.8.2, we give some elementary examples of (co)module-(co)algebras. Everything extends in an obvious way to right modules.

The main definition in this chapter is the following.

Definition 4.1.3. Let $\gamma : B \to H$ be a morphism of Hopf algebras. The morphism $\gamma : B \to H$ is a *crossed module of Hopf algebras* in case:

(i) (B, ϕ_B) is a left H-module which renders it a left H-module algebra and H-module coalgebra;

(ii) the map γ satisfies

$$\begin{aligned} \gamma\circ\phi_B &= \mu_H\circ(\mu_H\otimes S_H)\circ(\mathrm{id}_H\otimes\tau_{H,H})\circ(\triangle_H\otimes\gamma) \\ &= \mathrm{ad}_H\circ(\mathrm{id}_H\otimes\gamma), \end{aligned}$$

that is, γ is a morphism of H-modules where H carries the H-module structure given by the adjoint action ad_H;

(iii)

$$\begin{aligned}\phi_B \circ (\gamma \otimes \mathrm{id}_B) &= \mu_B \circ (\mu_B \otimes S_B) \circ (\mathrm{id}_B \otimes \tau_{B,B}) \circ (\Delta_B \otimes \mathrm{id}_B)\\ &= \mathrm{ad}_B.\end{aligned}$$

We call this last condition *Peiffer identity*.

Remark 4.1.4. Note that these conditions may be expressed in terms of elements $h \in H$ and $b, b' \in B$ in Sweedler notation more explicitly as:
(ii) $\gamma(h \cdot b) = h_1\gamma(b)S(h_2)$ and
(iii) $\gamma(b) \cdot b' = b_1 b' S(b_2)$.

Lemma 4.1.1. *Let A and H be Hopf algebras. Suppose that A is an H-module algebra and H-module coalgebra. Then we have for all $h \in H$ and all $a \in A$,*

$$h \cdot S(a) = S(h \cdot a).$$

Proof. In order to prove an identity about the antipode, we have to go back to the convolution algebra. The tensor product $H \otimes A$ of two Hopf algebras is a bialgebra, and with respect to this bialgebra structure, we have a convolution product on the space $\mathrm{Hom}(H \otimes A, A)$ of linear maps from $H \otimes A$ to A. Define now two maps

$$\nu : H \otimes A \to A, (h \otimes a) \mapsto h \cdot S(a),$$

and

$$\mu : H \otimes A \to A, (h \otimes a) \mapsto S(h \cdot a).$$

We will show that both ν and μ are convolution inverses to the module structure $\varphi : H \otimes A \to A$, and this will show by unicity of the inverse that they are equal. Indeed, let us compute

$$\begin{aligned}(\mu \star \varphi)(h \otimes a) &= \mu(h_1 \otimes a_1)\varphi(h_2 \otimes a_2)\\ &= S(h_1 \cdot a_1)(h_2 \cdot a_2)\\ &= S((h \cdot a)_1)(h \cdot a)_2\\ &= \epsilon(h \cdot a) = \epsilon_H(h)\epsilon_A(a) = \eta\epsilon(h \otimes a).\end{aligned}$$

Here, we have used that A is an H-module coalgebra: $(h_1 \cdot a_1) \otimes (h_2 \cdot a_2) = (h \cdot a)_1 \otimes (h \cdot a)_2$.

On the other hand,

$$\begin{aligned}(\nu \star \varphi)(h \otimes a) &= (h_1 \cdot S(a_1))(h_2 \cdot a_2)\\ &= h \cdot (S(a_1)a_2)\\ &= h \cdot (\epsilon_H(h)1) = \epsilon_H(h)\epsilon_A(a) = \eta\epsilon(h \otimes a),\end{aligned}$$

where we have used that A is an H-module algebra. The convolutions $\varphi \star \mu$ and $\varphi \star \nu$ are computed in the same way with the same result. □

Observe that in the special case of Hopf algebras in the category of sets, that is, groups, for the crossed module of groups $\mathrm{id}_G : G \to G$, this property is simply $(ghg^{-1})^{-1} = gh^{-1}g^{-1}$ for all $g, h \in G$. For a general crossed module of groups $\mu : M \to N$, this property follows from the fact that N acts by automorphisms on M.

This property imposes severe conditions on the Hopf algebra A in conjunction with the properties (i) to (iii) of a crossed module of Hopf algebras. Namely in case $A = H$, the adjoint action of a Hopf algebra H on itself does not in general satisfy $\mathrm{ad}_h(S(h')) = S(\mathrm{ad}_h(h'))$ for all $h, h' \in H$, but property (i) is related to the equality of these two terms; see Exercise 4.8.3.

Definition 4.1.5. A *morphism* $(\rho, \sigma) : (\gamma : B \to H) \to (\gamma' : B' \to H')$ between crossed modules of Hopf algebras $\gamma : B \to H$ and $\gamma' : B' \to H'$ is a pair of morphisms of Hopf algebras $\rho : B \to B'$ and $\sigma : H \to H'$ such that the following diagrams commute:

$$\begin{array}{ccc} B & \xrightarrow{\rho} & B' \\ \downarrow{\scriptstyle\gamma} & & \downarrow{\scriptstyle\gamma'} \\ H & \xrightarrow{\sigma} & H' \end{array} \qquad\qquad \begin{array}{ccc} B \otimes H & \xrightarrow{\rho\otimes\sigma} & B' \otimes H' \\ \downarrow{\scriptstyle\phi_B} & & \downarrow{\scriptstyle\phi'_B} \\ B & \xrightarrow{\rho} & B' \end{array}$$

It is easy to show that crossed modules of Hopf algebras form a category with respect to this notion of morphisms.

Remark 4.1.6. (a) Observe that the definition of a crossed module of Hopf algebras is not autodual: The Hopf algebra B carries only the structure of an H-module – the definition does not demand a comodule structure. On the other hand, the definition of a Hopf algebras *is* autodual.

(b) Observe that one cannot associate naively a four term exact sequence of Hopf algebras to a crossed module of Hopf algebras: Condition (ii) does not imply that the image of γ is an associative ideal (while it is always a coideal).

Remark 4.1.7. This definition differs from Definition 12 given in [30]: We do not make any assumption on cocommutativity nor impose constraints coming from compatibility with cocommutativity (condition (i) of Definition 12 in [30]). We do not ask the antipode S_B to be a morphism of H-modules either (condition (iii) of Definition 12 in [30]). In some sense, our definition is more the definition of a crossed module of bialgebras.

Let us denote $\mathtt{Lie}$, $\mathtt{Hopf}$, $\mathtt{Grp}$, $\mathtt{algGrp}$ the categories of $\mathbb{K}$-Lie algebras, respectively, $\mathbb{K}$-Hopf algebras, respectively, groups, respectively, (connected) algebraic groups defined over $\mathbb{C}$. For our purposes, considering algebraic groups will always imply that the ground field $\mathbb{K}$ is $\mathbb{C}$, the field of complex numbers. Let us furthermore denote by $\mathtt{ccHopf}$ and $\mathtt{cHopf}$ the full subcategories of $\mathtt{Hopf}$ consisting of cocommutative, respectively, commutative Hopf algebras. Recall the enveloping functor

$$U : \mathtt{Lie} \to \mathtt{ccHopf},$$

which associates to a Lie algebra $\mathfrak{g}$ its universal enveloping algebra $U\mathfrak{g}$; the functor of primitives

$$P : \mathtt{Hopf} \to \mathtt{Lie},$$

which associates to a Hopf algebra its Lie algebra of primitive elements (i. e., the $h \in H$ such that $\triangle h = 1 \otimes h + h \otimes 1$); the group algebra functor

$$\mathbb{K}[-] : \mathtt{Grp} \to \mathtt{ccHopf},$$

which associates to a group G its group-algebra $\mathbb{K}[G]$ which is a cocommutative Hopf algebra; the functor of group-like elements

$$G(-) : \mathtt{Hopf} \to \mathtt{Grp},$$

which associates to a Hopf algebra H with invertible antipode its group $G(H)$ of group-like elements, that is, of elements $h \in H$ with $\triangle h = h \otimes h$; the functor of regular functions

$$\mathbb{K}- : \mathtt{algGrp} \to \mathtt{cHopf},$$

which associates to a connected algebraic group G its Hopf algebra of regular functions $\mathbb{K}G$; the functor of characters

$$\chi : \mathtt{cHopf} \to \mathtt{algGrp},$$

which associates to a commutative Hopf algebra H (which is finitely generated as an algebra) the algebraic group of its characters $\chi(H)$, that is, of algebra morphisms from H to $\mathbb{K}$ – its underlying algebraic variety is built on the set of maximal ideals of the commutative algebra H.

4.2 Crossed modules of Lie and Hopf algebras

The goal of this section is to study the compatibility of the notion of crossed module with the standard functors U and P between the categories of Lie algebras and cocommutative Hopf algebras. The main theorem of this section shows how to associate a crossed module of Hopf algebras to a crossed module of Lie algebras.

Lemma 4.2.1. *An action of a Lie algebra $\mathfrak{g}$ by derivations (of the Lie bracket) on a Lie algebra $\mathfrak{h}$ extends uniquely to an action by derivations (of the associative product) of $U\mathfrak{g}$ on $U\mathfrak{h}$.*

Proof. Let D be a derivation of $\mathfrak{g}$. Note that as we want to extend D to a derivation $\tilde{D}$ of the associative algebra, we need to have $\tilde{D}(11) = \tilde{D}(1)1 + 1\tilde{D}(1)$ and, therefore, $\tilde{D}(1) = 0$, thus the map $\tilde{D}$ is already defined on $\mathbb{K} \subset U\mathfrak{g}$. On the other hand, the map $\tilde{D}$ is already defined on $\mathfrak{g} \subset U\mathfrak{g}$ by D. Suppose by induction that $\tilde{D}$ is already defined on $\mathfrak{g}^n \subset U\mathfrak{g}$. Then one can extend it to $\mathfrak{g}^{n+1} \subset U\mathfrak{g}$ by the derivation property:

$$\tilde{D}(xy) = \tilde{D}(x)y + x\tilde{D}(y),$$

for all $x \in \mathfrak{g}^n \subset U\mathfrak{g}$ and all $y \in \mathfrak{g}$. Observe that this extension is unique and vanishes on the ideal generated by the $x \otimes y - y \otimes x - [x, y]$ in $T\mathfrak{g}$.

Applying this scheme to the derivations given by all the elements $x \in \mathfrak{g}$, we obtain derivations $\tilde{D}_x \in \mathrm{Der}(U\mathfrak{h})$. We view this as a homomorphism of Lie algebras $\mathfrak{g} \to \mathrm{End}(U\mathfrak{h})$, $x \mapsto \tilde{D}_x$. By the universal property of the universal enveloping algebra, we deduce a homomorphism of associative algebras $U\mathfrak{g} \to \mathrm{End}(U\mathfrak{h})$. Its image consists (by construction) of derivations of the associative product. □

Lemma 4.2.2. *Let $\mathfrak{g}$ be a Lie algebra. An associative algebra A is an $U\mathfrak{g}$-module algebra if and only if the vector space A is a $\mathfrak{g}$-module such that elements of $\mathfrak{g}$ act by derivations (of the associative product of A).*

Proof. Suppose A is an $U\mathfrak{g}$-module algebra. Then the $U\mathfrak{g}$-module A is a $\mathfrak{g}$-module and the elements $x \in \mathfrak{g}$ satisfy equation (4.1), that is, for all $x \in \mathfrak{g}$ and all $a, b \in A$, we have

$$x \cdot (ab) = (x \cdot a)b + a(x \cdot b),$$

because x is primitive, that is, $\Delta(x) = 1 \otimes x + x \otimes 1$. Therefore, $\mathfrak{g}$ acts by derivations.

Conversely, let A be an algebra, which is a $\mathfrak{g}$-module where its elements act by derivations. A becomes a $U\mathfrak{g}$-module such that equation (4.2) holds trivially. We have to show equation (4.1). It holds by construction for the elements of $\mathfrak{g}$. As $U\mathfrak{g}$ is generated as an algebra by the products of elements of $\mathfrak{g}$, it is enough to show for all $x, y \in \mathfrak{g}$ and all $a, b \in A$ that $(xy) \cdot (ab) = ((xy)_1 \cdot a)((xy)_2 \cdot b)$. We compute

$$\begin{aligned}
(xy) \cdot (ab) &= x \cdot (y \cdot (ab)) \\
&= x \cdot ((y_1 \cdot a)(y_2 \cdot b)) \\
&= (x_1 \cdot (y_1 \cdot a))(x_2 \cdot (y_2 \cdot b)) \\
&= ((x_1 y_1) \cdot a)((x_2 y_2) \cdot b) \\
&= ((xy)_1 \cdot a)((xy)_2 \cdot b),
\end{aligned}$$

where we have used the bialgebra property from the fourth to the last line. □

Theorem 4.2.3. *The enveloping functor U extends to a functor (still denoted U) from the category of crossed modules of Lie algebras to the category of crossed modules of (cocommutative) Hopf algebras.*

Proof. Let $\mu : \mathfrak{m} \to \mathfrak{n}$ be a crossed module of Lie algebras. By functoriality, we get a homomorphism of associative algebras $\gamma := U\mu : U\mathfrak{m} \to U\mathfrak{n}$. By the previous two lemmas, $U\mathfrak{n}$ acts (by derivations of the associative product) on $U\mathfrak{m}$ and with this action, $U\mathfrak{m}$ becomes an $U\mathfrak{n}$-module algebra. Let us show that $U\mathfrak{m}$ is also an $U\mathfrak{n}$-module coalgebra, that is, that the coproduct $\Delta_{U\mathfrak{m}}$ and the counit $\epsilon_{U\mathfrak{m}}$ are $U\mathfrak{n}$-module homomorphisms. The conditions to be satisfied are explicitly written down in Equation (4.3).

Let $n \in \mathfrak{n} \subset U\mathfrak{n}$ and $m \in \mathfrak{m} \subset U\mathfrak{m}$. Then the action $n \cdot m$ is again primitive, that is,

$$(n \cdot m)_1 \otimes (n \cdot m)_2 = 1 \otimes (n \cdot m) + (n \cdot m) \otimes 1.$$

This equals $(n_1 \cdot m_1) \otimes (n_2 \cdot m_2)$, because the unit acts by the identity, the action of an element of $\mathfrak{n}$ on 1 is zero and both n and m are primitive. As both $U\mathfrak{n}$ and $U\mathfrak{m}$ are generated as algebras by primitive elements, the action property, the bialgebra property and Equation (4.1) show by induction the condition

$$(n \cdot m)_1 \otimes (n \cdot m)_2 = (n_1 \cdot m_1) \otimes (n_2 \cdot m_2)$$

for arbitrary $n \in U\mathfrak{n}$ and $m \in U\mathfrak{m}$. Indeed, the induction step with respect to $U\mathfrak{m}$, where we first assume n is primitive, reads

$$\begin{aligned}
(n \cdot (mm'))_1 \otimes (n \cdot (mm'))_2 &= ((n \cdot m)m' + m(n \cdot m'))_1 \otimes ((n \cdot m)m' + m(n \cdot m'))_2 \\
&= ((n \cdot m)_1 m'_1 + m_1(n \cdot m')_1) \otimes ((n \cdot m)_2 m'_2 + m_2(n \cdot m')_2) \\
&= ((n_1 \cdot m_1)m'_1 + m_1(n_1 \cdot m'_1)) \otimes ((n_2 \cdot m_2)m'_2 + m_2(n_2 \cdot m'_2)) \\
&= (n_1 \cdot (m_1 m'_1)) \otimes (n_2 \cdot (m_2 m'_2)).
\end{aligned}$$

Now the induction step with respect to $U\mathfrak{n}$, which then permits to pass to arbitrary n, reads

$$\begin{aligned}
(nn' \cdot m)_1 \otimes (nn' \cdot m)_2 &= (n \cdot (n' \cdot m))_1 \otimes (n \cdot (n' \cdot m))_2 \\
&= (n_1 \cdot (n' \cdot m)_1) \otimes (n_2 \cdot (n' \cdot m)_2) \\
&= (n_1 \cdot (n'_1 \cdot m_1)) \otimes (n_2 \cdot (n'_2 \cdot m_2)) \\
&= ((n_1 n'_1) \cdot m_1) \otimes ((n_2 n'_2) \cdot m_2).
\end{aligned}$$

The second condition in equation (4.3) is also easily verified. Both sides of

$$\epsilon_{U\mathfrak{m}}(n \cdot m) = \epsilon_{U\mathfrak{n}}(n)\epsilon_{U\mathfrak{m}}(m)$$

are zero unless $n \in \mathbb{K} \subset U\mathfrak{n}$ and $m \in \mathbb{K} \subset U\mathfrak{m}$. But in this case, the equation is true.

It remains to show properties (ii) and (iii). These two follow from the properties (a) and (b) of a crossed module. Indeed, on primitives, identity (ii) *is* identity (a) and identity (iii) *is* identity (b). The general case follows again from induction.

The first induction step for condition (ii) reads

$$\begin{aligned}\gamma(n\cdot(mm')) &= \gamma((n_1\cdot m)(n_2\cdot m'))\\ &= \gamma(n_1\cdot m)\gamma(n_2\cdot m')\\ &= n_1\gamma(m)S(n_2)n_3\gamma(m')S(n_4)\\ &= n_1\gamma(m)\gamma(m')S(n_2)\\ &= n_1\gamma(mm')S(n_2),\end{aligned}$$

where we have used $h_1S(h_2) = \epsilon(h)$ and $\epsilon(h_1)h_2 = h$ in a Hopf algebra H. Now the second induction step reads

$$\begin{aligned}\gamma(nn'\cdot m) &= \gamma(n\cdot(n'\cdot m))\\ &= n_1\gamma(n'\cdot m)S(n_2)\\ &= n_1n_1'\gamma(m)S(n_2')S(n_2)\\ &= (nn')_1\gamma(m)S((nn')_2).\end{aligned}$$

A similar reasoning with again two induction steps shows condition (iii).

One word about the morphism-side of the statement: A morphism of crossed modules $(\rho,\sigma) : (\mu : \mathfrak{m} \to \mathfrak{n}) \to (\mu' : \mathfrak{m}' \to \mathfrak{n}')$ is sent to a morphism of crossed modules of Hopf algebras $(U\rho, U\sigma) : (U\mu : U\mathfrak{m} \to U\mathfrak{n}) \to (U\mu' : U\mathfrak{m}' \to U\mathfrak{n}')$. This renders the first diagram in Definition 4.1.5 commutative. In order to render the second diagram commutative, one should use the natural isomorphism $\zeta_{\mathfrak{m},\mathfrak{n}} : U(\mathfrak{m}\times\mathfrak{n}) \to U\mathfrak{m}\otimes U\mathfrak{n}$ and the action $\phi_{U\mathfrak{m}} : U\mathfrak{n}\otimes U\mathfrak{m} \to U\mathfrak{m}$ given by $\phi_{U\mathfrak{m}} = \alpha\circ\zeta_{\mathfrak{m},\mathfrak{n}}^{-1}$, where α is the action of $\mathfrak{n}$ on $\mathfrak{m}$ from the crossed module. □

Theorem 4.2.4. *The functor of primitives P extends to a functor (still denoted P) from the category of crossed modules of Hopf algebras to the category of crossed modules of Lie algebras.*

Proof. Let $\gamma : B \to H$ be a crossed module of Hopf algebras (Definition 4.1.3). The $\mathbb{K}$-vector spaces of primitives $P(B)$ and $P(H)$ of B and H are Lie algebras, and the restriction of γ to $P(B)$ is a Lie algebra morphism $\gamma : P(B) \to P(H)$. By hypothesis, we have a map $\zeta : H\otimes B \to B$, which is an action of H on B. Restrict it to $\bar{\zeta} : H\otimes P(B) \to B$. $\bar{\zeta}$ takes its values in $P(B)$, because $(\star)$ the coproduct of B is a morphism of H-modules:

$$\begin{aligned}\Delta(h\cdot b) &\overset{(\star)}{=} \Delta(h)\cdot\Delta b\\ &= (h_1\otimes h_2)\cdot(b_1\otimes b_2)\\ &\overset{b\in P(B)}{=} (h_1\cdot 1)\otimes(h_2\cdot b) + (h_1\cdot b)\otimes(h_2\cdot 1)\end{aligned}$$

$$\begin{aligned} &= (\epsilon(h_1)\cdot 1)\otimes(h_2\cdot b)+(h_1\cdot b)\otimes(\epsilon(h_2)\cdot 1)\\ &= 1\otimes((\epsilon(h_1)h_2)\cdot b)+((\epsilon(h_2)h_1)\cdot b)\otimes 1\\ &= 1\otimes(h\cdot b)+(h\cdot b)\otimes 1. \end{aligned}$$

Then we may restrict $\tilde{\zeta}$ further to $\tilde{\zeta} : P(H)\otimes P(B) \to P(B)$. The action property of $\tilde{\zeta}$ is clear (and implies the property of a Lie algebra action), but we have to show that the action $\tilde{\zeta}$ is by derivations. This follows from the hypothesis that the multiplication of B is a morphism of H-modules, that is, equation (4.1), by taking all elements to be primitive.

Properties (ii) and (iii) of a crossed module of Hopf algebras imply, when restricted to the primitives, properties (a) and (b) of a crossed module of Lie algebras.

The morphism-side of the statement is straightforward. □

Let us deduce from the above theorems some equivalence between crossed modules of irreducible cocommutative Hopf algebras over a field of characteristic 0 and crossed modules of enveloping algebras. Recall that a nonzero coalgebra is called *irreducible* in case any two nonzero subcoalgebras have a nonzero intersection ([97], Definition p. 157). A Hopf algebra is called *irreducible* if its underlying coalgebra is. An irreducible coalgebra has 1 as its only group-like element.

Now recall that the functors U and P are equivalences of categories in characteristic 0 when one restricts the category of Hopf algebras to the full subcategory of irreducible cocommutative Hopf algebras (see [97] theorem 13.0.1, p. 274). Call a crossed module of Hopf algebras *irreducible* in case its component Hopf algebras are irreducible Hopf algebras. One deduces then from Theorem 4.2.3 the following.

Theorem 4.2.5. *An irreducible crossed module of cocommutative Hopf algebras is isomorphic to a crossed modules of enveloping algebras, that is, to a crossed module of Hopf algebras of the form $U(\mu : \mathfrak{m} \to \mathfrak{n})$, where $\mu : \mathfrak{m} \to \mathfrak{n}$ is a crossed module of Lie algebras.*

Remark 4.2.1. Let us comment on the possibility of a "standard cohomological framework" for crossed modules of Hopf algebras. In the case of Lie algebras, (discrete) groups or associative algebras, taking sections one may associate to a crossed module a 3-cohomology class, which expresses the equivalence class of the given crossed module. This scheme needs modification in the context of Hopf algebras. One way around this problem is to transfer using the functor U and Theorem 4.2.3 the equivalence relation for crossed modules of Lie algebras and their cohomology classes directly to Hopf algebras. This may not be the most canonical way to do so, but it sets at least compatibility conditions, which one might want to impose on some cohomological description of irreducible crossed modules of cocommutative Hopf algebras.

Remark 4.2.2. Let us remark that Baez and Crans [4] construct in their Section 6 a family $\{\mathfrak{g}_\hbar\}_\hbar$ of Lie 2-algebras which is a one-parameter deformation of the "trivial"

Lie 2-algebra $\mathfrak{g}$. Thus the family of crossed modules of Hopf algebras $U\mathfrak{g}_\hbar$, which may be associated to their one-parameter family by Theorem 4.2.3 may be regarded as a quantum 2-group, that is, a nontrivial deformation of Hopf 2-algebras.

In order to explain in which sense this deformation is nontrivial, we need the 2-categorical structure: Each semistrict Lie 2-algebra $\mathfrak{g}_\hbar$ with $\hbar \neq 0$ is nonstrict and only (nontrivially) equivalent to a strict one ("strictification"), while $\mathfrak{g}_0$ is strict.

4.3 Crossed modules of groups and Hopf algebras

We will now consider the problem to show that in case $\mu : M \to N$ is a crossed module of groups, the induced morphism between group algebras $\mu : \mathbb{K}[M] \to \mathbb{K}[N]$ is a crossed module of Hopf algebras. We start with some preliminaries.

Lemma 4.3.1. *Let the group N act on the group M by automorphisms. Then $\mathbb{K}[M]$ is a $\mathbb{K}[N]$-module algebra and $\mathbb{K}[N]$-module coalgebra.*

Proof. First of all, note that the action of N on M extends first to an action of N on $\mathbb{K}[M]$ by linearity. This can be viewed as a homomorphism of groups from N into the invertible endomorphisms of $\mathbb{K}[M]$, and by adjunction, this gives a homomorphism of algebras from $\mathbb{K}[N]$ to $\mathrm{End}(\mathbb{K}[M])$ which we interpret as an action of $\mathbb{K}[N]$ on $\mathbb{K}[M]$.

Now note that $\mathbb{K}[M]$ becomes a $\mathbb{K}[N]$-module coalgebra, because one always gets a module coalgebra by linearizing the action of a group on a set. This is clear for elements $n \in N$ and $m \in M$, as the commutativity of the diagram demands

$$(n \cdot m)_1 \otimes (n \cdot m)_2 = (n_1 \cdot m) \otimes (n_2 \cdot m),$$

which is true because all elements are group-like. The general statement is proven by double induction using that the $n \in N$ (resp. the $m \in M$) generate the algebra $\mathbb{K}[N]$ (resp. the algebra $\mathbb{K}[M]$).

The condition to be a module algebra means for all $n \in \mathbb{K}[N]$ and all $m, m' \in \mathbb{K}[M]$ that

$$n \cdot (mm') = (n_1 \cdot m)(n_2 \cdot m').$$

This is true for $n \in N$ and $m, m' \in M$, because the N-action on M is by automorphisms. The general case follows again from double induction. □

Theorem 4.3.2. *Given a crossed module of groups $\mu : M \to N$, the induced map between group algebras $\mu : \mathbb{K}[M] \to \mathbb{K}[N]$ is in a natural way a crossed module of Hopf algebras.*

Proof. We have seen in the above lemma that the action of N on M by automorphisms endows $\mathbb{K}[M]$ with an action of $\mathbb{K}[N]$ which renders it a module algebra and a module coalgebra.

It remains to show properties (ii) and (iii). Property (ii) reads for all $n \in \mathbb{K}[N]$ and all $m \in \mathbb{K}[M]$,

$$\mu(n \cdot m) = n_1 \mu(m) S(n_2).$$

This is exactly condition (a) in the crossed module of groups for $n \in N$ and $m \in M$. The general case can then be obtained by double induction. Property (iii) reads for all $m, m' \in \mathbb{K}[M]$,

$$\mu(m) \cdot m' = m_1 m' S(m_2).$$

Again, this is true for $m, m' \in M$ by the corresponding condition in the crossed module of groups and can be shown by double induction in the general case. □

Theorem 4.3.3. *Suppose given a crossed module of Hopf algebras $\gamma : B \to H$ with invertible antipodes. Then application of the functor of group-like elements to B and H induces a crossed module of groups $\gamma : G(B) \to G(H)$.*

Proof. Recall that the set of group-like elements $G(H)$ of a Hopf algebra H is a group in case H possesses an invertible antipode; see Proposition III.3.7, page 56, in [58]. Given a crossed module of Hopf algebras $\gamma : B \to H$, we obtain then by functoriality a homomorphism of groups $\gamma : G(B) \to G(H)$. The action of H on B induces a group action of $G(H)$ on B. As B is an H-module coalgebra, this action respects the group-like elements and induces thus an action of $G(H)$ on $G(B)$. This action is by automorphisms, because B is an H-module algebra. Properties (ii) and (iii) imply conditions (a) and (b). □

For infinite groups, the corresponding group algebras do not (in general) reflect all features of the given group. Therefore, one often passes to topological versions, like the C^*-algebra associated to a group. The correspondence between crossed modules of groups and crossed modules of their group algebras respects this kind of topological versions.

As an example, consider the group S^1 and the dense subgroup $\mathbb{Z}$, which is the image of the embedding $\mathbb{Z} \to S^1$, $k \mapsto e^{i\lambda k}$ where $\lambda / 2\pi$ is an irrational number. The quotient $S^1 / \mathbb{Z}$ is an abelian group, but carries the discrete topology, thus the associated C^*-algebra is trivial. It is explained in [9] (see also references therein) that a way to associate meaningful C^*-algebras to this example goes under the name *Hopfish algebras*. We will not go into the definition of a Hopfish algebra and refer to *loc. cit*.

The point we want to make here (we owe this remark to Chenchang Zhu) is that another way around the problem of associating a C^*-algebra to the quotient $S^1 / \mathbb{Z}$ is to take the crossed module of C^*-algebras associated to the crossed module of groups $\mathbb{Z} \to S^1$. Its corresponding 2-group (see Section 1.9) is then $S^1 \ltimes \mathbb{Z} \to S^1$ (which may be regarded as a groupoid, see Exercise 4.8.5). The upshot is that the corresponding crossed module of C^*-algebras of the groupoid $S^1 \ltimes \mathbb{Z} \to S^1$ replaces the C^*-algebra of the (pathological) quotient $S^1/\mathbb{Z}$. Note that this framework applies for any group G and for any normal subgroup N to the crossed module $N \to G$ or the corresponding 2-group/groupoid $G \ltimes N \to G$.

4.4 Crossed comodules of groups and Hopf algebras

The goal of this section is to study the compatibility of the notion of crossed module with the standard functors $\mathbb{K}-$ and χ between the categories of algebraic groups and Hopf algebras. As always in geometric situations, we will take the field of complex numbers $\mathbb{C}$ as the base field. We will need here the introduction of the new notion of *crossed comodule* of Hopf algebras (Definition 4.4.1) dual to the concept of crossed module of Hopf algebras. The main results of this section (Theorems 4.4.1 and 4.4.2) together with classical equivalence between algebraic groups and their rings of functions give a correspondence between commutative crossed comodules of Hopf algebras and crossed modules of connected algebraic groups.

Recall that a commutative Hopf algebra H, which is finitely generated as an algebra gives rise to an affine algebraic group; see [97], Section 6.3, page 123. The associated algebraic group is $\chi(H)$, the group of algebra homomorphisms $H \to \mathbb{C}$ or equivalently the group of maximal ideals of H. The equivalence is given by associating to a homomorphism $\phi : H \to \mathbb{C}$ its kernel $\operatorname{Ker}(\phi)$, which is a maximal ideal.

For a (left) coaction $\rho_L : L \to K \otimes L$ of K on L, we use Sweedler notation, that is,

$$\rho_L(l) = l^0 \otimes l^1 \in K \otimes L,$$

where the sum is understood.

Definition 4.4.1. Let $\zeta : K \to L$ be a morphism of Hopf algebras. The morphism ζ defines a *crossed comodule of Hopf algebras* in case:

(i) (L, ρ_L) is a K-comodule which renders it a K-comodule algebra and a K-comodule coalgebra;

(ii) $\rho_L \circ \zeta = (\mathrm{id}_K \otimes \zeta) \circ \mathrm{coad}_K$;

(iii) $(\zeta \otimes \mathrm{id}_L) \circ \rho_L = \mathrm{coad}_L$.

Definition 4.4.2. A *morphism* $(\rho, \sigma) : (\zeta : K \to L) \to (\zeta' : K' \to L')$ of crossed comodules of Hopf algebras $\zeta : K \to L$ and $\zeta' : K' \to L'$ is a pair of morphisms of Hopf algebras $\rho : K \to K'$ and $\sigma : L \to L'$ such that the following diagrams are commutative:

$$\begin{array}{ccc} K & \xrightarrow{\rho} & K' \\ \downarrow{\scriptstyle \zeta} & & \downarrow{\scriptstyle \zeta'} \\ H & \xrightarrow{\sigma} & H' \end{array} \qquad\qquad \begin{array}{ccc} L & \xrightarrow{\sigma} & L' \\ \downarrow{\scriptstyle \rho_L} & & \downarrow{\scriptstyle \rho_{L'}} \\ K \otimes L & \xrightarrow{\rho \otimes \sigma} & K' \otimes L' \end{array}$$

It is easy to show that crossed comodules form a category with this notion of morphisms.

We recall for (i) that, by definition, (L, ρ_L) is a left K-comodule coalgebra if $\triangle_L$ is a comodule morphism, that is, satisfies

$$(\mathrm{id}_K \otimes \triangle_L) \circ \rho_L \stackrel{\diamond}{=} \rho_{L\otimes L} \circ \triangle_L$$

with

$$\rho_{L\otimes L} := (\mu_{\mathbb{C}} \otimes \mathrm{id}_L \otimes \mathrm{id}_L) \circ (\mathrm{id}_K \otimes \tau \otimes \mathrm{id}_L) \circ (\rho_L \otimes \rho_L).$$

This means explicitly for $l \in L$ and $k \in K$,

$$l^0 \otimes (l^1)_1 \otimes (l^1)_2 = (l_1)^0 (l_2)^0 \otimes (l_1)^1 \otimes (l_2)^1.$$

In (ii) and (iii), coad_L and coad_K denote the adjoint coaction (and not the coadjoint action!) defined as

$$\mathrm{coad}_L := (\mu_L \otimes \mathrm{id}_L) \circ (\mathrm{id}_L \otimes \tau) \circ (\triangle_L \otimes S_L) \circ \triangle_L.$$

Explicitly, one has for all $l \in L$,

$$\mathrm{coad}_L(l) = l_1 S(l_3) \otimes l_2 \in L \otimes L.$$

The need for Definition 4.4.1 relies on the fact that the notion of a Hopf algebra is auto-dual, while that of a crossed module is not. The existence of both notions of crossed modules and comodules shows that there are (at least) two natural ways to categorify the notion of a Hopf algebra.

The next proposition generates natural examples of crossed comodules of Hopf algebras, namely when one considers the algebras of functions on a crossed module of algebraic groups.

Theorem 4.4.1. *The functor $\mathbb{C}-$ extends to a functor (still denoted $\mathbb{C}-$) from the category of crossed modules of connected algebraic groups to the category of crossed comodules of Hopf algebras.*

Proof. Recall that the functor $\mathbb{C}$ associates to a connected algebraic group G its commutative Hopf algebra $\mathbb{C}G$ of $\mathbb{C}$-valued regular functions on G. Let $\mu : M \to N$ be a crossed module of connected algebraic groups. We get a morphism of commutative Hopf algebras $\zeta := \mu^* : \mathbb{C}N \to \mathbb{C}M$ by precomposing a function with μ.

The action of N on M is transformed into a coaction of $\mathbb{C}N$ on $\mathbb{C}M$: $\rho_{\mathbb{C}M} : \mathbb{C}M \to \mathbb{C}N \otimes \mathbb{C}M$ is just the dual of the action map $N \times M \to M$, where we have used that $\mathbb{C}(N \times M) \cong \mathbb{C}N \otimes \mathbb{C}M$.

Now, $\mathbb{C}M$ is a $\mathbb{C}N$-comodule coalgebra, because of the fact that N acts on M by automorphisms. Indeed, observe that the coaction $\rho_{\mathbb{C}M} : \mathbb{C}M \to \mathbb{C}(N \times M)$ is given for a function f on M by $f \mapsto ((n,m) \mapsto f(n \cdot m))$. In the same vein, the coproduct is given by $f \mapsto ((m,m') \mapsto f(mm'))$. The commutativity of the comodule-coalgebra diagram is therefore the following equality between functions on $N \times M \times M$:

$$((n,m,m') \mapsto f(n \cdot (mm'))) = ((n,m,m') \mapsto f((n_1 \cdot m)(n_2 \cdot m'))).$$

This is true by the fact that $n_1 = n_2 = n$ (n group-like!) acts on M by automorphisms. Observe that, as all algebraic structures on $\mathbb{C}M$ and $\mathbb{C}N$ are defined by passing to the dual, the commutativity of the comodule-coalgebra diagram is exactly the dual (i. e., applying functions to) the property $n \cdot (mm') = (n \cdot m)(n \cdot m')$.

On the other hand, $\mathbb{C}M$ is a $\mathbb{C}N$-comodule algebra, because the multiplication on $\mathbb{C}M$ is the dual of the diagonal $M \to M \times M$, $m \mapsto (m,m)$. Indeed, the comodule-algebra diagram demands for $l, l' \in L$ the equality,

$$(ll')^0 \otimes (ll')^1 = l^0 l'^0 \otimes l^1 (l')^1.$$

Translated into function algebras by dualizing step-by-step each operation, for a given $f \otimes g \in \mathbb{C}(M \times M)$, this means the equality of the two functions $N \times M \to \mathbb{C}$,

$$((n,m) \mapsto f(n \cdot m)g(n \cdot m)) = ((n,m) \mapsto f(n \cdot m)g(n \cdot m)).$$

This is clearly true.

Finally, the identities (ii) and (iii) of crossed comodule of Hopf algebras are consequences (dualization) of the properties (a) and (b) of the crossed module $\mu : M \to N$.

The morphism-part of the statement is straightforward. □

Now, conversely, one can obtain crossed modules of groups from crossed comodules of Hopf algebras. This is the content of the following theorem.

Theorem 4.4.2. *The functor χ extends to a functor (still denoted χ) from the category of crossed comodules of commutative Hopf algebras to the category of crossed modules of groups.*

Its proof will be given after having introduced some notation and the preliminary Lemma 4.4.3.

For the rest of this section, K and L will be two commutative Hopf algebras, $\rho_L : L \to K \otimes L$ a linear map, $M := \chi[L]$, $N := \chi[K]$, $\eta_K, \gamma_K \in N$, $\phi_L, \psi_L \in M$.

One defines an "action" $\star_{\rho_L}$ of N on M by

$$\eta_K \star_{\rho_L} \phi_L := \mu_{\mathbb{C}} \circ (\eta_K \otimes \phi_L) \circ \rho_L.$$

In case $K = L$ and $\rho_L = \triangle_L$ (resp., $L = K$ and $\rho_L = \triangle_K$, resp., $L = K$ and $\rho_L = \mathrm{coad}_K$), we will simplify the notation $\star_{\rho_L}$ to $\star_L$ (resp., $\star_K$, resp., $\star_{\tilde{\rho}_K}$). From now on, (L, ρ_L) will be a left K-comodule algebra and coalgebra.

Lemma 4.4.3.
(α) $(M, \star_L)$ *is a group with inverse* $S_L^\star$;
(β) $\star_{\rho_L}$ *defines a group action of N on M;*
(γ) $\star_{\rho_L}$ *is an action by automorphisms;*
(δ) $\eta_K \star_{\tilde{\rho}_K} \gamma_K = \eta_K \star \gamma_K \star \eta_K^{-1}$.

Proof. (α) and (β) are proven using definitions and coassociativity of $\triangle_L$ and ρ_L. The proof of (γ), that is, N acts on M by automorphisms, is a little more involved and is based on the fact (⋄) that $\triangle_L$ is a comodule map:

$$\begin{aligned}
\eta_K \star_{\rho_L} (\phi_L \star_L \psi_L) &\overset{\star_{\rho_L}}{=} \mu_{\mathbb{C}}(\eta_K \otimes (\phi_L \star_L \psi_L)) \circ \rho_L \\
&\overset{\star_L}{=} \mu_{\mathbb{C}}(\mathrm{id}_k \otimes \mu_{\mathbb{C}}) \circ (\eta_K \otimes \phi_L \otimes \psi_L) \circ (\mathrm{id}_K \otimes \triangle_L) \circ \rho_L \\
&\overset{(\diamond)}{=} \mu_{\mathbb{C}}(\mathrm{id}_k \otimes \mu_{\mathbb{C}}) \circ (\eta_K \otimes \phi_L \otimes \psi_L) \circ \rho_{L \otimes L} \circ \triangle_L \\
&\overset{\rho_{L\otimes L}}{=} \mu_{\mathbb{C}}(\mathrm{id}_k \otimes \mu_{\mathbb{C}}) \circ (\eta_K \otimes \phi_L \otimes \psi_L) \\
&\qquad \circ (\mu_{\mathbb{C}} \otimes \mathrm{id}_L \otimes \mathrm{id}_L) \circ (\mathrm{id}_K \otimes \tau \otimes \mathrm{id}_L) \circ (\rho_L \otimes \rho_L) \circ \triangle_L \\
&\overset{\eta_K}{=} \mu_{\mathbb{C}}(\mathrm{id}_k \otimes \mu_{\mathbb{C}}) \circ (\mu_{\mathbb{C}}(\eta_K \otimes \eta_K) \otimes \phi_L \otimes \psi_L) \\
&\qquad \circ (\mathrm{id}_K \otimes \tau \otimes \mathrm{id}_L) \circ (\rho_L \otimes \rho_L) \circ \triangle_L \\
&\overset{\tau}{=} \mu_{\mathbb{C}}(\mu_{\mathbb{C}} \otimes \mu_{\mathbb{C}}) \circ (\eta_K \otimes \phi_L \otimes \eta_K \otimes \psi_L) \circ (\rho_L \otimes \rho_L) \circ \triangle_L \\
&\overset{\star_L}{=} (\mu_{\mathbb{C}} \circ (\eta_K \otimes \phi_L) \circ \rho_L) \star_L (\mu_{\mathbb{C}} \circ (\eta_K \otimes \psi_L) \circ \rho_L) \\
&\overset{\star_{\rho_L}}{=} (\eta_K \star_{\rho_L} \phi_L) \star_L (\eta_K \star_{\rho_L} \psi_L).
\end{aligned}$$

We finally prove (δ):

$$\begin{aligned}
\eta_K \star_{\rho_K} \gamma_K &\overset{\star_{\rho_K}}{=} \mu_{\mathbb{C}}(\eta_K \otimes \gamma_K) \circ \mathrm{coad}_K \\
&\overset{\mathrm{coad}_K}{=} \mu_{\mathbb{C}}(\eta_K \otimes \gamma_K) \circ (\mu_{\mathbb{C}} \otimes \mathrm{id}_K) \circ (\mathrm{id}_K \otimes \tau) \circ (\triangle_K \otimes S_K) \circ \triangle_K \\
&\overset{\mu_{\mathbb{C}}}{=} \mu_{\mathbb{C}}(\mu_{\mathbb{C}} \otimes \mathrm{id}_k) \circ (\eta_K \otimes \eta_K \otimes \gamma_K) \circ (\mathrm{id}_K \otimes \tau) \circ (\triangle_K \otimes S_K) \circ \triangle_K \\
&\overset{\tau}{=} \mu_{\mathbb{C}}(\mu_{\mathbb{C}} \otimes \mathrm{id}_k) \circ (\eta_K \otimes \gamma_K \otimes \eta_K) \circ (\triangle_K \otimes S_K) \circ \triangle_K \\
&\overset{\star_K}{=} \eta_K \star \gamma_K \star \eta_K^{-1}.
\end{aligned}$$
□

We can now turn to the proof of Theorem 4.4.2.

Proof. Let $\zeta : K \to L$ be a crossed comodule of commutative Hopf algebras K and L. By functoriality, we get a group homomorphism $\mu := \zeta^* : M \to N$. We already know from Lemma 4.4.3 that $\star_{\rho_L}$ is an action by automorphisms, so it only remains to show that identities (a) and (b) hold.

Identity (a) is obtained from axiom (ii):

$$\mu(\eta_K \star_{\rho_L} \psi_L) \overset{\mu}{=} (\eta_K \star_{\rho_L} \psi_L) \circ \zeta$$

$$
\begin{aligned}
&\overset{\star_{\rho_L}}{=} \mu_{\mathbb{C}}(\eta_K \otimes \psi_L) \circ \rho_L \circ \zeta \\
&\overset{\text{(ii)}}{=} \mu_{\mathbb{C}}(\eta_K \otimes \psi_L) \circ (\mathrm{id}_K \otimes \zeta) \circ \mathrm{coad}_K \\
&\overset{\mu}{=} \mu_{\mathbb{C}}(\eta_K \otimes \mu(\psi_L)) \circ \mathrm{coad}_K \\
&\overset{\star_{\tilde{\rho}_K}}{=} (\eta_K \star_{\rho_K} \mu(\psi_L)).
\end{aligned}
$$

Identity (b) is obtained from axiom (iii) and Lemma 4.4.3 (δ):

$$
\begin{aligned}
\mu(\phi_L) \star_{\rho_L} \psi_L &\overset{\mu}{=} (\phi_L \circ \zeta) \star_{\rho_L} \psi_L \\
&\overset{\star_{\rho_L}}{=} \mu_{\mathbb{C}}(\phi_L \otimes \psi_L) \circ (\zeta \otimes id_L) \circ \rho_L \\
&\overset{\text{(iii)}}{=} \mu_{\mathbb{C}}(\phi_L \otimes \psi_L) \circ \mathrm{coad}_L \\
&\overset{(\delta)}{=} \phi_L \star \psi_L \star \phi_L^{-1}.
\end{aligned}
$$

□

Corollary 4.4.4. *The functor χ extends to a functor from the category of crossed comodules of commutative Hopf algebras which are finitely generated as algebras to the category of crossed modules of affine algebraic groups.*

4.5 Integration of crossed modules of Lie algebras

In this section, we explain a general scheme of how to integrate crossed modules of Lie algebras. Note that not all Lie algebras do integrate into Lie groups. In general, only finite dimensional Lie algebras over $\mathbb{K} = \mathbb{R}$ or $\mathbb{C}$ integrate into Lie groups. This is Lie's third theorem; see Appendix D or directly [99], Theorem 3.15.1, page 230. It is therefore not expected that all crossed modules of Lie algebras integrate. Without entering into the discussion which infinite-dimensional Lie algebras integrate, one may restrict the integration procedure to crossed modules where all component Lie algebras are finite-dimensional. Furthermore, we restrict the base field to $\mathbb{R}$ or $\mathbb{C}$.

There are several ways to integrate a Lie algebra into a Lie group. We report here on two procedures passing through Hopf algebras. An integration procedure which yields the 1-connected real Lie group corresponding to a real Lie algebra will be reported on in Chapter 5.

4.5.1 Integration into formal groups

The two integration procedures we discuss here pass through Hopf algebras. The first integration process consists in associating to a finite dimensional Lie algebra $\mathfrak{g}$ its universal enveloping algebra $U\mathfrak{g}$. $U\mathfrak{g}$ is a cocommutative Hopf algebra. It is viewed as the space of "point distributions" at 1 in the group G integrating $\mathfrak{g}$, that is, as distributions on G with support in the point $1 \in G$; see [95] part II (LG), Section 5.5. This describes $U\mathfrak{g}$

as the continuous dual of the space of (germs of) functions on G (in $1 \in G$). In the formal setting, $U\mathfrak{g}$ can be seen as the continuous dual of the adic completion of the space of functions on some formal group law; see [95] part II, Section 5.6, LG 5.22. Therefore, in order to integrate $\mathfrak{g}$, one dualizes $U\mathfrak{g}$, which gives a completion of the Hopf algebra of formal functions, and finally recovers the group as group-like elements in this Hopf algebra of functions. One may adapt this framework such that the outcome is a formal group or an algebraic group (over $\mathbb{C}$); see [95], part II, Section 5.8, LG 5.37. This scheme of reasoning is illustrated in the following diagram:

$$\begin{array}{ccc}
\mathtt{Lie} & \underset{P}{\overset{U}{\longleftrightarrow}} & \mathtt{ccHopf} \\
\updownarrow & & \updownarrow \\
\mathtt{FGrp} & \underset{\mathbb{C}-}{\overset{G(-)}{\longleftrightarrow}} & \mathtt{cHopf}
\end{array}$$

In this diagram, FGrp is here the category of formal groups, the vertical arrows on the LHS denote integration and passage to the tangent space at 1 and the vertical arrows on the RHS denote completion and dualization.

Now this also works backwards. Starting with a formal group G, one takes the commutative Hopf algebra of formal functions. The dual of its completion is a cocommutative Hopf algebra, which is isomorphic to the universal enveloping algebra of the Lie algebra of G.

4.5.2 Integration into an algebraic group

Some modifications are necessary in order to adapt the above scheme to obtain genuine algebraic groups by integration. Consider the following diagram of functors:

$$\begin{array}{ccc}
\mathtt{Lie} & \overset{U}{\longrightarrow} & \mathtt{ccHopf} \\
\downarrow & & \downarrow{\scriptstyle -^{\circ}} \\
\mathtt{algGrp} & \underset{\chi}{\longleftarrow} & \mathtt{cHopf}
\end{array}$$

Here, algGrp is the category of connected affine algebraic groups over $\mathbb{C}$, and H° is the Hopf dual of a Hopf algebra H, that is, the Hopf algebra of matrix coefficients of finite-dimensional representations; see, for example, [61, 77].

A connected affine algebraic group G over $\mathbb{C}$ can be recovered in this way from its Lie algebra $\mathfrak{g} := \mathrm{Lie}(G)$ as $\chi(U\mathfrak{g}^{\circ})$ provided G is perfect, that is, $G = [G, G]$. More generally, if G has unipotent radical, then G is isomorphic to the characters on the subalgebra of basic representative functions on $U\mathfrak{g}$; see [51] for details. Note that the condition that the Hopf algebra is finitely generated as an algebra is satisfied by construction.

4.5.3 Integration of crossed modules

The above schemes of reasoning obviously also work for crossed modules of Lie algebras and integrates them thus (in case all components are finite dimensional) into crossed modules of formal or algebraic groups. The first step is to apply the functor U in order to obtain a crossed modules of Hopf algebras by Theorem 4.2.3. Then one passes to the continuous dual, which is naturally a crossed comodule of commutative Hopf algebras (see the previous section). Each constituent of this crossed comodule is the completion of the commutative Hopf algebra of functions on some formal group, or of germs of functions in an identity-neighborhood of some algebraic group G. Theorem 4.4.2 now implies that this crossed comodule gives rise to a crossed module of groups. One may adapt this scheme to formal groups or algebraic groups as described above, depending on the completion which we use. We summarize the above discussion in the following theorem.

Theorem 4.5.1. *Every crossed module of real or complex Lie algebras whose components are all finite dimensional Lie algebras can be integrated into a crossed module of formal groups whose associated crossed module of Lie algebras is the one we started with. Over $\mathbb{C}$, it can be integrated into a crossed module of algebraic groups whose associated crossed module of Lie algebras is the one we started with.*

Let us remark that Behrang Noohi has an integration procedure of crossed modules of Lie algebras which is closer to our constructions in Chapter 5; see [85]. Moreover, Noohi studies in detail the integration of morphisms.

Remark 4.5.1. Let us come back to the example to the crossed module of Lie algebras which represents the generator of $H^3(\mathfrak{g}, \mathbb{C})$, where $\mathfrak{g}$ is any simple complex finite dimensional Lie algebra. Here, we will associate a crossed module of Hopf algebras to it. This leads then to the definition of the enveloping 2-algebra of the string Lie algebra of Baez and Crans [4].

Denote by $\mathfrak{h}$ a Cartan subalgebra of $\mathfrak{g}$ and choose a Borel subalgebra $\mathfrak{b} \supset \mathfrak{h}$. Let $\mathfrak{n}$ be the nilpotent subalgebra of $\mathfrak{g}$ such that $\mathfrak{n} \oplus \mathfrak{h} = \mathfrak{b}$ as vector spaces. Denote by $M(\lambda)$ the Verma module of $\mathfrak{g}$ of highest weight λ. Namely $M(\lambda) = U\mathfrak{g} \otimes_{U\mathfrak{b}} \mathbb{C}_\lambda$, where $\mathbb{C}_\lambda$ is the one dimensional $\mathfrak{b}$-module given by the trivial action of $\mathfrak{n}$ and the $\mathfrak{h}$-action via $\lambda \in \mathfrak{h}^*$. $M(\lambda)$ possesses a unique maximal proper submodule $N(\lambda)$ and the quotient $L(\lambda)$ is therefore irreducible. Denote by $M(\lambda)^\sharp$, $N(\lambda)^\sharp$ and $L(\lambda)^\sharp$ the restricted duals of these graded $\mathfrak{g}$-modules. We have by definition a short exact sequence of $\mathfrak{g}$-modules:

$$0 \to L(\lambda)^\sharp \to M(\lambda)^\sharp \to N(\lambda)^\sharp \to 0,$$

which specializes to

$$0 \to \mathbb{C} \to M(0)^\sharp \to N(0)^\sharp \to 0$$

for $\lambda = 0$. On the other hand, there is a 2-cocycle $\alpha' \in C^2(\mathfrak{g}, N(0)^\sharp)$, which gives an abelian extension

$$0 \to N(0)^\sharp \to N(0)^\sharp \times_{\alpha'} \mathfrak{g} \to \mathfrak{g} \to 0.$$

We have seen in Section 2.5 that the splicing together of these two sequences gives a crossed module

$$\mu : N(0)^\sharp \to N(0)^\sharp \times_{\alpha'} \mathfrak{g},$$

which represents a generator of $H^3(\mathfrak{g}; \mathbb{C})$.

We have already seen another crossed module representing a generator of $H^3(\mathfrak{g}; \mathbb{C})$: Denoting by $P\mathfrak{g}$ and $\Omega\mathfrak{g}$ the path- and loop Lie algebra corresponding to $\mathfrak{g}$, there is a (general) extensions of Lie algebras

$$0 \to \Omega\mathfrak{g} \to P\mathfrak{g} \to \mathfrak{g} \to 0,$$

adapting the path-loop fibration to the Lie algebra setting. But a crossed module is nothing but a central extension of an ideal of some Lie algebra (together with a compatibility with the involved actions), and thus the standard central extension $\widehat{\Omega\mathfrak{g}}$ of $\Omega\mathfrak{g}$ gives rise to a crossed module

$$0 \to \mathbb{C} \to \widehat{\Omega\mathfrak{g}} \to P\mathfrak{g} \to \mathfrak{g} \to 0.$$

We have seen in Sections 1.7.2 and 2.8.1 that the corresponding class generates $H^3(\mathfrak{g}; \mathbb{C})$.

Direct application of Theorem 4.2.3 gives us a crossed module of Hopf algebras

$$\gamma : S(N(0)^\sharp) \to S(N(0)^\sharp) \otimes_{\widetilde{\alpha}'} U\mathfrak{g}. \tag{4.4}$$

Here, S denotes the symmetric algebra on a $\mathbb{K}$-vector space. The above mentioned crossed module corresponding to the path-loop fibration also gives rise to an enveloping algebra, but observe that it is not as algebraic as (4.4) and thus less meaningful (what is the enveloping algebra of a topological Lie algebra?) from the point of view of representation theory, while it has the advantage of being more geometric. More precisely, in the crossed module (4.4) we take enveloping algebras of Lie algebras which are *at most countably infinite dimensional* vector spaces, while the underlying vector spaces in the crossed module stemming from the path-loop fibration are not countably infinite dimensional.

Recall further from Chapter 2 that semistrict Lie 2-algebras are categorified Lie algebras with a functorial bracket, which is strictly antisymmetric, but satisfies the Jacobi identity only up to a Jacobiator. This Jacobiator gives a 3-cocycle θ of one of the underlying Lie algebras $\mathfrak{g}$ with values in some $\mathfrak{g}$-module V. The data $(\mathfrak{g}, V, [\theta])$ then completely specifies the equivalence class of the given semistrict Lie 2-algebra.

At the end of their article [4], Baez and Crans construct a family $\mathfrak{g}_\hbar$ of semi-strict Lie 2-algebras whose equivalence classes are given in this sense by the triplets $(\mathfrak{g}, \mathbb{C}, [\hbar\theta])$, where here θ is the Cartan cocycle, that is, $\theta = \langle [,], \rangle$ for $\langle , \rangle$ the Killing form of the simple complex finite dimensional Lie algebra $\mathfrak{g}$, and $\hbar$ is a scalar.

The construction which we cited here as an example works for any 3-cohomology class of a finite dimensional Lie algebra. Therefore, we deduce the following theorem directly from Theorem 4.2.3.

Theorem 4.5.2. *For any skeletal semistrict Lie 2-algebra corresponding to a triple $(\mathfrak{g}, V, [\theta] \in H^3(\mathfrak{g}, V))$, there is a crossed module of Hopf algebras*

$$\gamma : S(I) \to S(Q) \otimes_{\tilde{\alpha}''} U\mathfrak{g}, \tag{4.5}$$

which is a natural algebraic candidate for the enveloping algebra of the given Lie 2-algebra in the sense that all its underlying vector spaces are at most countably infinite dimensional.

Here, Q and I are $\mathfrak{g}$-modules such that I is injective and such that there is a short exact sequence of $\mathfrak{g}$-modules:

$$0 \to V \to I \to Q \to 0.$$

The class of the 2-cocycle $\alpha'' \in Z^2(\mathfrak{g}, Q)$ corresponds to the class of the given 3-cocycle θ under the corresponding connecting homomorphism. $\tilde{\alpha}''$ is the Hopf-algebra outcome of the construction (obtained by identifying the factors of $U(Q \times_{\alpha''} \mathfrak{g})$ with $S(Q)$ and $U\mathfrak{g}$, resp.). The above construction (4.4) is a special case. Recall from Section 2.6.2 that there cannot be a purely finite dimensional construction in general.

The properties of the preceding theorem are important, because they permit to establish a duality theory for the mentioned kind of enveloping algebras for skeletal semistrict Lie 2-algebras (up to equivalence of Lie 2-algebras; cf. [4]).

Remark 4.5.2. Let us finally discuss the integrability of the crossed module of Lie algebras

$$\mu : N(0)^\sharp \to N(0)^\sharp \times_{\alpha'} \mathfrak{g},$$

to which Theorem 4.5.1 does not apply. An integration of the more geometric crossed module $\Omega\mathfrak{g} \to P\mathfrak{g}$ is the object of [5]. In fact, the crossed module μ is not integrable (at least, not in a naïve way).

Indeed, we have seen that the crossed module μ which also represents the Cartan class in $H^3(\mathfrak{g}, \mathbb{C})$, is spliced together from an abelian extension

$$0 \to (U\mathfrak{g}^+)^* \to (U\mathfrak{g}^+)^* \times_\alpha \mathfrak{g} \to \mathfrak{g} \to 0,$$

and the short exact sequence of (adjoint) $\mathfrak{g}$-modules

$$0 \to \mathbb{C} \to (U\mathfrak{g})^* \to (U\mathfrak{g}^+)^* \to 0.$$

Let G be the connected, simply connected compact Lie group corresponding to $\mathfrak{g}$. The nonintegrability of μ is due to the fact that the $\mathfrak{g}$-module $U\mathfrak{g}$ (with the action by left multiplication) does not integrate into a G-module.

Even if it did, the integration would not have been this simple. In fact, the resulting crossed module of groups must contain a topologically nontrivial fiber bundle. If not, there would be a globally smooth section of the crossed module, and its 3-class would be presented by a globally smooth cocycle. This is impossible, because the smooth cohomology of G is trivial; see [53].

Notwithstanding this example, in principle crossed modules of the above kind which are spliced together from an abelian extension and a short exact sequence of modules may be integrated as follows—this gives a second integration procedure, which applies (in principle) to infinite dimensional Lie algebras. First, integrate the modules to locally convex G-modules, and suppose that the 2-cocycle from the abelian extension is continuous. Integrate it into a locally smooth group 2-cocycle using [80]. Obstructions to this step reside in $\pi_1(G)$ and $\pi_2(G)$ as specified in [80]. Finally, splice together the exact sequences to obtain a crossed module of groups.

4.6 Crossed modules of Hopf algebras and cat^1-Hopf algebras

In this section, we discuss the equivalence of categories between the category of cat^1-Hopf algebras and the category of crossed modules of Hopf algebras. For this, the notion of a semidirect product of Hopf algebras is essential.

4.6.1 Semidirect product of Hopf algebras

Let H be a Hopf algebra. The relevant algebraic structures for the semidirect product are the *smash product and smash coproduct*: Let $A = (A, \mu, \eta)$ be an H-module algebra via an action map $\phi : H \otimes A \to A$. Then

$$A\#H = (A \otimes H, \mu_{A\#_\phi H}, \eta_{A\#_\phi H}\eta_A \otimes \eta_H)$$

with product

$$\mu_{A\#_\phi H} := (\mu_A \otimes \mu_H) \circ (\mathrm{id}_A \otimes \phi \otimes \mathrm{id}_H \otimes \mathrm{id}_H) \circ (\mathrm{id}_A \otimes \mathrm{id}_H \otimes \tau \otimes \mathrm{id}_H) \circ (\mathrm{id}_A \otimes \Delta_H \otimes \mathrm{id}_A \otimes \mathrm{id}_H),$$

is an algebra, called the *smash product* of A and H. In down-to-earth terms, the product reads for $a, a' \in A$ and $h, h' \in H$,

$$(a \otimes h)(a' \otimes h') = a(h_1 \cdot a') \otimes h_2 h'.$$

Let $(C, \triangle, \epsilon)$ be an H-comodule via a coaction map $\rho : C \to H \otimes C$, denoted in Sweedler notation $\rho(c) = c^0 \otimes c^1$. Then

$$C \propto H := (C \otimes H, \triangle_{C\propto_\rho H}, \epsilon_{C\propto_\rho H} = \epsilon_C \otimes \epsilon_H)$$

with

$$\triangle_{C\propto_\rho H} := (\mathrm{id}_C \otimes \mu_H \otimes \mathrm{id}_C \otimes \mathrm{id}_H) \circ (\mathrm{id}_C \otimes \mathrm{id}_H \otimes \tau \otimes \mathrm{id}_H) \circ (\mathrm{id}_C \otimes \rho \otimes \mathrm{id}_H \otimes \mathrm{id}_H) \circ (\triangle_C \otimes \triangle_H),$$

is a coalgebra, called the *smash coproduct* of C and H. In terms of elements, we have for all $c \in C$ and all $h \in H$,

$$\triangle_{C\propto_\rho H}(c \otimes h) = (c_1 \otimes (c_2)^0 h_1) \otimes ((c_2)^1 \otimes h_2).$$

Radford [89] studied under which conditions a Hopf algebra H and an H-module algebra and H-comodule coalgebra B form together a semidirect product Hopf algebra. Such a pair (B, H) is called an *admissible pair* if $B \otimes H$ does form a bialgebra with the smash product and the smash coproduct. With this structure, $B \otimes H$ is then called the semidirect product of B and H.

Theorem 4.6.1 (Radford's recognition theorem). *Let H be a bialgebra over a field $\mathbb{K}$ and let B be an H-module algebra and H-comodule coalgebra. Let $\phi : H \otimes B \to B$ and $\rho : B \to H \otimes B$ be the H-module and H-comodule structure, respectively. Then the following assertions are equivalent:*
(a) $(B \otimes H, \mu_{B\#_\phi H}, \eta_{B\#_\phi H}, \triangle_{B\propto_\rho H}, \epsilon_{B\propto_\rho H})$ *is a bialgebra;*
(b) *B is an H-comodule algebra and an H-module coalgebra, ϵ_B is an algebra homomorphism, $\triangle_B(1_B) = 1_B \otimes 1_B$ and the following identities are satisfied for all $b, b' \in B$ and all $h \in H$:*
 (i) $\triangle(bb') = b_1((b_2)^0 \cdot b_1') \otimes (b_2)^1 b_2'$,
 (ii) $(h_1 \cdot b)^0 h_2 \otimes (h_1 \cdot b)^1 = h_1 b^0 \otimes (h_2 \cdot b^1)$.
(c) *ϵ_B and ρ are algebra homomorphisms, ϕ is a coalgebra homomorphism, we have $\triangle_B(1_B) = 1_B \otimes 1_B$, and (i) and (ii) are satisfied.*

Semidirect products are related to split short exact sequences. In Radford's isomorphism theorem, he explores when the splitting of an exact sequence gives rise to a semidirect product of Hopf algebras. In order to formulate it, we need the convolution product $\star$ in the endomorphism algebra $\mathrm{End}(A)$ of a bialgebra A. For $f, g \in \mathrm{End}(A)$, we have

$$f \star g := \mu_A \circ (f \otimes g) \circ \triangle_A.$$

With the convolution product, Radford constructs an idempotent $\Pi \in \mathrm{End}(A)$ whose image is the subalgebra B, which identifies A as the semidirect product of B and H.

We will also need the adjoint action ad_i with respect to a homomorphism of algebras $i : H \to A$ for a Hopf algebra H with antipode S, which is defined as

$$h \cdot a = \mathrm{ad}_i(h \otimes a) = i(h_1)\ a\ i(S(h_2)).$$

With respect to this action, A becomes an H-module algebra. Similarly, for the adjoint coaction co_π with respect to a homomorphism of coalgebras $\pi : C \to H$ where H is still a Hopf algebra. It is defined as

$$\mathrm{co}_\pi(c) = c^0 \otimes c^1 = \pi(c_1)S(\pi(c_3)) \otimes c_2.$$

Again, in this situation, C becomes an H-comodule coalgebra with respect to co_π.

Theorem 4.6.2 (Radford's isomorphism theorem). *Let H be a Hopf algebra over $\mathbb{K}$ with antipode S, and let A be a bialgebra. Let $i : H \to A$ and $\pi : A \to H$ be bialgebra homomorphisms such that $\pi \circ i = \mathrm{id}_H$. Put $\Pi = \mathrm{id}_A \star (i \circ S \circ \pi) \in \mathrm{End}(A)$, $B := \Pi(A)$ and let $j : B \to A$ be the inclusion. Then:*
(a) *B is a subalgebra of A and B carries a unique coalgebra structure such that Π is a coalgebra homomorphism.*
(b) *B is a H-submodule for ad_i (and thus an H-module algebra), and B is an H-subcomodule for co_π (and thus an H-comodule coalgebra).*
(c) *With these structures, (H, B) is an admissible pair.*
(d) *The map $B \otimes H \to A$, $b \otimes h \mapsto bi(h)$, is an isomorphism of bialgebras.*

Note that in particular, B is then an H-module algebra and coalgebra. For the proof of Radford's theorem, we refer to [89]. We will only mention that B is described explicitly as

$$B = \{b \in A \mid b_1 \otimes \pi(b_2) = b \otimes 1\}.$$

This property characterizes B as the equalizer in the following diagram:

$$B \xrightarrow{j} A \underset{(\mathrm{id}_A \otimes \pi) \circ \Delta_A}{\overset{\mathrm{id}_A \otimes \eta_H}{\rightrightarrows}} A \otimes H$$

Lemma 4.6.3. *The subalgebra B in Radford's theorem is also characterized as the coequalizer in the following diagram:*

$$A \otimes H \underset{\mu_A \circ (\mathrm{id}_A \otimes i)}{\overset{\mathrm{id}_A \otimes \epsilon_H}{\rightrightarrows}} A \xrightarrow{\Pi} B$$

Proof. In the proof of Radford's theorem, one explores the multiplicativity of the idempotent Π. One arrives at the formula:

$$\Pi(aa') = a_1 \Pi(a')(i \circ S \circ \pi)(a_2),$$

which is easily verified explicitly. In particular, we have

$$\Pi(ai(h)) = a_1\Pi(i(h))(i \circ S \circ \pi)(a_2) = \epsilon(h)\Pi(a),$$

because $\Pi(i(h)) = \epsilon(h)1_H$. This implies that the two compositions in the above diagram coincide. In order to prove the universal property, suppose that there is a space X with a map $\phi : A \to X$ giving equal compositions for the two arrows in the diagram. Define a map $\psi : B \to X$ by $\psi(\Pi(a)) = \phi(a)$ for all $a \in A$. This is well-defined, because in case $a \in B$, we have $a = \Pi(a)$ and, furthermore, $a \otimes 1 = a_1 \otimes \pi(a_2)$. Thus

$$\begin{aligned}\phi(a) &= \phi \circ \mu_A \circ (\mathrm{id}_A \otimes i \circ S)(a \otimes 1)\\ &= \phi \circ \mu_A(\mathrm{id}_A \otimes i \circ S)(a_1 \otimes \pi(a_2))\\ &= \phi(\Pi(a)).\end{aligned}$$

□

Let us summarize Radford's construction and at the same time change notation from Radford's notation to those of Fernandez Vilaboa et al. in [30]. The morphisms are now called $s : A \to H$ (instead of π), $e : H \to A$ (instead of i), $i_s : A_s \to H$ (instead of $j : B \to A$) and $p_s : A \to A_s$ (instead of $\Pi : A \to B$). Moreover, in the latter article, the authors work over a general symmetric monodical category $\mathcal{C}$.

Remark 4.6.1. Given morphisms of Hopf algebras $s : A \to H$ and $e : H \to A$ in $\mathcal{C}$ with $s \circ e = \mathrm{id}_H$, the endomorphism

$$\overline{p_s} = \mu_A \circ (\mathrm{id}_A \otimes (e \circ \lambda_H \circ s)) \circ \triangle_A : A \to A$$

is idempotent and there exists an object A_s and morphisms $i_s : A_s \to A$ and $p_s : A \to A_s$ such that $\overline{p_s} = i_s \circ p_s$ and $p_s \circ i_s = \mathrm{id}_{A_s}$. Here, λ_H denotes now the antipode of H and $\triangle_A$ the coproduct of A.

Furthermore, the morphism $i_s : A_s \to A$ is an equalizer for $\mathrm{id}_A \otimes \eta_H$ and $(\mathrm{id}_A \otimes s) \circ \triangle_A$, and the morphism $p_s : A \to A_s$ is a coequalizer for the morphisms $\mathrm{id}_A \otimes \epsilon_H$ and $\mu_A \circ (\mathrm{id}_A \otimes e)$.

Majid [74] and Bespalov [8] reinterpreted Radford's theorem in the following way. Majid identified in Proposition A.2 in [74] the subalgebra $B = A_s$ as a Hopf algebra in the braided category of Yetter–Drinfeld modules. While Majid works with a finiteness condition and the Drinfeld double Hopf algebra $D(H)$ of H, Bespalov works in the general setting of Yetter–Drinfeld modules over a symmetric monodical category $\mathcal{C}$ (see Theorems 4.1.2 and 4.1.3 as Corollaries of Theorem 4.2.1 in [8]).

Here, a *Yetter–Drinfeld module* X over a Hopf algebra H is an H-module and H-comodule such that the two structures are compatible in the sense that for all $h \in H$ and all $x, y \in X$,

$$(h_1 \cdot x)^0 h_2 \otimes (h_1 \cdot x)^1 = h_1 x^0 \otimes (h_2 \cdot x^1).$$

An example of Yetter–Drinfeld modules is given in Exercise 4.8.4. The category of Yetter–Drinfeld modules over a Hopf algebra H is braided by the Yetter–Drinfeld braiding

$$c(x \otimes y) = (x^0 \cdot y) \otimes x^1.$$

Note that in the special case where the coaction is trivial (i. e., $\rho_X = \eta_H \otimes \mathrm{id}_X$), the Yetter–Drinfeld braiding specializes to the tensor flip τ.

Now the subalgebra B from Radford's theorem is an algebra and a coalgebra, but the two structures do not form a bialgebra in the ordinary sense. Majid observed that they do form a bialgebra *in the category of Yetter–Drinfeld modules*, where B is a Yetter–Drinfeld module with respect to the adjoint action ad_i and the adjoint coaction co_π. In order to see this, one has to observe that property (i) in Radford's recognition theorem for semidirect products is exactly the property that product and coproduct satisfy the bialgebra axiom in the Yetter–Drinfeld category, while property (ii) is the Yetter–Drinfeld property.

Thus we complete Remark 4.6.1 by the following.

Remark 4.6.2. In the setting of Remark 4.6.1, A_s becomes a Hopf algebra in the braided category ${}_H^H\mathcal{C}$ of Yetter–Drinfeld modules over H with the structure maps:
(a) unit $\eta_{A_s} = p_s \circ \eta_A$;
(b) product $\mu_{A_s} = p_s \circ \mu_A \circ (i_s \otimes i_s)$;
(c) counit $\epsilon_{A_s} = \epsilon_A \circ i_s$;
(d) coproduct $\triangle_{A_s} = (p_s \otimes p_s) \circ \triangle_A \circ i_s$ (obtained in the proof of Radford's theorem by defining it such that $\triangle_{A_s} \circ p_s = (p_s \otimes p_s) \circ \triangle_A$);
(e) H-action

$$\varphi_{A_s} = p_s \circ \mu_A \circ (\mathrm{id}_A \otimes (\mu_A \circ \tau)) \circ (e \otimes (e \circ \lambda_H) \otimes i_s) \circ (\triangle_H \otimes \mathrm{id}_{A_s});$$

(f) H-coaction $\rho_{A_s} = (s \otimes p_s) \circ \triangle_A \circ i_s$, and
(g) antipode $\lambda_{A_s} = \varphi_{A_s} \circ (\mathrm{id}_H \otimes (p_s \circ \lambda_A \circ i_s)) \circ \rho_s$.

Furthermore, $\omega_A = \mu_a \circ (i_s \otimes e) : A_s \bowtie H \to A$ is an isomorphism of Hopf algebras in $\mathcal{C}$ with inverse $\omega_A^{-1} = (p_s \otimes s) \circ \triangle_A : A \to A_s \bowtie H$. Here, $A_s \bowtie H$ denotes the semidirect product of Hopf algebras with the above structure maps.

One can put the composition $e \circ \lambda_A \circ s$ also on the left-hand strand in the convolution product. This gives then an idempotent

$$\overline{{}_sp} = \mu_A \circ ((e \circ \lambda_H \circ s) \otimes \mathrm{id}_A) \circ \triangle_A : A \to A$$

and an object ${}_sA$ with morphisms ${}_si : {}_sA \to A$, ${}_sp : A \to {}_sA$ such that $\overline{{}_sp} = {}_si \circ {}_sp$ and ${}_sp \circ {}_si = \mathrm{id}_{{}_sA}$.

Remark 4.6.3. Note that in a cocommutative setting where H and A are cocommutative, as already Radford observes, the coaction ρ becomes trivial, that is, $\rho = \eta_H \otimes \mathrm{id}_B$. The authors in [30] consider more generally a Hopf algebra A in $\mathcal{C}$ such that $(s \otimes \mathrm{id}_a) \circ \tau \circ \triangle_A = (s \otimes \mathrm{id}_A) \circ \triangle_A$. This still implies that the coaction is trivial, the antipode is simply $\lambda_{A_s} = p_s \circ \lambda_A \circ i_s$ and the smash coproduct $\triangle_{A_s \propto_\rho H} = (\mathrm{id}_{A_s} \otimes \tau \otimes \mathrm{id}_H) \circ (\triangle_{A_s} \otimes \triangle_H)$. Moreover, H is a cocommutative Hopf algebra and A_s is an H-module coalgebra.

4.6.2 cat^1-Hopf algebras

We continue to work in a general symmetric monodical category $\mathcal{C}$ with split idempotents. This is not a serious restriction on the symmetric monoidal category $\mathcal{C}$, as it is shown in [8] that this can always be achieved. We denote by $\mathcal{H}$ the category of Hopf algebras and morphisms of Hopf algebras in $\mathcal{C}$.

Given two morphisms of Hopf algebras $f, g : A \to B$ in $\mathcal{C}$, a *fork* is a diagram

$$I \xrightarrow{i} A \overset{f}{\underset{g}{\rightrightarrows}} B$$

such that $f \circ i = g \circ i$. A fork is said to be *split* in case there exists $h : B \to A$ and $p : A \to I$ such that $h \circ f = \mathrm{id}_A$, $h \circ g = i \circ p$ and $p \circ i = \mathrm{id}_I$. Note that i is then an equalizer for f and g and that $\bar{p} = i \circ p : A \to A$ is a split idempotent. Conversely, the main example of split forks in our context are split idempotents $A \to A$ defining a subspace A_s such that the inclusion of A_s into A is an equalizer like, for example, $i_s : A_s \to A$ in Remark 4.6.1. One shows (see Lemma 4 in [30]) that in case A and B are Hopf algebras and f and g are morphisms of Hopf algebras, the object I in a split fork becomes a Hopf algebra and i a morphism of Hopf algebras just by factorizing the structure maps of A through equalizers.

As in [30], we will stick to a cocommutative framework. Cocommutativity will be built in the notion of cat^1-Hopf algebra and crossed module that we will use here. A reason for this are the simplification in coaction, coproduct and antipode that arise (see Remark 4.6.3), which will be crucial in the correspondence between cat^1-Hopf algebras and crossed modules.

Definition 4.6.4. Let A and H be two Hopf algebras in $\mathcal{H}$, $s, t : A \to H$ and $e : H \to A$ be morphisms of Hopf algebras. Then $\overline{A} := (A, H, s, t, e)$ is called a *pre-*cat^1*-Hopf algebra* if:

(a) $s \circ e = t \circ e = id_H$;
(b) $(s \otimes \mathrm{id}_A) \circ \tau \circ \triangle_A = (s \otimes \mathrm{id}_A) \circ \triangle_A$, and
(c) $(t \otimes \mathrm{id}_A) \circ \tau \circ \triangle_A = (t \otimes \mathrm{id}_A) \circ \triangle_A$.

Remark 4.6.5. Note that H is then necessarily a cocommutative Hopf algebra, while this need not be true for A.

Morphisms of pre-cat^1-Hopf algebras are defined in an obvious way as morphisms $f : A \to B$ for two pre-cat^1-Hopf algebras A and B over H, which commute with the structure maps. cat^1-Hopf algebras are then pre-cat^1-Hopf algebras together with a categorical composition. Note that this is different from the usual scheme where cat^1-objects are objects with two morphisms onto a normal subobject whose kernels commute. It is more in the spirit of a categorical Hopf algebra, that is, a category object in the category of Hopf algebras (over a fixed Hopf algebra H).

In order to define this categorical composition, we need a tensor product. This is again defined by idempotents and split forks.

Lemma 4.6.4. *If $\overline{A} = (A, H, s_A, t_A, e_A)$ and $\overline{B} = (B, H, s_B, t_B, e_B)$ are pre-cat^1-Hopf algebras over H, then:*
(a) *$\overline{A \otimes B} = (A \otimes B, H, \epsilon_A \otimes s_B, t_A \otimes \epsilon_B, (e_A \otimes e_B) \circ \Delta_H)$ is a pre-cat^1-Hopf algebra over H.*
(b) *There is a split fork*

$$A \otimes^H B \xrightarrow{j_{A,B}} A \otimes B \underset{\mathrm{id}_A \otimes ((t_B \otimes \mathrm{id}_B)\circ \Delta_B)}{\overset{((\mathrm{id}_A \otimes s_A)\circ \Delta_A)\otimes \mathrm{id}_B}{\rightrightarrows}} A \otimes H \otimes B$$

and $\overline{A \otimes^H B} = (A \otimes^H B, H, (\epsilon_A \otimes s_B)\circ j_{A,B}, (t_A \otimes \epsilon_B)\circ j_{A,B}, e_{A\otimes^H B})$ is a pre-cat^1-Hopf algebra over H where $e_{A\otimes^H B}$ is the factorization of $(e_A \otimes e_B) \circ \Delta_H$ through the equalizer $j_{A,B}$.

Proof. This is Lemma 7 in [30]. First of all, the assertion that $\overline{A \otimes B}$ is a pre-cat^1-Hopf algebra is proved by straightforward verifications. For the second assertion, let us only remark the following. The pre-cat^1-Hopf algebra $\overline{A \otimes^H B}$ is defined using an idempotent, namely (using the idempotent $\overline{p_s}$ associated to s_A)

$$\overline{p_{A,B}} = (\mu_A \otimes \mathrm{id}_B) \circ (\overline{p_s} \otimes (e_A \circ t_B) \otimes \mathrm{id}_B) \circ (\mathrm{id}_A \otimes \Delta_B) : A \otimes B \to A \otimes B.$$

More explicitly, for $a \otimes b \in A \otimes B$, the image under $\overline{p_{A,B}}$ is

$$\overline{p_{A,B}}(a \otimes b) = \overline{p_s}(a) e_A(t_B(b_1)) \otimes b_2.$$

Applying $\overline{p_{A,B}}$ two times, one needs to evaluate the endomorphism $\overline{p_s} \in \mathrm{End}(A)$ on a product. This can be computed:

$$\overline{p_s}(aa') = a_1 \overline{p_s}(a') e_A(\lambda_H(s_A(a_2))).$$

Therefore, we obtain

$$\overline{p_{A,B}}^2(a \otimes b) = \overline{p_s}(a)_1 \overline{p_s}(e_A(t_B(b_1))) e_A(\lambda_H(s_A(\overline{p_s}(a)_2))) e_A(t_B(b_2)) \otimes b_3.$$

But now one uses $\overline{p_s}(e_A(h)) = \epsilon_H(h)1$ for all $h \in H$ (see the proof of Lemma 4.6.3) in order to reduce the expression to

$$\overline{p_{A,B}}^2(a \otimes b) = \overline{p_s}(a)_1 e_A(\lambda_H(s_A(\overline{p_s}(a)_2))) e_A(t_B(b_1)) \otimes b_2.$$

Then we use the property that i_s is an equalizer (see Remark 4.6.1) for $\mathrm{id}_A \otimes \eta_A$ and $(\mathrm{id}_A \otimes s_A) \circ \triangle_A$ and the fact that $\overline{p_s}(a) = i_s(p_s(a))$ is an i_s-image to conclude

$$\overline{p_s}(a)_1 \otimes e_A(\lambda_H(s_A(\overline{p_s}(a)_2))) = \overline{p_s}(a) \otimes 1.$$

This shows finally that

$$\overline{p_{A,B}}^2(a \otimes b) = \overline{p_{A,B}}(a \otimes b)$$

as it should. Many other verifications for the proof are done in the proof of Lemma 7 in [30]. □

Fernandez et al. [30] go on showing that with this notion of a tensor product of two pre-cat¹-Hopf algebras over H, the category $\mathcal{PC}^1_H$ of pre-cat¹-Hopf algebras over H becomes a monoidal category.

Then, in Definition 9 of [30], cat¹-Hopf algebras are defined as monoids in the monoidal category $\mathcal{PC}^1_H$. More precisely, a pre-cat¹-Hopf algebra $\overline{A} = (A, H, s_A, t_A, e_A)$ is a cat¹*-Hopf algebra* in case there exists a morphism $m : \overline{A \otimes^H A} \to \overline{A}$ in $\mathcal{PC}^1_H$ such that $(\overline{A}, m, e_A)$ is an algebra in $\mathcal{PC}^1_H$.

Note that the unit constraint for the categorical composition m reads

$$m \circ (e \otimes^H \mathrm{id}_A) = m \circ (\mathrm{id}_A \otimes^H e) = \mathrm{id}_A,$$

where $e \otimes^H \mathrm{id}_A$ is the application of the bifunctor $- \otimes^H -$ to the morphisms $e : H \to A$ and id_A. As H is the unit for the tensor product, we have $e \otimes^H \mathrm{id}_A : H \otimes^H A \simeq A \to A \otimes^H A$.

It is remarkable that (as in the cases of groups, Lie algebras and associative algebras) the categorical composition m is here in the framework of Hopf algebras also already determined by the underlying structure. Namely, Fernandez et al. show in Proposition 10 [30] that in case m exists, it is necessarily of the form

$$m = \mu_A \circ (\overline{p_s} \otimes \mathrm{id}_A) \circ j_{A,A}.$$

Fernandez et al. develop a criterion in order to show that a given pre-cat¹-Hopf algebra is a Hopf algebra. The criterion reads the following.

Proposition 4.6.5. *A given pre-*cat¹*-Hopf algebra $\overline{A}$ is a* cat¹*-Hopf algebra if and only if*

$$\mu_A \circ ({}_t i \otimes i_s) = \mu_A \circ \tau \circ ({}_t i \otimes i_s).$$

4.6.3 Equivalence between crossed module and cat¹ Hopf algebras

In order to establish an equivalence of crossed modules of Hopf algebras with cat¹-Hopf algebras, Fernandez et al. [30] build the cocommutativity into the notion of crossed modules that they use.

Definition 4.6.6. Let $\gamma : B \to H$ be a morphism of Hopf algebras over the symmetric monoidal category $\mathcal{C}$. The morphism $\gamma : B \to H$ is called a *cocommutative crossed module of Hopf algebras* in case:
(i) $(\gamma \otimes \mathrm{id}_B) \circ \triangle_B = (\gamma \otimes \mathrm{id}_B) \circ \tau \circ \triangle_B$;
(ii) (B, ϕ_B) is a left H-module which renders it a left H-module algebra and H-module coalgebra;
(iii) $\lambda_B : B \to B$ is a morphism of H-modules;
(iv) $\gamma \circ \phi_B = \mathrm{ad}_H \circ (\mathrm{id}_H \otimes \gamma)$;
(v) $\phi_B \circ (\gamma \otimes \mathrm{id}_B) = \mathrm{ad}_B$.

Remark 4.6.7.
(a) Note that choosing $\mathcal{C}$ to be the category of sets with the cartesian product as monoidal structure, we get back the notion of a crossed module of groups.
(b) The notion of morphism of cocommutative crossed modules of Hopf algebras is the same as for general crossed modules of Hopf algebras.
(c) Condition (v) can be phrased equivalently as

$$\mu_B = \mu_B \circ (\phi_B \otimes \mathrm{id}_B) \circ (\gamma \otimes \tau) \circ (\triangle_B \otimes \mathrm{id}_B),$$

or, in terms of elements $b, b' \in B$,

$$(\gamma(b_1) \cdot b')b_2 = bb'.$$

Indeed, in one direction the identity $\gamma(b) \cdot b' = b_1 b' \lambda(b_2)$ can be used in $(\gamma(b_1) \cdot b')b_2 = bb'$ to transform the LHS which then gives bb' using that $\lambda(b_2)b_3 = \epsilon(b_2)$. In the other direction, replace $b_1 b'$ by $(\gamma(b_1) \cdot b')$ and get finally $\gamma(b) \cdot b'$ using again $b_2 \lambda(b_3) = \epsilon(b_2)$.

The main theorem of this section is the following theorem.

Theorem 4.6.6 (Fernandez et al.). *Let H be a cocommutative Hopf algebra. Then the categories $\mathcal{C}_H^1$ of* cat^1*-Hopf algebras over H and the category $\Xi\mathcal{M}od_H$ of cocommutative crossed modules over H are equivalent.*

Proof. This is Theorem 14 in [30] and we refer to it for details of the proof. Let us just sketch the big picture. We will construct functors $S : \mathcal{C}_H^1 \to \Xi\mathcal{M}od_H$ and $T : \Xi\mathcal{M}od_H \to \mathcal{C}_H^1$, which compose to give essential inverses of each other.

Given a cat^1-Hopf algebra $\overline{A} = (A, H, s, t, e, m)$ over H, we claim that $S(\overline{A}) := (A_s, H, \gamma := t \circ i_s)$ is a cocommutative crossed module over H. Note that in general A_s is first of all a left H-module algebra and H-comodule coalgebra. But here, because of the cocommutativity of the setting, A_s is a left H-coalgebra and the coaction is reduced to the trivial coaction $\rho_{A_s} = \eta_H \otimes \mathrm{id}_{A_s}$, as we remarked before in Remark 4.6.3. We will restrict to the verification of Conditions (iv) and (v), leaving the other conditions to the reader (but see [30]).

Condition (iv): Let $a \in A_s$ and $h \in H$. Recall from Radford's isomorphism theorem that A_s is an H-module for the adjoint action ad_e with $e : H \to A$. The result must be sent back to A_s via p_s. This is also expressed in item (e) of Remark 4.6.2 by the definition of φ_{A_s}. With this action, we have

$$\begin{aligned}\gamma(h \cdot a) &= t \circ i_s \circ p_s(e(h_1)i_s(a)e(\lambda_H(h_2)))\\ &= t(e(h_1)i_s(a)e(\lambda_H(h_2)))\\ &= h_1 t(i_s(a))\lambda_H(h_2)\\ &= \mathrm{ad}_H(h \otimes \gamma(a)).\end{aligned}$$

Here, we used that t is a morphism of algebras with $t \circ e = \mathrm{id}_H$.

Condition (v): Let $a, a' \in A_s$. Then we compute

$$\begin{aligned} i_s((t \circ i_s(a_1) \cdot a')a_2) &= i_s \circ p_s(e(t(i_s(a_1)))i_s(a')e(\lambda_H(t(i_s(a_2))))i_s(a_3))\\ &= e(t(i_s(a_1)))i_s(a')e(\lambda_H(t(i_s(a_2))))i_s(a_3)\\ &= e(t(i_s(a)_1))i_s(a')\overline{{}_t p}(i_s(a)_2)\\ &= e(t(i_s(a)_1))i_s(a')({}_t i \circ {}_t p)(i_s(a)_2)\\ &= e(t(i_s(a)_1))({}_t i \circ {}_t p)(i_s(a)_2)i_s(a')\\ &= e(t(i_s(a)_1))\overline{{}_t p}(i_s(a)_2)i_s(a')\\ &= e(t(i_s(a)_1))e(\lambda_H(t(i_s(a)_2)))i_s(a)_3 i_s(a')\\ &= (e \otimes e) \circ (t \otimes t)(i_s(a)_2 \otimes \lambda_A(i_s(a)_1))i_s(a)_3 i_s(a')\\ &= (e \circ t)(\epsilon(i_s(a)_1))i_s(a)_2 i_s(a')\\ &= i_s(a)i_s(a') = i_s(aa').\end{aligned}$$

Here, we used in the first line again the definition of φ_{A_s}, then the fact that $i_s \circ p_s$ is idempotent (and thus has no effect on elements of A_s), then the definition of $\overline{{}_t p}(a) = e(\lambda_H(t(a_1)))a_2$ (and that i_s is a morphism of coalgebras), then $\overline{{}_t p} = {}_t i \circ {}_t p$, then the equation

$$\mu_A \circ ({}_t i \otimes i_s) = \mu_A \circ \tau \circ ({}_t i \otimes i_s),$$

then we reassemble $\overline{{}_t p}$ and use its definition. Finally, extracting $e \otimes e$ and $t \otimes t$ from the first two terms, we see that they simplify to $\epsilon(i_s(a)_1)$, which is then reassembled with $i_s(a)_2$ to give the last line.

The upshot is thus

$$(t \circ i_s(a_1) \cdot a')a_2 = aa',$$

which has been shown to be equivalent to Condition (v) in Remark 4.6.7 (c).

The functor S is easily seen to be full and faithful. Conversely, given a crossed module of Hopf algebras (B, H, γ), we claim first that $(B, \varphi_B, \rho_B = \eta_H \otimes \mathrm{id}_B)$ is a left H-Yetter–Drinfeld module in ${}^H_H\mathcal{C}$. This is true because H is cocommutative.

Moreover, B is a Hopf algebra in ${}^H_H\mathcal{C}$ and by Radford's recognition theorem 4.6.1, the semidirect product $B \bowtie H$ defined by

$$B \bowtie H = (B \otimes H, \mu_{B\#_{\varphi_B}H}, \eta_B \otimes \eta_H, \triangle_{B\propto_\rho H}, \epsilon_B \otimes \epsilon_H, \lambda_{B\bowtie H}),$$

where we refer to [30] for the antipode, is a Hopf algebra in $\mathcal{C}$. Note that while the product is general and depends on the given φ_B, the coproduct reduces to

$$\triangle_{B\bowtie H} = (\mathrm{id}_B \otimes \tau \otimes \mathrm{id}_H) \circ (\triangle_B \otimes \triangle_H),$$

because of the triviality of the coaction.

The morphism $t := \mu_H \circ (\gamma \otimes \mathrm{id}_H) : B \bowtie H \to H$ is a morphism of Hopf algebras. For this, note first that for all $h \in H$ and all $b \in B$ we have

$$h\gamma(h) = \gamma(h_1 \cdot b)h_2.$$

Indeed, the RHS gives

$$\gamma(h_1 \cdot b)h_2 = h_1\gamma(b)\lambda_H(h_2)h_3 = h_1\gamma(b)\epsilon(h_2) = h\gamma(b).$$

This serves in showing that t is a morphism of algebras:

$$\begin{aligned} t(b \otimes h)t(b' \otimes h') &= \gamma(b)h\gamma(b')h' \\ &= \gamma(b)\gamma(h_1 \cdot b)h_2h' \\ &= t(\mu_{B\bowtie H}(b \otimes h, b' \otimes h')). \end{aligned}$$

For the other verifications that t is a morphism of Hopf algebras, we refer to [30].

Trivially, $s := \epsilon_B \otimes \mathrm{id}_H : B \bowtie H \to H$ and $e := \eta_H \otimes \mathrm{id}_H : H \to B \bowtie H$ are morphisms of Hopf algebras and $s \circ e = t \circ e = \mathrm{id}_H$, and we obtain a pre-cat^1-Hopf algebra $\overline{B \bowtie H} = (B \bowtie H, H, s, t, e)$.

The authors of [30] use then the criterion in Proposition 4.6.5, namely that

$$\mu_{B\bowtie H} \circ ({}_t i \otimes i_s) = \mu_{B\bowtie H} \circ \tau \circ ({}_t i \otimes i_s),$$

in order to show that $\overline{B \bowtie H}$ is indeed a cat^1-Hopf algebra. Radford's isomorphism theorem 4.6.2 together with $\gamma = t \circ i_s$ then shows that $S(\overline{B \bowtie H}) = (B, H, \gamma)$. The composition $T \circ S = \mathrm{id}$ is left to the reader. □

We propose now a different approach to this equivalence in Theorem 4.6.6, in the special case where the underlying symmetric monoidal category is just $\mathtt{Vect}_{\mathbb{K}}$, the category of $\mathbb{K}$-vector spaces. In order to connect cat^1-Hopf algebras and crossed modules of Hopf algebras, we propose to use the functors U and P and the known results on the relation between crossed modules of Lie algebras and cat^1-Lie algebras.

Theorem 4.6.7. *When restricting to the subcategory of irreducible cocommutative Hopf algebras, the categories of crossed modules of Hopf algebras and the category of pre-*cat^1*-Hopf algebra are equivalent.*

Proof. Given a pre-cat^1-Hopf algebra $s, t : A \to H$, $e : H \to A$ where H and A are irreducible cocommutative Hopf algebras (over the characteristic zero field $\mathbb{K}$), we have $A = U(P(A))$ and $H = U(P(H))$ (see [97] theorem 13.0.1, p. 274), and the morphisms are also induced from morphisms of Lie algebras. Denote by $\mathfrak{g}_1 := P(A)$ and $\mathfrak{g}_0 := P(H)$, then we have a precategory $\mathfrak{g}_1 \to \mathfrak{g}_0$ in the category of Lie algebras. But we have seen that the categorical composition is already encoded in the 2-vector space structure, thus $\mathfrak{g}_1 \to \mathfrak{g}_0$ becomes a genuine category object in the category of Lie algebras. It corresponds then to a crossed module of Lie algebras $\mu : \mathfrak{m} \to \mathfrak{n}$, and applying the functor U, we get the crossed module of Hopf algebras associated to the given pre-cat^1-Hopf algebra $s, t : A \to H$ by Theorem 4.2.3.

Conversely, given a crossed module of irreducible cocommutative Hopf algebras $\gamma : B \to H$, we can associate to it in the same way a pre-cat^1-Hopf algebra $s, t : A \to H$, $e : H \to A$. □

Remark 4.6.8. In the same way, one can show that the category of crossed comodules of commutative Hopf algebras and the category of commutative pre-cat^1-Hopf algebras (i. e., like in the Definition of a pre-cat^1-Hopf algebra, but with commutativity replacing the cocommutativity requirement) are equivalent when restricting to the subcategory of those Hopf algebras which are finitely generated as an algebra. It suffices to apply the functors $\mathbb{C}-$ and χ to translate the problem into groups, where we then apply Theorems 4.4.1 and 4.4.2.

Remark 4.6.9. In conjunction with Theorem 4.6.6, Theorem 4.6.7 shows that when one restricts the monoidal category $\mathcal{PC}^1_H$ to irreducible cocommutative Hopf algebras, every object becomes a monoid. In particular, this implies that the characterization of monoids given in Proposition 4.6.5 should be automatically satisfied in the irreducible setting.

4.7 Strict quantum 2-groups

We report in this section on a different approach to category objects in the category of Hopf algebras. The main reference is [75].

We start with a slightly more restrictive definition of a crossed module of Hopf algebras.

Definition 4.7.1. Let H be a Hopf algebra. The data of an *action-cocommutative crossed module* consists of:

(1) a Hopf algebra A in the category of left H-modules, that is, A is a left H-module via a map φ_A such that A is an H-module algebra and H-module coalgebra, with

the additional condition for all $h \in H$ and all $a \in A$,

$$h_1 \otimes (h_2 \cdot a) = h_2 \otimes (h_1 \cdot a); \tag{4.6}$$

(2) a morphism of Hopf algebras $d : A \to H$ such that

$$d \circ \varphi_A = \mathrm{ad}_H \circ (\mathrm{id}_H \otimes d);$$

(3) the Peiffer condition

$$\varphi_A \circ (d \otimes \mathrm{id}_A) = \mathrm{ad}_A.$$

The only new condition is equation (4.6), which we call *action cocommutativity*. Note that in the category of sets, an action-cocommutative crossed module is still the same as a crossed module of groups (exactly as cocommutative crossed modules or general crossed modules of Hopf algebras). Note that general crossed modules of Hopf algebras $\gamma : B \to H$ where B and H are cocommutative are examples of action-cocommutative crossed modules.

As in the previous section, we will use the semidirect product in order to construct a category object in the category of Hopf algebras from an action-cocommutative crossed module.

Lemma 4.7.1. *Given an action-cocommutative crossed module of Hopf algebras $d : A \to H$, the semidirect product $A \rtimes H$ of A and H is well-defined with the smash product*

$$(a \otimes h)(b \otimes g) = a(h_1 \cdot b) \otimes h_2 g,$$

and the tensor product coproduct

$$\Delta(a \otimes h) = (a_1 \otimes h_1) \otimes (a_2 \otimes h_2),$$

that is, with respect to the action φ_A and the trivial coaction $\rho_A = \eta_H \otimes \mathrm{id}_A$. We have morphisms of Hopf algebras

$$A \hookrightarrow A \rtimes H \underset{i}{\overset{s}{\rightleftarrows}} H$$

with $s(a \otimes h) = \epsilon(a)h$ and $i(h) = 1 \otimes h$. Moreover, there is a morphism of Hopf algebras $t : A \rtimes H \to H$ given by $t(a \otimes h) = d(a)h$ for all $a \in A$ and all $h \in H$.

Proof. According to Radford's recognition theorem, Theorem 4.6.1, we have to show that A is an H-comodule algebra and that Conditions (i) and (ii) are satisfied.

For the trivial coaction $\rho_A = \eta_H \otimes \mathrm{id}_A$, A is clearly an H-comodule algebra. Condition (i) reads for the trivial coaction simply $\Delta(aa') = a_1 a'_1 \otimes a_2 a'_2$, which is just the bialgebra condition for A. Condition (ii) becomes for a trivial coaction the action-cocommutativity condition in equation (4.6).

It remains to show that $t : A \rtimes H \to H$ is a morphism of Hopf algebras. We have shown this in the proof of Theorem 4.6.6. □

We will also need the cotensor product of comodules over a coalgebra.

Let C be a coalgebra, and consider a C-bicomodule M. This means that M carries a left- and a right C-coaction $\rho_L^M = \rho_L : M \to C \otimes M$ and $\rho_R^M = \rho_R : M \to M \otimes C$ which commute, that is,

$$(\mathrm{id}_C \otimes \rho_R) \circ \rho_L = (\rho_L \otimes \mathrm{id}_C) \circ \rho_R$$

upon identifying the final three-factor tensor product. For two such C-bicomodules M and N, we define their *cotensor product* $M \square N$ as the equalizer

$$M \square N := \left\{ \sum_i m_i \otimes n_i \mid \sum_I (\rho_R^M m_i) \otimes n_i = \sum_i m_i \otimes (\rho_L^N n_i) \right\} \subset M \otimes N.$$

Exactly as the tensor product of bimodules over an algebra, $M \square N$ becomes then a C-bicomodule with the remaining left and right C-coactions ρ_L^M and ρ_R^N.

Definition 4.7.2. An *embedded quantum groupoid* over a base coalgebra C (over the base field $\mathbb{K}$) consists of the following data:

(1) a counital coalgebra $M = (M, \triangle, \epsilon)$ over $\mathbb{K}$ with coalgebra maps $s, t : M \to C$ and $i : C \to M$ with $s \circ i = t \circ i = \mathrm{id}_C$ in the following diagram:

$$M \begin{smallmatrix} \xrightarrow{\;s\;} \\ \xleftarrow[\;i\;]{} \\ \xrightarrow[\;t\;]{} \end{smallmatrix} C;$$

(2) an associative nonunital algebra structure $m : M \otimes M \to M$ such that $\triangle$ and ϵ are algebra homomorphisms;

(3) the unity condition

$$m \circ (\mathrm{id}_M \otimes i) \circ (\mathrm{id} \otimes s) \circ \triangle = m \circ (i \otimes \mathrm{id}_M) \circ (t \otimes \mathrm{id}) \circ \triangle = \mathrm{id}_M,$$

and the source-target conditions

$$s \circ m = \epsilon \otimes s, \quad t \circ m = t \otimes \epsilon;$$

(4) an antipode $S : M \to M$ such that

$$(s \otimes S) \circ \triangle = \rho_L \circ S, \quad (S \otimes t) \circ \triangle = \rho_R \circ S,$$

and

$$m \circ (S \otimes \mathrm{id}_M) \circ \triangle = i \circ s, \quad m \circ (\mathrm{id}_M \otimes S) \circ \triangle = i \circ t.$$

The left and right C-coaction which we refer to in Condition (4) are obtained from the coproduct on M as

$$\rho_L := (t \otimes \mathrm{id}_M) \circ \triangle, \quad \rho_R := (\mathrm{id}_M \otimes s) \circ \triangle.$$

Now suppose that C is a Hopf algebra and that M, N are C-bicomodule algebras (i. e., both coactions are homomorphisms of algebras).

Lemma 4.7.2. *In case M and N are C-bicomodule algebras, the cotensor product $M\Box N$ becomes a C-bicomodule algebra with respect to the product*

$$(m \otimes n)(m' \otimes n') = mm' \otimes nn'.$$

Proof. In order to have a well-defined product, we have to show first of all that for all $m, m' \in M$ and all $n, n' \in N$, the above defined product $mm' \otimes nn'$ lies in $M\Box N$. We compute

$$\begin{aligned}
\rho_R(mm') \otimes nn' &= \rho_R(m)\rho_R(m') \otimes nn' \\
&= (\rho_R(m) \otimes n)(\rho_R(m') \otimes n') \\
&= (m \otimes \rho_L(n))(m' \otimes \rho_L(n')) \\
&= mm' \otimes \rho_L(n)\rho_L(n') \\
&= mm' \otimes \rho_L(nn').
\end{aligned}$$

Now we have to check that the remaining coactions ρ_L^M and ρ_R^N are morphisms of algebras with respect to this product. This follows at once from $\rho_L(mm') = \rho_L(m)\rho_L(m')$ and $\rho_R(nn') = \rho_R(n)\rho_R(n')$. □

Definition 4.7.3. A *strict quantum 2-group* is by definition a pair of Hopf algebras (H_1, H_0) linked by Hopf algebra maps in the diagram

$$H_1 \underset{t}{\overset{s}{\rightrightarrows}} H_0, \quad H_0 \xrightarrow{i} H_1$$

such that $s \circ i = t \circ i = \mathrm{id}_{H_0}$, together with an associative product $m : H_1 \otimes H_1 \to H_1$ forming an embedded quantum groupoid such that the quantum interchange law

$$m \circ \mu_{H_1\Box H_1} = \mu_{H_1} \circ (m \otimes m)$$

holds, both sides being maps $(H_1\Box H_1) \otimes (H_1\Box H_1) \to H_1$.

Here, $\mu_{H_1\Box H_1}$ is the above defined product on the cotensor product $H_1\Box H_1$ as bicomodule algebras.

The main result of the present section is the following theorem.

Theorem 4.7.3. *Given an action-cocommutative crossed module of Hopf algebras $d : A \to H$, the semidirect product construction gives a strict quantum 2-group*

$$A \rtimes H \underset{t}{\overset{s}{\rightrightarrows}} H, \quad H \xrightarrow{i} A \rtimes H$$

with categorical composition m and antipode given by

$$m(a \otimes h, b \otimes g) = \epsilon(h)ab \otimes g, \quad S(a \otimes h) = S(a_1) \otimes d(a_2)h$$

for all $a, b \in A$ and all $h, g \in H$, and furthermore $1 \otimes 1$ is a left unit for m.

Proof. First, we verify the conditions (1)–(4) in order to show that $A \rtimes H$ is an embedded quantum groupoid.

(1) We have already checked in Lemma 4.7.1 that

$$A \rtimes H \underset{t\ \ i}{\overset{s}{\rightleftarrows}} H$$

are Hopf algebra maps with $s \circ i = t \circ i = \mathrm{id}_H$.

(2) Let us check that the proposed product m on $A \rtimes H$ is associative and unital:

$$\begin{aligned} m(m(a \otimes h, b \otimes g), c \otimes f) &= \epsilon(h) m(ab \otimes g, c \otimes f) \\ &= \epsilon(h)\epsilon(g)((ab)c) \otimes f, \end{aligned}$$

and on the other hand

$$\begin{aligned} m(a \otimes h, m(b \otimes g, c \otimes f)) &= \epsilon(g) m(a \otimes h, bc \otimes f) \\ &= \epsilon(g)\epsilon(h)(a(bc)) \otimes f. \end{aligned}$$

Moreover, $m(1 \otimes 1, a \otimes h) = \epsilon(1) a \otimes h$, but $m(a \otimes h, 1 \otimes 1) = \epsilon(h) a \otimes 1$.

Now we check that the tensor coproduct $\triangle$ is an algebra homomorphism with respect to m:

$$\begin{aligned} \triangle(m(a \otimes h, b \otimes g)) &= \epsilon(h) \triangle (ab \otimes g) \\ &= \epsilon(h)((ab)_1 \otimes g_1) \otimes ((ab)_2 \otimes g_2) \\ &= \epsilon(h)((a_1 b_1) \otimes g_1) \otimes ((a_2 b_2) \otimes g_2) \\ &= \epsilon(h_1)\epsilon(h_2)((a_1 b_1) \otimes g_1) \otimes ((a_2 b_2) \otimes g_2) \\ &= (m \otimes m)((a_1 \otimes h_1) \otimes (b_1 \otimes g_1) \otimes (a_2 \otimes h_2) \otimes (b_2 \otimes g_2)) \\ &= (m \otimes m) \circ (1 \otimes \tau \otimes 1) \circ (\triangle \otimes \triangle)((a \otimes h) \otimes (b \otimes h)). \end{aligned}$$

Moreover, $\epsilon(\epsilon(h) ab \otimes g) = \epsilon(h)\epsilon(g)\epsilon(a)\epsilon(b)$ showing that the tensor counit is also an algebra homomorphism with respect to m.

Now we pass to the box tensor product. First of all, the box tenor product for $H_1 = A \rtimes H$ is by definition of the coactions

$$H_1 \square H_1 = \left\{ \sum a \otimes h \otimes b \otimes g \mid \sum a \otimes h_1 \otimes h_2 \otimes b \otimes g = \sum a \otimes h \otimes d(b_1) g_1 \otimes b_2 \otimes g_2 \right\},$$

because

$$(\mathrm{id}_{A \rtimes H} \otimes s)(a_1 \otimes h_1 \otimes a_2 \otimes h_2) = a_1 \otimes h_1 \otimes \epsilon(a_2) h_2 = a \otimes h_1 \otimes h_2$$

and $(t \otimes \mathrm{id}_{A \rtimes H})(b_1 \otimes g_1 \otimes b_2 \otimes g_2) = d(b_1) g_1 \otimes b_2 \otimes g_2$. Keeping always in mind that elements of $H_1 \square H_1$ are sums, we will now skip the sum-signs in order to simplify the notation and work only with homogeneous elements.

Applying $\mathrm{id}_A \otimes \epsilon_H \otimes \mathrm{id}_H \otimes \mathrm{id}_A \otimes \mathrm{id}_H$ to $a \otimes h_1 \otimes h_2 \otimes b \otimes g$, we obtain for elements in $H_1 \square H_1$

$$a \otimes h \otimes b \otimes g = a \otimes \epsilon(h)d(b_1)g_1 \otimes b_2 \otimes g_2, \tag{4.7}$$

and in the same vein, we obtain with other applications of ϵ in some spot

$$a \otimes \epsilon(b)h \otimes g = a \otimes \epsilon(h)d(b)g_1 \otimes g_2,$$

and

$$a \otimes \epsilon(b)\epsilon(g)h = a \otimes \epsilon(h)d(b)g. \tag{4.8}$$

Let us verify that the restriction of m to $H_1 \square H_1$ is a morphism of bicomodules. For the compatibility of M with the left coaction, we have to show that $(\mathrm{id}_{H_1} \otimes m) \circ (\rho_L \otimes \mathrm{id}_{H_1}) = \rho_L \circ m$, which follows from

$$\begin{aligned}(\mathrm{id}_{H_1} \otimes m)(t(a_1 \otimes h_1) \otimes (a_2 \otimes h_2), b \otimes g) &= d(a_1)h_1 \otimes \epsilon(h_2)a_2 b \otimes g \\ &= d(a_1)h \otimes a_2 b \otimes g\end{aligned}$$

and on the other hand,

$$\begin{aligned}\rho_L(m(a \otimes h, b \otimes g)) &= \rho_L(\epsilon(h)ab \otimes g) \\ &= \epsilon(h)d(a_1 b_1)g_1 \otimes a_2 b_2 \otimes g_2 \\ &= d(a_1)\epsilon(h)d(b_1)g_1 \otimes a_2 b_2 \otimes g_2 \\ &= d(a_1)h \otimes a_2 b \otimes g,\end{aligned}$$

where we have used equation (4.7) going from line 3 to line 4. For the compatibility with respect to the right coaction, we compute

$$\begin{aligned}(m \otimes \mathrm{id}_{H_1}) \circ (\mathrm{id}_{H_1} \otimes \rho_R)((a \otimes h) \otimes (b \otimes g)) &= (m \otimes \mathrm{id}_{H_1})(a \otimes h, b_1 \otimes g_1 \otimes \epsilon(b_2)g_2) \\ &= (m \otimes \mathrm{id}_{H_1})(a \otimes h, b \otimes g_1 \otimes g_2) \\ &= \epsilon(h)ab \otimes g_1 \otimes g_2,\end{aligned}$$

and on the other hand,

$$\begin{aligned}\rho_R(m(a \otimes h, b \otimes g)) &= \rho_R(\epsilon(h)ab \otimes g) \\ &= \epsilon(h)\epsilon(a_2 b_2)a_1 b_1 \otimes g_1 \otimes g_2 \\ &= \epsilon(h)ab \otimes g_1 \otimes g_2.\end{aligned}$$

(3) Now we verify the source-target conditions $t \circ m = t \otimes \epsilon$ and $s \circ m = \epsilon \otimes s$:

$$t(m(a \otimes h, b \otimes h)) = t(\epsilon(h)ab \otimes g)$$

$$
\begin{aligned}
&= \epsilon(h)d(ab)g \\
&= d(a)\epsilon(h)d(b)g \\
&= d(a)\epsilon(b)\epsilon(g)h \\
&= t(a \otimes h)\epsilon(b \otimes g),
\end{aligned}
$$

where we have used equation (4.8) from line 3 to line 4. In the same vein,

$$
\begin{aligned}
s(m(a \otimes h, b \otimes h)) &= s(\epsilon(h)ab \otimes g) \\
&= \epsilon(h)\epsilon(ab)g \\
&= \epsilon(h)\epsilon(a)\epsilon(b)g \\
&= \epsilon(a \otimes h)s(b \otimes g).
\end{aligned}
$$

Then we verify the unity condition $m \circ (\mathrm{id} \otimes i) \circ \rho_R = \mathrm{id}$ and $m \circ (i \otimes \mathrm{id}) \circ \rho_L = \mathrm{id}$:

$$
\begin{aligned}
(m \circ (\mathrm{id} \otimes i) \circ \rho_R)(a \otimes h) &= (m \circ (\mathrm{id} \otimes i))(a_1 \otimes h_1 \otimes \epsilon(a_2)h_2) \\
&= m(a_1 \otimes h_1, 1 \otimes \epsilon(a_2)h_2) \\
&= \epsilon(a_2)a_1 \otimes \epsilon(h_1)h_2 = a \otimes h,
\end{aligned}
$$

and

$$
\begin{aligned}
(m \circ (i \otimes \mathrm{id}) \circ \rho_L)(a \otimes h) &= (m \circ (i \otimes \mathrm{id}))(d(a_1)h_1 \otimes a_2 \otimes h_2) \\
&= m(1 \otimes d(a_1)h_1, a_2 \otimes h_2) \\
&= \epsilon(d(a_1)h_1)a_2 \otimes h_2 \\
&= \epsilon(a_1)a_2 \otimes \epsilon(h_1)h_2 = a \otimes h.
\end{aligned}
$$

(4) Next, we consider the antipode:

$$
\begin{aligned}
(s \otimes S) \circ \triangle(a \otimes h) &= (s \otimes S)((a_1 \otimes h_1) \otimes (a_2 \otimes h_2)) \\
&= \epsilon(a_1)h_1 \otimes S(a_2) \otimes d(a_3)h_2 \\
&= h_1 \otimes S(a_1) \otimes d(a_2)h_2 \\
&= d(\epsilon(a_2))h_1 \otimes S(a_1) \otimes d(a_3)h_2 \\
&= d(S(a_2))d(a_3)h_1 \otimes S(a_1) \otimes d(a_4)h_2 \\
&= (t \otimes \mathrm{id})(S(a_2) \otimes d(a_3)h_1) \otimes (S(a_1) \otimes d(a_4)h_2) \\
&= (t \otimes \mathrm{id})((S(a_1)_1 \otimes (d(a_2)h)_1) \otimes (S(a_1)_2 \otimes (d(a_2)h)_2)) \\
&= \rho_L(S(a_1) \otimes d(a_2)h) \\
&= (\rho_L \circ S)(a \otimes h).
\end{aligned}
$$

Similarly,

$$
(S \otimes t) \circ \triangle(a \otimes h) = (S \otimes t)((a_1 \otimes h_1) \otimes (a_2 \otimes h_2))
$$

$$
\begin{aligned}
&= S(a_1) \otimes d(a_2)h_1 \otimes d(a_3)h_2 \\
&= \epsilon(a_1)S(a_2) \otimes d(a_3)h_1 \otimes d(a_4)h_2 \\
&= S(a_2) \otimes d(a_3)h_1 \otimes \epsilon(S(a_1))d(a_4)h_2 \\
&= (\mathrm{id} \otimes s)(S(a_2) \otimes d(a_3)h_1 \otimes S(a_1) \otimes d(a_4)h_2) \\
&= (\mathrm{id} \otimes s)((S(a_1)_1 \otimes (d(a_2)h)_1) \otimes (S(a_1)_2 \otimes (d(a_2)h)_2)) \\
&= \rho_R(S(a_1) \otimes d(a_2)h) \\
&= (\rho_R \circ S)(a \otimes h).
\end{aligned}
$$

We compute furthermore

$$
\begin{aligned}
(m \circ (S \otimes \mathrm{id}) \circ \triangle)(a \otimes h) &= (m \circ (S \otimes \mathrm{id})((a_1 \otimes h_1) \otimes (a_2 \otimes h_2) \\
&= m(S(a_1) \otimes d(a_2)h_1, a_3 \otimes h_2) \\
&= \epsilon(d(a_2)h_1)\, S(a_1)a_3 \otimes h_2 \\
&= S(a_1)a_2 \otimes \epsilon(h_1)h_2 \\
&= \epsilon(a)1 \otimes h \\
&= i(\epsilon(a)h) = (i \circ s)(a \otimes h)
\end{aligned}
$$

and

$$
\begin{aligned}
(m \circ (\mathrm{id} \otimes S) \circ \triangle)(a \otimes h) &= (m \circ (\mathrm{id} \otimes S)((a_1 \otimes h_1) \otimes (a_2 \otimes h_2) \\
&= m(a_1 \otimes h_1, S(a_2) \otimes d(a_3)h_2) \\
&= \epsilon(h_1)a_1S(a_2) \otimes d(a_3)h_2 \\
&= 1 \otimes d(a)h \\
&= i(d(a)h) = (i \circ t)(a \otimes h).
\end{aligned}
$$

Observe that S interchanges t and s:

$$
(t \circ S)(a \otimes h) = t(S(a_1) \otimes d(a_2)h) = d(S(a_1))d(a_2)h = \epsilon(a)h = s(a \otimes h).
$$

The last step in the proof is the verification of the quantum interchange law:

$$
\begin{aligned}
m((a \otimes h)(a' \otimes h'), (b \otimes g)(b' \otimes g')) &= m(a(h_1 \cdot a') \otimes h_2h', b(g_1 \cdot b') \otimes g_2g') \\
&= \epsilon(h_2h')\, a(h_1 \cdot a')b(g_1 \cdot b') \otimes g_2g' \\
&= a(h \cdot a')b(g_1 \cdot b') \otimes g_2g'\epsilon(h') \\
&= a(d(b_1) \cdot (g_1 \cdot a'))b_2(g_2 \cdot b') \otimes g_3g'\epsilon(h')\epsilon(h) \\
&= ab_1(g_1 \cdot a')S(b_2)b_3(g_2 \cdot b') \otimes g_3g'\epsilon(h')\epsilon(h) \\
&= ab(g_1 \cdot a')(g_2 \cdot b') \otimes g_3g'\epsilon(h')\epsilon(h) \\
&= (ab \otimes g)(a'b' \otimes g')\epsilon(h')\epsilon(h) \\
&= m(a \otimes h, a' \otimes h')m(b \otimes g, b' \otimes g').
\end{aligned}
$$

Here, we used from line 3 to line 4 the property to be in $H_1 \square H_1$, more precisely, equation (4.7), and from line 4 to line 5 property (3) of a crossed module (Peiffer's identity). □

The remainder of this section will be used to present examples of strict quantum 2-groups.

Example 4.7.4. We have already seen how to transform a crossed module of Lie algebras $\mu : \mathfrak{m} \to \mathfrak{n}$ into a crossed module of Hopf algebras $\mu : U\mathfrak{m} \to U\mathfrak{n}$. Because of the cocommutativity of enveloping algebras, it is in particular an action-cocommutative crossed module of Hopf algebras. Thus the above theorem applies and gives rise to a strict quantum 2-group by the semidirect product construction $U\mathfrak{m} \rtimes U\mathfrak{n}$ for Hopf algebras. But it is easy to see that in this special case, $U\mathfrak{m} \rtimes U\mathfrak{n}$ is the enveloping algebra of the semidirect product Lie algebra $\mathfrak{m} \rtimes \mathfrak{n}$. Both $A = U\mathfrak{m}$ and $H = U\mathfrak{n}$ can be seen as Hopf subalgebras using the inclusion maps $a \mapsto a \otimes 1$ and $i : h \mapsto 1 \otimes h$.

The categorical composition m from Theorem 4.7.3 reads as follows. We have for $a, b \in \mathfrak{m}$ and $h, g \in \mathfrak{n}$,

$$m(a \otimes 1, b \otimes g) = ab \otimes g = (a \otimes 1)(b \otimes g), \quad m(1 \otimes h, b \otimes g) = 0.$$

Accordingly, there cannot exist a unit in $U\mathfrak{m} \rtimes U\mathfrak{n}$ with respect to the multiplication m. But $1 \otimes 1$ is a left unit.

Clearly, the same works for any pair of cocommutative Hopf algebras A, H where H acts on A respecting its Hopf algebra structure and where $d : A \to H$ is a Hopf algebra map, which satisfies Conditions (2) and (3) of Definition 4.7.1.

Example 4.7.5. Let $A = H$ be a cocommutative Hopf algebra, let H act on A by the adjoint action and $d = \mathrm{id}_H$. This is easily seen to form a crossed module in the sense of Definition 4.7.1: The adjoint action is always a morphism of algebras, but it is a morphism of coalgebras here only because we assumed our Hopf algebra to be cocommutative! We obtain thus a strict quantum 2-group with $H_0 = H$ and $H_1 = H \rtimes H$. The coproduct is trivial as before, but the product on $H \rtimes H$ is induced by the adjoint action and reads

$$(h \otimes g)(h' \otimes g') = h(g_1 h' S(g_2)) \otimes g_3 g'.$$

In fact, this semidirect product $H \rtimes H$ is isomorphic to the tensor product $H \otimes H$ as bialgebras. The isomorphism $\theta : H \rtimes H \to H \otimes H$ is defined by $\theta(h \otimes g) = hg_1 \otimes g_2$ with inverse $h \otimes g \mapsto hS(g_1) \otimes g_2$.

Using this isomorphism, we compute the maps on the second form of $H \otimes H$ to be

$$s(h \otimes g) = \epsilon(h)g, \quad t(h \otimes g) = h\epsilon(g), \quad i = \triangle,$$
$$m(h \otimes g, h' \otimes g') = hS(g)h' \otimes g', \quad S(h \otimes g) = g \otimes h.$$

The coactions on $H_1 = H \otimes H$ are computed to be $\rho_L = \Delta \otimes \mathrm{id}$ and $\rho_R = \mathrm{id} \otimes \Delta$. Therefore $H \square H = H \otimes \Delta(H) \otimes H$. Indeed, for $a \otimes h \otimes b \otimes h \in H \otimes H$, the condition to be in $H \square H$ is that

$$a \otimes h_1 \otimes h_2 \otimes b \otimes g = a \otimes h \otimes b_1 \otimes b_2 \otimes g.$$

By applying ϵ in the second position, we obtain

$$a \otimes h \otimes b \otimes g = a \otimes \Delta(\epsilon(h)b) \otimes g.$$

This reasoning also works with general elements of $H \square H$ which are sums of elementary tensors.

One sees that on $H \square H$, the categorical composition m reads in this description

$$m(a \otimes h_1, h_2 \otimes g) = a \otimes S(h_1)h_2 \otimes g = \epsilon(h)\, a \otimes g.$$

This presentation makes it easy to verify the quantum interchange law directly (while we know that it must be true by the general theory).

This adjoint action produces thus a strict quantum 2-group, which is analogous to the crossed module of groups $\mathrm{id}_G : G \to G$. The 2-group version of it, as we have seen, is

$$G \times G \underset{t \quad i}{\overset{s}{\rightrightarrows}} G,$$

where $s(g,h) = h$, $i(h) = (1,h)$ and $t(g,h) = d(g)h = gh$. One may translate the isomorphism with the tensor product in the above example on the level of Hopf algebras to group level. Then we obtain here a new 2-group $G \times G \to G$ where now $i(g) = (g,g)$ is the diagonal and s and t are the projections onto the first and second components. The categorical composition m is then given by $m((h,g),(h',g')) = (h,g')$ whenever $t(h,g) = g = h' = s(h',g')$. This is a special case of the product defined on the whole $G \times G$ by $\tilde{m}((h,g),(h',g')) = (hg^{-1}h', g')$ as corresponds to the formula $m(h \otimes g, h' \otimes g') = hS(g)h' \otimes g'$ for the categorical composition above.

Another class of examples of strict quantum 2-groups can be obtained from an action of a cocommutative Hopf algebra H on a commutative Hopf algebra A, rendering it an H-module algebra and H-module coalgebra. In order to obtain from these data an action-cocommutative crossed module of Hopf algebras, one needs a morphism of Hopf algebras $d : A \to H$ intertwining the given action with the adjoint action while Condition (3) in Definition 4.7.1 imposes that the image of d acts trivially.

Example 4.7.6. As an example of the preceding construction, take $H_0 = H$ a finite-dimensional cocommutative Hopf algebra, trivial morphism $d = \epsilon 1$ and $A = H^*$. There is a canonical coadjoint action of H on A (dualization of the adjoint action) and $H_1 =$

$A \rtimes H = D(H)$ is the quantum double Hopf algebra of H. In this class of examples we have $s = t$ (because d is trivial) and

$$s(a \otimes h)\epsilon(a)h, \quad i(h) = 1 \otimes h, \quad m(a \otimes h, b \otimes g)\epsilon(h)ab \otimes g, \quad S(a \otimes h) = S(a) \otimes h$$

according to Theorem 4.7.3.

Even more explicitly, one may take a finite group G, $A = \mathbb{K}G$, the Hopf algebra of functions on G, and $H = \mathbb{K}[G]$ the group algebra of G. The action of G on A is the adjoint action on the argument, and d is taken to be trivial. Then $H_1 = D(G)$ is the quantum double, $t = s$ is the counit projection and

$$m(a \otimes h, b \otimes g) = ab \otimes g$$

for all $h, g \in G$ and all $a, b \in \mathbb{K}G$. We can think of elements A of $D(G)$ as $\mathbb{K}[G]$-valued functions on G, that is, $A = \sum_{g \in G} a_g g$ with $a_g \in \mathbb{K}G$. Then for two such functions,

$$m(A, B) = \left(\sum_{g \in G} a_g\right)B$$

with the pointwise product of functions on G.

Similarly, we take $A = \mathbb{K}SU_2$ and $H = U(\mathfrak{su}_2)$. This makes sense, because one sees the enveloping algebra H as distributions supported in $1 \in SU_2$ on the Lie group SU_2 which then evaluate on elements of A, thus both are in natural duality. The action of H on A simply induced by the infinitesimal action corresponding to the action of SU_2 on its function algebra. The map d is still taken to be trivial.

H_1 consists of products of on the one hand $\mathfrak{su}_2$-valued functions and on the other hand $\mathbb{K}$-valued functions both on SU_2. We have for the categorical composition

$$m(a \otimes h, b \otimes g) = 0$$

as soon as $h \in \mathfrak{su}_2$ for any $a, b \in \mathbb{K}SU_2$ and any $g \in U(\mathfrak{su}_2)$. In other words, if A is $\mathfrak{su}_2$-valued and B arbitrary, $m(A, B) = 0$, while if A is $\mathbb{K}$-valued and B arbitrary, $m(A, B) = AB$ the pointwise product of functions.

The last example has noncommutative and noncocommutative H_1 and nontrivial d.

Example 4.7.7. Let G be a finite abelian group G and denote by $\widehat{G}$ the Pontryagin dual $\widehat{G}$ of G, that is, the group of continuous group homomorphisms from G to the unit circle S^1. Let $A = \bigoplus_{u \in \widehat{G}} A_u$ be a commutative $\widehat{G}$-graded Hopf algebra, where we write f_u for the component of $f \in A$ which lies in A_u. A is regarded as the Hopf algebra of functions on a some group M. We take $H_0 = \mathbb{K}\widehat{G}$, the Hopf algebra of functions on $\widehat{G}$. Elements of H_0 are sums of delta functions δ_u with $u \in \widehat{G}$. The map $d : A \to \mathbb{K}\widehat{G}$ is induced by the projection onto the label $u \in \widehat{G}$. In terms of the group M, d is induced by a map $\widehat{d} : \widehat{G} \to M^G$ from $\widehat{G}$ to the G-invariants M^G of M, the assignment of a label. The $\mathbb{K}\widehat{G}$-action on A is given by the grading and A becomes in this way a $\mathbb{K}\widehat{G}$-module algebra and coalgebra.

Passing to the strict quantum 2-group, we obtain $H_0 = \mathbb{K}\widehat{G}$ and $H_1 = A \rtimes \mathbb{K}\widehat{G}$, with the product

$$(f \otimes \delta_u)(h \otimes \delta_v) = fh_{u-v} \otimes \delta_v$$

for all $f, h \in A$ and all $u, v \in \widehat{G}$. Denoting by $e \in M$ the identity of the group M, we obtain

$$s(f \otimes \delta_u) = f(e)\delta_u, \quad t(f \otimes \delta_u) = f(\widehat{d}(u))\delta_u,$$
$$m(f \otimes \delta_u, h \otimes \delta_v) = \delta_{u,0} fh \otimes \delta_v, \quad S(f \otimes \delta_u) = L_{\widehat{d}(u)^{-1}}(S(f)) \otimes \delta_u,$$

where $L_s(f)(r) = f(sr)$ for all $s, r \in M$ and all $f \in A$ is the left translation by $s \in M$ on A and the antipode S on A, defined by $S(f)(r) = f(r^{-1})$, is given by inversion on M. The left and right coactions of $\mathbb{K}\widehat{G}$ are given by right and left actions of $\widehat{G}$ and these are

$$(f \otimes \delta_v) \cdot u = L_{\widehat{d}(u)}(f) \otimes \delta_{v-u}, \quad u \cdot (f \otimes \delta_v) = f \otimes \delta_{v-u}.$$

If we define $\psi : \mathbb{K}\widehat{G} \otimes A \to A \otimes \mathbb{K}\widehat{G}$ by

$$\psi(f \otimes \delta_u) = \delta_u \otimes L_{\widehat{d}(u)}(f)$$

for all $u \in \widehat{G}$ and all $f \in A$, then we arrive after some computation at

$$H_1 \square H_1 = (\mathrm{id} \otimes \psi \otimes \mathrm{id}) \circ (A \otimes A \otimes \Delta(H_0)).$$

To give a concrete, but infinite example, one may take $A = \mathbb{C}SU_2$ with the usual generators a, b, c, d for the matrix entries (thus subject to the relation $ad - bc = 1$). We take $\widehat{G} = \mathbb{Z}$ and the grading that corresponds to the S^1-action on SU_2 is given by the action of $S^1 \subset SU_2$ by conjugation from the right. This gives $|a| = |d| = 0$, $|b| = -2$ and $|c| = 2$. The subgroup of fixed elements is S^1, which we describe by identifying a complex number ω of modulus 1 with the diagonal matrix $\mathrm{diag}(\omega, \omega^{-1})$. The map $\widehat{d}$ consists in fixing an $\omega \in S^1$ such that $\widehat{d}(n) = \omega^n \in S^1 \subset SU_2$. Then $t(f \otimes \Delta_n) = f(\omega^n)\Delta_n$ and $s(f \otimes \Delta_n) = f(e)\Delta_n$. Note that $\mathbb{K}\mathbb{Z}$ is not a Hopf algebra due to the infinite sum in its coproduct, but most of the other structures still make sense. There are standard methods to handle the coproduct (cf. theory of locally compact quantum groups), but this will lead us too far afield. This kind of example can be generalized to general compact Lie groups M with $G \subset M$ the maximal torus and $\widehat{d}$ defined by $\mathrm{rank}(M)$ elements of G.

4.8 Exercises

Exercise 4.8.1. Let H be a Hopf algebra and $A = (A, \mu_A, \eta_A)$ be an algebra with a left coaction $\rho_A : A \to H \otimes A$ of H. Recall that A is an H-comodule algebra if and only if the following two diagrams are commutative:

$$\begin{array}{ccc} A \otimes A & \xrightarrow{\mu_A} & A \\ \downarrow{\scriptstyle \rho_{A\otimes A}} & & \downarrow{\scriptstyle \rho_A} \\ H \otimes A \otimes A & \xrightarrow{\mathrm{id}_H \otimes \mu_A} & H \otimes A \end{array}$$

where $\rho_{A\otimes A} = (\mu_H \otimes \mathrm{id} \otimes \mathrm{id}) \circ (\mathrm{id} \otimes \tau \otimes \mathrm{id}) \circ (\rho_A \otimes \rho_A)$.

$$\begin{array}{ccc} \mathbb{K} & \xrightarrow{\eta_A} & A \\ \downarrow{\scriptstyle \rho_{\mathbb{K}}} & & \downarrow{\scriptstyle \rho_A} \\ H \otimes \mathbb{K} & \xrightarrow{\mathrm{id}_H \otimes \eta_A} & H \otimes A \end{array}$$

where $\rho_{\mathbb{K}} = (\eta_H \otimes \mathrm{id}_{\mathbb{K}}) \circ \triangle_{\mathbb{K}}$. Express these two properties in terms of elements.

Similarly, let $(C, \triangle_C, \epsilon_C)$ be a coalgebra with a left coaction $\rho_C : C \to H \otimes C$ by H. Recall that C is an H-comodule coalgebra if and only if the following two diagrams are commutative:

$$\begin{array}{ccc} C & \xrightarrow{\triangle_C} & C \otimes C \\ \downarrow{\scriptstyle \rho_C} & & \downarrow{\scriptstyle \rho_{C\otimes C}} \\ H \otimes C & \xrightarrow{\mathrm{id}_H \otimes \triangle_C} & H \otimes C \otimes C \end{array}$$

and

$$\begin{array}{ccc} C & \xrightarrow{\epsilon_C} & \mathbb{K} \\ \downarrow{\scriptstyle \rho_C} & & \downarrow{\scriptstyle \rho_{\mathbb{K}}} \\ H \otimes C & \xrightarrow{\mathrm{id}_H \otimes \epsilon_C} & H \otimes \mathbb{K} \end{array}$$

where $\rho_{\mathbb{K}} := (\eta_H \otimes \mathrm{id}_{\mathbb{K}}) \circ \triangle_{\mathbb{K}}$. Express these two properties in terms of elements.

Exercise 4.8.2. This exercise gives some elementary examples of module-algebras and comodule-coalgebras.

(a) Show that any coalgebra is a comodule-coalgebra with respect to the coaction given by the coproduct and similarly that every algebra is a module-algebra for the action given by the product.

(b) Show that any algebra A is a module-algebra over the enveloping algebra of the Lie algebra $\mathfrak{Der}(A)$ of derivations of the associative algebra A and over the bialgebra $\mathbb{K}[G]$ where $G = \mathrm{Aut}(A)$ is the group of algebra automorphisms of A.

(c) Let H be a Hopf algebra. Show that the adjoint comodule structure defined by

$$\rho_H : H \otimes H \otimes H, \quad h \mapsto h_1 S(h_3) \otimes h_2$$

endows H with the structure of an H-comodule-coalgebra.

Exercise 4.8.3. In this exercise, we consider the adjoint action of a Hopf algebra H on itself, that is,

$$\mathrm{ad} : H \otimes H \to H, \quad h \otimes a \mapsto h_1 a S(h_2).$$

(a) Show that H is an H-module algebra and coalgebra in case H is cocommutative. Find an example to illustrate that in general H is not an H-module coalgebra.
(b) Show that $\mathrm{ad}_h(S(a)) \neq S(\mathrm{ad}_h(a))$ implies once again that in general H is not an H-module algebra and coalgebra.

Exercise 4.8.4. Show that the Yetter–Drinfeld modules in the monoidal category of sets (with the cartesian product) are exactly the *augmented racks*,that is, given by the data of a group G, a G-set X and a map $p : X \to G$ such that for all $x \in X$ and all $g \in G$

$$p(g \cdot x) = gp(x)g^{-1}.$$

Exercise 4.8.5. Suppose given a 2-group, that is, the data of two groups G_0 and G_1 together with group homomorphisms $s, t : G_1 \to G_0$, $i : G_0 \to G_1$, and $m : G_1 \times_{G_0} G_1 \to G_1$. Show that the 2-group $s, t : G_1 \to G_0$ is in particular a groupoid, that is, a category where every arrow is invertible.

4.9 Bibliographical notes

Our standard reference for Hopf algebras is Kassel's book [58]. We took, for example, the notions of (co)module-(co)algebra form there. The definition of a crossed module of Hopf algebras is taken from Frégier–Wagemann [31]. To our knowledge, the earliest definition of a crossed module of Hopf algebras is due to Fernandez Vilaboa, Lopez Lopez and Villanueva Novoa [30]. Lemma 4.1.1 is taken from [58].

Sections 4.2, 4.3 and 4.4 are taken from [31]. Sections 4.5.1 and 4.5.2 are taken from Krähmer–Wagemann [62]. Section 4.5.3 is new.

Section 4.6.1 is taken from Radford's article [89]. Sections 4.6.2 and 4.6.3 are taken from [30]. Section 4.7 is taken from Majid's article [75].

5 Crossed modules with geometric structure

This chapter is concerned with crossed modules of groups and Lie algebras with geometric structure. First of all, we will concentrate on crossed modules of Lie groups and the elementary integration theory of crossed modules of Lie algebras into crossed modules of Lie groups. Then we will discuss crossed modules with other geometric structures such as crossed modules of Lie–Rinehart algebras and crossed modules in Malcev theory. All these subjects draw heavily on the previous chapters.

5.1 Crossed modules of Lie groups

In the same way as crossed modules of groups are linked to 3-cohomology, crossed modules of Lie groups are linked to 3-cohomology. Precautions have to be taken though, because in order to assign 3-cocycles to crossed modules of Lie groups, the latter have to be split as sequences of manifolds. In order to weaken this very serious constraint, we will work with locally split sequences of manifolds (i. e., locally trivial fiber bundles) and consequently with locally smooth 3-cocycles. The definition of locally smooth cohomology is recalled in Appendix D. The first section draws heavily on Chapter 1 concerning the algebraic side.

All Lie groups in this section have underlying finite dimensional, second countable, separated smooth (i. e., $\mathcal{C}^{\infty}$) manifolds, except the three examples in the section about integration. We will suppose that all Lie groups of dimension greater or equal to one are connected. We suppose that the base field $\mathbb{K}$ is $\mathbb{R}$ or $\mathbb{C}$ in all geometric situations in this chapter.

5.1.1 Smooth outer actions

Let L be a Lie group. The group of all Lie group automorphisms of L is denoted $\mathrm{Aut}(L)$. Conjugation by elements of L gives a map $C_L : L \to \mathrm{Inn}(L)$ onto the normal subgroup of inner automorphisms $\mathrm{Inn}(L)$, which consists of automorphisms which are conjugations conj_l, that is, $l' \mapsto \mathrm{conj}_l(l') := l l' l^{-1}$, by elements l of L. The corresponding quotient is denoted $\mathrm{Out}(L)$.

In Appendix D we have introduced locally smooth cochains and locally smooth cohomology with values in a smooth G-module A associated to a Lie group G. In this chapter, we will also need non-abelian cochains and cohomology groups, that is, having values in a non-abelian group L which carries a G-action. These are not defined in all degrees, but in degree 1 everything works out nicely.

Definition 5.1.1. Let G and L be Lie groups and $S : G \to \mathrm{Aut}(L)$ be a homomorphism defining a smooth action of G on L. A locally smooth map $f : G \to L$ is called a 1-*cocycle*

https://doi.org/10.1515/9783110750959-005

if for all $g, g' \in G$,

$$f(gg') = f(g)S(g)(f(g')).$$

The set of (non-abelian, locally smooth) 1-cocycles with respect to the smooth action S is denoted $Z^1_s(G, L)_S$. The group $L = C^0_s(G, L)$ acts on $Z^1_s(G, L)_S$ by setting for all $g \in G$ and all $l \in L$,

$$(l \cdot f)(g) := lf(g)S(g)(l)^{-1},$$

and the set of orbits of L in $Z^1_s(G, L)_S$ is the cohomology set

$$H^1_s(G, L)_S := Z^1_s(G, L)_S \,/\, L.$$

Exactly as in Chapter 1, we will need outer actions in order to discuss crossed modules. In the present locally smooth setting, the corresponding notion is given in the following definition.

Definition 5.1.2. Let G and L be Lie groups. We define $C^1_s(G, \mathrm{Aut}(L))$ as the set of all maps $S : G \to \mathrm{Aut}(L)$ with $S(1) = \mathrm{id}_L$ and such that there exists an open identity neighborhood $U \subset G$ such that the map

$$U \times L \to L, \quad (g, l) \mapsto S(g)(l)$$

is smooth. We call a map $S \in C^1_s(G, \mathrm{Aut}(L))$ a *smooth outer action* of G on L if there exists $\beta : G^2 \to L$, smooth in an identity neighborhood of G^2, such that for all $g, g' \in G$,

$$S(g)S(g')S(gg')^{-1} = \mathrm{conj}_{\beta(g,g')}. \tag{5.1}$$

Note that the equality takes place in $\mathrm{Aut}(L)$. Observe that this is the locally-smooth version of the non-abelian 2-cocycles in Definition 1.4.4. On the set of outer actions, we define as before an equivalence relation by

$$S \sim S' \Leftrightarrow \exists h \in C^1_s(G, L) : \; S' = (C_L \circ h) \cdot S.$$

More explicitly, S is *equivalent* to S' if there exists a locally smooth 1-cochain h such that for all $g \in G$, $S'(g) = \mathrm{conj}_{h(g)} S(g)$, that is, in case $S'(g)S(g)^{-1}$ is conjugation by an element $h(g)$.

We write $[S]$ for an equivalence class of outer actions represented by S and call $[S]$ a *smooth G-kernel*.

Recall from Chapter 1 the notation $\delta_S : G \times G \to L$,

$$\delta_S(g, g') := S(g)S(g')S(gg')^{-1}.$$

Then equation (5.1) may be written as $\delta_S = C_L \circ \beta$ or $\delta_S = \operatorname{conj} \beta$. For $f \in C^1_s(G, L)$, we also define $\delta_f : G \times G \to L$ by $\delta_f(g, g') := f(g)f(g')f(gg')^{-1}$ for all $g, g' \in G$. With respect to the pointwise product in $C^1_s(G, L)$, we then have for $h, f \in C^1_s(G, L)$:

$$\begin{aligned} \delta_{hf}(g, g') &= h(g)f(g)h(g')f(g')f(gg')^{-1}h(gg')^{-1} \\ &= h(g)\operatorname{conj}_{f(g)}(h(g'))\delta_f(g, g')h(gg')^{-1}. \end{aligned} \tag{5.2}$$

Remark 5.1.3. (a) A smooth outer action $S : G \to \operatorname{Aut}(L)$ need not be a group homomorphism, but $\delta_S = C_L \circ \beta$ implies that S induces a group homomorphism $s = Q_L \circ S : G \to \operatorname{Out}(L)$, where $Q_L : \operatorname{Aut}(L) \to \operatorname{Out}(L)$ denotes the quotient map. Clearly, s depends only on the class $[S]$ of S, because $Q_L \circ C_L = 1$, the constant map.

(b) If G and L are discrete groups, then for each homomorphism $s : G \to \operatorname{Out}(L)$, there exists a map $S : G \to \operatorname{Aut}(L)$ with $S(1) = \mathrm{id}_L$ and $Q_L \circ S = s$ and a map $\beta : G \times G \to L$ with $\delta_S = C_L \circ \beta$. In this case, all outer actions are smooth, the smoothness conditions on S and β are vacuous and $S \sim S'$ is equivalent to $Q_L \circ S = Q_L \circ S'$. As a consequence, a smooth G-kernel is in this case just a group homomorphism $s : G \to \operatorname{Out}(L)$, that is, an ordinary G-kernel.

(c) If $S \in C^1_s(G, \operatorname{Aut}(L))$ is a homomorphism of groups, then we may choose $\beta = 1$, the constant map, and S defines a smooth action of G on L. Indeed, by definition the action is smooth on a set of the form $U \times L$. Smoothness on all of $G \times L$ follows thus by left-translation by $g' \in G$ using the fact that all automorphisms $S(g)$ are smooth and thanks to $S(g'g) = S(g')S(g)$.

(d) Suppose $S_1, S_2 \in C^1_s(G, \operatorname{Aut}(L))$ and that the map

$$U \times L \to L^2, \quad (g, l) \mapsto (S_1(g)(l), S_2(g)(l))$$

is smooth. Then the map $U \times L \to L$, $(g, l) \mapsto S_1(g)S_2(g)(l)$ is smooth as a composition of smooth maps. Therefore, $C^1_s(G, \operatorname{Aut}(L))$ is closed under pointwise products.

(e) If $S : G \to \operatorname{Aut}(L)$ is a smooth outer action, then we obtain in particular

$$C_L(\beta(g, g^{-1})) = \delta_S(g, g^{-1}) = S(g)S(g^{-1}),$$

which leads to

$$S(g)^{-1} = S(g^{-1})C_L(\beta(g, g^{-1})^{-1}).$$

Hence there exists also an identity neighborhood $U \subset G$ for which the map

$$U \times L \to L, \quad (g, l) \mapsto S(g)^{-1}(l) = S(g^{-1})\beta(g, g^{-1})^{-1} l \beta(g, g^{-1})$$

is smooth.

Recall the definition of an initial subgroup of a Lie group; see Definition II.6.1, page 345, in [82].

Definition 5.1.4. An injective morphism $i : H \hookrightarrow G$ of Lie groups is called an *initial Lie subgroup* if $T_1 i : T_1 H = \mathfrak{h} \to T_1 G = \mathfrak{g}$ is injective and for all smooth maps $f : M \to G$ from a smooth manifold M to G with $\mathrm{Im}(f) \subset H$, the corresponding map $i^{-1} \circ f : M \to H$ is smooth.

According to Rem. II.6.5 (b), page 346, in [82] in finite dimensions, all subgroups are initial.

Lemma 5.1.1. *Each smooth outer action $S : G \to \mathrm{Aut}(L)$ defines on the center $Z(L)$ the structure of a smooth G-module by $g \cdot z := S_Z(g)(z) := S(g)(z)$ for all $g \in G$ and all $z \in Z(L)$.*

Proof. An outer action is an action up to inner automorphisms. As inner automorphisms act trivially on the center $Z(L)$, S defines a genuine action of G on $Z(L)$. Let $U \subset G$ be an open identity neighborhood such that the map $U \times L \to L$, $(g, l) \mapsto S(g)(l)$ is smooth. As $Z(L)$ is an initial Lie subgroup of L, restriction and corestriction define a smooth map

$$U \times Z(L) \to Z(L), \quad g \cdot z = S_Z(g)(z).$$

Via S, each element $g \in G$ acts on $Z(L)$ as a Lie group automorphism, thus the smoothness of the action on all of $G \times L$ follows by left-translation from the smoothness on $U \times L$. □

With respect to this module structure, we have the usual definitions of locally smooth Lie group cohomology of G with values in $Z(L)$; see Appendix D. The cohomology groups are then denoted by

$$H^p_s(G, Z(L))_S = Z^p_s(G, Z(L))_S \,/\, d_S(C^{p-1}_s(G, Z(L))),$$

indicating the reference to the smooth outer action S. As the action of G on $Z(L)$ defined by the smooth outer action S depends only on the equivalence class $[S]$, we also write

$$Z^p_s(G, Z(L))_{[S]} := Z^p_s(G, Z(L))_S,$$

and

$$H^p_s(G, Z(L))_{[S]} := H^p_s(G, Z(L))_S.$$

Lemma 5.1.2. *Let G and L be Lie groups and consider the group $C^1_s(G, L)$ with respect to the pointwise multiplication. This group acts on the set $C^1_s(G, \mathrm{Aut}(L))$ via $h \cdot S := \mathrm{conj}_h \circ S$. This means explicitly for all $g \in G$: $(h \cdot S)(g) = \mathrm{conj}_{h(g)} \circ S(g)$. The group $C^1_s(G, L)$ acts thus on the product set*

$$C^1_s(G, \mathrm{Aut}(L)) \times C^2_s(G, L) \quad \text{via } h \cdot (S, \beta) := (h \cdot S, h * S\beta) \tag{5.3}$$

where for all $g, g' \in G$,

$$(h * S\beta)(g, g') := h(g)S(g)(h(g'))\beta(g, g')h(gg')^{-1}.$$

The stabilizer of (S, β) *is given by*

$$C_s^1(G, L)_{(S,\beta)} = Z_s^1(G, Z(L))_S,$$

which depends only on $[S]$, *but not on* β. *Moreover, the following assertions hold:*

(a) *The subset* $\{(S, \beta) \mid \delta_S = C_L \circ \beta\}$ *is invariant.*

(b) *If* $\delta_S = C_L \circ \beta$, *then* $\mathrm{Im}(d_S\beta) \subset Z(L)$ *holds for*

$$(d_S\beta)(g, g', g'') = S(g)(\beta(g', g''))\beta(g, g'g'')\beta(gg', g'')^{-1}\beta(g, g')^{-1}.$$

(c) *If* $\delta_S = C_L \circ \beta$ *and* $h \cdot (S, \beta) = (S', \beta')$, *then* $d_{S'}\beta' = d_S\beta$.

(d) *If* $\delta_S = C_L \circ \beta$, *then* $d_S\beta \in Z_s^3(G, Z(L))_S$.

(e) *For a smooth outer action* S *with* $\delta_S = C_L \circ \beta$, *the cohomology class* $[d_S\beta] \in H_s^3(G, Z(L))_S$ *depends only on the equivalence class* $[S]$.

Proof. Equation (5.3) defines a group action thanks to $(hh') \cdot S = h \cdot (h' \cdot S)$ and thanks to

$$\begin{aligned}
&(h *_{h' \cdot S} (h' *_S \beta))(g, g') \\
&= h(g)(h' \cdot S)(g)(h'(g'))(h' *_S \beta)(g, g')h(gg')^{-1} \\
&= h(g)h'(g)S(g)(h(g'))h'(g)^{-1}h'(g)S(g)(h'(g'))\beta(g, g')h'(gg')^{-1}h(gg')^{-1} \\
&= (hh')(g)S(g)(hh'(g'))\beta(g, g')(hh')(gg')^{-1} \\
&= ((hh') *_S \beta)(g, g'),
\end{aligned}$$

where $g, g' \in G$ and $h, h' \in C_s^1(G, L)$. To compute the stabilizer of (S, β), observe first that the condition $h \cdot S = S$ is equivalent to $h \in C_s^1(G, Z(L))$. Indeed, for all $g \in G$, $\mathrm{conj}_{h(g)} \circ S(g) = S(g)$ implies $\mathrm{conj}_{h(g)} = \mathrm{id}$, that is, $h(g) \in Z(L)$ for all $g \in G$. Then $h *_S \beta = \beta \cdot d_S h$, and this equals β if and only if h is a cocycle. Therefore, $C_s^1(G, L)_{(S,\beta)} = Z_s^1(G, Z(L))_S$.

(a) If (S, β) satisfies $\delta_S = C_L \circ \beta$, then we obtain for $h \in C_s^1(G, L)$ with Formula (5.2) the relation

$$\begin{aligned}
\delta_{h \cdot S}(g, g') &= \delta_{\mathrm{conj}_h \circ S}(g, g') \\
&= \mathrm{conj}_{h(g)} \mathrm{conj}_{S(g)}(\mathrm{conj}_{h(g')})\delta_S(g, g') \mathrm{conj}_{h(gg')^{-1}} \\
&= \mathrm{conj}_{h(g)} \mathrm{conj}_{S(g)(h(g'))} \mathrm{conj}_{\beta(g,g')} \mathrm{conj}_{h(gg')^{-1}} \\
&= \mathrm{conj}_{h(g)S(g)(h(g'))\beta(g,g')h(gg')^{-1}} \\
&= \mathrm{conj}_{h *_S \beta}(g, g').
\end{aligned}$$

(b) This follows from

$$\begin{aligned}
&\mathrm{conj}_{d_S\beta}(g,g',g'') \\
&= \mathrm{conj}_{S(g)(\beta(g',g''))\beta(g,g'g'')\beta(gg',g'')^{-1}\beta(g,g')^{-1}} \\
&= S(g)\,\mathrm{conj}_{\beta(g',g'')}\,S(g)^{-1}\,\mathrm{conj}_{\beta(g,g'g'')}\,\mathrm{conj}_{\beta(gg',g'')^{-1}}\,\mathrm{conj}_{\beta(g,g')^{-1}} \\
&= S(g)\delta_S(g',g'')S(g)^{-1}\delta_S(g,g'g'')\delta_S(gg',g'')^{-1}\delta_S(g,g')^{-1} \\
&= S(g)(S(g')S(g'')S(g'g'')^{-1})S(g)^{-1}S(g)S(g'g'')S(gg'g'')^{-1}S(gg'g'') \\
&\quad\cdot S(g'')^{-1}S(gg')^{-1}S(gg')S(g')^{-1}S(g)^{-1} \\
&= 1.
\end{aligned}$$

(c) In the following computation, we will use frequently that $d_S\beta$ and $d_{S'}\beta'$ have values in $Z(L)$ and thus commute with all elements of L. Remark that this implies that if $d_{S'}\beta'(g,g',g'') = \mathrm{conj}_l(A)$ for some term A and $l \in L$, then we may conclude $d_{S'}\beta'(g,g',g'') = A$. We use underbraces to indicate which parts of the formula cancel each other:

$$\begin{aligned}
&d_{S'}\beta'(g,g',g'') \\
&= S'(g)(\beta'(g',g''))\beta'(g,g'g'')\beta'(gg',g'')^{-1}\beta'(g,g')^{-1} \\
&= (\mathrm{conj}_{h(g)}\circ S(g))(\beta'(g',g''))\beta'(g,g'g'')\beta'(gg',g'')^{-1}\beta'(g,g')^{-1} \\
&= \underbrace{h(g)}_{1}\,S(g)[h(g')S(g')(h(g''))\beta(g',g'')\underbrace{h(g'g'')^{-1}}_{5}]\,\underbrace{h(g)^{-1}}_{2}\,\underbrace{h(g)}_{2} \\
&\quad\underbrace{S(g)(h(g'g''))}_{5}\,\beta(g,g'g'')\,\underbrace{h(gg'g'')^{-1}}_{3}\,\underbrace{h(gg'g'')}_{3}\,\beta(gg',g'')^{-1} \\
&\quad S(gg')(h(g''))^{-1}\,\underbrace{h(gg')^{-1}}_{4}\,\underbrace{h(gg')}_{4}\,\beta(g,g')^{-1}S(g)(h(g'))^{-1}\,\underbrace{h(g)^{-1}}_{1} \\
&= \underbrace{S(g)(h(g'))}_{1}\,S(g)\circ S(g')(h(g''))[S(g)(\beta(g',g''))\beta(g,g'g'')\beta(gg',g'')^{-1}] \\
&\quad S(gg')(h(g''))^{-1}\beta(g,g')^{-1}\,\underbrace{S(g)(h(g'))^{-1}}_{1} \\
&= S(gg')(h(g''))[d_S\beta(g,g',g'')\beta(g,g')]S(gg')(h(g''))^{-1}\beta(g,g')^{-1} \\
&= d_S\beta(g,g',g'').
\end{aligned}$$

(d) This has been shown in Lemma 1.2.3.

(e) This has been shown in Lemma 1.2.4. □

Definition 5.1.5. Let $S \in C^1_s(G, \mathrm{Aut}(L))$ be a smooth outer action and pick an $\beta \in C^2_s(G, L)$ such that $\delta_S = \mathrm{conj}_\beta$. The cohomology class

$$\chi_s(S) := [d_S\beta] \in H^3_s(G, Z(L))_S$$

does not depend on the choice of β and is constant on the equivalence class of S, so we may write $\chi_S([S]) = \chi_S(S)$. We call $\chi_S(S)$ the *characteristic class* of S.

5.1.2 Extensions of Lie groups

We will now discuss extensions of Lie groups. This section builds on Appendix A and on Section 1.4 where extensions of (discrete) groups were discussed.

Definition 5.1.6. Let G be a Lie group. A subgroup H of G is called a *split Lie subgroup* if it carries a Lie group structure for which the inclusion map $i_H : H \hookrightarrow G$ is a morphism of Lie groups and the right action of H on G defined by the restriction of the multiplication map of G to a map $G \times H \to G$ defines a smooth H-principal bundle. This means that the coset space G/H is a smooth manifold and that the quotient map $\pi : G \to G/H$ has smooth local sections.

Remark 5.1.7. Since the Lie algebra $\mathfrak{h}$ of a Lie subgroup H of a Lie group G has a closed complement in $\mathfrak{g}$, every Lie subgroup of a finite dimensional Lie group G is split. This need not be the case in infinite dimensions.

Recall the notion of an initial subgroup from Definition 5.1.4.

Lemma 5.1.3. *Each split Lie subgroup of G is initial.*

Proof. The condition that H is a split Lie subgroup implies that there exists an open subset U of some $\mathbb{R}^n$ and a smooth map $\sigma : U \to G$ such that

$$U \times H \to G, \quad (x, h) \mapsto \sigma(x)h$$

is a diffeomorphism onto an open subset of G. We conclude that for every manifold M, a map $f : M \to H$ is smooth if and only if it is smooth as a map to G, respectively, the open subset $\sigma(U)H$ of G. □

Another way of showing this is to remark that in finite dimensions, all subgroups are initial; see Remark II.6.5 (b), page 346, in [82].

Recall the notion of an extensions of groups in Definition 1.4.1. The following definition is a Lie group version of this.

Definition 5.1.8. An *extension of Lie groups* is a short exact sequence

$$1 \to L \xrightarrow{i} \widehat{G} \xrightarrow{q} G \to 1$$

of Lie group morphisms, for which $L \cong \mathrm{Ker}(q)$ is a split Lie subgroup. This means that $\widehat{G}$ is a smooth L-principal bundle over G and $G \cong \widehat{G}/L$. In the following, we shall identify L with the normal subgroup $i(L) \subset G$.

We call two extension $L \hookrightarrow \widehat{G}_1 \twoheadrightarrow G$ and $L \hookrightarrow \widehat{G}_2 \twoheadrightarrow G$ of the Lie group G by the Lie group L *equivalent* if there exists a Lie group morphism $\varphi : \widehat{G}_1 \to \widehat{G}_2$ such that the following diagram commutes:

$$\begin{array}{ccccccccc} 1 & \longrightarrow & L & \xrightarrow{i_1} & \widehat{G}_1 & \xrightarrow{q_1} & G & \longrightarrow & 1 \\ & & \downarrow{\scriptstyle \mathrm{id}_L} & & \downarrow{\scriptstyle \varphi} & & \downarrow{\scriptstyle \mathrm{id}_G} & & \\ 1 & \longrightarrow & L & \xrightarrow{i_2} & \widehat{G}_2 & \xrightarrow{q_2} & G & \longrightarrow & 1 \end{array}$$

It is easy to see that any such morphism φ is an isomorphism of Lie groups. We write $\mathrm{Ext}(G, L)$ for the set of equivalence classes of Lie group extensions of G by L.

We call an extension $q : \widehat{G} \to G$ with kernel $\mathrm{Ker}(q) = L$ *split* if there exists a Lie group morphism $\sigma : G \to \widehat{G}$ with $q \circ \sigma = \mathrm{id}_G$. In this case, the map $L \rtimes_S G \to \widehat{G}$, $(l, g) \mapsto l\sigma(g)$ is an isomorphism, where the semidirect product is defined by the homomorphism

$$S := C_L \circ \sigma : G \to \mathrm{Aut}(L),$$

where $C_L : \widehat{G} \to \mathrm{Aut}(L)$ is the conjugation action of $\widehat{G}$ on the normal subgroup L. The map S is a smooth action of G on L.

Example 5.1.9. In order to understand the need of locally smooth cocycles as opposed to (globally) smooth cocycles, let us examine the following example: It is well-known that the sequence

$$0 \to \mathbb{Z}/2\mathbb{Z} \to SU_2(\mathbb{C}) \to SO(3, \mathbb{R}) \to 1$$

expresses that $SU_2(\mathbb{C})$ is the universal covering of $SO(3, \mathbb{R})$, and thus $\pi_1(SO(3, \mathbb{R})) = \mathbb{Z}/2\mathbb{Z}$, because $SU_2(\mathbb{C}) \cong S^3$ is simply connected. It can be obtained by interpreting $SU_2(\mathbb{C})$ in terms of quaternions of norm 1, which act on $\mathbb{R}i \oplus \mathbb{R}j \oplus \mathbb{R}k$ preserving the norm.

The sequence is an extension of Lie groups, which cannot be described by a smooth 2-cocycle, because it is a nontrivial fibration and does not admit (global) smooth sections. It can, however, be described by a locally smooth 2-cocycle, and thus enters into the theory set up in this chapter.

Remark 5.1.10. We give a description of Lie group extensions $L \hookrightarrow \widehat{G} \twoheadrightarrow G$ in terms of data associated to G and L; cf. Remark 1.4.3. Let $q : \widehat{G} \to G$ be a Lie group extension of G by L. By assumption, the map q has a smooth local section. Hence there exists a locally smooth global section $\sigma : G \to \widehat{G}$, which is *normalized* in the sense that $\sigma(1) = 1$, that is, $\sigma \in C^1_s(G, \widehat{G})$. Then the map

$$\Phi : L \times G \to \widehat{G}, \quad (l, g) \mapsto l\sigma(g)$$

is a bijection, which restricts to a diffeomorphism on an identity neighborhood. In general, Φ is not continuous, but we may nevertheless use it to identify $\widehat{G}$ with the product set $L \times G$ endowed with the multiplication

$$(l,g)(l',g') = (lS(g)(l')\beta(g,g'),gg'), \tag{5.4}$$

where

$$S := C_L \circ \sigma : G \to \mathrm{Aut}(L) \quad \text{with } C_L : \widehat{G} \to \mathrm{Aut}(L), C_L(g)(l) = glg^{-1} \tag{5.5}$$

and

$$\beta : G \times G \to L, \quad (g,g') \mapsto \sigma(g)\sigma(g')\sigma(gg')^{-1}. \tag{5.6}$$

Note that β is smooth in an identity neighborhood and the map $G \times L \to L$, $(g,l) \mapsto S(g)(l)$ is smooth on a set of the form $U \times L$, where U is an identity neighborhood of G. The maps S and β satisfy the relations

$$\sigma(g)\sigma(g') = \beta(g,g')\sigma(gg'), \tag{5.7}$$
$$S(g)S(g') = C_L(\beta(g,g'))S(gg'), \quad \text{i.e. } \delta_S = C_L \circ \beta, \tag{5.8}$$

and

$$\beta(g,g')\beta(gg',g'') = S(g)(\beta(g',g''))\beta(g,g'g''), \quad \text{i.e. } d_S\beta = 1. \tag{5.9}$$

Definition 5.1.11. The elements of the set

$$Z_s^2(G,L) := \{(S,\beta) \in C_s^1(G,\mathrm{Aut}(L)) \times C_s^2(G,L) \mid \delta_S = C_L \circ \beta,\ d_S\beta = 1\}$$

are called *smooth factor systems* for the pair (G,L). Note that the set $Z_s^2(G,L)$ is the disjoint union

$$Z_s^2(G,L) = \bigcup_S Z_s^2(G,L)_S := \bigcup_S \{\beta \in C_s^2(G,L) \mid \delta_S = C_L \circ \beta,\ d_S\beta = 1\}.$$

In this way, we will view in this non-abelian framework 2-cocycles as pairs (S,β) rather than merely as maps β. In case L is abelian one can fix an S and deal with the sets $Z_s^2(G,L)_S$ individually, but this is impossible for non-abelian L.

Lemma 5.1.4. *Let $S \in C^1(G,\mathrm{Aut}(L))$ and $\beta \in C^2(G,L)$ with $\delta_S = \mathrm{conj}_\beta$. The set $L \times G$ becomes a group with respect to the product*

$$(l,g)(l',g') = (lS(g)(l')\beta(g,g'),gg')$$

if and only if $d_S\beta = 1$.

Proof. This has been proven in Lemma 1.4.1. □

The preceding discrete lemma leads in the Lie group framework to the following.

Proposition 5.1.5. *(a) If (S,β) is a smooth factor system, then $L \times_{(S,\beta)} G$ carries a unique structure of a Lie group for which the map*

$$L \times G \to L \times_{(S,\beta)} G, \quad (l,g) \mapsto (l,g)$$

is smooth on a set of the form $L \times U$ where U is an open identity neighborhood in G. Then

$$q : L \times_{(S,\beta)} G \to G, \quad (l,g) \mapsto g$$

is a Lie group extension of G by L.

(b) Each Lie group extension $q : \widehat{G} \to G$ of G by L gives rise to smooth factor systems by choosing a locally smooth normalized section $\sigma : G \to \widehat{G}$ and defining $(S,\beta) := (C_L \circ \sigma, \delta_\sigma)$. Then the given extension $q : \widehat{G} \to G$ is equivalent to $L \times_{(S,\beta)} G$.

Proof. The proof uses elements already proven in Chapter 1 in the discrete case.

(a) Let $U_G \subset G$ be an open symmetric 1-neighborhood such that the maps

$$U_G \times L \to L, \quad (g,l) \mapsto S(g)(l) \quad \text{and} \quad \omega|_{U_G \times U_G}$$

are smooth. We consider the set

$$U := L \times U_G = q^{-1}(U_G) \subset \widehat{G} := L \times_{(S,\beta)} G.$$

Then $U = U^{-1}$ because q is a group homomorphism (recall that

$$(l,g)^{-1} = (S(g)^{-1}(l^{-1}\beta(g,g^{-1})^{-1}), g^{-1})$$

from Lemma 1.4.1). We endow U with the product manifold structure inherited from $L \times U_G$. Since the multiplication $m_G|_{U_G \times U_G} : U_G \times U_G \to G$ is continuous, there exists an open symmetric identity neighborhood $V_G \subset U_G$ with $V_G V_G \subset U_G$. Then the set $V := L \times V_G$ is an open subset of U such that the multiplication map

$$V \times V \to U, \quad ((l,g),(l',g')) \mapsto (lS(g)(l')\beta(g,g'), gg')$$

is smooth. The inversion

$$U \to U, \quad (l,g) \mapsto (\beta(g^{-1},g)^{-1} S(g^{-1})(l^{-1}), g^{-1})$$

(cf. the proof of Lemma 1.4.1) is smooth. For $(l,g) \in \widehat{G}$, let $V_g \subset U_G$ be an open identity neighborhood with $c_g(V_g) \subset U_G$. Then $c_{(l,g)}(q^{-1}(V_g)) \subset U$ by the explicit formula for conjugation, that is,

$$(l,g)(l',g')(l,g)^{-1} = (lS(g)(l')\beta(g,g')\beta(gg'g^{-1},g)^{-1} S(gg'g^{-1})^{-1}(l^{-1}), gg'g^{-1}),$$

as can be easily deduced from Lemma 1.4.1. That the conjugation map $c_{(l,g)}$: $q^{-1}(V_g) \to U$ is smooth in an identity neighborhood follows immediately from the special case

$$(1,g)(l',g')(1,g)^{-1} = (S(g)(l')\beta(g,g')\beta(gg'g^{-1},g)^{-1}, gg'g^{-1})$$

of the preceding formula, because the function $g \mapsto \beta(g,g')\beta(gg'g^{-1},g)^{-1}$ is smooth in an identity neighborhood thanks to the following lemma as G is connected. Finally, Theorem D.2.1 in Appendix D implies that $\widehat{G}$ carries a unique Lie group structure for which the inclusion map $U = L \times U_G \hookrightarrow \widehat{G}$ is a local diffeomorphism onto an identity neighborhood of $\widehat{G}$. It is clear that with respect to this Lie group structure on $\widehat{G}$, the map $q : \widehat{G} \to G$ is a smooth L-principal bundle because the map $V_G \to \widehat{G}$, $g \mapsto (1,g)$ defines a section of q, which is smooth on an identity neighborhood in G (that may be smaller that V_G).

(b) Assume, conversely, that $q : \widehat{G} \to G$ is a Lie group extension of G by L. Then there exists an open 1-neighborhood $U_G \subset G$ and a smooth section $\sigma : U_G \to \widehat{G}$ of the map $q : \widehat{G} \to G$. We extend σ arbitrarily to a global section $G \to \widehat{G}$ and put $(S,\beta) := (C_L \circ \sigma, \delta_\sigma)$. Now (S,β) is a smooth factor system, so that we can use Lemma 5.1.4 and (a) to obtain a Lie group structure on $L \times_{(S,\beta)} G$ for which there exists an identity neighborhood $V_G \subset G$ such that the multiplication map

$$L \times V_G \to L \times_{(S,\beta)} G, \quad (l,v) \mapsto (l,1)(1,v) = (l,v)$$

is smooth. This implies that the group isomorphism $L \times_{(S,\beta)} G \to \widehat{G}$ is a local diffeomorphism, hence an isomorphism of Lie groups. □

It remains to prove that for connected Lie groups (all our Lie groups of dimension ≥ 1 are supposed to be connected!) the map $g \mapsto \beta(g,g')\beta(gg'g^{-1},g)^{-1}$ is smooth in an identity neighborhood of G.

Lemma 5.1.6. *Let G be a (connected) Lie group and suppose $S \in C^1_s(G, \mathrm{Aut}(L))$ and $\beta \in C^2_s(G,L)$ satisfy*

$$\delta_S = \mathrm{conj}_\beta, \quad \text{and} \quad d_S\beta = 1.$$

Then the map $\beta_g : g' \mapsto \beta(g,g')\beta(gg'g^{-1},g)^{-1}$ is smooth in an identity neighborhood of G, and in particular $(S,\beta) \in Z^2_s(G,L)$.

Proof. That the map $\beta_g : g' \mapsto \beta(g,g')\beta(gg'g^{-1},g)^{-1}$ is smooth in an identity neighborhood of G means that conjugation with $(1,g)$ is smooth in an identity neighborhood, because

$$(1,g)(l',g')(1,g)^{-1} = (S(g)(l')\beta_g(g'), gg'g^{-1}).$$

The set of all elements $x \in L \times_{(S,\beta)} G$ for which the conjugation c_x is smooth in an identity neighborhood is a subsemigroup containing L and $\{1\} \times W$ for some identity neighborhood W in G. If G is connected, it is generated by W as a semigroup, so that $L \times_{(S,\beta)} G$ is generated by $L \times W$. Therefore, the map $\beta_g : g' \mapsto \beta(g,g')\beta(gg'g^{-1},g)^{-1}$ is smooth in an identity neighborhood of G. □

The following discussion is very close to the discrete case developed in Section 1.4. In particular, the following proposition is a locally smooth version of Lemma 1.4.2.

Proposition 5.1.7. *For two smooth factor systems $(S,\beta), (S',\beta') \in Z_s^2(G,L)$, the following are equivalent:*
(a) *$L \times_{(S,\beta)} G$ and $L \times_{(S',\beta')} G$ are equivalent extensions of G by L.*
(b) *There exists an $h \in C_s^1(G,L)$ with $h \cdot (S,\beta) = (S',\beta')$.*

If these conditions are satisfied, then the map

$$\varphi : L \times_{(S,\beta)} G \to L \times_{(S',\beta')} G, \quad (l,g) \mapsto (lh(g),g)$$

is an equivalence of extensions and all equivalences of extensions $L \times_{(S,\beta)} G \to L \times_{(S',\beta')} G$ are of this form.

Proof. The proof of Lemma 1.4.2 is easily adapted. The constructed bijective map φ is smooth in an identity neighborhood which implies that it is an isomorphism of Lie groups. □

Corollary 5.1.8. *The map*

$$Z_s^2(G,L) \to \mathrm{Ext}(G,L), \quad (S,\beta) \mapsto [L \times_{(S,\beta)} G]$$

induces a bijection

$$H_s^2(G,L) := Z_s^2(G,L)/C_s^1(G,L) \to \mathrm{Ext}(G,L).$$

The preceding proposition implies in particular that

$$L \times_{(S,\beta)} G \sim L \times_{(S',\beta')} G$$

implies $[S] = [S']$, that is, that equivalent extensions correspond to the same smooth G-kernel. In the following, we will write $\mathrm{Ext}(G,L)_{[S]}$ for the set of equivalence classes of N-extensions of G corresponding to the G-kernel $[S]$.

Theorem 5.1.9. *Let S be an outer action of G on L with $\mathrm{Ext}(G,L)_{[S]} \neq \emptyset$. Then each extension class in $\mathrm{Ext}(G,L)_{[S]}$ can be represented by a Lie group of the form $L \times_{(S,\beta)} G$. Any other Lie group extension $L \times_{(S,\beta')} G$ representing an element of $\mathrm{Ext}(G,L)_{[S]}$ satisfies*

$$\beta'\beta^{-1} \in Z^2(G,Z(L))_{[S]},$$

and the groups $L \times_{(S,\beta)} G$ and $L \times_{(S,\beta')} G$ define equivalent extensions if and only if

$$\beta'\beta^{-1} \in B^2(G,Z(L))_{[S]}.$$

Proof. The proof is essentially like the proof of Theorem 1.4.4. □

Corollary 5.1.10. *For a G-kernel* $[S]$ *such that* $\mathrm{Ext}(G,L)_{[S]} \neq \emptyset$, *the map*

$$\begin{aligned} H^2(G,Z(L)) \times \mathrm{Ext}(G,L)_{[S]} &\to \mathrm{Ext}(G,L)_{[S]} \\ (\epsilon, [L \times_{(S,\beta)} G]) &\mapsto [L \times_{(S,\beta\cdot\epsilon)} G] \end{aligned}$$

is a well-defined simply transitive action.

Remark 5.1.12. In the special case where $L = A$ is an abelian Lie group, we recover the classical theory. Indeed, conjugation is trivial on A and a smooth factor system (S,β) for the pair (G,A) consists of a smooth action $S : G \to \mathrm{Aut}(A)$ and an element $\beta \in C_s^2(G,A)$. In this case, d_S is the usual group coboundary operator corresponding to the (smooth) action of G on A. Therefore, (S,β) defines a Lie group $A \times_{(S,\beta)} G$ if and only if $d_S\beta = 0$, that is, β is a smooth 2-cocycle. The corresponding Lie group can be written $A \times_\beta G$ with the multiplication

$$(a,g)(a',g') = (a + g\cdot a' + \beta(g,g'), gg').$$

Further $S \sim S'$ if and only if $S = S'$. Hence a smooth G-kernel $[S]$ is the same as a smooth G-module structure on A and $\mathrm{Ext}(G,A)_S$ is the set of equivalence classes of A-extensions of G with underlying G-module structure given by S. Corollary 5.1.10 reduces then to the classical bijection

$$H_s^2(G,A) \to \mathrm{Ext}(G,A)_S, \quad [\beta] \mapsto [A \times_\beta G].$$

In the following theorem, we obtain a cohomological criterion for the nonemptiness of $\mathrm{Ext}(G,L)_S$.

Theorem 5.1.11. *Let S be a smooth outer action of G on L. Then we have*

$$[d_S\beta] = 1 \quad \Leftrightarrow \quad \mathrm{Ext}(G,L)_{[S]} \neq \emptyset.$$

Proof. If there exists an extension corresponding to $[S]$, then we may assume by Proposition 5.1.5 that it is of the form $L \times_{(S,\beta)} G$ for some factor system (S,β). But then we have $d_S\beta = 1$, and thus $[d_S\beta] = 1$.

Suppose conversely that S is an outer action with $[d_S\beta] = 1$. Then there exists $\beta \in C_s^2(G,L)$ with $\delta_S = \mathrm{conj}_\beta$ and some $h \in C_s^2(G,Z(L))_S$ with $d_S\beta = d_S h^{-1}$ (because $[d_S\beta] = 1$), so that $\beta' := \beta \cdot h \in C^2(G,L)$ satisfies $d_S\beta' = d_S\beta d_S h = 1$ and $\delta_S = \mathrm{conj}_{\beta'}$ (because the factor h does not play a role in the conjugation). Hence (S,β') is a smooth factor system and Proposition 5.1.5 implies the existence of a group extension $L \times_{(S,\beta')} G$ of G by L corresponding to $[S]$. □

We end this section with a discussion of *split Lie group extensions*, that is, those extensions $q : \widehat{G} \to G$ for which there exists a smooth group homomorphism $\sigma : G \to \widehat{G}$ with $q \circ \sigma = \mathrm{id}_G$. This implies that $\widehat{G}$ is isomorphic to the semidirect product $L \rtimes_S G$ with respect to the smooth action $S := \mathrm{conj}_\sigma$ of G on L.

Proposition 5.1.12. *The extension $\widehat{G} = L \times_{(S,\beta)} G$ splits if and only if there exists an $h \in C^1_s(G, L)$ such that*

$$h * \beta = 1.$$

In this case, S is a smooth action of G on L. With respect to this action, $h * \beta = 1$ is equivalent to $d_S h = \beta^{-1}$.

Consider the set

$$Z^1_s(G, L, Z(L))_S := \{f \in C^1_s(G, L) \mid \operatorname{Im}(d_S f) \subset Z(L)\}$$

and the action of G on $\operatorname{Aut}(L)$ by $S'(g)(\varphi) := S(g) \circ \varphi \circ S(g)^{-1}$. Then the abelian group $C^1_s(G, Z(L))$ acts on $C^1_s(G, L)$ by pointwise multiplication and we have an injective map

$$Z^1_s(G, L, Z(L))_S / C^1_s(G, Z(L)) \to Z^1(G, \operatorname{Aut}(L))_{S'}, \quad [f] \mapsto C_L \circ f$$

of the set of $C^1_s(G, Z(L))$-orbits in $Z^1_s(G, L, Z(L))_S$ onto the set of all 1-cocycles $G \to C_L(L) \subset \operatorname{Aut}(L)$ that can be lifted to elements of $C^1_s(G, L)$.

Moreover,

$$[\beta] \cdot [L \times_{(S,1)} G] = [L \times_{(S,\beta)} G] \in \operatorname{Ext}(G, L)_{[S]} \quad \text{splits} \quad \Leftrightarrow [\beta^{-1}] \in \operatorname{Im}(\delta),$$

where δ is the natural map

$$\delta : Z^1_s(G, L, Z(L))_S / C^1_s(G, Z(L)) \to H^2_s(G, Z(L))_S, \quad [f] \mapsto [d_S f].$$

Proof. Since the orbit of the factor system (S, β) under the group $C^1_s(G, L)$ consists of all other factor systems describing the same extension, that is, corresponding to different choices of a section $G \to \widehat{G}$ (see Proposition 5.1.5), the extension $\widehat{G}$ splits if and only if this orbit contains a factor system $(S', \beta') = h \cdot (S, \beta)$ for which the corresponding section $\sigma' : G \to \widehat{G}$, $g \mapsto (h(g), g)$ is a group homomorphism, that is, $\delta_{\sigma'} = 1$. Hence the first assertion follows from $\delta_{\sigma'} = h *_S \beta$.

Suppose now that β has values in $Z(L)$. Then $\delta_S = C_L \circ \beta = 1$ shows that S is a representation. Moreover, $h *_S \beta = d_S h \cdot \beta$ vanishes if and only if $d_S h = \beta^{-1}$.

To verify the remaining assertions, we first observe that for $f \in Z^1_s(G, L, Z(L))_S$ and $h \in C^1_s(G, Z(L))$ we have

$$\operatorname{Im}(d_S(fh)) = \operatorname{Im}(d_S(f)\, d_S(h)) \subset Z(L),$$

so that the subset $Z^1_s(G, L, Z(L))_S$ of $C^1_s(G, L)$ is invariant under the action of $C^1_s(G, Z(L))$ by pointwise multiplication. It is clear that for two elements $f, f' \in Z^1_s(G, L, Z(L))_S$ the condition $\operatorname{conj}_f = \operatorname{conj}_{f'}$ is equivalent to $f' \in f \cdot C^1_s(L, Z(L))$. This leads immediately to the asserted injectivity statement. Moreover, for each $f \in Z^1_s(G, L, Z(L))_S$ the map $\operatorname{conj}_f : G \to \operatorname{Aut}(L)$ is a 1-cocycle because $d_{S'}(C_L \circ f) = C_L \circ d_S f = 1$, and the latter relation shows that $Z_s^1(G, L, Z(L))_S$ is precisely the set of all elements of $C_s^1(G, L)$ for which $C_L \circ f = \mathrm{conj}_f$ is a 1-cocycle.

Finally, we observe that the map δ is well-defined, and that $[\beta^{-1}] \in \mathrm{Im}(\delta)$ is equivalent to the existence of some $h \in Z_s^1(G, L, Z(L))_S$ with $d_S h = \beta^{-1}$. Since β has values in $Z(L)$, the observations from above show that this is equivalent to the splitting of $L \times_{(S,\beta)} G$. □

Remark 5.1.13. Suppose that the extension $\widehat{G} = L \times_{(S,\beta)} G$ splits, so that we may assume without loss of generality that $\beta = 1$.

(a) For $h \in C_s^1(G, L)$ we have then

$$h \cdot (S, 1) = (h \cdot S, h *_S 1) = (h \cdot S, d_S h)$$

with $d_S h(g, g') := h(g)S(g)(h(g'))h(gg')^{-1}$. We conclude that

$$\{h \in C_s^1(G, L) \mid h *_S 1 = 1\} = Z_s^1(G, L)_S$$

is the set of 1-cocycles for the smooth action of G on L defined by the homomorphism $S : G \to \mathrm{Aut}(L)$. Therefore, the orbit of $(S, 1)$ under $C_s^1(G, L)$ may contain different elements of the form $(S', 1)$. The representations $S' : G \to \mathrm{Aut}(L)$ arising in this way have the form

$$S_f := \mathrm{conj}_f \cdot S \quad \text{for } f \in Z_s^1(G, L)_S.$$

If L is abelian, all representations S_f are the same, so that the $C_s^1(G, L)$-orbit of $(S, 1)$ contains no other element of the form $(S', 1)$. But if $f \in Z_s^1(G, L)_S$ is a cocycle whose values do not lie in the center $Z(L)$ of L, then $S_f \neq S$. Nevertheless, the two extensions $L \rtimes_S G$ and $L \rtimes_{S_f} G$ are equivalent.

(b) A condition which is slightly weaker than the condition $d_S h = 1$ is that $S_h = (C_L \circ h) \cdot S$ is a group homomorphism, which in view of $\delta_{h \cdot S} = C_L \circ (d_S h)$ (cf. proof of Lemma 5.1.2 (a)) is equivalent to $\mathrm{Im}(d_S h) \subset Z(L)$. This means that $C_L \circ h \in Z_s^1(G, \mathrm{Aut}(L))$ with respect to the action of G on $\mathrm{Aut}(L)$ given by $g \cdot \varphi := S(g)\varphi S(g)^{-1}$. Then $d_S d_S h = d_{S_h} d_S h = d_{S_h}(h *_S 1) = 1$ (Lemma 5.1.2 (d)) implies that $d_S h \in Z_s^2(G, Z(L))_S$, so that on the level of extension classes we have

$$[L \rtimes_S G] = [L \times_{(h \cdot S, d_S h)} G] = [d_S h] \cdot [L \times_{(S_h, 1)} G] = [d_S h] \cdot [L \rtimes_{S_h} G].$$

This shows that the semidirect products

$$[L \rtimes_S G] \quad \text{and} \quad [L \rtimes_{S_h} G]$$

are equivalent extensions of G by L if and only if the class $[d_S h] \in H_s^2(G, Z(L))_S$ vanishes.

Corollary 5.1.13. *The quotient group $L_{ad} := L/Z(L)$ carries a canonical Lie group structure for which $q_L : L \to L_{ad}$ defines a central extension of L_{ad} by $Z(L)$. Furthermore, let $S : G \to \mathrm{Aut}(L)$ be a smooth action of G on L and consider the induced action $\overline{S}$ of G on L_{ad}. Then there is an exact sequence of groups, respectively, pointed sets*

$$Z(L)^G \hookrightarrow L^G \to L_{\mathrm{ad}}^G \to H_s^1(G,Z(L))_S \to H_s^1(G,L)_S \to H_s^1(G,L_{\mathrm{ad}})_{\overline{S}} \xrightarrow{\delta} H_s^2(G,Z(L))_S,$$

where for $f \in Z_s^1(G,L_{\mathrm{ad}})_S$ we put

$$\delta([f]) = [d_S\widehat{f}]$$

for some $\widehat{f} \in Z_s^1(G,L,Z(L))_S$ satisfying $q_L \circ \widehat{f} = f$ and

$$\mathrm{Ext}(G,L)_{[S]}^{\mathrm{split}} = -\mathrm{Im}(\delta)\cdot[L \rtimes_S G],$$

where $\mathrm{Ext}(G,L)_{[S]}^{\mathrm{split}}$ is the set of split extension classes.

Proof. For the 7-term sequence and the description of the connecting homomorphism δ corresponding to the Lie group extension $Z(L) \hookrightarrow L \twoheadrightarrow L_{\mathrm{ad}}$, see Exercise 5.4.4.

If the cocycle $f : G \to L$ has values in $Z(L)$, then $f\cdot(S,1) = (S,1)$. Thus $Z_s^1(G,Z(L))_S$ corresponds to the stabilizer of $(S,1) \in Z_s^2(G,L)$ and the map

$$Z_s^1(G,L,Z(L))_S \to Z_s^2(G,L), \quad f \mapsto f\cdot(S,1)$$

factors through an injective map

$$Z_s^1(G,L,Z(L))/Z_s^1(G,Z(L))_S \cong Z_s^1(G,L_{\mathrm{ad}})_S, \quad [f] \mapsto f\cdot(S,1)$$

because for each $h \in Z_s^1(G,L_{\mathrm{ad}})_S$, there is an $\widehat{h} \in Z_s^1(G,L,Z(L))_S$ with $q_L \circ \widehat{h} = h$ and $\widehat{h}\cdot(S,1) = (\widehat{h}\cdot S, d_S\widehat{h})$.

Since $d_S\widehat{h} \in Z_s^2(G,Z(L))_S$, this factor system corresponds to the extension class

$$[d_S\widehat{h}]\cdot[L \rtimes_S G] = \delta(h)\cdot[L \rtimes_S G] \in \mathrm{Ext}(G,L)_{[S]}.$$

This class equals $[L \rtimes_S G]$ if and only if $\delta(h) = 0$, which is equivalent to $[h]$ being contained in the image of $H_s^1(G,L)_S$ under the connecting homomorphism δ. The remaining assertion follows from Remark 5.1.13 (b) and Proposition 5.1.12. □

5.1.3 Crossed modules of Lie groups

This section deals with the smooth version of a crossed module of groups, that is, crossed modules of Lie groups. All algebraic steps taken here are very close to the procedure in Chapter 1. The only change with respect to Chapter 1 is that we now have to justify smoothness conditions.

Definition 5.1.14. A morphism of Lie groups $\mu : M \to N$ together with a homomorphism $\widehat{S} : N \to \mathrm{Aut}(M)$ defining a smooth action $\widehat{S} : N \times M \to M$, $(n, m) \mapsto \widehat{S}(n)(m)$ of N on M is called a *crossed module of Lie groups* if the following conditions are satisfied:
(a) $\mu \circ \widehat{S}(n) = \mathrm{conj}_{\mu(n)} \circ \mu$ for all $n \in N$.
(b) $\widehat{S} \circ \mu = C_M : M \to \mathrm{Aut}(M)$ is the conjugation action.
(c) $\mathrm{Ker}(\mu)$ is a split Lie subgroup of M and $\mathrm{Im}(\mu)$ is a split Lie subgroup of N for which μ induces an isomorphism $M/\mathrm{Ker}(\mu) \to \mathrm{Im}(\mu)$.

Observe that conditions (a) and (b) express that $\mu : M \to N$ is in particular a crossed module of groups. In particular, the conditions (a) and (b) imply as before that $\mathrm{Im}(\mu) \subset N$ is a normal subgroup, that $\mathrm{Ker}(\mu) \subset M$ lies in the center $Z(M)$ of M and that N acts smoothly on $\mathrm{Ker}(\mu)$. Similarly, as in Chapter 1, standard examples of crossed modules of Lie groups include normal split subgroups and central extensions.

Like in Chapter 1.3, we change now our point of view and consider a crossed module as a central extension $\widehat{L} \to L$ of a normal split Lie subgroup $L \subset \widehat{G}$ of a Lie group $\widehat{G}$ for which the conjugation action of $\widehat{G}$ on L lifts to a smooth action on $\widehat{L}$.

Conversely, if $L \subset \widehat{G}$ is a split normal subgroup and $\mu : \widehat{L} \to L$ is a central extension of Lie groups of L by Z, then we have a natural smooth L-module structure on $\widehat{L}$ given by $l \cdot l' := \widehat{C}_L(l)(l') := ll'l^{-1}$. To obtain the structure of a crossed module for $\mu : \widehat{L} \to \widehat{G}$, one has to extend the action of L on $\widehat{L}$ to a smooth $\widehat{G}$-action $\widehat{G} \times \widehat{L} \to \widehat{L}$, $(g, l) \mapsto g \cdot l$ for which μ is equivariant.

From now on, we assume that this data is given. We will assume without loss of generality that $\widehat{L} = Z \times_\omega L$ for some locally smooth 2-cocycle $\omega \in Z^2_s(L, Z)$ and write the $\widehat{G}$-module structure as $(g, z) \mapsto g \cdot z$. Then the $\widehat{G}$-module structure on $\widehat{L} = Z \times_\omega L$ can be written as

$$g \cdot (z, l) = (g \cdot z + \theta(g, g \cdot l), g \cdot l), \tag{5.10}$$

where the function $\theta : \widehat{G} \to C^1_s(L, Z)$ is a map for which $\widetilde{\theta} : \widehat{G} \times L \to Z$, $(g, l) \mapsto \theta(g)(l)$ is smooth on an open neighborhood of $\widehat{G} \times \{1\}$ in $\widehat{G} \times L$. As in Lemma 1.3.1, the fact that equation (5.10) defines a group action of $\widehat{G}$ on $\widehat{L}$ means two things: First of all, θ must be a cocycle with respect to the natural action of $\widehat{G}$ on $C^1_s(L, Z)$ by $(g \cdot \alpha)(l) := g \cdot \alpha(g^{-1} l g)$. The second requirement is that with respect to the natural action of $\widehat{G}$ on $C^2_s(L, Z)$ by $(g \cdot \beta)(l_1, l_2) := g \cdot \beta(g^{-1} l_1 g, g^{-1} l_2 g)$ and with respect to the trivial action of L on Z, we have

$$d\theta(g) = g \cdot \omega - \omega = d_S(\omega)(g) \quad \text{for all } g \in \widehat{G}. \tag{5.11}$$

Therefore, the data we are now working with is given by $\omega \in Z^2_s(L, Z)$ with $\widehat{L} = Z \times_\omega L$ and $\theta \in Z^1_s(\widehat{G}, C^1_s(L, Z))_f$ for which the corresponding smooth action $\widehat{S}$ of $\widehat{G}$ on $\widehat{L}$ turns the homomorphism $\mu : \widehat{L} \to \widehat{G}$, $(z, l) \mapsto l$ into a crossed module.

Lemma 5.1.14. *For a cocycle $\theta : \widehat{G} \to C^1_s(L, Z)$ satisfying condition (5.11), the following are equivalent:*

(a) *The corresponding action of* $\widehat{G}$ *on* $\widehat{L}$ *(see (5.10)) is smooth.*
(b) *The function* $\widetilde{\theta} : \widehat{G} \times L \to Z$, $(g, l) \mapsto \theta(g)(l)$ *is smooth in an identity neighborhood and for each* $l \in L$ *the function*

$$\theta_l : \widehat{G} \to Z, \quad g \mapsto \theta(g)(g \cdot l) + \omega(g \cdot l, l^{-1})$$

is smooth in an identity neighborhood of $\widehat{G}$.

Proof. First of all, the action map $\widehat{G} \times \widehat{L} \to \widehat{L}$ is smooth in an identity neighborhood if and only if the map $\widetilde{\theta} : \widehat{G} \times L \to Z$ is smooth in an identity neighborhood. Indeed, if the action is smooth in an identity neighborhood, the function $(g, l) \mapsto \theta(g)(g \cdot l)$ has this property and the map $\widehat{G} \times L \to \widehat{G} \times L$, $(g, l) \mapsto (g, g^{-1} \cdot l)$ is a diffeomorphism fixing $(1, 1)$. Therefore, $\widetilde{\theta}$ is smooth in an identity neighborhood. The converse is clear.

As a second step, the action has smooth orbit maps (i. e., for every fixed $l \in \widehat{L}$, the map $\widehat{G} \to \widehat{L}$, $g \mapsto g \cdot l$) if and only if for all $l \in L$, the map θ_l is smooth in an identity neighborhood of $\widehat{G}$. Indeed, since $\widehat{G}$ acts by Lie group automorphisms, the orbit maps are smooth if and only if they are smooth in an identity neighborhood. It suffices to consider orbit maps of elements of the form $(0, l)$ because $\widehat{G}$ acts smoothly on Z. For $l \in L$, the orbit map is smooth in an identity neighborhood of $\widehat{G}$ if and only if the map $g \mapsto (g \cdot (0, l))(0, l)^{-1}$ is smooth in an identity neighborhood. The explicit expression is

$$\begin{aligned}(g \cdot (0, l))(0, l)^{-1} &= (\theta(g)(g \cdot l), g \cdot l)(-\omega(g, g^{-1}), g^{-1}) \\ &= (\theta(g, g \cdot l) - \omega(l, l^{-1}) + \omega(g \cdot l, l^{-1}), (g \cdot l)l^{-1}).\end{aligned}$$

As ω is a locally smooth 2-cocycle, the smoothness of the orbit maps is thus equivalent to the fact that

$$g \mapsto \theta(g, g \cdot l) + \omega(g \cdot l, l^{-1})$$

is smooth in an identity neighborhood for all $l \in L$.

Suppose in a last step that the automorphic action of $\widehat{G}$ on $\widehat{L}$ is smooth. Then by the first step $\widetilde{\theta}$ is smooth in an identity neighborhood and by the second step, the maps θ_l are smooth in an identity neighborhood for all $l \in L$. Conversely, if (b) is satisfied, each $g \in \widehat{G}$ acts by a group automorphism which is smooth in an identity neighborhood, hence by a Lie group automorphism, that is, the action is smooth. □

Remark 5.1.15. In fact, the requirement that the orbit maps are smooth (or, equivalently, the condition that the maps θ_l are smooth in an identity neighborhood for all $l \in L$) follows from the local smoothness of the action, because all our Lie groups of dimension ≥ 1 are supposed to be connected. Indeed, the set of all elements of $\widehat{L}$ with smooth orbit map is a subgroup containing Z. If the action is locally smooth, then this subgroup also contains an identity neighborhood, hence it contains the whole group $Z \times_\omega L$ by connectedness of L.

The natural action of L on $\widehat{L}$ is given by

$$l \cdot (z', l') := (z, l)(z', l')(z, l)^{-1} = (z + \omega(l, l') - \omega(ll'l^{-1}, l), ll'l^{-1}),$$

where we note that the right-hand side does not depend on $z \in Z$. Compare this to the definition of θ, that is,

$$g \cdot (z, l) = \big(g \cdot z + \theta(g, g \cdot l), g \cdot l\big).$$

As in equations (1.1), (1.2) and (1.3), this comparison leads to the condition that the restriction of θ to L must be $\widetilde{\omega}$:

Definition 5.1.16. We define

$$\widetilde{\omega}(l)(l') := \omega(l, l') - \omega(l, l^{-1}l'l),$$

and observe that (5.10) defines a smooth action extending the given action of L if $\theta|_L = \widetilde{\omega}$, that is, that the pair (ω, θ) is an element of the group

$$D := \big\{(\omega, \theta) \in Z^2_s(L, Z) \times Z^1_s(\widetilde{G}, C^1_s(L, Z)) \mid d\theta = d_S(\omega),\ \theta|_L = \widetilde{\omega}\big\}.$$

That the set of all these elements forms indeed a subgroup follows from Lemma 5.1.14. We also write

$$D(\omega) := \big\{\theta \in Z^2_s(\widetilde{G}, C^1_s(L, Z)) \mid (\omega, \theta) \in D\big\} = \mathrm{pr}_2(D)$$

for the set of all smooth actions of $\widetilde{G}$ on $\widehat{L}$ extending the given action of L.

The following lemma provides a description of the set $D(\omega)$ as a homogeneous space.

Lemma 5.1.15. *Let $q : \widehat{G} \to G := \widehat{G}/L$ denote the quotient map, $\widehat{L} = Z \times_\omega L$ a central Lie group extension of L by Z and a smooth $\widehat{G}$-action on $\widehat{L}$ turning $\mu : \widehat{L} \to L \subset \widehat{G}$ into a smooth crossed module. For any $(\omega, \theta) \in D$ and $\beta \in Z^1_s(G, Z^1_s(L, Z))_0$, we have*

$$\beta * (\omega, \theta) := \big(\omega, \theta + q^*\beta\big) \in D.$$

This defines for each $\omega \in \mathrm{pr}_1(D) \subset Z^2_s(L, Z)$ an action of $Z^1_s(G, Z^1_s(L, Z))_0$ on the set $D(\omega)$, which is simply transitive.

Proof. That $\beta * (\omega, \theta) \in D$ follows from

$$d(\theta + q^*\beta) = d\theta + dq^*\beta = d\theta + q^*(d\beta) = d\theta = d_S\omega.$$

If, conversely, $(\omega, \theta), (\omega, \theta') \in D$, then the difference $\theta' - \theta$ has values in $Z^1_s(L, Z)$, because $d\theta = d\theta' = d_S\omega$, and vanishes on L, because $\theta|_L = \theta'|_L = \widetilde{\omega}$. Hence the difference $\theta' - \theta$ is of the form $q^*\beta$ for some $\beta \in Z^1_s(G, Z^1_s(L, Z))_0$. This proves the lemma. □

As in Chapter 1.3, we go on and construct a map $Q : D \to H^3_s(G,Z)$ for $G := \widehat{G}/L$ such that $Q(\omega,\theta) = 0$ characterizes the extendability of ω to a cocycle $\bar{\omega} \in Z^2_s(\widehat{G},Z)$ (where we assume that we are given a smooth action of $\widehat{G}$ on Z) satisfying $\bar{\omega}|_{\widehat{G}\times L} = \widetilde{\omega}$. In the following, we fix a pair $(\omega,\theta) \in D$ and consider the corresponding smooth action of $\widehat{G}$ on $\widehat{L}$ given by

$$\widehat{S}(g)(z,l) := (g\cdot z + \theta(g)(g\cdot l), g\cdot l)$$

extending the given action on Z.

In order to define the map Q, let $\rho : G \to \widehat{G}$ be a locally smooth section with $\rho(1) = 1$ of the quotient map $\widehat{G} \to G$. Define then the lift of the smooth action by

$$S := \widehat{S}\circ\rho : G \to \mathrm{Aut}(\widehat{L}).$$

There exists a $\beta \in C^2_s(G,\widehat{L})$ with $q_L\circ\beta = \alpha$ where $q_L : \widehat{L} \to L$ is the quotient map. Then

$$\delta_S = \widehat{S}\circ\alpha = \widehat{S}\circ q_L\circ\beta = \mathrm{conj}\circ\beta,$$

where conj is the conjugation action in $\widehat{L}$. Thus S is a smooth outer action of G on $\widehat{L}$.

Then we have as before (cf. Proposition 1.3.2)

Proposition 5.1.16. *The map*

$$\widetilde{Q}(\omega,\theta,\rho,\beta) := d_S\beta : G^3 \to \widehat{L}, \quad (x,y,z) \mapsto S(x)(\beta(y,z))\beta(x,yz)\beta(xy,z)^{-1}\beta(x,y)^{-1}$$

has values in Z and is a 3-cocycle in $Z^3_s(G,Z)$. Its cohomology class $Q(\omega,\theta) := [d_S\beta] \in H^3(G,Z)$ does not depend on the choice of β and ρ.

Proof. We refer to Lemmas 1.2.2, 1.2.3 and 1.2.4, because the local smoothness conditions do not alter the reasoning. □

Lemma 5.1.17. *The map $Q : D \to H^3_s(G,Z)$ is a group homomorphism.*

Proof. Let $(\theta,\omega) \in D$. We choose as before $\beta \in C^2_s(G,\widehat{L})$ with $q_L\circ\beta = \alpha$, where ρ is a locally smooth section of the quotient map $\widehat{G} \to G$ and α its default to be a group homomorphism. We have to see how $\widetilde{Q}(\omega,\theta,\rho,\beta)$ depends on (ω,θ). In product coordinates $\widehat{L} = Z\times_\omega L$, any β with $q_L\circ\beta = \alpha$ can be written as $\beta = (\beta_Z,\alpha)$, independently of ω and θ.

Writing $g\cdot l = glg^{-1}$ for the conjugation action of $\widehat{G}$ on L, we now obtain

$$\begin{aligned} S(g)\cdot\beta(g',g'') &= \widehat{S}(\rho(g))(\beta_Z(g',g''),\alpha(g',g'')) \\ &= (g\cdot\beta_Z(g',g'') + \theta(\rho(g))(\rho(g)\cdot\alpha(g',g'')), \rho(g)\cdot\alpha(g',g'')). \end{aligned}$$

Therefore,

$$\begin{aligned}(S(g)\cdot\beta(g',g''))\beta(g,g'g'') &= (g\cdot\beta_Z(g',g'')+\beta_Z(g,g'g'')+\theta(\rho(g))(\rho(g)\cdot\alpha(g',g''))\\&\quad+\omega(\rho(g)\cdot\alpha(g',g''),\alpha(g,g'g'')),(\rho(g)\cdot\alpha(g',g''))\alpha(g,g'g'')),\end{aligned}$$

and similarly

$$\begin{aligned}\beta(g,g')\beta(gg',g'') = (\beta_Z(g,g')+\beta_Z(gg',g'')\\ +\omega(\alpha(g,g'),\alpha(gg',g'')),\alpha(g,g')\alpha(gg',g'')).\end{aligned}$$

Since $\mathrm{Im}(d_S\beta)\subset Z$, this implies that

$$\begin{aligned}(d_S\beta)(g,g',g'') &= (d_S\beta_Z)(g,g',g'')+\theta(\rho(g))(\rho(g)\cdot\alpha(g',g''))\\&\quad+\omega(\rho(g)\cdot\alpha(g',g''),\alpha(g,g'g''))-\omega(\alpha(g,g'),\alpha(gg',g'')).\end{aligned}$$

Hence the cohomology class of $d_S\beta$ depends additively on the pair (ω,θ), that is, $Q : D\to H^3_s(G,Z)$ is a group homomorphism. □

5.1.4 Significance of the zero class in $H^3(G,Z)$

We now continue in the framework of the previous section to show that $Q(\omega,\theta)=0$ characterizes the extendability of ω to a cocycle $\tilde{\omega}\in Z^2_s(\widehat{G},Z)$.

Theorem 5.1.18. *Let $\omega\in Z^2_s(L,Z)$ define the central extension $\widehat{L}=Z\times_\omega L$, and $\theta\in Z^1_s(\widehat{G},C^1_s(L,Z))_\omega$ describe the action as in equation (5.10) such that $\mu:\widehat{L}\to\widehat{G}$, $(z,l)\mapsto l$ is a crossed module of groups. Further choose a section ρ and let α and β be defined as before. Then the following are equivalent:*

(1) *$Q(\omega,\theta)=[d^M\beta]=0$ in $H^3(G,Z)$.*
(2) *There exists an extension*

$$0\to Z\to\widehat{\widehat{G}}\xrightarrow{q}\widehat{G}\to 1$$

and a $\widehat{G}$-equivariant equivalence $\widehat{L}\to q^{-1}(L)$ of central Z-extensions of L.
(3) *There exists a cocycle $\omega_{\widehat{G}}\in Z^2(\widehat{G},Z)$ with $\omega_{\widehat{G}}|_{L\times L}=\omega$ and $\tilde{\omega}_{\widehat{G}}|_{\widehat{G}\times L}=\theta$.*

Proof. (1)⇒(2): Suppose that $[d^M\beta]=0$, that is, that $d^M\beta=d^M\chi$ for some $\chi\in C^2(G,Z)$. Then $\beta':=\beta\chi^{-1}$ satisfies on the one hand $\mu\circ\beta'=\alpha$, because $\mu\circ\chi^{-1}=1$ as χ has values in Z. On the other hand, $d^M\beta'=d^M\beta-d^M\chi=0$. Denoting by S the outer action, the pair (S,β') permits to define the group $\widehat{\widehat{G}}=\widehat{L}\times_{(S,\beta')}G$ via the product

$$(l,g)(l',g'):=(lS(g)(l')\beta'(g,g'),gg').$$

This group $\widehat{\widehat{G}}$ is an extension of G by $\widehat{L}$. Furthermore, we can view $(\mathrm{conj} \circ \rho, \alpha)$ as a factor system for G and L, which leads to the description of G as $L \times G$, endowed with the product

$$(l,g)(l',g') = (l\,\mathrm{conj}_{\rho(g)}(l')\alpha(g,g'), gg').$$

Hence the map

$$q : \widehat{\widehat{G}} = \widehat{L} \times G = (Z \times_\omega L) \times G \to \widehat{G} = L \times G, ((z,l),g) \mapsto (l,g)$$

defines a group extension of $\widehat{G}$ by Z containing $\widehat{L}$. It remains to verify that the action of $\widehat{G}$ on $\widehat{L}$, induced by the conjugation action of $\widehat{\widehat{G}}$ for which Z acts trivially, coincides with the action given by θ.

It follows from the construction that the conjugation action of $\widehat{\widehat{G}}$ on $\widehat{L} = Z \times_\omega L$ is given by

$$\begin{aligned}(l,g)(z,l')(l,g)^{-1} &= \mathrm{conj}_l \circ S(g)(z,l') \\ &= \mathrm{conj}_l(g \cdot z + \theta(g)(\mathrm{conj}_g(l')), \mathrm{conj}_g(l')) \\ &= (g \cdot z + \theta(g)(\mathrm{conj}_g(l')) + \theta(l)(\mathrm{conj}_l \circ \mathrm{conj}_g(l')), \mathrm{conj}_l \circ \mathrm{conj}_g(l')) \\ &= (g \cdot z + \theta((l,g))(\mathrm{conj}_l \circ \mathrm{conj}_g(l')), \mathrm{conj}_{(l,g)}(l')) \\ &= (l,g) \cdot (z,l').\end{aligned}$$

This completes the proof of (2).

(2)⇒(3): Let $q : \widehat{\widehat{G}} \to \widehat{G}$ be an extension as in (2). We may without loss of generality assume that $\widehat{\widehat{G}} = Z \times_{\omega_{\widehat{G}}} \widehat{G}$ for some $\omega_{\widehat{G}} \in Z^2(\widehat{G}, Z)$ such that $\omega_{\widehat{G}}$ extends ω, that is, $\widehat{L} = Z \times_\omega L$ is contained in $\widehat{\widehat{G}}$. The condition that the inclusion $\widehat{L} \hookrightarrow \widehat{\widehat{G}}$ is $\widehat{G}$-equivariant means that $\tilde{\omega}_{\widehat{G}}|_{\widehat{G}\times L} = \theta$.

(3)⇒(1): Let $\omega_{\widehat{G}} \in Z^2(\widehat{G}, Z)$ be a cocycle as in (3) and let

$$q_{\widehat{G}} : \widehat{\widehat{G}} := Z \times_{\omega_{\widehat{G}}} \widehat{G} \to \widehat{G}$$

be the corresponding extension. The condition $\omega_{\widehat{G}}|_{L\times L} = \omega$ means that the group $\widehat{L} = Z \times_\omega L$ is a subgroup of $\widehat{G}$, and the second condition $\tilde{\omega}_{\widehat{G}}|_{\widehat{G}\times L} = \theta$ ensures that the conjugation action of $\widehat{\widehat{G}}$ on $\widehat{L}$ induces the action of $\widehat{G}$ on $\widehat{L}$ defined by θ.

Now take the section $\widehat{\rho} : G \to \widehat{G}$ defined by $g \mapsto (0, \rho(g))$. Then its corresponding failure to be a group homomorphism $\alpha_{\widehat{\rho}}$ satisfies $\mu \circ \alpha_{\widehat{\rho}} = \alpha$, where α is the failure for the section ρ. Therefore, $\alpha_{\widehat{\rho}}$ may be taken as β, and, observing that the outer action is just the composition of the conjugation action and $\widehat{\rho}$, we obtain

$$\begin{aligned}d^M \alpha_{\widehat{\rho}}(x,y,z) &= \widehat{\rho}(x)\widehat{\rho}(y)\widehat{\rho}(z)\widehat{\rho}(yz)^{-1}\widehat{\rho}(x)^{-1}\widehat{\rho}(x)\widehat{\rho}(yz)\widehat{\rho}(xyz)^{-1}\widehat{\rho}(xyz)\widehat{\rho}(z)^{-1} \\ &\quad \times \widehat{\rho}(xy)^{-1}\widehat{\rho}(xy)\widehat{\rho}(y)^{-1}\widehat{\rho}(x)^{-1} \\ &= 1.\end{aligned}$$

We deduce $\tilde{Q}(\omega, \theta, \rho, \alpha_{\widehat{\rho}}) = d^M \alpha_{\widehat{\rho}} = 1$, and thus $Q(\omega, \theta) = [d^M \alpha_{\widehat{\rho}}] = 0$. □

5.1.5 The smooth obstruction crossed module

Like in Section 1.4.3 we define now for a given smooth outer action an obstruction crossed module, which expresses the obstruction to the existence of a smooth general extension with this given smooth outer action.

Definition 5.1.17. Let S be a smooth outer action of G on L and define $\beta \in C_s^2(G,L)$ by $\delta_S = \mathrm{conj}_\beta$; cf. Remark 1.4.5. We then have $d_S\beta \in Z_s^3(G,Z(L))_{[S]}$; see Lemma 5.1.2. The corresponding cohomology class

$$[d_S\beta] \in H_s^3(G,Z(L))_{[S]}$$

was called characteristic class of S, but will also be called the *obstruction class* of the outer action S in the present context.

We have already computed that taking a different β with $\delta_S = \mathrm{conj}_\beta$, we have $\epsilon := \beta'\beta^{-1} \in C_s^2(G,Z(L))$ and $d_S\beta$ and $d_S\beta'$ differ by the coboundary $d_S\epsilon$, thus the class $[d_S\beta]$ does not depend on the choice of β.

For equivalent S and S', that is, $S \sim S'$, there exists $h \in C_s^1(G,L)$ with $S' = h \cdot S$, and we may choose β and β' such that $h *_S \beta = \beta'$. We have seen in Lemma 5.1.2 that $d_S\beta = d_{S'}\beta'$. The upshot of this discussion is that the class $[d_S\beta]$ depends only on the equivalence class of S, that is, on the smooth G-kernel S.

Up to now, we have studied the situation where a general extension of G by L already existed. Let us see now how a certain crossed module answers the *existence question*, that is, given a smooth outer action, does there exist a general extension of Lie groups, which realizes this outer action? For this, we will associate to a smooth outer action $S : G \to \mathrm{Aut}(L)$ a crossed module of Lie groups $\mu : L \to G^S$.

Indeed, let $S : G \to \mathrm{Aut}(L)$ be a smooth outer action and let $\beta \in C_s^2(G,L)$ with $\delta_S = \mathrm{conj}_\beta$. Consider the normal subgroup $L \subset \mathrm{Aut}(L)$ of automorphisms given by conjugation with some element of L, and denote by $\mathrm{ad} : L \to \mathrm{conj}_L \subset \mathrm{Aut}(L)$ the quotient map $L \to L/Z(L) = \mathrm{conj}_L =: L_{\mathrm{ad}}$. On the product set,

$$G^S := L_{\mathrm{ad}} \times G$$

we define the group product

$$\begin{aligned}(\mathrm{conj}_l, x)(\mathrm{conj}_{l'}, x') &:= (\mathrm{conj}_{lS(x)(l')\beta(x,x')}, xx') \\ &= (\mathrm{conj}_l\,\mathrm{conj}_{S(x)(l')}\,\delta_S(x,x'), xx').\end{aligned}$$

Note that the second form of the product shows that it does not depend on the choice of β. Define

$$S_1 : G \to \mathrm{Aut}(L_{\mathrm{ad}}), \quad S_1(g)(\mathrm{conj}_l) := \mathrm{conj}_{S(g)(l)}\,.$$

This defines a smooth outer action on L_{ad} with $\delta_{S_1} = \mathrm{conj}_{\delta_S}$, if we consider δ_S as L_{ad}-valued.

Lemma 5.1.19. *The above defined product renders G^S a group with the following properties:*

(a) *It carries a natural Lie group structure for which the map $q_S : G^S \to G, (\mathrm{conj}_l, x) \mapsto x$ defines a group extension of G by L_{ad}. The equivalence class of this extension depends only on the G-kernel $[S]$.*

(b) *The map $\eta : G^S \to \mathrm{Aut}(L), (\mathrm{conj}_l, x) \mapsto \mathrm{conj}_l \circ S(x)$ defines a smooth action of G^S on L.*

(c) *The map $\mu : L \to G^S, l \mapsto (\mathrm{conj}_l, 1)$ defines a Lie group crossed module with $\mathrm{Ker}(\mu) = Z(L)$ and $\mathrm{Coker}(\mu) = G^S/L_{\mathrm{ad}} \cong G$, and S is the corresponding smooth outer action of $G \cong G^S/L_{\mathrm{ad}}$ on L.*

Proof.

(a) Let us show that (S_1, δ_{S_1}) is a smooth factor system for G and L_{ad}. Since we define the Lie group structure on $L_{\mathrm{ad}} = C_L(L)$ in such a way that $C_L : L \to L_{\mathrm{ad}}$ is the quotient map, the relation $\delta_S = C_L \circ \beta$ implies $\delta_S \in C_s^2(G, L_{\mathrm{ad}})$ and we have $\delta_{S_1} = \mathrm{conj}_{\delta_S}$.

If $U \subset G$ is an open identity neighborhood such that the map $U \times L \to L, (g, l) \mapsto S(g) \cdot l$ is smooth, then the map

$$U \times L_{\mathrm{ad}}, \quad (g, \mathrm{conj}_l) \mapsto \mathrm{conj}_{S(g) \cdot l} = S(g) \, \mathrm{conj}_l \, S(g)^{-1} = S_1(g)(\mathrm{conj}_l)$$

is also smooth.

The local smoothness of $\delta_S = \mathrm{conj}_\beta$ follows from the local smoothness of β.

Let $x, x', x'' \in G$. Then compute

$$\begin{aligned}
&S_1(x)(\delta_S(x', x''))\delta_S(x, x'x'') \\
&\quad = S(x)(S(x')S(x'')S(x'x'')^{-1})S(x)^{-1}S(x)S(x'x'')S(xx'x'')^{-1} \\
&\quad = S(x)S(x')S(x'')S(xx'x'')^{-1} \\
&\quad = S(x)S(x')S(xx')^{-1}S(xx')S(x'')S(xx'x'')^{-1} \\
&\quad = \delta_S(x, x')\delta_S(xx', x'').
\end{aligned}$$

This shows that (S_1, δ_{S_1}) is a factor system and, therefore, by Proposition 5.1.5 that G^S is a Lie group and that $q_S : G^S \to G$ is a Lie group extension of G by L_{ad}.

To see that the equivalence class of the extension $G^S = L_{\mathrm{ad}} \times_{(S_1, \delta_S)} G$ of G by L_{ad} depends only on the equivalence class $[S]$, let $\alpha \in C_s^1(G, L)$ and $S' := \mathrm{conj}_\alpha \circ S$. Then

$$S_1' := \mathrm{conj}_h \circ S_1 = h \cdot S_1$$

for $h := \mathrm{conj}_\alpha \in C_s^1(G, L_{\mathrm{ad}})$. We obtain

$$\delta_{S'}(x, x') = \mathrm{conj}_{\alpha(x)S(x)(\alpha(x'))} \circ \delta_S(x, x') \circ \mathrm{conj}_{\alpha(xx')^{-1}} = (h *_{S_1} \delta_S)(x, x').$$

According to Proposition 5.1.7, the map

$$\varphi : G^{S'} \to G^S, \quad (\mathrm{conj}_l, x) \mapsto (\mathrm{conj}_{l\alpha(x)}, x) = (\mathrm{conj}_l \circ h(x), x)$$

is an equivalence of extensions of G by L_{ad}.

(b) Let us first show that η is a representation:

$$\begin{aligned}\eta(\mathrm{conj}_l, x)\eta(\mathrm{conj}\, l', x') &= \mathrm{conj}_l \circ S(x) \circ \mathrm{conj}_{l'} \circ S(x')\\ &= \mathrm{conj}_l \circ \mathrm{conj}_{S(x)(l')} \circ S(x) \circ S(x')\\ &= \mathrm{conj}_{lS(x)(l')} \circ \delta_S(x, x') \circ S(xx')\\ &= \eta(\mathrm{conj}_{lS(x)(l')} \circ \delta_S(x, x'), xx')\\ &= \eta((\mathrm{conj}_l, x)(\mathrm{conj}_{l'}, x')).\end{aligned}$$

Since $\eta(G^S)$ consists of smooth automorphisms of L, the smoothness of the representation follows from the local smoothness and the smoothness of the orbit maps; local smoothness is a consequence of the local smoothness of the map

$$(L_{\mathrm{ad}} \times G) \times L \to L, \quad ((\mathrm{conj}_l, S(g)), l') \mapsto \mathrm{conj}_l \circ S(g) \cdot l'$$

which, in addition, shows that the action of G^S on L is a map $G^S \times L \to L$ that is smooth on a set of the form $U^S \times L$, where U^S is an identity neighborhood in G^S. Since G^S acts on L by smooth maps, this implies the smoothness of the action.

(c) This follows directly from (a) and (b). □

Remark 5.1.18. On the one hand, this construction gives the *obstruction crossed module* $\mu : L \to G^S$ as the canonical central extension

$$0 \to Z(L) \to L \to L_{\mathrm{ad}} \to 1$$

of the ideal L_{ad} in G^S. On the other hand (see the following lemma), the crossed module $\mu : L \to G^S$ can also be seen as the pullback crossed module of the canonical crossed module

$$0 \to Z(L) \to L \to \mathrm{Aut}(L) \to \mathrm{Out}(L) \to 1$$

by the group homomorphism $\psi : G \to \mathrm{Out}(L)$ obtained by $\psi = q \circ S$ using the quotient map $q : \mathrm{Aut}(L) \to \mathrm{Out}(L)$.

Denote by $q_S : G^S \to G$ the projection.

Lemma 5.1.20. *The map*

$$\psi_S = (\eta, q_S) : G^S \to \mathrm{Aut}(L) \times G$$

is injective and yields an isomorphism of groups

$$G^S \cong \{(\varphi, x) \in \mathrm{Aut}(L) \times G \mid S(x) \in \varphi \circ \mathrm{conj}_L\}.$$

Proof. Since $\mathrm{Ker}(q_S) = L_{\mathrm{ad}}$ and $\mathrm{Ker}(\eta) \cap L_{\mathrm{ad}} = \{1\}$, the map ψ_S is an injective homomorphism of groups.

For each element $(\mathrm{conj}_l, x) \in G^S$, we have

$$\psi_S(\mathrm{conj}_l, x) = (\mathrm{conj}_l \circ S(x), x),$$

which proves the left-to-right inclusion $\subseteq$. But for any pair $(\varphi, x) \in \mathrm{Aut}(L) \times G$ with $\varphi \in S(x) \circ \mathrm{conj}_L$, there exists an element $l \in L$ with $\varphi = \mathrm{conj}_l \circ S(x)$, which implies that (φ, x) is the ψ_S-image of (conj_l, x). □

The next lemma sheds some light on the passage from a given Lie group extension of G by L with smooth outer action S to the corresponding Lie group extension $q_S : G^S \to G$ of G by $Z(L)$.

Lemma 5.1.21. *Let $q : \widehat{G} \to G$ be a Lie group extension of G by L with associated smooth G-kernel $[S]$ and denote (as usual) by* conj *the conjugation map* $\mathrm{conj} : \widehat{G} \to \mathrm{Aut}(L)$. *Then $\widehat{G}/Z(L) \cong G^S$ and the map*

$$\psi = (\mathrm{conj} \circ (S \circ q)^{-1}, q) : \widehat{G} \to \mathrm{Aut}(L) \times G, \quad x \mapsto (\mathrm{conj}_x \circ S(q(x))^{-1}, q(x))$$

defines a Lie group extension

$$0 \to Z(L) \to \widehat{G} \xrightarrow{\psi} G^S \to 1.$$

This assignment has the following properties:

(a) *If $q_j : \widehat{G}_j \to G$, $j = 1, 2$, are equivalent extensions of G by L, then $\psi_j : \widehat{G}_j \to G^S$ are equivalent extensions of G^S by $Z(L)$. We thus obtain a map*

$$\tau : \mathrm{Ext}(G, L)_{[S]} \to \mathrm{Ext}(G^S, Z(L))_{[\eta]}.$$

(b) *A Lie group extension $\psi : \widehat{G^S} \to G^S$ of G^S by $Z(L)$ comes from an extension of G by L corresponding to $[S]$ if and only if there exists a G^S-equivariant equivalence*

$$\alpha : L \to \psi^{-1}(L_{\mathrm{ad}}) \subseteq \widehat{G^S}$$

of central extensions of L_{ad} by $Z(L)$. Note that $G^S \cong \widehat{G^S}/Z(L)$ acts on $\psi^{-1}(L_{\mathrm{ad}})$ because $Z(L)$ acts trivially on this group.

Proof. Since the extension $\widehat{G}$ corresponds to $[S]$, it is equivalent to an extension of the type $L \times_{(S,\beta)} G$ for some smooth factor system (S, β) by Proposition 5.1.5. This means that

$$\mathrm{conj}_{(l,x)} = \mathrm{conj}_l \circ S(x) = \mathrm{conj}_l \circ S(q(l, x)), \tag{5.12}$$

so that $\psi(l,x) = (\mathrm{conj}_l, x)$. Now the explicit formulas for the multiplication in $L \times_{(S,\beta)} G$ and G^S imply that ψ is a surjective homomorphism of Lie groups:

$$\begin{aligned}\psi((l,x)(l',x')) &= \psi(lS(x)(l')\beta(x,x'), xx')\\ &= (\mathrm{conj}_{lS(x)(l')\beta(x,x')}, xx')\\ &= (\mathrm{conj}_{lS(x)(l')}\,\delta_S(x,x'), xx')\\ &= (\mathrm{conj}_l, x)(\mathrm{conj}_{l'}, x')\\ &= \psi(l,x)\psi(l',x').\end{aligned}$$

Its kernel is $Z(L)$, a split Lie subgroup of L, and thus ψ defines a Lie group extension of G^S by $Z(L)$.

(a) If $\varphi : \widehat{G}_1 \to \widehat{G}_2$ is an equivalence of L-extensions of G, then we have for the conjugation actions conj^j of $\widehat{G}_j$ on L for $j = 1,2$ the relation $\mathrm{conj}^2 \circ \varphi = \mathrm{conj}^1$ because $\varphi|_L = \mathrm{id}_L$. Therefore, the quotient maps

$$\psi_j = (\mathrm{conj}^j \circ (S \circ q_j)^{-1}, q_j) : \widehat{G}_j \to G^S$$

satisfy

$$\begin{aligned}\psi_2 \circ \varphi &= ((\mathrm{conj}^2 \circ \varphi) \circ (S \circ q_2 \circ \varphi)^{-1}, q_2 \circ \varphi)\\ &= (\mathrm{conj}^1 \circ (S \circ q_1)^{-1}, q_1) = \psi_1.\end{aligned}$$

This means that $\varphi : \widehat{G}_1 \to \widehat{G}_2$ is also an equivalence of extensions of G^S by $Z(L)$.

(b) Suppose first that the extension $\psi : \widehat{G^S} \to G^S$ by $Z(L)$ comes from the L-extension $q : \widehat{G} \to G$ corresponding to $[S]$ in the sense that $\widehat{G} = \widehat{G^S}$ with $\widehat{G}/Z(L) \cong G^S$. We may assume that $\widehat{G} = Z(L) \times_{(S,\beta)} G$ by Proposition 5.1.5. Then equation (5.12) shows that $(l,x) \in \widehat{G}$ acts on L by $\mathrm{conj}_l \circ S(x) = \eta(\mathrm{conj}_l, x)$. Therefore, the inclusion $L \hookrightarrow \widehat{G}$ onto the subset $\psi^{-1}(L_{\mathrm{ad}})$ is equivariant with respect to the action of G^S and, therefore, in particular for the action of L_{ad}, so that it is an equivalence of central extensions of L_{ad} by $Z(L)$.

Suppose, conversely, that $\psi : \widehat{G^S} \to G^S$ is an extension of G^S by $Z(L)$ for which there exists a G^S-equivariant equivalence $\alpha : L \to \psi^{-1}(L_{\mathrm{ad}})$ of central extensions of L_{ad} by $Z(L)$. Then

$$\widehat{G^S}/\alpha(L) = \widehat{G^S}/\psi^{-1}(L_{\mathrm{ad}}) \cong G^S/L_{\mathrm{ad}} \cong G,$$

so that we obtain an extension of G by L using the quotient map $q = q_S \circ \psi : \widehat{G^S} \to G$ with kernel $\psi^{-1}(L_{\mathrm{ad}})$. The action of $G^S \cong \widehat{G^S}/Z(L)$ on L induced by the conjugation action of $\widehat{G^S}$ on L coincides with the given action

$$\eta : G^S \to \mathrm{Aut}(L), \quad (\mathrm{conj}_l, x) \mapsto \mathrm{conj}_l \circ S(x)$$

of G^S on L, because α is G^S-equivariant. Therefore, the smooth outer action associated to the extension $q : \widehat{G^S} \to G$ is $[S]$.

For the conjugation action conj of $\widehat{G}^S$ on L, we have $\mathrm{conj} = \eta \circ \psi$ and $q = q_S \circ \psi$, so that the corresponding map $\widehat{G}^S \to G^S$ coincides with ψ. This means that $\psi : \widehat{G}^S \to G^S$ comes from the extension $q : \widehat{G}^S \to G$ in the sense described above. □

Let an outer action $[S]$ be given. Recall the obstruction crossed module $\mu : L \to G^S$. Its characteristic class is $[d_S\beta] \in H^3(G, Z(L))_{[S]}$.

The following theorem shows that the cohomology class $[d_S\beta] \in H^3_s(G, Z(L))_{[S]}$ associated to $[S]$ is exactly the obstruction against the existence of a Lie group extension of G by L realizing this smooth outer action. The first part has already been shown in Theorem 5.1.11. The second part of the theorem concerns the realizability as an *abelian* extension (but of the more complicated group G^S).

Theorem 5.1.22. *Let S be a smooth outer action of G on L.*
(a) *Then, concerning non-abelian Lie group extensions, we have*

$$[d_S\beta] = 0 \quad \Leftrightarrow \quad \mathrm{Ext}(G, L)_{[S]} \neq \emptyset.$$

(b) *Furthermore, we have that the central extension*

$$0 \to Z(L) \to L \to L_{\mathrm{ad}} \to 1$$

can be embedded in a G^S-equivariant way into an abelian extension

$$0 \to Z(L) \to \widehat{G}^S \to G^S \to 1$$

if and only if $[d_S\beta] = 0$.

Proof.
(a) This has already been shown in Theorem 5.1.11.
(b) We have seen in Lemma 5.1.19 that the map $\mu : L \to G^S$ defines a Lie group crossed module with smooth outer action S. According to Theorem 5.1.18, the condition $[d_S\beta] = 0$ is equivalent to the possibility of embedding the central extension

$$0 \to Z(L) \to L \to L_{\mathrm{ad}} \to 1$$

in a G^S-equivariant way into an abelian extension

$$0 \to Z(L) \to \widehat{G}^S \to G^S \to 1.$$ □

5.1.6 Integration of crossed modules of Lie algebras

In this section, we will only describe the most elementary integration theory of crossed modules of Lie algebras. We will always suppose that the base field is $\mathbb{R}$ or $\mathbb{C}$. For more refined versions we refer, for example, to [85]. For the following lemma; cf. [46].

Lemma 5.1.23. *Let G be a 1-connected Lie group with Lie algebra $T_1G = \mathfrak{g}$. Then we have an isomorphism of identity components of automorphism groups*

$$\varphi : \mathrm{Aut}_0(G) \cong \mathrm{Aut}_0(\mathfrak{g}).$$

The isomorphism φ sends conj_g *into* Ad_g *for all $g \in G$.*

Proof. For an automorphism $\phi \in \mathrm{Aut}_0(G)$, the induced linear map $T_1\phi$ sends T_1G to T_1G. By ordinary Lie theory, $T_1\phi$ preserves the bracket. Then the same is true for ϕ^{-1}, thus $T_1\phi$ is an automorphism. Conversely, given an automorphism of Lie algebras $\psi : \mathfrak{g} \to \mathfrak{g}$, it can be lifted to a homomorphism of Lie groups $\widehat{\psi} : G \to G$ because G is 1-connected. The same applies to its inverse, thus ψ lifts to an automorphism of Lie groups $\widehat{\psi} : G \to G$.

The conjugation by $g \in G$ sends $1 \mapsto 1$, thus $T_1 \mathrm{conj}_g : T_1G \to T_1G$. By definition, $\mathrm{Ad}_g = T_1 \mathrm{conj}_g$. □

Proposition 5.1.24. *Let $\mu : M \to N$ be a crossed module of Lie groups. Denote by $\mathfrak{m} := T_1M$ and $\mathfrak{n} := T_1N$ the associated Lie algebras. Then the induced linear map $T_1\mu : \mathfrak{m} \to \mathfrak{n}$ is a crossed module of Lie algebras, called the derived crossed module of Lie algebras corresponding to $\mu : M \to N$.*

Proof. The linear map $T_1\mu : \mathfrak{m} \to \mathfrak{n}$ is a homomorphism of Lie algebras by ordinary Lie theory. The smooth action $\widehat{S} : N \to \mathrm{Aut}(M)$ of N on M by automorphisms induces $T_1\widehat{S} : \mathfrak{n} \to T_1 \mathrm{Aut}(M)$. The image takes values in $T_1 \mathrm{Aut}_0(M)$, the tangent space at id_M of the identity component of the Lie group $\mathrm{Aut}(M)$. As in Lemma 5.1.23, there is a smooth homomorphism $\varphi : \mathrm{Aut}_0(M) \to \mathrm{Aut}_0(\mathfrak{m})$ between identity components of the Lie groups of automorphisms of M and $\mathfrak{m}$. Here, φ need not be an isomorphism as M is not necessarily 1-connected. Nevertheless, $T_1\varphi$ sends $T_1\widehat{S}(n)$ for $n \in \mathfrak{n}$ to an infinitesimal automorphism, that is, a derivation of $\mathfrak{m}$. In this way, the Lie algebra $\mathfrak{n}$ acts via $T_1\widehat{S}$ on the Lie algebra $\mathfrak{m}$ by derivations.

It remains to check the two requirements (a) and (b) of a crossed module of Lie algebras. Both are obtained from the corresponding statements on group level by applying the functor $T_1(-)$: We have for $n \in \mathfrak{n}$,

$$T_1(\mu \circ \widehat{S})(n) = T_1\mu \circ T_1\widehat{S}(n) = T_1(\mathrm{conj})(n) \circ T_1\mu = \mathrm{ad}_n \circ T_1\mu,$$

and this translates into $T_1\mu(n \cdot m) = [n, T_1\mu(m)]$ for all $m \in \mathfrak{m}$.

For condition (b), we observe that Ad corresponds to ad under the passage $T_1\varphi : \mathrm{Aut}_0(\mathfrak{m}) \to \mathrm{Der}(\mathfrak{m})$:

$$T_1\varphi(T_1(\widehat{S} \circ \mu)) = T_1\varphi(T_1\widehat{S} \circ T_1\mu) = T_1\varphi(T_1 \mathrm{conj}) = T_1\varphi(\mathrm{Ad}) = \mathrm{ad},$$

and this translates into $T_1\mu(m) \cdot m' = [m, m']$ for all $m, m' \in \mathfrak{m}$. □

Proposition 5.1.25. *Let $\mu : \mathfrak{m} \to \mathfrak{n}$ be a crossed module of finite dimensional Lie algebras. Then there is a unique (up to isomorphism) crossed module of 1-connected Lie groups $\hat{\mu} : M \to N$ whose derived crossed module is $\mu : \mathfrak{m} \to \mathfrak{n}$.*

Proof. The finite dimensional Lie algebras $\mathfrak{m}$ and $\mathfrak{n}$ integrate to 1-connected Lie groups M and N, respectively, by Lie's third theorem; see Appendix D. The Lie groups M and N are unique up to isomorphism of Lie groups by the requirement of 1-connectedness. The homomorphism of Lie algebras $\mu : \mathfrak{m} \to \mathfrak{n}$ integrates into a homomorphism of Lie groups $\hat{\mu} : M \to N$ by Lie's second theorem because M is 1-connected; see Appendix D.

The action of an element of $\mathfrak{n}$ by derivations on $\mathfrak{m}$ provides a homomorphism of Lie algebras $\mathfrak{n} \to \mathrm{Der}(\mathfrak{m})$. By 1-connectedness of N, it integrates to a homomorphism of Lie groups $N \to \mathrm{Aut}_0(\mathfrak{m})$. Lemma 5.1.23 shows (again using 1-connectedness) that the composition with φ^{-1} leads to a homomorphism of Lie groups $\widehat{S} : N \to \mathrm{Aut}_0(M)$, which is a smooth action of N on M by automorphisms.

It remains to check the requirements (a), (b) and (c) of a crossed module of Lie groups. Conditions (a) and (b) follow directly from the fact that (1-connected) integration is functorial. Condition (c) follows from the fact that Lie subalgebras integrate into Lie subgroups, and Lie subgroups are always split in finite dimensions; see Remark 5.1.7. □

Remark 5.1.19. Recall that (globally) smooth cohomology of a finite dimensional connected compact Lie group vanishes. This is Hu's theorem [53]. On the other hand, locally smooth and (globally) smooth cohomology are isomorphic in case of contractible coefficients, for example, for coefficients in a real- or complex vector space, see [102] Proposition 1.5. Therefore, we do not have nontrivial examples of crossed modules of compact connected finite dimensional Lie groups such that the module $V = \mathrm{Ker}(\mu)$ is a real or complex vector space.

Example 5.1.20. The following is an example of a nonintegrable crossed module of Lie algebras. Indeed, it is enough to find a nonintegrable Lie algebra $\mathfrak{g}$, that is, a Lie algebra for which there does not exist a Lie group G such that $T_1G \cong \mathfrak{g}$ as Lie algebras. It is clear that we must look for examples in infinite dimensions and we will require G to be a locally convex Lie group, that is, a group which is also a locally convex manifold such that group product and inversion are smooth. In case $\mathfrak{g}$ is a nonintegrable Lie algebra, the crossed module $\mathrm{id}_{\mathfrak{g}} : \mathfrak{g} \to \mathfrak{g}$ is a crossed module of Lie algebras which cannot be integrated into a crossed module of Lie groups.

A standard example of a nonintegrable Lie algebra is the following. Suppose $\mathfrak{g}$ is a locally convex, *locally exponential* Lie algebra, that is, the Baker–Campbell–Hausdorff formula endows a circular locally convex 0-neighborhood of $\mathfrak{g}$ with the structure of a local Lie group; see [82] Definition IV.2.1, page 370. The *period group* $\Pi(\mathfrak{g})$ is the image of the period homomorphism

$$\mathrm{per}_{\mathfrak{g}} : \pi_2(G_{\mathrm{ad}}) \to \mathfrak{z}(\mathfrak{g}),$$

where $\mathfrak{g}$ is viewed as the central extension of $\mathfrak{g}_{\mathrm{ad}} := \mathfrak{g}/\mathfrak{z}(\mathfrak{g})$ by its center $\mathfrak{z}(\mathfrak{g})$. The locally exponential Lie algebra $\mathfrak{g}_{\mathrm{ad}}$ always integrates into a Lie group $G_{\mathrm{ad}} \subset \mathrm{Aut}(\mathfrak{g})$. Integration of the 2-cocycle defining the central extension over elements of $\pi_2(G_{\mathrm{ad}})$ yields then the period map.

Suppose further that the period group $\Pi(\mathfrak{g}) \cong \mathbb{Z}$, namely $\Pi(\mathfrak{g}) = \mathbb{Z}d$. Let $\theta \in \mathbb{R} \setminus \mathbb{Q}$. Then $\mathfrak{z} := \{(x, \theta x) \mid x \in \mathfrak{z}(\mathfrak{g})\}$ is a central ideal of $\mathfrak{g} \times \mathfrak{g}$, so that $\mathfrak{h} := \mathfrak{g} \times \mathfrak{g}/\mathfrak{z}$ is locally exponential. One computes

$$\Pi(\mathfrak{h}) \cong \mathbb{Z}[(d,0)] + \mathbb{Z}[(0,d)] = (\mathbb{Z} + \mathbb{Z}\theta)[(d,0)].$$

As this group is not discrete, there is no Lie group G with $T_1 G \cong \mathfrak{h}$; see *loc. cit.*

It remains to find an example of a locally convex, locally exponential Lie algebra $\mathfrak{g}$ with period group $\Pi(\mathfrak{g}) \cong \mathbb{Z}$, example due to van Est–Korthagen. The Lie algebra $\mathfrak{g}$ is the central extension of the Banach–Lie algebra $\mathcal{C}^1(S^1, \mathfrak{su}_2(\mathbb{C}))$ of $\mathcal{C}^1$-differentiable maps from the circle S^1 to the Lie algebra $\mathfrak{su}_2(\mathbb{C})$, by the center $\mathbb{R}$ defined by the 2-cocycle

$$\omega(f,g) := \int_0^1 \mathrm{tr}(f(t)g'(t))\,dt,$$

where we identify functions on $S^1 \cong \mathbb{R}/\mathbb{Z}$ with 1-periodic functions on $\mathbb{R}$. Then $\mathfrak{g}_{\mathrm{ad}} = \mathcal{C}^1(S^1, \mathfrak{su}_2(\mathbb{C}))$ and $G_{\mathrm{ad}} = \mathcal{C}^1(S^1, SU_2(\mathbb{C}))$ leads to $\pi_2(G_{\mathrm{ad}}) \cong \pi_3(SU_2(\mathbb{C})) \cong \pi_3(S^3) \cong \mathbb{Z}$. Now one shows that $\mathrm{per}_{\mathfrak{g}}$ is nontrivial to verify that $\Pi(\mathfrak{g}) \cong \mathbb{Z}$.

Another (simpler) example of a locally convex, locally exponential Lie algebra $\mathfrak{g}$ with period group $\Pi(\mathfrak{g}) \cong \mathbb{Z}$ is the Lie algebra $\mathfrak{u}(\mathcal{H})$ of the Lie group of unitary operators $U(\mathcal{H})$ on an infinite-dimensional complex Hilbert space $\mathcal{H}$; see *loc. cit.* Indeed, in general the period group $\Pi(\mathfrak{g})$ of the Lie algebra $\mathfrak{g}$ of a 1-connected locally exponential Fréchet Lie group G is given by $\pi_1(Z(G))$, that is, the fundamental group of the center of G. This applies to $U(\mathcal{H})$, because by Kuiper's theorem $U(\mathcal{H})$ has the weak homotopy type of a point. As the center $Z(U(\mathcal{H}))$ is the circle S^1, we obtain $\Pi(\mathfrak{u}(\mathcal{H})) \cong \mathbb{Z}$.

Example 5.1.21. The string group after Baez–Crans–Stevensson–Schreiber (see Theorem 1.7.3) is an example of a crossed module of infinite dimensional Lie groups obtained as the integration of a crossed module of Lie algebras, namely as the integration of the crossed module in Section 2.8.1; see Remark 2.8.1.

Example 5.1.22. We have seen in Section 2.8.2 that there is a another crossed module of Lie algebras representing the Cartan class $[\theta] \in H^3(\mathfrak{g}, \mathbb{C})$ in the third cohomology of a complex simple finite dimensional Lie algebra $\mathfrak{g}$; see Remark 2.8.10.

It is interesting to observe that while the crossed module of Lie algebras in Section 2.8.1 is integrable, the crossed module in Remark 2.8.10 is not integrable, while representing the same cohomology class. Indeed, let us show that the crossed module

$$0 \to \mathbb{C} = L(0)^\vee \to M(0)^\vee \to N(0)^\vee \times_\alpha \mathfrak{g} \to \mathfrak{g} \to 0$$

is not integrable. Here, $\mathfrak{g}$ is a finite dimensional complex simple Lie algebra, $M(\lambda)$ its Verma module of highest weight λ, $N(\lambda)$ its greatest proper submodule and $L(\lambda)$ its simple quotient, while $\alpha \in Z^2(\mathfrak{g}, N(0)^\vee)$ is a certain 2-cocycle (such that $\partial[\alpha] = [\theta]$).

The main point why this crossed module does not integrate is that the representations involved are not integrable. An infinite dimensional representation of $\mathfrak{g}$ is integrable to a representation of the 1-connected Lie group G integrating $\mathfrak{g}$ only if the elements f_i and e_i in the Cartan basis of $\mathfrak{g}$ act locally nilpotently. This is not the case for the action on $M(0)^*$ and $N(0)^*$ here.

5.2 Crossed modules in Malcev theory

There is a different way to associate to a group (not necessarily a Lie group!) a Lie algebra, the *Magnus Lie algebra* below in Proposition 5.2.1. It is purely algebraic in nature, but can serve to place an abstract group in a certain geometric context and is therefore is treated in this chapter. This section is due to Kapranov [56].

Given a group G, recall the *lower central series* of successive commutator subgroups associated to G:

$$G = \gamma_1(G) \subset \gamma_2(G) \subset \dots, \quad \gamma_{r+1}(G) = [G, \gamma_r(G)].$$

Recall that for two subgroups A and B of G, $[A, B]$ is the subgroup generated by the commutators $[a, b] = aba^{-1}b^{-1}$ for all $a \in A$ and all $b \in B$. Recall furthermore that a group G is called *nilpotent* if the lower central series terminates in $\{1\}$ after a finite number of steps.

Each $\gamma_{r+1}(G)$ is a normal subgroup in $\gamma_r(G)$ with abelian quotient. Let $\mathbb{K}$ be a field of characteristic zero. Then the $\mathbb{K}$-vector space

$$L(G) := \bigoplus_{r=1}^{\infty} (\gamma_r(G)/\gamma_{r+1}(G)) \otimes_{\mathbb{Z}} \mathbb{K}$$

carries a bracket induced by the group commutator in G. $L(G)$ becomes a $\mathbb{K}$-Lie algebra with this bracket, known as the *Magnus Lie algebra* associated to G.

Proposition 5.2.1. *The $\mathbb{K}$-vector space $L(G)$ is a Lie algebra with respect to the bracket induced by the group commutator.*

Proof. We will only sketch the proof and refer for details to [14] Chapter II, Section 4, numbers 4–5, pages 40–45. Moreover, we will adopt only in this proof the Bourbaki convention that commutators are $[a, b] = a^{-1}b^{-1}ab$ and conjugations are $\mathrm{conj}_b(a) = b^{-1}ab$. Outside this proof, we will stick to our usual convention. We encourage the interested reader to translate the reasonings and formulae in this proof to our usual convention for commutators and conjugations.

By definition, we have $[\gamma_r(G), \gamma_s(G)] \subset \gamma_{r+s}(G)$. This implies that the group commutator induces a well-defined map

$$[,] : \gamma_r(G)/\gamma_{r+1}(G) \times \gamma_s(G)/\gamma_{s+1}(G) \to \gamma_{r+s}(G)/\gamma_{r+s+1}(G).$$

The $\mathbb{K}$-bilinearity comes from the following commutator identity (shown in [13] I, Section 6, number 2, page A I.66, formule (4 bis)):

$$[xx', y] = (x')^{-1}[x, y]x'[x', y].$$

For $a \in \gamma_{r+s}(G)$, we have $a^{-1}(x')^{-1}ax' = [a, x'] \in \gamma_{r+s+1}(G)$, and applying this to the commutator $a = [x, y]$, we obtain that $[x, y] = \mathrm{conj}_{x'}[x, y]$ modulo $\gamma_{r+s+1}(G)$. The above formula implies thus

$$[xx', y] = [x, y][x', y] \quad \text{modulo} \quad \gamma_{r+s+1}(G),$$

that is, the additivity of the Lie bracket in $L(G)$ in the first variable. Observe $[y, x] = [x, y]^{-1}$. This implies together with the above identity

$$[x, yy'] = [x, y][x, y'] \quad \text{modulo} \quad \gamma_{r+s+1}(G),$$

that is, the additivity in the second variable. In total, we have shown bilinearity of the Lie bracket in $L(G)$.

As $[x, x] = 1$, we have that the bracket in $L(G)$ is antisymmetric. The Jacobi identity follows in the same way as the additivity from [13] I page 66, formule (15):

$$[\mathrm{conj}_y(x), [y, z]][\mathrm{conj}_z(y), [z, x]][\mathrm{conj}_x(z), [x, y]] = 1.$$ □

Now let $\mu : M \to N$ be a crossed module of groups. We associate to M the *N-lower central sequence*

$$M = \gamma_1(N, M) \subset \gamma_2(N, M) \subset \cdots,$$

where $\gamma_{r+1}(N, M)$ is the normal subgroup in M generated as a normal subgroup by the elements of the form

$$\eta(n)(m)\, m^{-1} \quad \text{for } n \in N \text{ and } m \in \gamma_r(M, N),$$

where $\eta : N \to \mathrm{Aut}(M)$ is the action of N on M given in the crossed module. A crossed module $\mu : M \to N$ is called *nilpotent* if both series $(\gamma_r(N))_{r\geq 1}$ and $(\gamma_r(N, M))_{r\geq 1}$ terminate in $\{1\}$ after a finite number of steps.

Proposition 5.2.2. (a) *Both*

$$\mu|_{\gamma_r(N,M)} : \gamma_r(N, M) \to \gamma_r(N) \quad \text{and} \quad \bar{\mu} : M/\gamma_r(N, M) \to N/\gamma_r(N)$$

are crossed modules of groups with respect to the induced structure.

(b) *The successive quotients $\gamma_r(N,M)/\gamma_{r+1}(N,M)$ are abelian, and the group commutator in M induces a Lie bracket on*

$$L(N,M) := \bigoplus_{r=1}^{\infty}(\gamma_r(N,M)/\gamma_{r+1}(N,M)) \otimes_{\mathbb{Z}} \mathbb{K}$$

rendering $L(N,M)$ a $\mathbb{K}$-Lie algebra. The group homomorphism $\mu : M \to N$ induces a homomorphism of Lie algebras

$$L(\mu) : L(N,M) \to L(N).$$

(c) *The formula*

$$\gamma_p(N) \times \gamma_q(N,M) \ni (n,m) \mapsto \eta(n)(m)\, m^{-1} \in \gamma_{p+q}(N,M) \tag{5.13}$$

defines an action of $L(N)$ on $L(N,M)$ by derivations and makes

$$L(\mu; M \to N) = (L(\mu) : L(N,M) \to L(N))$$

into a crossed module of Lie algebras.

Proof. (a) Concerning $\mu|_{\gamma_r(N,M)} : \gamma_r(N,M) \to \gamma_r(N)$, the properties of a crossed module are satisfied by restriction of those of $\mu : M \to N$. Indeed, $\mu|_{\gamma_r(N,M)}$ is a homomorphism of groups, because $\mu(\gamma_r(N,M)) \subset \gamma_r(N)$:

$$\mu(\eta(n)(m)\, m^{-1}) = n\mu(m)n^{-1}\mu(m)^{-1} \in \gamma_r(N,M)$$

for $m \in \gamma_{r-1}(N,M)$ and $n \in N$. Further $\gamma_r(N)$ acts by automorphisms on $\gamma_r(N,M)$ because $\gamma_r(N,M)$ is stable under the N-action on M:

$$\eta(n)(\eta(n')(m)\, m^{-1}) = \eta(n)(\eta(n')(m))(\eta(n)(m))^{-1} \in \gamma_r(N,M).$$

The axioms are clear, because they are true in $\mu : M \to N$.
Concerning $\overline{\mu} : M/\gamma_r(N,M) \to N/\gamma_r(N)$, the reasoning is very similar. The subgroup $\gamma_r(N,M)$ in M is normal by definition. The induced morphism is well-defined, because $\mu(\gamma_r(N,M)) \subset \gamma_r(N)$. The action, induced by the action of N on M, is well-defined because $\gamma_r(N,M)$ is stable under the N-action on M. Again, the axioms are true because they are true in $\mu : M \to N$.

(b) The subgroup $\gamma_{r+1}(N,M)$ in $\gamma_r(N,M)$ is normal by restriction. As $\gamma_{r+1}(N,M)$ contains the derived subgroup of $\gamma_r(N,M)$, the quotient $\gamma_r(N,M)/\gamma_{r+1}(N,M)$ is abelian. The fact that the group commutator induces a Lie bracket on $L(N,M)$ follows exactly as in Proposition 5.2.1. The fact that $\mu : M \to N$ is a group homomorphism implies that the induces map $L(\mu)$ from $L(N,M)$ to $L(N)$ is a homomorphism of Lie algebras.

(c) Let us denote $\eta(n)(m)m^{-1} =: \zeta(n)(m)$. First, we show that the (Lie algebra) action is by derivations, that is, $\zeta(n)[m,m'] = [\zeta(n)(m),m'][m,\zeta(n)(m')]$. For $n \in \gamma_r(N)$, $m \in \gamma_p(N,M)$ and $m' \in \gamma_q(N,M)$, this equality takes place in $\gamma_{p+q+r}(N,M)$ and should be true modulo $\gamma_{p+q+r+1}(N,M)$. As we work modulo $\gamma_{p+q+r+1}(N,M)$, $\zeta(n)(m) \in \gamma_{p+r}(N,M)$ commutes with $\zeta(n)(m') \in \gamma_{q+r}(N,M)$, because their commutator has values in $\gamma_{p+q+2r}(N,M)$ and $r \geq 1$. But, for example, m and m' do not commute in general.

We compute

$$\begin{aligned}&[\zeta(n)(m),m'][m,\zeta(n)(m')]\\&\quad = (n\cdot m)m^{-1}m'm(n\cdot m)^{-1}(m')^{-1}m(n\cdot m')(m')^{-1}m^{-1}m'(n\cdot m')^{-1}.\end{aligned}$$

As $m(n\cdot m)^{-1} \in \gamma_{p+r}(N,M)$, $m^{-1}m'm(n\cdot m)^{-1}(m')^{-1}m \in \gamma_{p+r}(N,M)$ as well. This elements commutes thus with $(n\cdot m')(m')^{-1} \in \gamma_{q+r}(N,M)$, and we obtain

$$\begin{aligned}&[\zeta(n)(m),m'][m,\zeta(n)(m')]\\&\quad = (n\cdot m)(n\cdot m')(m')^{-1}m^{-1}m'm(n\cdot m)^{-1}mm^{-1}(n\cdot m')^{-1}\end{aligned}$$

after cancellations (and then inserting mm^{-1} again). In this last line, we identify the commutator $(m')^{-1}m^{-1}m'm \in \gamma_{p+q}(N,M)$, and this commutator commutes with $(n\cdot m^{-1})m \in \gamma_{p+r}(N,M)$, where we have used that the group action is by automorphisms to write $(n\cdot m)^{-1} = n\cdot m^{-1}$. We obtain

$$\begin{aligned}&[\zeta(n)(m),m'][m,\zeta(n)(m')]\\&\quad = (n\cdot m)(n\cdot m')(n\cdot m^{-1})m(m')^{-1}m^{-1}m'(n\cdot m')^{-1}\end{aligned}$$

after cancellation. Commute once again in the same way to put the commutator $m(m')^{-1}m^{-1}m'$ at the very end. We obtain

$$\begin{aligned}&[\zeta(n)(m),m'][m,\zeta(n)(m')]\\&\quad = (n\cdot m)(n\cdot m')(n\cdot m^{-1})(n\cdot m')^{-1}m'm(m')^{-1}m^{-1}m'(m')^{-1}\\&\quad = (n\cdot m)(n\cdot m')(n\cdot m^{-1})(n\cdot m')^{-1}m'm(m')^{-1}m^{-1} = \zeta(n)[m,m'].\end{aligned}$$

The next step is to show under which conditions, the compositions of group automorphisms $\phi,\psi \in \mathrm{Aut}(M)$ is sent to the sum of the corresponding derivations $f(\phi), f(\psi)$, defined by $m \mapsto f(\phi)(m) := \phi(m)m^{-1}$. We have

$$\begin{aligned}\psi(\phi(m))m^{-1} &= \psi(\phi(m))\psi(m^{-1})\psi(m)m^{-1}\\&= \psi(\phi(m)m^{-1})\psi(m)m^{-1}\\&= \psi(f(\phi)(m))f(\psi)(m).\end{aligned}$$

Observe that this formula is close to the chain rule for the derivative of a composition, though different. One can go one step further:

$$\psi(\phi(m))m^{-1} = \psi(f(\phi)(m))\,(f(\phi)(m))^{-1} f(\phi)(m) f(\psi)(m).$$

This formula means that in case the degrees of ϕ, ψ and m are such that the term

$$f(\psi)(f(\phi)(m)) = \psi(f(\phi)(m))\,(f(\phi)(m))^{-1}$$

is zero in the quotient, the derivation corresponding to the composition of the two automorphisms (i. e., $f(\psi \circ \phi)(m) = \psi(\phi(m))m^{-1}$) is equal to the sum of the derivations (i. e., $f(\phi)(m)f(\psi)(m)$). We will apply this to the automorphisms ψ defined by $\psi(m) = (nn')\cdot m$ and ϕ defined by $\phi(m) = (n'n)^{-1}\cdot m$ for two elements $n, n' \in N$ in order to show that ζ is a Lie algebra action, that is,

$$\zeta[n,n'](m) = \zeta(n)(\zeta(n')(m))(\zeta(n')(\zeta(n)(m)))^{-1}. \tag{5.14}$$

Once again, it is essential to consider the degrees of the different terms and to determine in which degree this equality lives and modulo which degree it has to be true. Therefore, take $n \in \gamma_p(N)$, $n' \in \gamma_q(N)$ and $m \in \gamma_r(N, M)$. Then equality (5.14) takes place in $\gamma_{p+q+r}(N, M)$ and we consider it modulo $\gamma_{p+q+r+1}(N, M)$.
We compute

$$\begin{aligned}
\zeta[n,n'](m) &= n\cdot(n'\cdot(n^{-1}\cdot((n')^{-1}\cdot m)))m^{-1}\\
&= nn'\cdot((n'n)^{-1}\cdot m)m^{-1}\\
&= nn'\cdot((n'n)^{-1}\cdot m)(nn'\cdot m^{-1})(nn'\cdot m)m^{-1}\\
&= nn'\cdot(((n'n)^{-1}\cdot m)m^{-1})(nn'\cdot m)m^{-1}\\
&= nn'\cdot(\zeta((n'n)^{-1})(m))(nn'\cdot m)m^{-1}\\
&= nn'\cdot(\zeta((n'n)^{-1})(m))(\zeta((n'n)^{-1})(m))^{-1}\zeta((n'n)^{-1})(m)(nn'\cdot m)m^{-1}\\
&= \zeta(nn')(\zeta((n'n)^{-1})(m))\zeta((n'n)^{-1})(m)\zeta(nn')(m).
\end{aligned}$$

Now observe that by the choice of degrees, the term $\zeta(nn')(\zeta((n'n)^{-1})(m))$ is of degree $2p + 2q + r$, thus zero in the quotient. We therefore obtain modulo $\gamma_{p+q+r+1}(N, M)$:

$$\zeta[n,n'](m) = \zeta((n'n)^{-1})(m)\zeta(nn')(m),$$

and the two terms $\zeta((n'n)^{-1})(m)$ and $\zeta(nn')(m)$ commute. By the above considerations, the composition of automorphisms leads to the sum of derivations and this means that the derivation $\zeta((n'n)^{-1})(-)$ corresponding to the inverse of the derivation $\zeta(n'n)(m)$ must be simply the opposite of $\zeta(n'n)(m)$.

On the other hand,

$$\begin{aligned}
&\zeta(n)(\zeta(n')(m))(\zeta(n')(\zeta(n)(m)))^{-1}\\
&= (nn'\cdot m)(n\cdot m^{-1})m(n'\cdot m^{-1})(n\cdot m)m^{-1}(n'\cdot m)(n'n\cdot m^{-1})\\
&= (nn'\cdot m)m^{-1}m(n\cdot m^{-1})m(n'\cdot m^{-1})(n\cdot m)m^{-1}(n'\cdot m)m^{-1}m(n'n\cdot m^{-1})\\
&= (nn'\cdot m)m^{-1}m(n'n\cdot m^{-1})\\
&= \zeta(nn')(m)(\zeta(n'n)(m))^{-1}.
\end{aligned}$$

Here, we have used that

$$m(n\cdot m^{-1})m(n'\cdot m^{-1})(n\cdot m)m^{-1}(n'\cdot m)m^{-1} \in \gamma_{p+q+2r}(N,M),$$

which means that this term is zero in the quotient. Observe further that in the final expression the two terms $\zeta(nn')(m)$ and $(\zeta(n'n)(m))^{-1}$ commute with each other and give the same terms as for $\zeta[n,n'](m)$. This shows that the ζ is a Lie algebra action.

It remains to show conditions (a) and (b) of a crossed module of Lie algebras, but these are in fact immediate from the conditions (a) and (b) of the crossed module of groups. Indeed,

$$\begin{aligned}
L(\mu)(\zeta(n)(m)) &= \mu((n\cdot m)m^{-1})\\
&= n\mu(m)n^{-1}\mu(m)^{-1}\\
&= [n,\mu(m)].
\end{aligned}$$

In the same vein,

$$\begin{aligned}
\zeta(L(\mu)(m))(m') &= (\mu(m)\cdot m')(m')^{-1}\\
&= mm'm^{-1}(m')^{-1}\\
&= [m,m'].
\end{aligned}$$

□

In this framework, one may not only describe the passage from the group G to its Magnus Lie algebra $L(G)$, but also an integration theory passing from the Lie algebra to a Lie group, which will rather be an algebraic group here. This theory is due to Malcev [76] and works over a field $\mathbb{K}$ of characteristic zero. Let us first describe it for a finite-dimensional nilpotent Lie algebra $\mathfrak{g}$. The Lie algebra $\mathfrak{g}$ is called *nilpotent* if its lower central series

$$\mathfrak{g} =: \gamma_1(\mathfrak{g}) \supset \gamma_2(\mathfrak{g}) \supset \cdots \supset \gamma_{r+1}(\mathfrak{g}) := [\mathfrak{g},\gamma_r(\mathfrak{g})]$$

terminates in $\gamma_N(\mathfrak{g}) = \{0\}$ for some $N > 1$.

Malcev theory [76] constructs a unipotent algebraic group $\exp(\mathfrak{g})$ for the finite-dimensional nilpotent Lie algebra $\mathfrak{g}$. Recall (see, e. g., Chapter XVI in [58]) that the augmentation ideal $I \subset U\mathfrak{g}$ of the universal enveloping algebra $U\mathfrak{g}$ of a Lie algebra $\mathfrak{g}$ defines the I-adic completion $\widehat{U\mathfrak{g}}$ on $U\mathfrak{g}$. By definition, $\widehat{U\mathfrak{g}}$ is the inverse limit of the quotients $U\mathfrak{g}/I^n$ with respect to the surjections arising from the inclusions $I^n \subset I^{n-1}$. As $\mathfrak{g}$ is nilpotent, we have $\bigcap_{n\geq 1} I^n = \{0\}$, and thus $U\mathfrak{g}$ embeds into $\widehat{U\mathfrak{g}}$. As an inverse limit, $\widehat{U\mathfrak{g}}$ carries the inverse limit topology, which is called in this context I-adic topology (described in *loc. cit.*), and $\widehat{U\mathfrak{g}}$ becomes a topological Hopf algebra. The group $\exp(\mathfrak{g})$ (more precisely, the group of $\mathbb{K}$ points $\exp(\mathfrak{g})(\mathbb{K})$ of the algebraic group $\exp(\mathfrak{g})$) is the group of group-like elements in $\widehat{U\mathfrak{g}}$, that is, of those $x \in \widehat{U\mathfrak{g}}$ such that $\triangle(x) = x \otimes x$. As a set, it consists precisely of the elements of the form $\exp(x) = \sum_{i=0}^{\infty} \frac{x^i}{i!}$ for $x \in \mathfrak{g}$. This construction is functorial and associates to a morphism $f : \mathfrak{g} \to \mathfrak{g}'$ of finite-dimensional nilpotent Lie algebras a morphism $\exp(f) : \exp(\mathfrak{g}) \to \exp(\mathfrak{g}')$ of algebraic groups.

For a general Lie algebra $\mathfrak{g}$, one defines its pro-nilpotent completion to be $\widehat{\mathfrak{g}} := \varprojlim \mathfrak{k}$ where $\mathfrak{k}$ runs over the finite-dimensional nilpotent quotients of $\mathfrak{g}$. We have then the pro-algebraic group $\exp(\widehat{\mathfrak{g}}) = \varprojlim \exp(\mathfrak{k})$.

Before we apply this theory to crossed modules, let us identify for a crossed module of Lie algebras $\mu : \mathfrak{m} \to \mathfrak{n}$ the terms of the lower central series of the differential graded Lie algebra $\mu : \mathfrak{m} \to \mathfrak{n}$ with the terms in the $\mathfrak{n}$-lower central series of $\mathfrak{m}$. Indeed, recall that a *differential graded Lie algebra* $\mathfrak{g}^\bullet$ is a graded vector space $\mathfrak{g} = \bigoplus_{l\leq 0} \mathfrak{g}^l$ with a graded bracket $[\mathfrak{g}^l, \mathfrak{g}^m] \subset \mathfrak{g}^{l+m}$ which satisfies graded skewsymmetry and the graded Jacobi identity together with a differential $d : \mathfrak{g}^l \to \mathfrak{g}^{l+1}$, which is a graded derivation of $[,]$ of degree 1 and satisfies $d \circ d = 0$. The lower central series of a differential graded Lie algebra is defined as before to be

$$\mathfrak{g}^\bullet =: \gamma_1(\mathfrak{g}^\bullet) \supset \gamma_2(\mathfrak{g}^\bullet) \supset \cdots \supset \gamma_{r+1}(\mathfrak{g}^\bullet) := [\mathfrak{g}^\bullet, \gamma_r(\mathfrak{g}^\bullet)].$$

Obviously, the $\gamma_r(\mathfrak{g}^\bullet)$ are differential graded Lie subalgebras (even ideals!), and thus differential graded Lie algebras in their own right. A differential graded Lie algebra is called nilpotent in case there exists $N > 1$ such that $\gamma_N(\mathfrak{g}^\bullet) = \{0\}$.

On the other hand, for a crossed module of Lie algebras $\mu : \mathfrak{m} \to \mathfrak{n}$ the $\mathfrak{n}$-lower central series of $\mathfrak{m}$ is defined to be

$$\mathfrak{m} =: \gamma_1(\mathfrak{n}, \mathfrak{m}) \supset \gamma_2(\mathfrak{n}, \mathfrak{m}) \supset \cdots \supset \gamma_{r+1}(\mathfrak{n}, \mathfrak{m}) := \eta(\mathfrak{n})(\gamma_r(\mathfrak{n}, \mathfrak{m})).$$

The crossed module of Lie algebras $\mu : \mathfrak{m} \to \mathfrak{n}$ may be viewed as a differential graded Lie algebra $\mathfrak{g}_\mu^\bullet$ with $\mathfrak{n} =: \mathfrak{g}_\mu^0$ and $\mathfrak{m} =: \mathfrak{g}_\mu^{-1}$, $d := \mu$ and the bracket of an element of $\mathfrak{g}_\mu^0$ and an element of $\mathfrak{g}_\mu^{-1}$ given by the action of $\mathfrak{n}$ on $\mathfrak{m}$. The following proposition compares the terms of the lower central series of the differential graded Lie algebra $(\mu : \mathfrak{m} \to \mathfrak{n}) = \mathfrak{g}_\mu^\bullet$ with crossed submodules constructed from the $\mathfrak{n}$-lower central series of $\mathfrak{m}$.

Proposition 5.2.3. *We have the equality*

$$\gamma_r(\mathfrak{g}^\bullet_\mu) = (\mu|_{\gamma_r(\mathfrak{n},\mathfrak{m})} : \gamma_r(\mathfrak{n},\mathfrak{m}) \to \gamma_r(\mathfrak{n}))$$

between the crossed module $\gamma_r(\mathfrak{g}^\bullet_\mu)$ *(given by the differential graded Lie subalgebra* $\gamma_r(\mathfrak{g}^\bullet_\mu)$*) and the crossed module given by the restriction of* μ *to* $\gamma_r(\mathfrak{n},\mathfrak{m})$ *(which takes values in* $\gamma_r(\mathfrak{n})$*).*

Proof. In the differential graded Lie subalgebra $\gamma_r(\mathfrak{g}^\bullet_\mu)$, there are terms which are entirely (sums of) brackets of elements in $\mathfrak{g}^0$ – these terms constitute $\gamma_r(\mathfrak{g}^0_\mu) = \gamma_r(\mathfrak{n})$. Otherwise, all other nonzero terms are (sums of) brackets of the form $[x,y]$ with $x \in \mathfrak{g}^0$ and $y \in \gamma_{r-1}(\mathfrak{g}^\bullet_\mu)$ for degree reasons. Note that for these $[x,y] = x \cdot y$ by definition of the bracket in $\mathfrak{g}^\bullet_\mu$. Thus we have by induction

$$\gamma_r(\mathfrak{n},\mathfrak{m}) = \eta(\mathfrak{n})(\gamma_{r-1}(\mathfrak{n},\mathfrak{m})) = [\mathfrak{g}^0_\mu, \gamma_{r-1}(\mathfrak{n},\mathfrak{m})] = [\mathfrak{g}^0_\mu, \gamma_{r-1}(\mathfrak{g}^\bullet_\mu)] = \gamma_r(\mathfrak{g}^\bullet_\mu).$$ □

A crossed module of Lie algebras $\mu : \mathfrak{m} \to \mathfrak{n}$ will be called *nilpotent* in case the corresponding differential graded Lie algebra $\mathfrak{g}^\bullet_\mu$ is nilpotent, that is, if there exists $N > 1$ such that $\gamma_N(\mathfrak{g}^\bullet_\mu) = 0$.

Proposition 5.2.4. *Let* $\mu : \mathfrak{m} \to \mathfrak{n}$ *be a crossed module of Lie algebras. Then we have:*
(a) *If* $\mu : \mathfrak{m} \to \mathfrak{n}$ *is nilpotent, then both* $\mathfrak{m}$ *and* $\mathfrak{n}$ *are nilpotent Lie algebras.*
(b) *Assume* $\mu : \mathfrak{m} \to \mathfrak{n}$ *is finite dimensional. Then* $\mu : \mathfrak{m} \to \mathfrak{n}$ *is nilpotent if and only if* $\mathfrak{n}$ *is a nilpotent Lie algebra.*

Proof. (a) The nilpotency of $\mathfrak{g}^\bullet_\mu$ implies in particular the nilpotency of $\mathfrak{g}^0_\mu = \mathfrak{n}$. On the other hand, it also implies the nilpotency of the bracket $[m,m'] = \mu(m) \cdot m'$ for $m, m' \in \mathfrak{m}$, thus the nilpotency of $\mathfrak{m}$.
(b) As in (a), the nilpotency of $\mathfrak{g}^\bullet_\mu$ implies in particular the nilpotency of $\mathfrak{n}$. Conversely, by Engel's theorem for the nilpotent Lie algebra $\mathfrak{n}$ acting on the finite-dimensional vector space $\mathfrak{m}$, there exists $N > 1$ with $\gamma_N(\mathfrak{n},\mathfrak{m}) = 0$. □

Theorem 5.2.5. *Let* $\mu : \mathfrak{m} \to \mathfrak{n}$ *be a nilpotent crossed module of finite-dimensional Lie algebras. Then*

$$\exp(\mu) := (\exp(\mu) : \exp(\mathfrak{m}) \to \exp(\mathfrak{n}))$$

has a natural structure of a crossed module of algebraic groups over $\mathbb{K}$*. This crossed module is nilpotent. The correspondence* $(\mu : \mathfrak{m} \to \mathfrak{n}) \mapsto \exp(\mu)$ *establishes an equivalence of categories between the category of nilpotent crossed modules of finite-dimensional Lie* $\mathbb{K}$*-algebras and the category of nilpotent crossed modules of algebraic groups over* $\mathbb{K}$*.*

Proof. In view of the preceding results, the theorem follows from the fact that Malcev's construction [76] $\mathfrak{g} \mapsto \exp(\mathfrak{g})$ is an equivalence of categories between the category of nilpotent finite-dimensional Lie $\mathbb{K}$-algebras and the category of unipotent algebraic groups over $\mathbb{K}$; cf. Chapter 8 in [33]. □

5.3 Crossed modules of Lie–Rinehart algebras

The crossed modules, which will be discussed in this section, are not crossed modules of Lie groups. They are nevertheless related to geometric structure, because the basic data of a Lie–Rinehart algebra includes a commutative algebra, which has an evident geometric meaning in the realm of affine algebraic geometry. The basic reference for this section is [16].

Definition 5.3.1. Let A be a commutative $\mathbb{K}$-algebra. A *Lie–Rinehart algebra* is an A-module $\mathcal{L}$ endowed with the structure of a $\mathbb{K}$-Lie algebra and a morphism of A-modules and $\mathbb{K}$-Lie algebras

$$\mathrm{anc} : \mathcal{L} \to \mathrm{Der}(A),$$

called the *anchor*, such that for all $X, Y \in \mathcal{L}$ and all $a \in A$, we have

$$[X, aY] = a[X, Y] + \mathrm{anc}(X)(a)\, Y.$$

Example 5.3.2. (a) The most basic example of a Lie–Rinehart algebra is the Lie algebra and A-module of all derivations $\mathrm{Der}(A)$ itself. The anchor is the identity map and all axioms are easily checked.

(b) The other extreme example of a Lie–Rinehart algebra is a $\mathbb{K}$-Lie algebra, where we have $A = \mathbb{K}$ and $\mathrm{Der}_{\mathbb{K}}(\mathbb{K}) = \{0\}$.

(c) From a more geometric point of view, the first example can be viewed as the Lie algebra of vector fields on a manifold. This becomes very clear, for example, in the case $A = C^{\infty}(M)$ for a manifold M. More generally, any Lie algebroid L gives rise to an example of a Lie–Rinehart algebra. Indeed, L is a vector bundle over a manifold M whose space of sections $\Gamma(L)$ is an A-module for $A = C^{\infty}(M)$ and a $\mathbb{K} = \mathbb{R}$-Lie algebra, which satisfies the conditions. These are exactly the examples which arise from finitely-generated projective A-modules.

(d) Observe that A-Lie algebras, that is, Lie algebras whose Lie bracket is A-linear, are examples of Lie–Rinehart algebras. These are exactly the examples where the anchor is trivial. Geometrically, they correspond the Lie algebra bundles.

Morphisms of Lie–Rinehart algebras $\varphi : \mathcal{L}_1 \to \mathcal{L}_2$ are morphisms of A-modules and $\mathbb{K}$-algebras such that $\mathrm{anc}_1 = \mathrm{anc}_2 \circ \varphi$.

Definition 5.3.3. Let $\mathcal{L}$ be a Lie–Rinehart algebra and M be an A-module. We will say that $\mathcal{L}$ *acts* on M or that M is an $\mathcal{L}$-*module* if there is a $\mathbb{K}$-linear map

$$\mathcal{L} \otimes M \to M, \quad (X, m) \mapsto X \cdot m$$

for all $X \in \mathcal{L}$ and all $m \in M$ such that:

(a) $[X, Y] \cdot m = X \cdot (Y \cdot m) - Y \cdot (X \cdot m)$ for all $X, Y \in \mathcal{L}$ and all $m \in M$;

(b) $(aX) \cdot m = a(X \cdot m)$ for all $X \in \mathcal{L}$, $m \in M$ and all $a \in A$;

(c) $X \cdot (am) = a(X \cdot m) + \mathrm{anc}(X)(a)\, m$ for all $X \in \mathcal{L}$, $m \in M$ and all $a \in A$.

Observe that condition (a) means that M carries the action of the $\mathbb{K}$-Lie algebra $\mathcal{L}$, while condition (b) means the A-linearity of this action, viewed as a map $\mathcal{L} \to \mathrm{End}_{\mathbb{K}}(M)$. Sometimes M is called a zeroth-order module, because the map $\mathcal{L} \to \mathrm{End}_{\mathbb{K}}(M)$ is a zeroth-order differential operator. Condition (c) means that the operators $(X\cdot) \in \mathrm{End}_{\mathbb{K}}(M)$ are first-order differential operators with symbol anc.

Definition 5.3.4. Let $\mathcal{L}$ be a Lie–Rinehart algebra and $\mathcal{R}$ be an A-Lie algebra. We say that $\mathcal{L}$ *acts* on $\mathcal{R}$ *by derivations* in case $\mathcal{R}$ is an $\mathcal{L}$-module such that for all $X \in \mathcal{L}$ and all $r, r' \in \mathcal{R}$,

$$X \cdot [r, r'] = [X \cdot r, r'] + [r, X \cdot r'].$$

Definition 5.3.5. A *crossed module of Lie–Rinehart algebras* $\mu : \mathcal{R} \to \mathcal{L}$ is a morphism of Lie–Rinehart algebras between an A-Lie algebra $\mathcal{R}$ and a Lie–Rinehart algebra $\mathcal{L}$ together with an action of $\mathcal{L}$ on $\mathcal{R}$ by derivations such that:

(a) $\mu(X \cdot r) = [X, \mu(r)]$ for all $X \in \mathcal{L}$ and all $r \in \mathcal{R}$;
(b) $\mu(r) \cdot r' = [r, r']$ for all $r, r' \in \mathcal{R}$;
(c) the composition $\mathcal{R} \stackrel{\mu}{\to} \mathcal{L} \stackrel{\mathrm{anc}}{\to} \mathrm{Der}(A)$ is zero.

Example 5.3.6. We have the usual classes of examples, adapted to the Lie–Rinehart setting:

(a) The kernel of a morphism of Lie–Rinehart algebras is a crossed module. Defining a Lie–Rinehart ideal in the same way (namely as a Lie–Rinehart subalgebra $\mathcal{N} \subset \mathcal{L}$, which is an ideal in the sense of $\mathbb{K}$-Lie algebras such that the composition $\mathcal{N} \subset \mathcal{L} \stackrel{\mathrm{anc}}{\to} \mathrm{Der}(A)$ is zero), the inclusion map of an ideal is an example of a crossed module, too.
(b) For any $\mathcal{L}$-module $\mathcal{R}$, the zero map $\mathcal{R} \to \mathcal{L}$ is a crossed module.
(c) Let $\mu : \mathcal{R} \to \mathcal{L}$ be a surjective morphism of Lie–Rinehart algebras whose kernel lies in the center $Z(\mathcal{R})$ of $\mathcal{R}$, that is, the intersection of the center of $\mathcal{R}$ as a $\mathbb{K}$-Lie algebra and the kernel of anc. Then $\mu : \mathcal{R} \to \mathcal{L}$ is a crossed module of Lie–Rinehart algebras. The action is given by $X \cdot r := [r', r]$ where $\mu(r') = X$ for all $X \in \mathcal{L}$ and $r, r' \in \mathcal{R}$.

Let $\mu : \mathcal{R} \to \mathcal{L}$ be a crossed module of Lie–Rinehart algebras. Notice that $\mathrm{Im}(\mu)$ is a $\mathbb{K}$-Lie algebra ideal and an A-submodule of $\mathcal{L}$, thus $\mathrm{Coker}(\mu) = \mathcal{L} / \mathrm{Im}(\mu)$ is a Lie–Rinehart algebra. Furthermore, $\mathrm{Ker}(\mu)$ is an abelian and central ideal of $\mathcal{R}$ and the action of $\mathcal{L}$ on $\mathcal{R}$ induces a $\mathrm{Coker}(\mu)$-module structure on $\mathrm{Ker}(\mu)$. In the same sense as before, one introduces a notion of morphism and an equivalence relation on the set of crossed modules of Lie–Rinehart algebras $\mathrm{crmod}(\mathcal{L}, M)$ with fixed cokernel $\mathcal{L}$ and fixed kernel M.

Recall from Appendix D that the cohomology of Lie–Rinehart algebras $H^*(\mathcal{L}, M)$ of a Lie–Rinehart algebra $\mathcal{L}$ with values in an $\mathcal{L}$-module M is given by the usual

Chevalley–Eilenberg cohomology of the Lie algebra underlying $\mathcal{L}$, but with cochains which are linear over A. The main result of this section reads then the following.

Theorem 5.3.1. *For any Lie–Rinehart algebra $\mathcal{L}$ which is projective as an A-module and any $\mathcal{L}$-module M, there exists a natural bijection between* $\mathrm{crmod}(\mathcal{L}, M)$ *and* $H^3(\mathcal{L}, M)$.

As in Remark 2.9.5, this absolute version of the theorem is deduced from the following relative version. The only thing to know for this is that there are enough injective modules in the category of Lie–Rinehart modules and Lie–Rinehart cohomology vanishes on injective modules. This follows from the fact that Lie–Rinehart cohomology is an Ext-functor for a Lie–Rinehart algebra $\mathcal{L}$ which is projective over A; see Proposition 1.14 in [55]. The relative version has the advantage of treating objects which live one level lower in cohomological degree and are thus easier to manipulate.

Theorem 5.3.2. *Suppose $\mathcal{L}$ is a Lie–Rinehart algebra which is projective as an A-module. Let $\pi : \mathcal{L} \to \mathcal{G}$ be a surjective morphism of Lie–Rinehart algebras, which admits an A-linear section. Then there exists a natural bijection between* $\mathrm{crmod}(\mathcal{L}, \mathcal{G}, M)$ *and* $H^3(\mathcal{L}, \mathcal{G}, M)$.

Proof. The proof of Theorem 5.3.2 is very similar to the proof of Theorem 2.9.1. We only have to prove the A-linearity of cochains at different stages of the proof.

Let $\mu : \mathcal{R} \to \mathcal{L}$ be a crossed module with quotient morphism $\pi : \mathcal{L} \to \mathcal{G}$ and with $\mathrm{Ker}(\mu)$ identified with M as a $\mathcal{G}$-module. We associate to the crossed module μ a 2-cochain f with values in M associated to the A-linear section $s : \mathcal{G} \to \mathcal{L}$ of π and an A-linear section $\sigma : \mathrm{Im}(\mu) = \mathrm{Ker}(\pi) \to \mathcal{R}$ of μ. Observe that for all $X, Y \in \mathcal{G}$, we set

$$g(X, Y) = \sigma([s(X), s(Y)] - s[X, Y]) \in \mathcal{R}.$$

This corresponds to our $\beta(X, Y)$ in equation (2.1) of Chapter 2. In the present context, the cochain g is A-linear, that is, $g(aX, Y) = ag(X, Y) = g(X, aY)$. For example, the first equality follows from

$$\begin{aligned} g(aX, Y) &- ag(X, Y) \\ &= \sigma([s(aX), s(Y)] - s[aX, Y]) - a\sigma([s(X), s(Y)] - s[X, Y]) \\ &= -\sigma(\mathrm{anc}_{\mathcal{L}}(s(Y))(a)s(X)) + \sigma(s(\mathrm{anc}_{\mathcal{G}}(Y)(a)X)) \\ &= -\sigma(\mathrm{anc}_{\mathcal{L}}(s(Y))(a)s(X) - \mathrm{anc}_{\mathcal{G}}(Y)(a)s(X)) = 0, \end{aligned}$$

because $\mathrm{anc}_{\mathcal{L}}(s(Y)) = \mathrm{anc}_{\mathcal{G}}(Y)$ as s is a section of π which is a morphism of Lie–Rinehart algebras ($Z(a) = \pi(Z)(a)$ implies $s(Y)(a) = (\pi \circ s)(Y)(a) = Y(a)$).

Furthermore, we set for all $Z \in \mathcal{L}$,

$$\Psi(Z) = \sigma(Z - s \circ \pi(Z)) \in \mathcal{R}.$$

With these notation, put

$$f(Z, Z') = g(\pi(Z), \pi(Z')) - Z' \cdot \Psi(Z) + Z \cdot \Psi(Z') - [\Psi(Z), \Psi(Z')] - \Psi[Z, Z'].$$

As in the proof of Theorem 2.9.1, this 2-cochain on $\mathcal{L}$ has values in M. We have to prove here that it is A-linear, that is,

$$f(aZ,Z') = af(Z,Z') = f(Z,aZ')$$

for all $Z,Z' \in \mathcal{L}$ and all $a \in A$. Observe that not only π and Ψ are A-linear, but $g(\pi(Z),\pi(Z'))$ is A-bilinear and $[\Psi(Z),\Psi(Z')]$ is $\mathbb{K}$-bilinear in (Z,Z'), because g is A-bilinear and the bracket in $\mathcal{R}$ is $\mathbb{K}$-bilinear. Furthermore, the action of $\mathcal{L}$ on $\mathcal{R}$ is A-linear in the sense that the corresponding map $\mathcal{L} \to \mathrm{End}_{\mathbb{K}}(\mathcal{R})$ is A-linear. Thus we have

$$\begin{aligned} &f(aZ,Z') - af(Z,Z') \\ &\quad = -[Z',a\Psi(Z)] - \Psi([aZ,Z']) + a[Z',\Psi(Z)] + a\Psi([Z,Z']) \\ &\quad = -\mathrm{anc}(Z')(a)\Psi(Z) + \mathrm{anc}(Z')(a)\Psi(Z) = 0. \end{aligned}$$

The complex of relative Lie–Rinehart algebra cohomology is by definition the quotient complex

$$0 \to C^*(\mathcal{G},M) \overset{\pi^*}{\to} C^*(\mathcal{L},M) \overset{\kappa^*}{\to} C^*(\mathcal{G},\mathcal{L},M) \to 0.$$

The relative cocycle we associate to the crossed module μ is now $\kappa^* f \in C^2(\mathcal{G},\mathcal{L},M)$. The cohomology in $C^2(\mathcal{G},\mathcal{L},M)$ is denoted $H^3(\mathcal{G},\mathcal{L},M)$. In order to show that $\kappa^* f$ is a cocycle, we introduce as before a cochain k defined by

$$k(X,Y,Z) = \sum_{\text{cycl.}} g(X,[Y,Z]) + \sum_{\text{cycl.}} s(X)\cdot g(Y,Z) \in M.$$

Now the situation is the following:

$$\begin{array}{ccccc} C^2(\mathcal{G},M) & \xrightarrow{\pi^*} & C^2(\mathcal{L},M) & \xrightarrow{\kappa^*} & C^2(\mathcal{G},\mathcal{L},M) \\ \downarrow d & & \downarrow d & & \downarrow d \\ C^3(\mathcal{G},M) & \xrightarrow{\pi^*} & C^3(\mathcal{L},M) & \xrightarrow{\kappa^*} & C^3(\mathcal{G},\mathcal{L},M) \end{array}$$

As in Chapter 2, a direct computation shows $df = v^* k$. This identity then implies then $d\kappa^* f = \kappa^* df = \kappa^* \pi^* f = 0$ and, therefore, $\kappa^* f$ is a cocycle. As in Chapter 2, one shows that the class of $\kappa^* f$ does not depend on the choice of the sections s and σ and that relatively equivalent crossed modules give rise to the same cohomology class.

All we have done this far can be resumed in the existence of a well-defined map

$$\mathrm{crmod}(\mathcal{G},\mathcal{L},M) \to H^3(\mathcal{G},\mathcal{L},M), \quad [\mu : \mathcal{R} \to \mathcal{L}] \mapsto [\kappa^* f].$$

Conversely, suppose given a cocycle in $C^2(\mathcal{G},\mathcal{L},M)$ which we lift to a cochain $f \in C^2(\mathcal{L},M)$. As $\kappa^* f$ is a cocycle, we have a cochain $k \in C^3(\mathcal{G},M)$ such that $df = \pi^* k$. In particular, the restriction of f to $\mathrm{Ker}(\pi)$ gives a cocycle in $C^2(\mathrm{Ker}(\pi), M)$. We get thus a Lie–Rinehart algebra structure on the A-module direct sum $\mathcal{R} = M \oplus \mathrm{Ker}(\pi)$, which makes it a central extension using the bracket

$$[(m,l),(m',l')] = (f(l_1,l_2),[l_1,l_2]).$$

Observe that $\mathcal{R}$ is with this bracket an A-Lie algebra, because of the A-linearity of f and the fact that the restriction of f onto $\mathrm{Ker}(\pi) \times \mathrm{Ker}(\pi)$ is a cocycle. Restriction onto $\mathcal{L} \times \mathrm{Ker}(\pi)$, we obtain from f an action of $\mathcal{L}$ on $\mathcal{R}$ by the formula

$$Z \cdot (m,l) = (\pi(Z) \cdot m + f(Z,l),[Z,l]);$$

cf. equation (2.4). One easily verifies that with these data, the map $\mu : \mathcal{R} \to \mathcal{L}$, given by $(m,l) \mapsto l$, is a crossed module. The addition of a coboundary to f does not affect the (relative) equivalence class of this crossed module.

We thus get a well-defined map

$$H^3(\mathcal{G},\mathcal{L},M) \to \mathrm{crmod}(\mathcal{G},\mathcal{L},M), \quad [\kappa^* f] \mapsto [\mu : \mathcal{R} \to \mathcal{L}].$$

By construction, we obtain as associated cohomology class to this crossed module the class of f. In the other direction, the two maps also compose to the identity. This is shown by constructing explicitly a morphism of crossed modules from the given crossed module to the one which we have constructed from its cohomology class. □

5.4 Exercises

Exercise 5.4.1. In this exercise, we give an example of a nontrivial Lie group extension of the Lie group $G = \mathbb{R}^2$ by the abelian Lie group $A = \mathbb{R}$; cf. [83] Remark 2.16.

(a) Show that the smooth cohomology of G, $H^2_s(G,A)$, is isomorphic to the continuous cohomology of its Lie algebra $\mathfrak{g}$ with values in $\mathfrak{a} = \mathbb{R}$, that is,

$$H^2_s(G,A) \cong H^2_{\mathrm{cont}}(\mathfrak{g},\mathfrak{a}).$$

(b) Compute $H^2_{\mathrm{cont}}(\mathfrak{g},\mathfrak{a})$ explicitly, noting that all cochains are continuous. Distinguish the cases of trivial and nontrivial action.

(c) Conclude that

$$H^2_s(G,A) \cong H^2_{\mathrm{cont}}(\mathfrak{g},\mathfrak{a}) \cong \begin{cases} \mathbb{R} & \text{for } \mathfrak{g} \cdot \mathfrak{a} = \{0\} \\ \{0\} & \text{for } \mathfrak{g} \cdot \mathfrak{a} \neq \{0\}. \end{cases}$$

Exercise 5.4.2. In this exercise, we illustrate Corollary 5.1.10 by showing that the action of $H^2_s(G,Z(N))$ on $\mathrm{Ext}(G,N)_{[S]}$ can be expressed as the Baer product with respect to the extension of G by $Z(N)$; cf. [83], Remark 2.15.

(a) Recall that the center $Z(N)$ of a Lie group N is an initial subgroup and that a smooth outer action of G on N induces on $Z(N)$ the structure S_Z of a smooth G-module. Deduce

$$H_s^2(G, Z(N))_S \cong \mathrm{Ext}(G, Z(N))_{S_Z}.$$

(b) Let now $Z(N) \to \widehat{G}_1 \overset{q_1}{\to} G$ be a Lie group extension of G by the smooth G-module $Z(N)$ and $N \to \widehat{G}_2 \overset{q_2}{\to} G$ be a Lie group extension of G by N corresponding to the G-kernel $[S]$. Show that the group

$$H := q_1^* \widehat{G}_2 = \{(g_1, g_2) \in \widehat{G}_1 \times \widehat{G}_2 \mid q_1(g_1) = q_2(g_2)\}$$

is a Lie group extension of G by $Z(N) \times N$. Show that its subgroup $\triangle_Z := \{(z, z^{-1}) \mid z \in Z(N)\}$ is a central split Lie subgroup of N.

(c) Show that the *Baer product* $\widehat{G} := H / \triangle_Z$ is a Lie group extension of G by $N \cong (Z(N) \times N) / \triangle_Z$.

(d) Show that if $\widehat{G}_1 \cong Z(N) \times_f G$ and $\widehat{G}_2 \cong N \times_{(S,\omega)} G$, then

$$H \cong (Z(N) \times N) \times_{((S_Z,S),(f,\omega))} G$$

and that $\widehat{G} \cong N \times_{(S, f\cdot\omega)} G$.

This implies that the action of the abelian group $H_s^2(G, Z(N))$ on $\mathrm{Ext}(G, N)_{[S]}$ can be expressed as the Baer product with respect to the extension of G by $Z(N)$.

Exercise 5.4.3. Computation of $\mathrm{Im}(\delta)$ of the map

$$\delta : Z_s^1(G, N, Z(N))_S / C_s^1(G, Z(N)) \to H_s^2(G, Z(N))_S, \quad [f] \mapsto [d_S f]$$

in the special case $S = 1$; cf. Remark 2.19 in [83]. We consider the groups $N = \widetilde{\mathrm{Sl}}_2(\mathbb{R})$, the universal covering group of $\mathrm{Sl}_2(\mathbb{R})$ and the discrete group G with $2g$ generators $\alpha_1, \ldots, \alpha_{2g}$ subject to the commutator relation

$$[\alpha_1, \alpha_2] \cdots [\alpha_{2g-1}, \alpha_{2g}] = 1$$

with $[\alpha_k, \alpha_l] := \alpha_k \alpha_l \alpha_k^{-1} \alpha_l^{-1}$.

(a) Show that the center $Z(N)$ is isomorphic to $\mathbb{Z}$, and thus $N_{\mathrm{ad}} := N / Z(N) \cong \mathrm{PSl}_2(\mathbb{R})$. Therefore, a homomorphism $h : G \to N_a d$ corresponds to a $(2g)$-tuple of points $(x_1, \ldots, x_{2g}) \in \mathrm{PSl}_2(\mathbb{R})^{2g}$ satisfying

$$[x_1, x_2] \cdots [x_{2g-1}, x_{2g}] = 1.$$

(b) Using that G is the fundamental group $\pi_1(\Sigma_g)$ of a compact connected orientable surface Σ_g of genus g and that $H_2(\Sigma_g) \cong \mathbb{Z}$, show that

$$H^2(G, Z(N)) \cong H^2(\Sigma_g, Z(N)) \cong \mathrm{Hom}(H_2(\Sigma_g), Z(N)) \cong Z(N).$$

This describes the locally smooth cohomology $H_s^2(G, Z(N))_S$, because G and $Z(N)$ are discrete.

(c) Deduce from the above that $\mathrm{Im}(\delta)$ consists of the set

$$\{[y_1, y_2] \cdots [y_{2g-1}, y_{2g}] \mid (y_1, \ldots, y_{2g}) \in N^{2g}, \quad [y_1, y_2] \cdots [y_{2g-1}, y_{2g}] \in Z(N)\}.$$

For more information about this, see Remark 2.19 in [83].

Exercise 5.4.4. This exercise is about the 7-term exact sequence in Corollary 5.1.13. Recall the context. The quotient group $L_{\mathrm{ad}} := L/Z(L)$ carries a canonical Lie group structure for which $q_L : L \to L_{\mathrm{ad}}$ defines a central extension of L_{ad} by $Z(L)$. Furthermore, let $S : G \to \mathrm{Aut}(L)$ be a smooth action of G on L and consider the induced action $\overline{S}$ of G on L_{ad}. Then we want to show that there is an exact sequence of groups, respectively, pointed sets

$$Z(L)^G \hookrightarrow L^G \to L_{\mathrm{ad}}^G \to H_s^1(G, Z(L))_S \to H_s^1(G, L)_S \to H_s^1(G, L_{\mathrm{ad}})_{\overline{S}} \xrightarrow{\delta} H_s^2(G, Z(L))_S,$$

where for $f \in Z_s^1(G, L_{\mathrm{ad}})_S$ we put

$$\delta([f]) = [d_S \widehat{f}]$$

for some $\widehat{f} \in Z_s^1(G, L, Z(L))_S$ satisfying $q_L \circ \widehat{f} = f$ and

$$\mathrm{Ext}(G, L)_{[S]}^{\mathrm{split}} = -\mathrm{Im}(\delta) \cdot [L \rtimes_S G],$$

where $\mathrm{Ext}(G, L)_{[S]}^{\mathrm{split}}$ is the set of split extension classes.

(a) Show that $Z(L)^G \hookrightarrow L^G \to L_{\mathrm{ad}}^G$ is exact.

(b) Define a map $L_{\mathrm{ad}}^G \to C_s^1(G, Z(L))_S$ by $l \mapsto (x \mapsto (S(x)(l)l^{-1}))$. Show that it takes values in $Z_s^1(G, Z(L))_S$. Show exactness of the sequence at L_{ad}^G and $H_s^1(G, Z(L))_S$.

(c) Show exactness of $H_s^1(G, Z(L))_S \to H_s^1(G, L)_S \to H_s^1(G, L_{\mathrm{ad}})_{\overline{S}}$.

(d) Show exactness of the sequence at $H_s^1(G, L_{\mathrm{ad}})_{\overline{S}}$.

5.5 Bibliographical notes

Section 5.1 is taken from Neeb's article [83]. The scope of Neeb's article are infinite dimensional Lie groups. In order to simplify things in this book, we reduced and adapted Neeb's article to finite dimensional Lie groups. This implies that we do not need to consider the *strongly smooth cohomology* $H_{ss}^*(G, N)$. On the other hand, this does not make the theory empty, as explained in [102]. We encourage the interested reader to go on to study the infinite dimensional case with the help of Neeb's article. Section 5.1.6 is new, drawing on (within other things) Neeb's article [82].

Section 5.2 is taken from Kapranov's (unpublished?) article [56]. Section 5.3 is taken from Casas–Ladra–Pirashvili [16].

6 Crossed modules of racks

In this chapter, we generalize crossed modules of groups to crossed modules of racks. Racks are generalizations of groups whose axioms capture the essential properties of group conjugation and algebraically encode two of the Reidemeister moves. Because of the latter, they have proven useful in defining link and knot invariants.

It turns out that Lie racks, Leibniz algebras and rack bialgebras display similar relations as Lie groups, Lie algebras and bialgebras. One goal of this section is to lift these relations to crossed modules, generalizing the relations between crossed modules of Lie groups, crossed modules of Lie algebras and crossed modules of bi/Hopf algebras discussed in the previous sections. There are two competing versions of a crossed module of racks, the one-sided version and the two-sided version. While the one-sided version is explored in its relation to categorical racks and topology, the two-sided version is more adapted to the relation with Leibniz algebras and rack bialgebras.

6.1 Racks and related notions

6.1.1 Racks and Leibniz algebras

We begin by recalling the notion of a rack, which results from axiomatizing the properties of group conjugation.

Definition 6.1.1. A (left) *rack* consists of a set X equipped with a binary operation denoted $(x,y) \mapsto x \rhd y$ such that for all x, y, and $z \in X$, the map $y \mapsto x \rhd y$ is bijective and

$$x \rhd (y \rhd z) = (x \rhd y) \rhd (x \rhd z).$$

In the same way, a right rack consists of a set X equipped with a binary operation denoted $(x,y) \mapsto x \lhd y$ such that for all x, y, and $z \in X$, the map $x \mapsto x \lhd y$ is bijective and

$$(x \lhd y) \lhd z = (x \lhd z) \lhd (y \lhd z).$$

One can always transform a left rack into a right rack (and vice versa) by sending the bijective map $y \mapsto x \rhd y$ to its inverse (which is then denoted $z \mapsto z \lhd x$). In this book, we will mainly work with left racks, but in fact, any left rack carries also a right rack operation.

Proposition 6.1.1. *A rack can be defined equivalently as a set X with two binary operations $(x,y) \mapsto x \rhd y$ and $(x,y) \mapsto x \lhd y$ such that for all $x, y, z \in X$:*

(a) $x \rhd (y \rhd z) = (x \rhd y) \rhd (x \rhd z)$;

https://doi.org/10.1515/9783110750959-006

(b) $(x \lhd y) \lhd z = (x \lhd z) \lhd (y \lhd z)$;
(c) $(x \rhd y) \lhd x = y$;
(d) $y \rhd (x \lhd y) = x$.

Proof. Note that identities (c) and (d) express the fact that the two binary operations are inverses of each other. It therefore remains to show the following: Given a (left) rack $(X, \rhd)$, identity (b) holds for the right rack product $(x, y) \mapsto x \lhd y$. Indeed, let $x, y, z \in X$, then we have on the one hand,

$$((x \rhd (y \rhd z)) \lhd x) \lhd y = z,$$

and on the other hand,

$$(((x \rhd y) \rhd (x \rhd z)) \lhd (x \rhd y)) \lhd x = z.$$

This gives for $w := x \rhd (y \rhd z) = (x \rhd y) \rhd (x \rhd z)$ the relation

$$(w \lhd x) \lhd y = (w \lhd (x \rhd y)) \lhd x. \tag{6.1}$$

Now replace $y := u \lhd v$ and $x = v$. Then $x \rhd y$ becomes $v \rhd (u \lhd v) = u$ and, therefore, equation (6.1) becomes

$$(w \lhd v) \lhd (u \lhd v) = (w \lhd u) \lhd v.$$ □

The conjugation in a group G gives rise to a (left) rack operation given by $(g, h) \mapsto ghg^{-1}$ and a right rack operation by $(g, h) \mapsto g \lhd h := h^{-1}gh$.

The notion of a unit leads to pointed racks.

Definition 6.1.2. A *pointed rack* $(X, \rhd, 1)$ consists of a set X equipped with a binary operation $\rhd$ and an element $1 \in X$ satisfying:
(a) $x \rhd (y \rhd z) = (x \rhd y) \rhd (x \rhd z)$;
(b) For all $a, b \in X$, there exists a unique $x \in X$ such that $a \rhd x = b$;
(c) $1 \rhd x = x$ and $x \rhd 1 = 1$ for all $x \in X$.

Once again, the conjugation rack of a group is an example of a pointed rack. For formal reasons, we will denote the conjugation rack underlying the group G by $\mathrm{Conj}(G)$. Denote by Racks the category of racks, that is, the category whose objects are racks and whose morphisms are rack homomorphisms as defined below. Then Conj is a functor from the category of groups Grp to Racks.

Definition 6.1.3. Let R and S be two racks. A *morphism of racks* is a map $\mu : R \to S$ such that

$$\mu(r \rhd r') = \mu(r) \rhd \mu(r')$$

for all $r \in R$. In the usual way, we will speak about iso- and automorphisms of racks.

Definition 6.1.4. Let R be a rack. The *associated group* to R, denoted $\mathrm{As}(R)$, is the quotient of the free group $F(R)$ on the set R by the normal subgroup generated by the elements $xy^{-1}x^{-1}(x \rhd y)$ for all $x, y \in R$. Denote the canonical morphism of racks by $i : R \to \mathrm{As}(R)$.

The importance of the associated group $\mathrm{As}(R)$ of a rack R comes from the following universal mapping property.

Lemma 6.1.2. *Let R be a rack and G be a group. For any morphism of racks $f : R \to \mathrm{Conj}(G)$, there exists a unique group morphism $g : \mathrm{As}(R) \to G$ such that $g \circ i = f$.*

Proof. By freeness of the free group on the set R, a morphism of racks $R \to \mathrm{Conj}(G)$ induces a group homomorphism $F(R) \to G$. This morphism sends all elements $xy^{-1}x^{-1}(x \rhd y)$ to $1 \in G$, because both the commutator in $F(R)$ and the rack product are sent to the commutator in G. The first assertion is true because $F(R) \to G$ is a group homomorphism, and the second assertion is true because $R \to \mathrm{Conj}(G)$ is a rack morphism. Thus $F(R) \to G$ factors through a morphism $\mathrm{As}(R) \to G$. This group homomorphism is unique, because it is fixed on the generators of $\mathrm{As}(R)$. □

From this, one can deduce that the functor $\mathrm{As} : \mathsf{Racks} \to \mathsf{Grp}$ from the category of racks to the category of groups is left adjoint to the functor $\mathrm{Conj} : \mathsf{Grp} \to \mathsf{Racks}$ which associates to a group its underlying conjugation rack.

Remark 6.1.5. In fact, the unit of the adjunction is just the map i. By standard arguments, the unit of the adjunction is injective, but only as a map

$$i : \mathrm{Conj}(G) \to \mathrm{Conj}(\mathrm{As}(\mathrm{Conj}(G)))$$

for a group G.

We observe that the compositions $\mathrm{Conj}(\mathrm{As}(R))$ for a racks R and $\mathrm{As}(\mathrm{Conj}(G))$ for a group G are, in general, far from being equal to R or G, respectively. For example, for an abelian group A, the conjugation rack $\mathrm{Conj}(A)$ is the set A with the trivial rack product, while $\mathrm{As}(\mathrm{Conj}(A))$ is the free abelian group on the set A.

Definition 6.1.6. Let G be a group and X be a G-set. We say that X together with a map $p : X \to G$ is an *augmented rack* in case it satisfies the *augmentation identity*, that is,

$$p(g \cdot x) = g\,p(x)\,g^{-1}$$

for all $g \in G$ and all $x \in X$.

We observe that for any augmented rack $p : X \to G$, one may define a rack operation on X as $x \rhd x' := p(x) \cdot x'$ for all $x, x' \in X$. Then the map p becomes an equivariant map (with respect to the given G-action on X and the conjugation action on the group G) and a morphism of racks. Augmented racks are in fact the Yetter–Drinfel'd modules over the Hopf algebra G (in the symmetric monoidal category of sets), or in other words, the Drinfel'd center of the symmetric monoidal category of G-modules; see [34] and Exercise 4.8.4.

Example 6.1.7. There are many examples of augmented racks. For example, for each rack R, the canonical morphisms $R \to \mathrm{Aut}(R)$ and $i : R \to \mathrm{As}(R)$ are augmented racks.

Definition 6.1.8. Let R be a rack and X be a set. We say that R *acts* on X or that X is an *R-set* in case there are bijections $(r\cdot) : X \to X$ for all $r \in R$ such that

$$r \cdot (r' \cdot x) = (r \rhd r') \cdot (r \cdot x)$$

for all $x \in X$ and all $r, r' \in R$.

Lemma 6.1.3. *An action of R on X is equivalent to a morphism of racks $\mu : R \to \mathrm{Bij}(X)$ with values in the conjugation rack underlying the group of bijections on X.*

Proof. Indeed, the defining equation in Definition 6.1.3 can be written

$$(r\cdot) \circ (r'\cdot) \circ (r\cdot)^{-1}(y) = (r \rhd r') \cdot (y)$$

for all $r, r' \in R$ and all $y \in X$. This shows that the rack product in R is sent to the rack product in $\mathrm{Conj}(\mathrm{Bij}(X))$. □

Remark 6.1.9. Exactly as in Proposition 6.1.1, one can induce from a left action of R on X a right action of R on X by passing to the inverse: $x \cdot r := (r \cdot -)^{-1}(x)$. This right action satisfies the property of a right rack action, that is,

$$(x \cdot r'') \cdot r = (x \cdot r) \cdot (r'' \lhd r).$$

Indeed, we have for all $x \in X$ and all $r, r' \in R$ that

$$\begin{aligned}(x \cdot r) \cdot r' &= (r' \cdot -)^{-1}((r \cdot -)^{-1}(x)) = ((r \cdot -) \circ (r' \cdot -))^{-1}(x) \\ &= (((r \rhd r') \cdot -) \circ (r \cdot -))^{-1}(x) = (r \cdot -)^{-1} \circ ((r \rhd r') \cdot -)^{-1}(x) \\ &= (x \cdot (r \rhd r')) \cdot r.\end{aligned}$$

Now replace $r' := r'' \lhd r$ to get $r \rhd r' = r \rhd (r'' \lhd r) = r''$. This gives

$$(x \cdot r'') \cdot r = (x \cdot r) \cdot (r'' \lhd r).$$

Later for the two-sided version of a crossed module, we will need the notion of a bimodule over a rack. This notion is attached to a rack with an involution. The given involution on the rack R is supposed to be sent to the involution on the module X (which is in the definition below fixed to be the inversion $\mathrm{Bij}(X) \ni \phi \mapsto \phi^{-1}$).

Definition 6.1.10. Let R be a rack and X be a set. Suppose given left and right operators $(r\cdot), (\cdot r) \in \mathrm{Bij}(X)$ for all $r \in R$. X is called an *R-bimodule* if for all $r, r' \in R$ and all $x \in X$, we have

(RRX) $r \cdot (r' \cdot x) = (r \rhd r') \cdot (r \cdot x)$;
(RXR) $r \cdot (x \cdot r') = (r \cdot x) \cdot (r \rhd r')$;
(XRR) $(r \cdot x) \cdot r' = (x(\cdot r)^{-1}) \cdot r'$.

In case R is pointed, one demands that 1 acts as the identity.

Remark 6.1.11. (a) The meaning of these relations will become clear when studying the link between rack bimodules, bimodules of rack bialgebras and Leibniz bimodules. In fact, the relation (XRR) which corresponds to the relation (MLL) for Leibniz algebras, cannot be written as such for racks. Therefore, we use here an *integrated version* of the relation $(r \cdot v) \cdot r' = -(v \cdot r) \cdot r'$, which corresponds to (MLL) in the realm of Leibniz bimodules.

(b) It is clear that forgetting the right operation turns a bimodule into an R-set. Conversely, given a left action of a rack on a set and taking the right operation to be the inverse of the left operation gives a bimodule structure.

(c) The presence of the inversion $(-)^{-1}$ in axiom (XRR) has to be understood in the following way. A rack comes with an involution $R \to R$, which sends $r \rhd -$ to $(r \rhd -)^{-1}$. In general, this involution should be a piece of the structure of a rack. In the same way, the rack underlying the group of bijections $\mathrm{Bij}(X)$ comes with such an involution. Then one demands for the bimodule structure that the involution of R should be sent to the involution of $\mathrm{Bij}(X)$.

The following structure is the analogue of the semidirect product of a group G by a G-module. As it is only "half of the structure" (the term concerning x is missing), it is termed *hemi-semidirect product*; see [59].

Definition 6.1.12. Let R be a rack and X be an R-set. The *hemi-semidirect product rack* consists of the set $X \times R$ equipped with the rack product

$$(x, r) \rhd (x', r') := (r \cdot x', r \rhd r')$$

for all $x, x' \in X$ and all $r, r' \in R$.

The following definition is the Lie-group-version of a rack.

Definition 6.1.13. A *Lie rack* is a pointed rack M with the structure of a smooth manifold such that for all $x, y \in M$, the rack operation $(x, y) \mapsto x \rhd y$ is a smooth map $M \times M \to M$ and the map $y \mapsto x \rhd y$ is a diffeomorphisms of M.

Remark 6.1.14. Note that here again we can introduce a right operation $x \mapsto x \lhd y$, which is also a diffeomorphism and satisfies the properties (a)–(d) displayed in Proposition 6.1.1.

A *morphisms of Lie racks* $\phi : M \to M'$ is a map of pointed manifolds satisfying for all $x, y \in M$ the condition $\phi(x \rhd y) = \phi(x) \rhd' \phi(y)$. The class of all Lie racks forms a category called `LieRack`.

Let us now come to Leibniz algebras: Leibniz algebras have been invented by A. M. Blokh [10] in 1965, and then rediscovered by J.-L. Loday in 1992 in the search of an explanation for the absence of periodicity in algebraic K-Theory [65, p. 323, equation (10.6.1.1)'].

Definition 6.1.15. A *Leibniz algebra* (over $\mathbb{K}$) is a $\mathbb{K}$-vector space $\mathfrak{h}$ equipped with a linear map $[\,,\,] : \mathfrak{h} \otimes \mathfrak{h} \to \mathfrak{h}$, written $x \otimes y \mapsto [x, y]$ such that the (left) Leibniz identity holds for all $x, y, z \in \mathfrak{h}$,

$$[x,[y,z]] = [[x,y],z] + [y,[x,z]]. \tag{6.2}$$

A morphism of Leibniz algebras $f : \mathfrak{h} \to \mathfrak{h}'$ is a $\mathbb{K}$-linear map preserving brackets, that is, for all $x, y \in \mathfrak{h}$ we have $f([x,y]) = [f(x), f(y)]'$. Note first that each Lie algebra is a Leibniz algebra giving rise to a functor i from the category `LieAlg` of Lie algebras to the category `Leib` of Leibniz algebras.

The following relation to Leibniz algebras is due to M. Kinyon [60].

Proposition 6.1.4. *Let $(M, 1, \rhd)$ be a Lie rack and $\mathfrak{h} = T_1M$. Define the following bracket $[\,,\,]$ on $\mathfrak{h}$ by*

$$[x,y] = \frac{\partial}{\partial t} T_1 L_{a(t)}(y)\Big|_{t=0}, \tag{6.3}$$

where $t \mapsto a(t)$ is any smooth curve defined on an open real interval containing 0 satisfying $a(0) = 1$, $(da/dt)(0) = x \in \mathfrak{h}$ and $L_{a(t)} = a(t) \rhd -$ means the left translation by $a(t)$ (with respect to the rack product $\rhd$). Then we have the following:
(a) *$(\mathfrak{h}, [\,,\,])$ is a real Leibniz algebra.*
(b) *Let $\phi : (M, 1, \rhd) \to (M', 1', \rhd')$ be a morphism of Lie racks. Then $T_1\phi : \mathfrak{h} \to \mathfrak{h}'$ is a morphism of Leibniz algebras.*

Proof. (a) Since for each $a \in M$, we have $L_a(1) = 1$, it follows that the tangent map T_1L_a maps the tangent space T_1M to T_1M. Therefore, the curve $t \mapsto T_1L_{a(t)}$ is a curve of $\mathbb{R}$-linear maps $T_1M \to T_1M$, whence equation (6.3) defines a well-defined real bilinear map $\mathfrak{h} \times \mathfrak{h} \to \mathfrak{h}$.

Let $x, y, z \in \mathfrak{h}$, and let $t \mapsto a(t)$ and $t \mapsto b(t)$ two smooth curves of an open interval (containing 0) into M such that $a(0) = 1 = b(0)$ and $(da/dt)(0) = x$, $(db/dt)(0) = y$. We compute

$$\begin{aligned}
&[x,[y,z]]\\
&= \frac{\partial^2}{\partial s\partial t}(T_1L_{a(s)}(T_1L_{b(t)}(z)))\Big|_{s,t=0} = \frac{\partial^2}{\partial s\partial t}T_1(L_{a(s)} \circ L_{b(t)})(z)\Big|_{s,t=0}\\
&= \frac{\partial^2}{\partial s\partial t}T_1(L_{a(s)\rhd b(t)} \circ L_{a(s)})(z)\Big|_{s,t=0} = \frac{\partial^2}{\partial s\partial t}(T_1L_{a(s)\rhd b(t)}(T_1L_{a(s)}(z)))\Big|_{s,t=0}\\
&= \frac{\partial^2}{\partial s\partial t}T_1L_{a(s)\rhd b(t)}\Big|_{s,t=0}(T_1L_{a(0)}(z)) + \frac{\partial}{\partial t}T_1L_{a(0)\rhd b(t)}\Big|_{t=0}\left(\frac{\partial}{\partial s}(T_1L_{a(s)}(z))\right)\Big|_{s=0}.
\end{aligned}$$

Since $a(0) = 1$, we have $T_1 L_{a(0)}(z) = z$ and $a(0) \rhd b(t) = b(t)$, whence the last term equals $[y, [x, z]]$. Since for each s the curve $t \mapsto a(s) \rhd b(t)$ is equal to 1 at $t = 0$, we get

$$\frac{\partial^2}{\partial s \partial t} T_1 L_{a(s) \rhd b(t)}\Big|_{s,t=0} (T_1 L_{a(0)}(z)) = \left[\frac{\partial}{\partial s} T_1 L_{a(s)}(y)\Big|_{s=0}, z \right] = [[x, y], z]$$

proving the Leibniz identity.

(b) Since ϕ maps 1 to $1'$, its tangent map $T_1\phi$ maps $T_1 M$ to $T_{1'} M'$. We get for all $x, y \in \mathfrak{h} = T_1 M$, where $t \mapsto a(t)$ is a smooth curve in M with $a(0) = 1$ and $(da/dt)(0) = x$:

$$\begin{aligned} T_1\phi([x, y]) &= T_1\phi\Big(\frac{\partial}{\partial t} T_1 L_{a(t)}(y)\Big|_{t=0} \Big) = \frac{\partial}{\partial t} (T_1(\phi \circ L_{a(t)})(y))\Big|_{t=0} \\ &= \frac{\partial}{\partial t} (T_1(L'_{\phi(a(t))} \circ \phi)(y))\Big|_{t=0} = \frac{\partial}{\partial t} T_{1'} L'_{\phi(a(t))}\Big|_{t=0} (T_1\phi(y)) \\ &= [T_1\phi(x), T_1\phi(y)]. \end{aligned}$$

□

Remark 6.1.16. As explained in Remark 6.1.14, the Lie rack M can also be seen as a right Lie rack. This right Lie rack gives rise by a similar procedure to a (right) Leibniz bracket $[,]_R$ on $\mathfrak{h} = T_1 M$. This right Leibniz bracket is related to the left Leibniz bracket $[,]$ on $T_1 M$, which we constructed in the theorem by $[x, y]_R = -[y, x]$ as is easily verified.

Let $\texttt{Leib}_{fd}$ denote the category of finite-dimensional real Leibniz algebras. The preceding proposition shows that there is a functor $T_*\mathcal{R} : \texttt{LieRack} \to \texttt{Leib}_{fd}$, which associates to any Lie rack M its tangent space $T_1 M$ at the distinguished point $1 \in M$ equipped with the Leibniz bracket equation (6.3).

Furthermore, recall that each Leibniz algebra has two canonical subspaces

$$Q(\mathfrak{h}) := \left\{ x \in \mathfrak{h} \mid \exists\, N \in \mathbb{N} \setminus \{0\},\ \exists\, \lambda_1, \ldots, \lambda_N \in \mathbb{K},\ \exists\, x_1, \ldots, x_N \right.$$
$$\left. \text{such that } x = \sum_{r=1}^{N} \lambda_r [x_r, x_r] \right\}, \tag{6.4}$$

$$\mathfrak{z}(\mathfrak{h}) := \{ x \in \mathfrak{h} \mid \forall\, y \in \mathfrak{h} :\ [x, y] = 0 \}. \tag{6.5}$$

It is not hard to deduce from the Leibniz identity that both $Q(\mathfrak{h})$ and $\mathfrak{z}(\mathfrak{h})$ are two-sided abelian ideals of $(\mathfrak{h}, [\ ,\ ])$, that $Q(\mathfrak{h}) \subset \mathfrak{z}(\mathfrak{h})$, and that the quotient Leibniz algebras

$$\overline{\mathfrak{h}} := \mathfrak{h}/Q(\mathfrak{h}) \quad \text{and} \quad \mathfrak{h}/\mathfrak{z}(\mathfrak{h}) \tag{6.6}$$

are Lie algebras. The ideal $Q(\mathfrak{h})$ is called the *ideal of squares* and $\mathfrak{z}(\mathfrak{h})$ is called the *left center* of $\mathfrak{h}$. Since the ideal $Q(\mathfrak{h})$ is clearly mapped into the ideal $Q(\mathfrak{h}')$ by any morphism of Leibniz algebras $\mathfrak{h} \to \mathfrak{h}'$ (which is a priori not the case for $\mathfrak{z}(\mathfrak{h})$!), there is an obvious functor $\mathfrak{h} \to \overline{\mathfrak{h}}$ from the category of all Leibniz algebras to the category of all Lie algebras.

It is easy to observe that in both cases of the above Lie algebras $\overline{\mathfrak{h}}$ and $\mathfrak{h}/\mathfrak{z}(\mathfrak{h})$, the following structure is present.

Definition 6.1.17. A quintuple $(\mathfrak{h}, p, \mathfrak{g}, [\,,\,]_{\mathfrak{g}}, \dot{\rho})$ is called an *augmented Leibniz algebra* if the following holds:

1. $(\mathfrak{g}, [\,,\,]_{\mathfrak{g}})$ is a Lie algebra.
2. $\mathfrak{h}$ is a $\mathbb{K}$-vector space which is a left $\mathfrak{g}$-module via the linear map $\dot{\rho} : \mathfrak{g} \otimes \mathfrak{h} \to \mathfrak{h}$ written $\dot{\rho}_x(h) = x \cdot h$ for all $x \in \mathfrak{g}$ and $h \in \mathfrak{h}$.
3. $p : \mathfrak{h} \to \mathfrak{g}$ is a morphism of $\mathfrak{g}$-modules, that is, for all $x \in \mathfrak{g}$ and $h \in \mathfrak{h}$,

$$p(x \cdot h) = [x, p(h)]_{\mathfrak{g}}. \tag{6.7}$$

A *morphism of augmented Leibniz algebras* $(\mathfrak{h}, p, \mathfrak{g}, [\,,\,]_{\mathfrak{g}}, \dot{\rho}) \to (\mathfrak{h}', p', \mathfrak{g}', [\,,\,]'_{\mathfrak{g}}, \dot{\rho}')$ is a pair (F, f) of linear maps where $f : \mathfrak{g} \to \mathfrak{g}'$ is a morphism of Lie algebras, $F : \mathfrak{h} \to \mathfrak{h}$ is a morphism of Lie algebra modules over f, that is, for all $h \in \mathfrak{h}$ and $x \in \mathfrak{g}$,

$$F(x \cdot h) = f(x) \cdot F(h). \tag{6.8}$$

Moreover, the obvious diagram commutes, that is,

$$p' \circ F = f \circ p. \tag{6.9}$$

The following properties are immediate from the definitions.

Proposition 6.1.5. *Let $(\mathfrak{h}, p, \mathfrak{g}, [\,,\,]_{\mathfrak{g}}, \dot{\rho})$ be an augmented Leibniz algebra. Define the following bracket on $\mathfrak{h}$:*

$$[x, y]_{\mathfrak{h}} := p(x) \cdot y. \tag{6.10}$$

(a) *$(\mathfrak{h}, [,]_{\mathfrak{h}})$ is a Leibniz algebra on which $\mathfrak{g}$ acts as derivations. If (F, f) is a morphism of augmented Leibniz algebras, then F is a morphism of Leibniz algebras.*
(b) *The kernel of p, $\mathrm{Ker}(p)$, is a $\mathfrak{g}$-invariant two-sided abelian ideal of $\mathfrak{h}$ satisfying $Q(\mathfrak{h}) \subset \mathrm{Ker}(p) \subset \mathfrak{z}(\mathfrak{h})$.*
(c) *The image of p, $\mathrm{Im}(p)$, is an ideal of the Lie algebra $\mathfrak{g}$.*

Proof. We just check the Leibniz identity: Let $x, y, z \in \mathfrak{h}$, then writing $[\,,\,]_{\mathfrak{h}} = [\,,\,]$,

$$\begin{aligned}[x, [y, z]] &= p(x) \cdot (p(y) \cdot z) \\ &= p(x) \cdot (p(y) \cdot z) - p(y) \cdot (p(x) \cdot z) + p(y) \cdot (p(x) \cdot z) \\ &= [p(x), p(y)]_{\mathfrak{g}} \cdot z + [y, [x, z]] \overset{(6.7)}{=} p(p(x) \cdot y) \cdot z + [y, [x, z]] \\ &= [[x, y], z] + [y, [x, z]].\end{aligned}$$ □

It follows that the class of augmented Leibniz algebras forms a category LeibA. There is a forgetful functor from LeibA to Leib associating to $(\mathfrak{h}, p, \mathfrak{g}, [\,,\,]_{\mathfrak{g}}, \dot{\rho})$ the Leibniz algebra $(\mathfrak{h}, [\,,\,]_{\mathfrak{h}})$ where the Leibniz bracket $[\,,\,]_{\mathfrak{h}}$ is defined in equation (6.10).

On the other hand, there is a functor from Leib to LeibA associating to each Leibniz algebra $(\mathfrak{h}, [\,,\,])$ the augmented Leibniz algebra $(\mathfrak{h}, p, \bar{\mathfrak{h}}, [\,,\,]_{\bar{\mathfrak{h}}}, \mathrm{ad}')$ where $p : \mathfrak{h} \to \bar{\mathfrak{h}}$ is the canonical projection and the representation ad' of the Lie algebra $\bar{\mathfrak{h}}$ on the Leibniz algebra $\mathfrak{h}$ is defined by (for all $x, y \in \mathfrak{h}$)

$$\mathrm{ad}'_{p(x)}(y) := \mathrm{ad}_x(y) = [x, y]. \tag{6.11}$$

Remark 6.1.18. There exists an inverse to the construction in Proposition 6.1.4, namely an integration process, which integrates finite-dimensional real (augmented) Leibniz algebras into (augmented) Lie racks. In order to make this meaningful, one also imposes that the integration process is such that finite-dimensional Lie algebras are integrated into the standard simply-connected Lie group associated to a finite-dimensional real Lie algebra. Such an integration procedure exists (see [11]), but for the moment, the only known procedure is not functorial in general. It is functorial only for morphisms of Leibniz algebras, which *reduce* the spectrum of the ad-operators.

6.1.2 Racks, Leibniz algebras and rack bialgebras

We now come to the notion which is analogous to the notion of the group algebra in the framework of racks. First, let us recall some basic notions about coalgebras. We shall assume in this subsection that $\mathbb{Q} \subset \mathbb{K}$.

Let C be a $\mathbb{K}$-vector space. Recall that a linear map $\Delta : C \to C \otimes C$ is called a *coassociative comultiplication* in case $(\Delta \otimes \mathrm{id}_C) \circ \Delta = (\mathrm{id}_C \otimes \Delta) \circ \Delta$, and the pair (C, Δ) is called a (coassociative) *coalgebra*. In Sweedler's notation, the coproduct on elements is written as

$$\Delta(a) = a_1 \otimes a_2.$$

Recall that this notation implies a sum over all tensors which form $\Delta(a)$.

Let (C', Δ') be another coalgebra. The coalgebra (C, Δ) is called *cocommutative* if $\tau \circ \Delta = \Delta$ where $\tau : C \otimes C \to C \otimes C$ denotes the canonical flip map. Recall furthermore that a linear map $\epsilon : C \to \mathbb{K}$ is called a *counit* for the coalgebra (C, Δ) in case $(\epsilon \otimes \mathrm{id}_C) \circ \Delta = (\mathrm{id}_C \otimes \epsilon) \circ \Delta = \mathrm{id}_C$. The triple (C, Δ, ϵ) is called a *counital coalgebra*. Moreover, a counital coalgebra (C, Δ, ϵ) equipped with an element 1 is called *coaugmented* if $\Delta(1) = 1 \otimes 1$ and $\epsilon(1) = 1 \in \mathbb{K}$. Recall that a morphism $\phi : (C, \Delta, \epsilon, 1) \to (C', \Delta', \epsilon', 1')$ of counital coaugmented coalgebras is a $\mathbb{K}$-linear map satisfying $(\phi \otimes \phi) \circ \Delta = \Delta' \circ \phi$, $\epsilon' \circ \phi = \epsilon$, and $\phi(1) = 1'$. Moreover, for any counital coaugmented coalgebra the $\mathbb{K}$-submodule of all *primitive elements* is defined by

$$P(C) := \{x \in C \mid \Delta(x) = x \otimes 1 + 1 \otimes x\}. \tag{6.12}$$

The analogue of the notion of a bialgebra in the framework of racks is the notion of a rack bialgebra.

Definition 6.1.19. A *rack bialgebra* $(B, \triangle, \epsilon, 1, \mu)$ is a coassociative, counital, coaugmented coalgebra $(B, \triangle, \epsilon, 1)$ together with a product $\rhd : B \times B \to B$, which is a morphism of coalgebras (and satisfies in particular $1 \rhd 1 = 1$) such that the following identities hold for all $a, b, c \in B$:

$$1 \rhd a = a, \tag{6.13}$$

$$a \rhd 1 = \epsilon(a)1, \tag{6.14}$$

$$a \rhd (b \rhd c) = (a_1 \rhd b) \rhd (a_2 \rhd c). \tag{6.15}$$

The last condition (6.15) is called the *self-distributivity* condition.

Note that we do not demand that the coalgebra B should be cocommutative. Note furthermore that the self-distributivity condition here is the *linearized* version of the self-distributivity of a rack.

Example 6.1.20. Any coassociative, counital, coaugmented coalgebra $(C, \triangle, \epsilon, 1)$ carries a *trivial rack bialgebra structure* defined by the left-trivial multiplication

$$a \rhd b := \epsilon(a)b, \tag{6.16}$$

which in addition is easily seen to be associative and left-unital, but in general not unital.

Example 6.1.21. Let $(H, \triangle_H, \epsilon_H, \mu_H, 1_H, S)$ be a cocommutative Hopf algebra over $\mathbb{K}$. Then it is easy to see (see also the particular case $B = H$ and $\Phi = \mathrm{id}_H$ of Proposition 6.1.6) that the new product $\rhd : H \otimes H \to H$ defined by the usual *adjoint representation*

$$h \rhd h' := \mathrm{ad}_h(h') := h_1 h' S(h_2), \tag{6.17}$$

equips the coassociative, counital, coaugmented coalgebra $(H, \triangle_H, \epsilon_H, 1_H)$ with a rack bialgebra structure.

In general, the adjoint representation does not preserve the coalgebra structure if no cocommutativity is assumed; cf. Exercise 4.8.3.

Example 6.1.22. Let $(X, 1)$ be a pointed rack. Then there is a natural rack bialgebra structure on the vector space $\mathbb{K}[X]$ which has the elements of X as a basis. $\mathbb{K}[X]$ carries the usual coalgebra structure such that all $x \in X$ are set-like: $\triangle(x) = x \otimes x$ for all $x \in X$. The product $\rhd$ is then the linearization of the rack product. By functoriality, $\rhd$ is compatible with $\triangle$ and 1. Observe that this construction differs slightly from the construction in [15], Section 3.1.

As in the case of Leibniz algebras and Lie racks, there is an associated *augmented* structure:

Definition 6.1.23. An *augmented rack bialgebra* is a quadruple (B, p, H, ℓ) consisting of a coassociative, counital, coaugmented coalgebra $(B, \triangle, \epsilon, 1)$, of a cocommutative Hopf algebra $(H, \triangle_H, \epsilon_H, 1_H, \mu_H, S)$, of a morphism of coalgebras $p : B \to H$, and of a left action $\ell : H \otimes B \to B$ of H on B, which is a morphism of coalgebras (i. e., B is a H-module-coalgebra) such that for all $h \in H$ and $a \in B$

$$h.1 = \epsilon_H(h)1, \tag{6.18}$$

$$p(h \cdot a) = \mathrm{ad}_h(p(a)), \tag{6.19}$$

where ad denotes the usual adjoint representation for Hopf algebras; see, for example, equation (6.17).

We shall define a *morphism* $(B, p, H, \ell) \to (B', p', H', \ell')$ of augmented rack bialgebras to be a pair (ϕ, ψ) of k-linear maps where $\phi : (B, \triangle, \epsilon, 1) \to (B', \triangle', \epsilon', 1')$ is a morphism of coalgebras, and $\psi : H \to H'$ is a morphism of Hopf algebras such that the obvious diagrams commute:

$$p' \circ \phi = \psi \circ p \quad \text{and} \quad \ell' \circ (\psi \otimes \phi) = \phi \circ \ell. \tag{6.20}$$

An immediate consequence of this definition is the following.

Proposition 6.1.6. *Let (B, p, H, ℓ) be an augmented rack bialgebra. Then the coassociative, counital, coaugmented coalgebra $(B, \epsilon, 1)$ will become a rack bialgebra by means of the product*

$$a \rhd b := p(a) \cdot b \tag{6.21}$$

for all $a, b \in B$. In particular, each cocommutative Hopf algebra H becomes an augmented rack bialgebra via $(H, \mathrm{id}_H, H, \mathrm{ad})$. In general, for each augmented rack bialgebra the map $p : B \to H$ is a morphism of rack bialgebras.

Proof. We check first that $\rhd$ is a morphism of coalgebras $B \otimes B \to B$: Let $a, b \in B$, then we have, using the fact that the action ℓ and the maps p are coalgebra morphisms:

$$\begin{aligned} \triangle(a \rhd b) &= \triangle(p(a) \cdot b) = (p(a)_1 \cdot b_1) \otimes (p(a)_2 \cdot b_2) \\ &= (p(a_1) \cdot b_1) \otimes (p(a_2) \cdot b_2) \\ &= (a_1 \rhd b_1) \otimes (a_2 \rhd b_2), \end{aligned}$$

whence $\rhd$ is a morphism of coalgebras. Clearly,

$$\epsilon(a \rhd b) = \epsilon(p(a) \cdot b) = \epsilon_H(p(a))\epsilon(b) = \epsilon(a)\epsilon(b),$$

whence $\rhd$ preserves counits. Let us next compute both sides of the self-distributivity identity (6.15). We have for all $a, b, c \in B$,

$$a \rhd (b \rhd c) = p(a).(p(b) \cdot c) = (p(a)p(b)) \cdot c,$$

and

$$(a_1 \rhd b) \rhd (a_2 \rhd c) = (p(a_1) \cdot b) \rhd (p(a_2) \cdot c)$$
$$= (p(p(a_1) \cdot b)) \cdot (p(a_2) \cdot c)$$
$$= (p(p(a_1) \cdot b)p(a_2)) \cdot c,$$

and we compute, using the fact that p is a morphism of coalgebras,

$$\begin{aligned} p(p(a_1) \cdot b)p(a_2) &= p(p(a)_1 \cdot b)p(a)_2 \\ &\overset{(6.19)}{=} \mathrm{ad}_{p(a)_1}(p(b))p(a)_2 \\ &= p(a)_1 p(b)\, S(p(a)_2)p(a)_3 \\ &= p(a)_1 p(b)\epsilon_H(p(a)_2) \\ &= p(a)p(b), \end{aligned}$$

which proves the self-distributivity identity. Moreover, we have

$$1_B \rhd a = p(1).a = 1_H.a = a,$$

and

$$a \rhd 1 = p(a).1 \overset{(6.18)}{=} \epsilon_H(p(a))1 = \epsilon_B(a)1.$$

This shows that the coassociative, counital, coaugmented coalgebra B becomes a rack bialgebra. □

Example 6.1.24. Exactly in the same way as a pointed rack gives rise to a rack bialgebra $\mathbb{K}[X]$, an augmented pointed rack $p : X \to G$ gives rise to an augmented rack bialgebra $p : \mathbb{K}[X] \to \mathbb{K}[G]$.

The link to Leibniz algebras is contained in the following.

Proposition 6.1.7. *Let $(B, \triangle, \epsilon, 1, \mu)$ be a rack bialgebra. Then its subspace of all primitive elements, $P(B) =: \mathfrak{h}$, is a subalgebra with respect to the product $\rhd$ satisfying the Leibniz identity*

$$x \rhd (y \rhd z) = (x \rhd y) \rhd z + y \rhd (x \rhd z) \tag{6.22}$$

for all $x, y, z \in \mathfrak{h} = P(B)$. Hence the pair $(\mathfrak{h}, [\ ,\ ])$ with $[x, y] := x \rhd y$ for all $x, y \in \mathfrak{h}$ is a Leibniz algebra over k. Moreover, every morphism of rack bialgebras maps primitive elements to primitive elements, and thus induces a morphism of Leibniz algebras.

Proof. Let $x \in \mathfrak{h}$ and $a \in B$. Since μ is a morphism of coalgebras and x is primitive, we get

$$\triangle(a \rhd x) = (a_1 \rhd x) \otimes (a_2 \rhd 1) + (a_1 \rhd 1) \otimes (a_2 \rhd x)$$

$$\overset{(6.14)}{=} ((a_1\epsilon(a_2)) \rhd x) \otimes 1 + 1 \otimes ((\epsilon(a_1)a_2) \rhd x)$$
$$= (a \rhd x) \otimes 1 \; + \; 1 \otimes (a \rhd x),$$

whence $a \rhd x$ is primitive.

It follows that $\mathfrak{h}$ is a subalgebra with respect to $\rhd$. Let $x, y, z \in \mathfrak{h}$. Then since x is primitive, it follows from $\Delta(x) = x \otimes 1 + 1 \otimes x$ and the self-distributivity identity (6.15) that

$$x \rhd (y \rhd z) = (x \rhd y) \rhd (1 \rhd z) + (1 \rhd y) \rhd (x \rhd z) \overset{(6.13)}{=} (x \rhd y) \rhd z + y \rhd (x \rhd z),$$

proving the left Leibniz identity. The morphism statement is clear, since each morphism of rack bialgebras is a morphism of coalgebras and preserves primitives. □

As an immediate consequence, we get that the functor P induces a functor from the category of all rack bialgebras to the category of all Leibniz algebras over $\mathbb{K}$.

Remark 6.1.25. Define *set-like elements* to be elements a in a rack bialgebra B such that $\Delta(a) = a \otimes a$. Thanks to the fact that $\rhd$ is a morphism of coalgebras, the set of set-like elements Slike(B) is closed under $\rhd$. In fact, Slike(B) is a rack, and one obtains in this way a functor Slike : `RackBialg` $\to$ `Racks`.

Proposition 6.1.8. *The functor of set-likes* Slike : `RackBialg` $\to$ `Racks` *has the functor* $\mathbb{K}[-]$: `Racks` $\to$ `RackBialg` *(see Example* 6.1.22*) as its left-adjoint.*

Proof. This follows from the adjointness of the same functors, seen as functors between the categories of pointed sets and of coassociative, counital, coaugmented, cocommutative coalgebras, observing that the coalgebra morphism induced by a morphism of racks respects the rack product. □

Observe that the restriction of Slike : `RackBialg` $\to$ `Racks` to the subcategory of cocommutative Hopf algebras `Hopf` (where the Hopf algebra is given the rack product defined in equation (6.17)) gives the usual functor of group-like elements.

We will now associate to a Leibniz algebra $\mathfrak{h}$ a rack bialgebra, that is, construct an inverse process to Proposition 6.1.7. First of all, recall that each Leibniz algebra has two canonical ideals $Q(\mathfrak{h})$ and $\mathfrak{z}(\mathfrak{h})$ (see equations (6.4) and (6.5)) with $Q(\mathfrak{h}) \subset \mathfrak{z}(\mathfrak{h})$ such that the corresponding quotient Leibniz algebras are Lie algebras.

In order to perform the following constructions of rack bialgebras for any given Leibniz algebra $\mathfrak{h}$, choose first a two-sided ideal $\mathfrak{z} \subset \mathfrak{h}$ such that

$$Q(\mathfrak{h}) \subset \mathfrak{z} \subset \mathfrak{z}(\mathfrak{h}), \tag{6.23}$$

let $\mathfrak{g}$ denote the quotient Lie algebra $\mathfrak{h}/\mathfrak{z}$, and let $p : \mathfrak{h} \to \mathfrak{g}$ be the natural projection.

The Lie algebra $\mathfrak{g}$ naturally acts as derivations on $\mathfrak{h}$ by means of (for all $x, y \in \mathfrak{h}$)

$$p(x) \cdot y := [x, y] =: \mathrm{ad}_x(y) \tag{6.24}$$

because $\mathfrak{z} \subset \mathfrak{z}(\mathfrak{h})$. Note that

$$\mathfrak{h}/\mathfrak{z}(\mathfrak{h}) \cong \{\mathrm{ad}_x \in \mathrm{Hom}_K(\mathfrak{h}, \mathfrak{h}) \mid x \in \mathfrak{h}\}, \tag{6.25}$$

as Lie algebras. Then $(\mathfrak{h}, p, \mathfrak{g}, [,]_{\mathfrak{g}}, \cdot)$ is an augmented Leibniz algebra.

Consider now the coassociative, counital, coaugmented, cocommutative coalgebra $(B = S(\mathfrak{h}), \Delta, \epsilon, 1)$, which is actually a commutative cocommutative Hopf algebra over $\mathbb{K}$ with respect to the symmetric multiplication. The linear map $p : \mathfrak{h} \to \mathfrak{g}$ induces a unique morphism of Hopf algebras

$$\tilde{p} : S(\mathfrak{h}) \to S(\mathfrak{g}) \tag{6.26}$$

satisfying

$$\tilde{p}(x_1 \cdot \cdots \cdot x_k) = p(x_1) \cdot \cdots \cdot p(x_k) \tag{6.27}$$

for any nonnegative integer k and $x_1, \ldots, x_k \in \mathfrak{h}$. In other words, the association $S : V \to S(V)$ is a functor from the category of all $\mathbb{K}$-vector spaces to the category of all commutative unital coassociative, counital, coaugmented, cocommutative coalgebras. Consider the universal enveloping algebra $U\mathfrak{g}$ of the Lie algebra $\mathfrak{g}$. Since we now assume $\mathbb{Q} \subset \mathbb{K}$, the Poincaré–Birkhoff–Witt theorem holds (see, e. g., [22], Sections 2.1–2.4, pp. 69–84). More precisely, the symmetrisation map $\omega : S(\mathfrak{g}) \to U\mathfrak{g}$, defined by

$$\omega(1_{S(\mathfrak{g})}) = 1_{U\mathfrak{g}}, \quad \text{and} \quad \omega(x_1 \cdot \cdots \cdot x_k) = \frac{1}{k!} \sum_{\sigma \in S_k} x_{\sigma(1)} \cdots x_{\sigma(k)}, \tag{6.28}$$

is an isomorphism of coassociative, counital, coaugmented, cocommutative coalgebras (in general not of associative algebras). We now need an action of the Hopf algebra $H = U\mathfrak{g}$ on B, and an intertwining map $p : B \to U\mathfrak{g}$. In order to get this, we first look at $\mathfrak{g}$-modules: The $\mathbb{K}$-module $\mathfrak{h}$ is a $\mathfrak{g}$-module by means of equation (6.24), the Lie algebra $\mathfrak{g}$ is a $\mathfrak{g}$-module via its adjoint representation and the linear map $p : \mathfrak{h} \to \mathfrak{g}$ is a morphism of $\mathfrak{g}$-modules since p is a morphism of Leibniz algebras. Now $S(\mathfrak{h})$ and $S(\mathfrak{g})$ are $\mathfrak{g}$-modules in the usual way, that is, for all $k \in \mathbb{N} \setminus \{0\}$, $x, x_1, \ldots, x_k \in \mathfrak{g}$, and $h_1 \ldots, h_k \in \mathfrak{h}$,

$$x \cdot (h_1 \cdot \cdots \cdot h_k) := \sum_{r=1}^{k} h_1 \cdot \cdots \cdot (x \cdot h_r) \cdot \cdots \cdot h_k, \tag{6.29}$$

$$x \cdot (x_1 \cdot \cdots \cdot x_k) := \sum_{r=1}^{k} x_1 \cdot \cdots \cdot [x.x_r] \cdot \cdots \cdot x_k, \tag{6.30}$$

and of course $x \cdot 1_{S(\mathfrak{h})} = 0$ and $x \cdot 1_{S(\mathfrak{g})} = 0$. Recall that $U\mathfrak{g}$ is a $\mathfrak{g}$-module via the adjoint representation $\mathrm{ad}_x(u) = x \cdot u = xu - ux$ for all $x \in \mathfrak{g}$ and all $u \in U(\mathfrak{g})$. It is easy to see that the map $\tilde{p}$ in equation (6.27) is a morphism of $\mathfrak{g}$-modules, and it is well known that the symmetrization map ω (6.28) is also a morphism of $\mathfrak{g}$-modules; see, for example, [22, p. 82, Proposition 2.4.10]. Define the $\mathbb{K}$-linear map $p : S(\mathfrak{h}) \to U\mathfrak{g}$ by the composition

$$p := \omega \circ \tilde{p}. \tag{6.31}$$

Then p is a map of coassociative, counital, coaugmented, cocommutative coalgebras and a map of $\mathfrak{g}$-modules. Thanks to the universal property of the universal enveloping algebra, it follows that $S(\mathfrak{h})$ and $U\mathfrak{g}$ are left $U\mathfrak{g}$-modules, via (for all $x_1, \ldots, x_k \in \mathfrak{g}$, and for all $a \in S(\mathfrak{h})$)

$$(x_1 \cdots x_k) \cdot a = x_1.(x_2.(\cdots x_k \cdot a) \cdots) \tag{6.32}$$

and the usual adjoint representation (6.17) (for all $u \in U\mathfrak{g}$)

$$\mathrm{ad}_{x_1 \cdots x_k}(u) = (\mathrm{ad}_{x_1} \circ \cdots \circ \mathrm{ad}_{x_k})(u), \tag{6.33}$$

and that p intertwines the $U\mathfrak{g}$-action on $C = S(\mathfrak{h})$ with the adjoint action of $U\mathfrak{g}$ on itself. Finally, it is a routine check using the above identities (6.29) and (6.17) that $S(\mathfrak{h})$ becomes a module coalgebra. We can resume the preceding considerations in the following theorem.

Theorem 6.1.9. *Let $\mathfrak{h}$ be a Leibniz algebra over $\mathbb{K}$, let $\mathfrak{z}$ be a two-sided ideal of $\mathfrak{h}$ such that $Q(\mathfrak{h}) \subset \mathfrak{z} \subset \mathfrak{z}(\mathfrak{h})$, let $\mathfrak{g}$ denote the quotient Lie algebra $\mathfrak{g} := \mathfrak{h}/\mathfrak{z}$, and let $p : \mathfrak{h} \to \mathfrak{g}$ be the canonical projection.*

1. *Then there is a canonical $U(\mathfrak{g})$-action ℓ on the coassociative, counital, coaugmented, cocommutative coalgebra $B := S(\mathfrak{h})$ (making it into a module coalgebra leaving invariant 1) and a canonical lift of p to a morphism of coalgebras, $p : S(\mathfrak{h}) \to U\mathfrak{g}$ such that equation (6.19) holds.*
   Hence the quadruple $(S(\mathfrak{h}), p, U\mathfrak{g}, \ell)$ is an augmented rack bialgebra whose associated Leibniz algebra is equal to $\mathfrak{h}$ and this is true independently of the choice of $\mathfrak{z}$. The resulting rack product $\rhd$ of $S(\mathfrak{h})$ is also independent on the choice of $\mathfrak{z}$ and is explicitly given as follows for all positive integers k, l and $x_1, \ldots, x_k, y_1, \ldots, y_l \in \mathfrak{h}$:

$$(x_1 \cdot \cdots \cdot x_k) \rhd (y_1 \cdot \cdots \cdot y_l) = \frac{1}{k!} \sum_{\sigma \in S_k} (\mathrm{ad}^S_{x_{\sigma(1)}} \circ \cdots \circ \mathrm{ad}^S_{x_{\sigma(k)}})(y_1 \cdot \cdots \cdot y_l), \tag{6.34}$$

   where ad^S_x denotes the action of the Lie algebra $\mathfrak{h}/\mathfrak{z}(\mathfrak{h})$ (see equation (6.25)) on $S(\mathfrak{h})$ according to equation (6.29).
2. *In case $\mathfrak{z} = Q(\mathfrak{h})$, the construction mentioned in (a) is a functor $\mathfrak{h} \to \mathsf{UAR}^\infty(\mathfrak{h})$ from the category of all Leibniz algebras to the category of all augmented rack bialgebras associating to $\mathfrak{h}$ the rack bialgebra*

$$\mathsf{UAR}^\infty(\mathfrak{h}) := (S(\mathfrak{h}), \Phi, U(\mathfrak{g}), \ell)$$

   and to each morphism f of Leibniz algebras the pair $(S(f), U(\bar{f}))$ where $\bar{f}$ is the induced Lie algebra morphism.

Proof. A great deal of the statements has already been proven in the discussion before the theorem. Note that for all $x, y \in \mathfrak{h}$ we have by definition

$$[x,y] = p(x) \cdot y = x \rhd y,$$

independently of the chosen ideal $\mathfrak{z}$. Moreover, we compute

$$\begin{aligned}(x_1 \cdot \dots \cdot x_k) &\rhd (y_1 \cdot \dots \cdot y_l)\\ &= ((\omega \circ \tilde{p})(x_1 \cdot \dots \cdot x_k)) \cdot (y_1 \cdot \dots \cdot y_l)\\ &= \frac{1}{k!} \sum_{\sigma \in S_k} (p(x_{\sigma(1)}) \cdots p(x_{\sigma(k)})) \cdot (y_1 \cdot \dots \cdot y_l),\end{aligned}$$

which gives the desired formula since for all $x \in \mathfrak{h}$ and $a \in S(\mathfrak{h})$, we have

$$p(x) \cdot a = \mathrm{ad}_x^s(a).$$

Let us then show functoriality. For this, let $f : \mathfrak{h} \to \mathfrak{h}'$ be a morphism of Leibniz algebras, and let $\overline{f} : \overline{\mathfrak{h}} \to \overline{\mathfrak{h}'}$ be the induced morphism of Lie algebras. Hence we get

$$p' \circ f = \overline{f} \circ p, \tag{6.35}$$

where $p' : \mathfrak{h}' \to \overline{\mathfrak{h}'}$ denotes the corresponding projection modulo $Q(\mathfrak{h}')$. Let $S(f) : S(\mathfrak{h}) \to S(\mathfrak{h}')$, $S(\overline{f}) : S(\overline{\mathfrak{h}}) \to S(\overline{\mathfrak{h}'})$, and $U(\overline{f}) : U(\overline{\mathfrak{h}}) \to U(\overline{\mathfrak{h}'})$ be the induced maps of Hopf algebras, that is, $S(f)$ (resp., $S(\overline{f})$) satisfies equation (6.27) (with p replaced by f (resp., by $\overline{f}$)), and $U(\overline{f})$ satisfies

$$U(\overline{f})(x_1 \cdots x_k) = \overline{f}(x_1) \cdots \overline{f}(x_k)$$

for all positive integers k and $x_1, \dots, x_k \in \overline{\mathfrak{h}}$. If $\omega : S(\overline{\mathfrak{h}}) \to U(\overline{\mathfrak{h}})$ and $\omega' : S(\overline{\mathfrak{h}'}) \to U(\overline{\mathfrak{h}'})$ denote the corresponding symmetrisation maps (6.28), then it is easy to see from the definitions that

$$\omega' \circ S(\overline{f}) = U(\overline{f}) \circ \omega.$$

Equation (6.35) implies

$$\tilde{p}' \circ S(f) = S(p') \circ S(f) = S(\overline{f}) \circ S(p) = S(\overline{f}) \circ \tilde{p},$$

and composing from the left with ω' yields the equation

$$p' \circ S(f) = U(\overline{f}) \circ p. \tag{6.36}$$

Moreover for all $h, h' \in \mathfrak{h}$ we have, since f is a morphism of Leibniz algebras,

$$f(p(h) \cdot h') = f([h,h']) = [f(h), f(h')]' = p'(f(h)) \cdot f(h') = \overline{f}(p(h)) \cdot f(h'),$$

hence for all $x \in \overline{h}$

$$f(x \cdot h) = \overline{f}(x) \cdot f(h),$$

and upon using equation (6.29), we get for all $a \in S(\mathfrak{h})$

$$S(f)(x \cdot a) = \overline{f}(x) \cdot S(f)(a),$$

showing finally for all $u \in U\mathfrak{h}$ and all $a \in S(\mathfrak{h})$,

$$S(f)(u \cdot a) = U(\overline{f})(u) \cdot S(f)(a). \tag{6.37}$$

Associating to every Leibniz algebra $\mathfrak{h}$ the above defined augmented rack bialgebra $(S(\mathfrak{h}), \Phi, U\overline{\mathfrak{h}}, \ell)$, and to every morphism $\psi : \mathfrak{h} \to \mathfrak{h}'$ of Leibniz algebras the pair of linear maps $(S(\psi), U(\overline{\psi}))$, we can easily check that $S(\psi)$ is a morphism of coalgebras, $U(\psi)$ is a morphism of Hopf algebras, such that the two relevant diagrams (6.20) commute which easily follows from (6.36) and (6.37). The rest of the functorial properties is a routine check. □

Remark 6.1.26. This theorem should be compared to Proposition 3.5 in [15]. In [15], the authors construct a rack bialgebra structure on $N := \mathbb{K} \oplus \mathfrak{h}$. The rack product is given by the bracket of the Leibniz algebra. We work in the preceding theorem with the whole symmetric algebra on the Leibniz algebra. Thus, in some sense, we extend their Proposition 3.5 “to all orders.”

The above rack bialgebra associated to a Leibniz algebra $\mathfrak{h}$ can be seen as one version of an *enveloping algebra* of $\mathfrak{h}$.

Let us summarize. Given a rack R, there is a rack bialgebra $\mathbb{K}[R]$, which is the linearization of the rack R. Given a Lie rack M, its tangent space to the unit 1 is endowed with the structure of a (natural, in general nontrivial) Leibniz algebra. Leibniz algebras have “enveloping algebras,” which are rack bialgebras, and every rack bialgebra has as its primitives a Leibniz algebra and as its set-likes a pointed rack. There are some other links between these structures which we do not discuss, for example, the point-distributions on a Lie rack supported in 1 form a rack bialgebra. In total, the structures of rack, rack bialgebra and Leibniz algebra enjoy (largely speaking) the same links as the structures of group, bi/Hopf algebra and Lie algebra.

6.2 Crossed modules of racks

Definition 6.2.1. Let R and S be racks. We say that S *acts* on R *by automorphisms* in case there is an action of S on R such that

$$s \cdot (r \rhd r') = (s \cdot r) \rhd (s \cdot r')$$

for all $s \in S$ and all $r, r' \in R$.

Definition 6.2.2. A *(one-sided) crossed module of racks* is a morphism of racks $\mu : R \to S$ together with an action of S on R by automorphisms such that:
(a) μ is equivariant, that is, $\mu(s \cdot r) = s \rhd \mu(r)$ for all $s \in S$, $r \in R$ and
(b) Peiffer's identity is satisfied, that is, $\mu(r) \cdot r' = r \rhd r'$ for all $r, r' \in R$.

We now provide some classes of examples.

Example 6.2.3. Exactly as in the case of crossed modules of groups, there is for every rack R the adjoint crossed module of racks:

$$\mathrm{Ad} : R \to \mathrm{Aut}(R),$$

where $\mathrm{Ad}(r) := (s \mapsto r \rhd s)$. With this definition, $\mathrm{Ad}(r)$ is a bijection, and self-distributivity implies that it is also an automorphism:

$$\mathrm{Ad}(r)(s_1 \rhd s_2) = r \rhd (s_1 \rhd s_2) = (r \rhd s_1) \rhd (r \rhd s_2) = \mathrm{Ad}(r)(s_1) \rhd \mathrm{Ad}(r)(s_2).$$

The automorphism group $\mathrm{Aut}(R)$ of R acts by $\phi \cdot r := \phi(r)$ for all $r \in R$ and all $\phi \in \mathrm{Aut}(R)$. In fact, this is an augmented rack:

$$\mathrm{Ad}(\phi \cdot r) = \mathrm{Ad}(\phi(r)) = \phi \circ \mathrm{Ad}(r) \circ \phi^{-1}.$$

The induced rack structure on R via Ad (i. e., $r \rhd s := \mathrm{Ad}(r) \cdot s$) and the original rack structure on R coincide.

Example 6.2.4. Let $\mu : M \to N$ be a crossed module of groups. Passing to the associated conjugation racks of M and N, we obtain a crossed module of racks. Indeed, the group morphism $N \to \mathrm{Bij}(M)$ gives rise to a morphism of racks, and thus we have a rack action of the conjugation rack N on the conjugation rack M. Moreover, we have

$$n \cdot (m \rhd m') = n \cdot (mm'm^{-1}) = (n \cdot m)(n \cdot m')(n \cdot m)^{-1} = (n \cdot m) \rhd (n \cdot m'),$$

for all $n \in N$ and all $m, m' \in M$, where we have used the fact that the group N acts on M by automorphisms. Finally, the equivariance condition for μ and the Peiffer identity follow from the analogous conditions for the crossed module of groups.

Remark 6.2.5. We have already seen in Chapter 1 a mechanism to construct *explicit* examples of crossed modules of groups, and thus, *a fortiori*, of racks. Indeed, we have seen how to construct crossed modules in an explicit way from cohomology classes $[\theta] \in H^3(G, V)$ for some group G and a G-module V. Choose an injective presentation of V, that is, a short exact sequence

$$0 \to V \to I \to Q \to 0, \tag{6.38}$$

where I is an injective G-module. The long exact sequence in cohomology contains the connecting map

$$\partial : H^2(G, Q) \to H^3(G, V),$$

which by injectivity of I is an isomorphism. There exists thus a unique class $[\alpha] \in H^2(G,Q)$ with $\partial[\alpha] = \theta$. To $[\alpha]$, one may associate an abelian extension

$$0 \to Q \to Q \times_\alpha G \to G \to 0,$$

and this extension can be spliced together with the short exact sequence (6.38) to give a crossed module

$$0 \to V \to I \to Q \times_\alpha G \to G \to 0.$$

Under the isomorphism between equivalence classes of crossed modules of groups with kernel V and cokernel G and $H^3(G,V)$, this crossed module corresponds to $[\theta]$. We have seen in Chapter 1 explicit examples of this construction, which thus yield explicit crossed modules of racks.

Example 6.2.6. An augmented rack $p : X \to G$ is an example of a crossed module of racks. Indeed, we have already remarked that X may be equipped with a rack operation making p an equivariant morphism of racks. Then G acts on the rack X, because the group morphism $G \to \mathrm{Bij}(X)$ is also a morphism of conjugation racks (cf. Lemma 6.1.3). Moreover, G acts by automorphisms, because

$$\begin{aligned} g\cdot(x \rhd x') &= g\cdot(p(x)\cdot x') \\ &= (gp(x)g^{-1}g)\cdot x' \\ &= p(g\cdot x)\cdot(g\cdot x') \\ &= (g\cdot x) \rhd (g\cdot x') \end{aligned}$$

for all $g \in G$ and all $x, x' \in X$. We have already mentioned that the equivariance of p comes from the augmentation identity. The Peiffer identity comes from the definition of the rack operation on X.

Example 6.2.7. There is also the notion of a *generalized augmented rack*, that is, an augmented rack of racks instead of groups. For this, let R be a rack and X be an R-module (in the sense of Definition 6.1.8). Suppose there is a map $p : X \to R$ which satisfies the *generalized augmentation identity*, that is,

$$p(r\cdot x) = r \rhd p(x),$$

for all $r \in R$ and all $x \in X$. Then this generalized augmented rack defines a crossed module of racks. Namely, X becomes a rack with the product

$$x \rhd y := p(x)\cdot y,$$

for all $x, y \in X$. For this, note first that p is a morphism of racks:

$$p(x \rhd y) = p(p(x)\cdot y) = p(x) \rhd p(y).$$

We verify the rack identity using the definition of the rack product, the fact that p is a morphism and the fact that $\cdot$ is a rack action:

$$\begin{aligned} x \rhd (y \rhd z) &= p(x) \cdot (p(y) \cdot z) \\ &= (p(x) \rhd p(y)) \cdot (p(x) \cdot z) \\ &= p(x \rhd y) \cdot (x \rhd z) \\ &= (x \rhd y) \rhd (x \rhd z). \end{aligned}$$

Furthermore, R acts by automorphisms on the rack X using the definition of the rack product, the generalized augmentation identity and the fact that $\cdot$ is a rack action:

$$\begin{aligned} r \cdot (x \rhd y) &= r \cdot (p(x) \cdot y) \\ &= (r \rhd p(x)) \cdot (r \cdot y) \\ &= p(r \cdot x) \cdot (r \cdot y) \\ &= (r \cdot x) \rhd (r \cdot y). \end{aligned}$$

It is clear that p is equivariant by the generalized augmentation identity, and that the Peiffer identity follows from the definition of the rack product on X. Thus in conclusion, $p : X \to R$ is a crossed module of racks.

These three classes of examples are ordered here with growing generality, that is, crossed modules of groups are particular augmented racks, and augmented racks are particular generalized augmented racks. It turns out that the last class of examples is equivalent to crossed modules of racks:

Proposition 6.2.1. *There is a one-to-one correspondence between crossed modules of racks and generalized augmented racks.*

Proof. In Example 6.2.7, we gave the construction of a crossed module of racks from a generalized augmented rack. Conversely, given a crossed module of racks $\mu : R \to S$, forgetting the rack structure on R leaves us with a generalized augmented rack. Clearly, the two constructions are inverse to each other. □

Thanks to this proposition, we will very often regard crossed modules simply as a rack R, an R-module X and an equivariant map $p : X \to R$.

Proposition 6.2.2. *The functor* $\mathrm{As} : \mathtt{Racks} \to \mathtt{Grp}$ *sends crossed modules of racks to crossed modules of groups.*

Proof. Let $p : R \to S$ be a crossed module of racks. Then by functoriality, p extends to a group homomorphism $p : \mathrm{As}(R) \to \mathrm{As}(S)$, which extends the map p on elements of word length one to words of arbitrary length.

The rack action $R \times S \to R$ gives rise to a rack morphism $S \to \mathrm{Bij}(R)$ that extends by the universal property to a group homomorphism $\mathrm{As}(S) \to \mathrm{Bij}(R)$. The action of

As(S) on R is then extended to As(R) demanding that it should be an action by group automorphisms:

$$s \cdot (r_1 r_2) = (s \cdot r_1)(s \cdot r_2),$$

for all $r_1, r_2 \in \mathrm{As}(R)$ and all $s \in \mathrm{As}(S)$. This equation first extends the action of As(S) to an action on the free group $F(R)$, but as $s \cdot (r_1 \rhd r_2) = (s \cdot r_1) \rhd (s \cdot r_2)$, this passes to the quotient As(R) of $F(R)$.

The map p is equivariant, that is, $p(s \cdot r) = s \rhd p(r) = sp(r)s^{-1}$ for all $r \in \mathrm{As}(R)$ and all $s \in \mathrm{As}(S)$, because this statement holds for elements of word length one, and extends to all elements $r \in \mathrm{As}(R)$ by the fact that p is a group homomorphism and the action is an action by automorphisms. It extends finally to all $s \in \mathrm{As}(S)$ using the action property.

The Peiffer identity is shown in a similar way. □

One can do the replacement of racks by the associated groups also partially, that is, replace in a crossed module $p : X \to R$ the rack R by the group As(R). Note that the statements of the following two propositions compose to give back the statement of the previous proposition.

Proposition 6.2.3. *Given a crossed module of racks, the corresponding map $p : X \to \mathrm{As}(R)$ is an augmented rack.*

Proof. The rack homomorphism $p : X \to R$ gives rise by composition to a map $p : X \to \mathrm{As}(R)$. The action of R on X (given by a rack homomorphism $R \to \mathrm{Bij}(X)$) extends by the universal property to a group action of As(R) on X. The identity

$$p(r \cdot x) = rp(x)r^{-1}$$

is true for elements r of word length one in As(R) (and all elements $x \in X$), because of $p(r \cdot x) = r \rhd p(x)$. This then extends to all elements using the action property. □

In the same way, one can replace the G-set X in an augmented rack, regarded as a rack, by the associated group As(X).

Proposition 6.2.4. *Given an augmented rack $p : X \to G$, the induced map $p : \mathrm{As}(X) \to G$ is a crossed module of groups.*

Proof. Regarding X as a rack, the rack morphism $p : X \to \mathrm{Conj}(G)$ gives rise by the universal property to a group homomorphism $p : \mathrm{As}(X) \to G$. The next step is to extend the action of G on X to an action of G on As(X) by automorphisms. For this, one imposes

$$g \cdot (x_1 x_2) = (g \cdot x_1)(g \cdot x_2)$$

for all $x_1, x_2 \in X$ and all $g \in G$ in order to extend the action on X to an action on the free group $F(X)$. Then, using $g \cdot (x_1 \rhd x_2) = (g \cdot x_1) \rhd (g \cdot x_2)$, it becomes an action by automorphisms on the quotient $\mathrm{As}(X)$ of $F(X)$.

One shows the equivariance of p and the Peiffer identity as in the proof of Proposition 6.2.2. □

In conclusion, we have three classes of examples:

$$\mathrm{crmod}(\mathsf{Grp}) \subset \mathsf{augm\ Racks} \subset \mathrm{crmod}(\mathsf{Racks}).$$

Furthermore, we have the pair of adjoint functors

$$\mathrm{Conj} : \mathsf{Grp} \rightleftarrows \mathsf{Racks} : \mathrm{As},$$

which extend to functors going back and forth between these classes. The (elementary) composition functors, which arise have components of the form

$$\mathrm{Conj}(\mathrm{As}(R))$$

for a rack R, or

$$\mathrm{As}(\mathrm{Conj}(G))$$

for a group G. In general, $\mathrm{As}(\mathrm{Conj}(G))$ is far from being G and $\mathrm{Conj}(\mathrm{As}(R))$ is far from being R. This is the information loss one suffers by going from crossed modules of racks to augmented racks, or from augmented racks to crossed modules of groups.

In the following, we will also need a two-sided version of a crossed module of racks, that is, a version where R is an S-bimodule. Recall that the notion of bimodule is attached to a rack with involution, and the corresponding involutions on R and S are supposed to be sent to each other. One may think of these involutions as being the passage $\phi \mapsto \phi^{-1}$ in the bijections of R (resp., S).

Definition 6.2.8. Let $\mu : R \to S$ be a morphism of racks. The morphism μ is called a *(two-sided) crossed module of racks* in case R is an S-bimodule with a left and a right operation by automorphisms such that:
(a) $\mu(s \cdot r) = s \rhd \mu(r)$ and $\mu(r \cdot s) = \mu(r) \rhd s$.
(b) $\mu(r) \cdot r' = r \rhd r' = r \cdot \mu(r')$.

It is clear that forgetting the right operation we obtain a one-sided crossed module from a two-sided crossed module, and extending a given left action by taking the inverses to a right operation, we obtain a two-sided crossed module from a one-sided crossed module.

6.3 Crossed modules of rack bialgebras

Rack bialgebras are a linearized version of a rack. We therefore linearize the notions leading to crossed modules of racks in order to define crossed modules of rack bialgebras.

Definition 6.3.1. We say that a rack bialgebra R, defined over a field $\mathbb{K}$, *acts* on a $\mathbb{K}$-vector space V in case for any $r \in R$, there exists an endomorphism $r\cdot \in \mathrm{End}(V)$ such that for all $r, r' \in R$ and all $v \in V$,

$$r \cdot (r' \cdot v) = (r_1 \rhd r') \cdot (r_2 \cdot v).$$

Recall that we use Sweedler notation and the notation $\Delta(r) = r_1 \otimes r_2$ includes a sum. In case V is a rack bialgebra with unit 1_V, we will always add by convention that for all $r \in R$, $r \cdot 1_V = \epsilon_R(r) 1_V$.

Remark 6.3.2. (a) As an example, let a group G act linearly on a $\mathbb{K}$-vector space V. Then the rack bialgebra $\mathbb{K}[\mathrm{Conj}(G)]$ of the conjugation rack $\mathrm{Conj}(G)$ acts on V in the above described manner. More generally, if a rack X acts linearly on V, then $\mathbb{K}[X]$ acts in the above manner on V. As all elements of X are group-like, the above identity boils down to the rack-action-identity

$$r \cdot (r' \cdot v) = (r \rhd r') \cdot (r \cdot v).$$

(b) If a Lie algebra $\mathfrak{g}$ acts on V, then the rack bialgebra underlying the universal enveloping algebra $U\mathfrak{g}$ acts in the above manner on V. In this case, the above identity boils down for elements $x, y \in \mathfrak{g}$ to

$$x \cdot (y \cdot v) = y \cdot (x \cdot v) + [x, y] \cdot v,$$

because all elements $x, y \in \mathfrak{g}$ are primitive.

As for racks, we also have the notion of a bimodule, attached to a rack bialgebra with involution. We will always consider this involution to be $R \ni r \mapsto -r$.

Definition 6.3.3. Let R be a rack bialgebra. An *R-bimodule* is a $\mathbb{K}$-vector space V with operations of R from the left and from the right, which satisfy for all $r, r' \in R$ and all $v \in V$:

(1) $r \cdot (r' \cdot v) = (r_1 \rhd r') \cdot (r_2 \cdot v)$;
(2) $r \cdot (v \cdot r') = (r_1 \cdot v) \cdot (r_2 \rhd r')$;
(3) $(r \cdot v) \cdot r' = -(v \cdot r) \cdot r'$ holds only for $r, r' \notin \mathbb{K}1$.

Furthermore, 1 is supposed to act as the identity from the left and from the right.

Remark 6.3.4. (a) Let us explain the origin of this definition. Let $\mathfrak{h}$ be a Leibniz algebra. For the rack bialgebra $R := \mathfrak{h} \oplus \mathbb{K}1$, the above definition of an R-bimodule gives exactly a Leibniz bimodule; see Definition 6.4.1. Indeed, equations (1) and (2) are easily seen to be equations (LLM) and (LML) of a Leibniz bimodule. The last equation, equation (MLL), is more difficult to translate. But it is equivalent (using (LML)) to the condition $(r \cdot v) \cdot r' = -(v \cdot r) \cdot r'$. This last condition can be taken as it stands for rack bialgebras.

(b) Observe that one may always take $v \cdot r := -r \cdot v$ in order to transform an action of a rack bialgebra into the structure of a bimodule. In the same way, one may always take $v \cdot r := 0$ in order to transform a left action of a rack bialgebra into the structure of a bimodule. This is analoguous to the symmetric and antisymmetric Leibniz bimodule structures associated to a Lie module.

Definition 6.3.5. A *(one-sided) crossed module of rack bialgebras* is a morphism of rack bialgebras $\mu : R \to S$ together with an action of S on R such that R becomes an S-module algebra and an S-module coalgebra and for all $s \in S$ and all $r, r' \in R$, we have:

(a) $\mu(s \cdot r) = s \rhd \mu(r)$;

(b) $\mu(r) \cdot r' = r \rhd r'$.

There is also a two-sided version.

Definition 6.3.6. A *(two-sided) crossed module of rack bialgebras* is a morphism of rack bialgebras $\mu : R \to S$ together with the structure of an S-bimodule on R such that R becomes an S-module algebra and an S-module coalgebra and for all $s \in S$ and all $r, r' \in R$, we have:

(a) $\mu(s \cdot r) = s \rhd \mu(r)$ and $\mu(r \cdot s) = \mu(r) \rhd s$;

(b) $\mu(r) \cdot r' = r \rhd r' = r \cdot \mu(r')$.

Let us remind the reader that R is an S-module algebra in case for all $s \in S$ and all $r, r' \in R$

$$s \cdot (r \rhd r') = (s_1 \cdot r) \rhd (s_2 \cdot r'),$$

and

$$s \cdot 1_R = \epsilon_S(s)\, 1_R.$$

In the same way, R is an S-module coalgebra in case for all $s \in S$ and all $r \in R$, we have

$$(s \cdot r)_1 \otimes (s \cdot r)_2 = (s_1 \cdot r_1) \otimes (s_2 \cdot r_2),$$

and

$$\epsilon_R(s \cdot r) = \epsilon_S(s)\, \epsilon_R(r).$$

Let us develop two examples of one-sided crossed modules of rack bialgebras. Discussing later the relations between the different types of crossed modules, we will see that two-sided crossed modules of racks and of Leibniz algebras give examples of two-sided crossed modules of rack bialgebras.

Proposition 6.3.1. *Let $p : X \to G$ be an augmented rack. Then p linearizes to $p : \mathbb{K}[X] \to \mathbb{K}[G]$ which is a (one-sided) crossed module of rack bialgebras.*

Proof. We have already seen that the linearized rack $\mathbb{K}[X]$ is a rack bialgebra, and the group algebra $\mathbb{K}[G]$ can be viewed as a rack bialgebra with respect to

$$g \rhd h := ghg^{-1},$$

for all $g, h \in G$. This rack product, here defined on generators, is then extended by linearity to all of $\mathbb{K}[G]$. It is the same to view G as a conjugation rack and then take the linearization of the rack. By functoriality of the linearization, p induces a linear map $p : \mathbb{K}[X] \to \mathbb{K}[G]$. The group action of G on the set X induces an action of the rack bialgebra $\mathbb{K}[G]$ on $\mathbb{K}[X]$. Indeed, the property

$$g \cdot (h \cdot x) = ghg^{-1} \cdot (g \cdot x) = (g \rhd h) \cdot (g \cdot x)$$

is satisfied on a basis, thus everywhere. Note that here all elements are group-like or set-like. Now $\mathbb{K}[X]$ becomes a $\mathbb{K}[G]$-module algebra, because for $g \in G$ and $x, y \in X$, we have

$$\begin{aligned} g \cdot (x \rhd y) &= g \cdot (p(x) \cdot y) \\ &= (g \rhd p(x)) \cdot (g \cdot y) \\ &= (gp(x)g^{-1}) \cdot (g \cdot y) \\ &= p(g \cdot x) \cdot (g \cdot y) \\ &= (g \cdot x) \rhd (g \cdot y). \end{aligned}$$

As all elements are group-like or set-like, the property of being a $\mathbb{K}[G]$-module coalgebra is trivially satisfied. Property (a) reads

$$p(g \cdot x) = gp(x)g^{-1} = g \rhd p(x),$$

and property (b) is the definition (on a basis) of the rack product on $\mathbb{K}[X]$. □

We have seen that a cocommutative Hopf algebra is a rack bialgebra with respect to the adjoint action. In order for a crossed module of cocommutative Hopf algebras $\gamma : B \to H$ to induce a crossed module of rack bialgebras, we need that the action is also compatible with the antipode. As B is already an H-module algebra and coalgebra, the condition we need is that B is a Hopf algebra in the category of H-modules, that is, an H-module Hopf algebra. This implies for the antipode for all $h \in H$ and all $b \in B$ that

$$h \cdot S(b) = S(h \cdot b).$$

Proposition 6.3.2. *Let $\gamma : B \to H$ be a crossed module of cocommutative Hopf algebras such that B becomes an H-module Hopf algebra. Then $\gamma : B \to H$ is a (one-sided) crossed module of rack bialgebras.*

Proof. The Hopf algebras B and H become rack bialgebras with respect to the adjoint action, that is, for all $b, b' \in B$,

$$b \rhd b' := \mathrm{ad}_b(b') := b_1 b' S(b_2).$$

The morphism $\gamma : B \to H$ is then a morphism of rack bialgebras, because a morphism of Hopf algebras is a morphism of algebras and of coalgebras, which commutes with the antipodes. The H-module structure is an action of rack bialgebras, because we have for all $h, h' \in H$ and all $b \in B$ on the one hand,

$$h \cdot (h' \cdot b) = hh' \cdot b,$$

and on the other hand,

$$(h_1 \rhd h') \cdot (h_2 \cdot b) = h_1 h' S(h_2) h_3 \cdot b = h_1 h' \epsilon(h_2) \cdot b = hh' \cdot b.$$

The H-module Hopf algebra B is an H-module coalgebra, but we have to show that it is also an H-module algebra with respect to the adjoint action. Here, we use the property of being an H-module algebra to go from the first to the second line, cocommutativity and the compatibility with the antipode to go from the second to the third line and the property H-module coalgebra to go from the third to the fourth line:

$$\begin{aligned} h \cdot (b \rhd b') &= h \cdot (b_1 b' S(b_2)) \\ &= (h_1 \cdot b_1)(h_2 \cdot b')(h_3 \cdot S(b_2)) \\ &= (h_1 \cdot b_1)(h_3 \cdot b') S(h_2 \cdot b_2) \\ &= (h \cdot b)_1 (h_3 \cdot b') S(h \cdot b)_2 \\ &= (h_1 \cdot b) \rhd (h_2 \cdot b'). \end{aligned}$$

It remains to show the properties (a) and (b). But both properties of a crossed module of Hopf algebras (i. e., $\gamma \circ \phi_B = \mathrm{ad}_H(\mathrm{id}_B \otimes \gamma)$ and $\phi_B \circ (\gamma \otimes \mathrm{id}_H) = \mathrm{ad}_B$) are exactly the corresponding properties for rack bialgebras. □

6.4 Links between the different types of crossed modules

The goal of this subsection is to establish the links between two-sided crossed modules that can be inferred from the links between racks, Leibniz algebras and rack bialgebras.

6.4.1 Crossed modules of Lie racks and Leibniz algebras

Let us start with Leibniz algebras and more precisely with the notion of an action of a Leibniz algebra or of a Leibniz bimodule.

Definition 6.4.1. Let $\mathfrak{h}$ be a Leibniz algebra. A vector space M is a (left) *Leibniz $\mathfrak{h}$-bimodule* in case there exist maps $\mathfrak{h}\times M \to M$, written $(x,m)\mapsto x\cdot m$, and $M\times\mathfrak{h}\to M$, written $(m,x)\mapsto m\cdot x$ such that for all $x,y\in\mathfrak{h}$ and all $m\in M$
(LLM) $x\cdot(y\cdot m) = [x,y]\cdot m + y\cdot(x\cdot m)$;
(LML) $x\cdot(m\cdot y) = (x\cdot m)\cdot y + m\cdot[x,y]$;
(MLL) $m\cdot[x,y] = (m\cdot x)\cdot y + x\cdot(m\cdot y)$.

Observe that while (LLM) asserts that M becomes a left $\mathfrak{h}$-module (in the sense of Lie algebras) where the ideal of squares $\mathrm{Leib}(\mathfrak{h})$ acts trivially, equation (MLL) does not assert that M becomes a right $\mathfrak{h}$-module. Observe further that (under the assumption of (LML)) the condition (MLL) is equivalent to $(m\cdot x)\cdot y = -(x\cdot m)\cdot y$ for all $x,y\in\mathfrak{h}$ and all $m\in M$.

Remark 6.4.2. The equations (LLM), (LML) and (MLL) are in fact three realizations of the (left) Leibniz identity $[x,[y,z]] = [[x,y],z]+[y,[x,z]]$, obtained by replacing (x,y,z) by (x,y,m) (for LLM), (x,m,y) (for LML) and (m,x,y) (for MLL). These equations arise naturally if one asks the direct sum of a (left) Leibniz algebra $\mathfrak{h}$ and a (left) Leibniz bimodule M to become again a (left) Leibniz algebra; see [66]. Pay attention to the fact that we consider here left Leibniz algebras, while Loday and Pirashvili usually consider right Leibniz algebras (and right Leibniz bimodules).

Definition 6.4.3. Let $\mathfrak{h}$ and $\mathfrak{m}$ be Leibniz algebras. $\mathfrak{h}$ is said to *act* on $\mathfrak{m}$ *by derivations* in case $\mathfrak{m}$ is a Leibniz $\mathfrak{h}$-bimodule such that the following compatibility relations hold between the bracket on $\mathfrak{m}$ and the two actions:
- $x\cdot[m,m'] = [x\cdot m,m'] + [m,x\cdot m']$;
- $[m,x\cdot m'] = [m\cdot x,m'] + x\cdot[m,m']$;
- $[m,m'\cdot x] = [m,m']\cdot x + [m',m\cdot x]$;

for all $m,m'\in\mathfrak{m}$ and all $x\in\mathfrak{h}$.

Remark 6.4.4. These three equations arise again as (LLM), (LML) and (MLL) with (x,y,m) replaced by (x,m,m').

Leibniz algebras are intrinsically two-sided, so dealing with Leibniz algebras, there is only a two-sided version of a crossed module.

Definition 6.4.5. A *(two-sided) crossed module of Leibniz algebras* is a morphism of Leibniz algebras $\mu:\mathfrak{m}\to\mathfrak{n}$ with an action of $\mathfrak{n}$ on $\mathfrak{m}$ by derivations such that:
(a) $\mu(n\cdot m) = [n,\mu(m)]$ and $\mu(m\cdot n) = [\mu(m),n]$;
(b) $\mu(m)\cdot m' = [m,m'] = m\cdot\mu(m')$;

for all $n\in\mathfrak{n}$ and all $m,m'\in\mathfrak{m}$.

Theorem 6.4.1. *The tangent Leibniz algebras of a two-sided crossed module of Lie racks form a crossed module of Leibniz algebras.*

Proof. The main tool for the proof is Proposition 6.1.4. By functoriality, the morphism of Lie racks $\mu : R \to S$ gives rise to a morphism of Leibniz algebras $T_1\mu : \mathfrak{r} \to \mathfrak{s}$ where $\mathfrak{r} := T_1 R$ and $\mathfrak{s} := T_1 S$. The smooth operations of S on R gives rise by passing to the tangent spaces to a left and a right operation of $\mathfrak{s}$ on $\mathfrak{r}$. Let us show that these provide a Leibniz bimodule structure of $\mathfrak{r}$.

Let $\alpha :]-\epsilon,\epsilon[\to U \subset S$ and $\beta :]-\epsilon,\epsilon[\to U \subset S$ be smooth curves in a neighborhood U of $1 \in S$ such that $\alpha(0) = \beta(0) = 1$ and $\frac{d}{dt}|_{t=0}\alpha(t) = x \in \mathfrak{s}$ and $\frac{d}{ds}|_{s=0}\beta(s) = y \in \mathfrak{s}$. We will denote the operations of S on R and the subsequent operations of $\mathfrak{s}$ on $\mathfrak{r}$ which are obtained as their differentials at 1 both in the same way. For the differentials at 1 of the rack product, we use the notation $\frac{d}{ds}|_{s=0}\alpha(t) \rhd \beta(s) = \mathrm{Ad}_{\alpha(t)} y$ and $\frac{d}{dt}|_{t=0}\frac{d}{ds}|_{s=0}\alpha(t) \rhd \beta(s) = [x,y]$ by definition of the Leibniz bracket on $\mathfrak{s}$. We obtain in this way for all $r \in R$ by derivation of

$$\alpha(t)\cdot(\beta(s)\cdot r) = (\alpha(t) \rhd \beta(s))\cdot(\alpha(t)\cdot r)$$

first with respect to s at $s = 0$

$$\alpha(t)\cdot(y\cdot r) = \mathrm{Ad}_{\alpha(t)}(y)\cdot(\alpha(t)\cdot r)$$

and then with respect to t at $t = 0$

$$x\cdot(y\cdot r) = [x,y]\cdot r + y\cdot(x\cdot r).$$

This is condition (LLM). In the same way, the condition

$$\alpha(t)\cdot(r\cdot\beta(s)) = (\alpha(t)\cdot r)\cdot(\alpha(t) \rhd \beta(s))$$

gives by deriving first with respect to s at $s = 0$ and then with respect to t at $t = 0$

$$x\cdot(r\cdot y) = (x\cdot r)\cdot y + r\cdot[x,y].$$

This is condition (LML) for a Leibniz bimodule. Observe now that passing to differentials sends the involutions of R and S (which were supposed to be $\phi \mapsto \phi^{-1}$) to the involution of $\mathfrak{r}$ and $\mathfrak{s}$ (which we set to be $x \mapsto -x$). By deriving then

$$(\alpha(t)\cdot r)\cdot\beta(s) = \left(r(\cdot\alpha(t))^{-1}\right)\cdot\beta(s)$$

with respect to s and t at $s = 0$ and $t = 0$, we obtain finally

$$(x\cdot r)\cdot y = -(r\cdot x)\cdot y.$$

This is equivalent (under the assumption of (LML)) to (MLL).

The fact that S operates on R from the left and from the right by automorphisms translates into infinitesimal operations of $T_1 S$ from the left and from the right on the Leibniz algebra $\mathfrak{r}$ by derivations. It is clear that $T_1\mu$ is equivariant with respect to both actions by (a) and satisfies Peiffer's identity for both actions by (b). □

Let us comment on the two-sided nature of the construction.

Remark 6.4.6. Recall from Remark 6.1.14 that a Lie rack can be considered to have two smooth operations $(x,y) \mapsto x \rhd y$ and $(x,y) \mapsto x \lhd y$, which satisfy the properties displayed in Proposition 6.1.1. Both left and right translation maps $y \mapsto x \rhd y$ and $x \mapsto x \lhd y$ are diffeomorphisms. Left and right translations give rise by Proposition 6.1.4 to a left and a right Leibniz bracket on $T_1 R$ and $T_1 S$. These two are related as explained in Remark 6.1.16.

Remark 6.4.7. For the moment, there is no theorem stating the converse, that is, that crossed modules of (finite dimensional, real) Leibniz algebras can be integrated into crossed modules of Lie racks. The principal difficulty here is that the integration procedure of Leibniz algebras into Lie racks alluded to earlier (see [11]) is not functorial, thus we do not know for the moment how the morphism of Leibniz algebras underlying the crossed module will induce a morphism of Lie racks. The main issue is that a morphism of Leibniz algebras is not necessarily ad-spectrum reducing (a property that is needed in our construction in order to induce a morphism of Lie racks).

6.4.2 Crossed modules of rack bialgebras and Leibniz algebras

We have already seen several examples for crossed modules of rack bialgebras. Here comes an example from Leibniz algebras. Let $p : \mathfrak{h} \to \mathfrak{g}$ be an augmented Leibniz algebra, that is, the Lie algebra $\mathfrak{g}$ acts on the vector space $\mathfrak{h}$ and p is a linear map, which satisfies $p(x \cdot h) = [x, p(h)]$ for all $x \in \mathfrak{g}$ and all $h \in \mathfrak{h}$. An augmented Leibniz algebra gives rise to a crossed module of rack bialgebras.

Proposition 6.4.2. *The augmented rack bialgebra $p : S(\mathfrak{h}) \to U\mathfrak{g}$ constructed in Theorem 6.1.9 is a one-sided crossed module of rack bialgebras.*

Proof. We have seen that $p : S(\mathfrak{h}) \to U\mathfrak{g}$ is a morphism of coalgebras, and the rack product on $S(\mathfrak{h})$ is constructed such that it becomes a morphism of rack bialgebras. Moreover, $p(u \cdot h) = \mathrm{ad}_u(p(h))$ for all $u \in U\mathfrak{g}$ and all $h \in S(\mathfrak{h})$. Furthermore, we have seen that $S(\mathfrak{h})$ is a $U\mathfrak{g}$-module and a $U\mathfrak{g}$-module coalgebra. Let us show that the action is an action of the rack bialgebra $U\mathfrak{g}$ on the rack bialgebra $S(\mathfrak{h})$. For all $u, u' \in U\mathfrak{g}$ and all $h \in S(\mathfrak{h})$, we have on the one hand,

$$u \cdot (u' \cdot h) = uu' \cdot h,$$

and on the other hand,

$$\begin{aligned}(u_1 \rhd u') \cdot (u_2 \cdot h) &= \mathrm{ad}_{u_1}(u') \cdot (u_2 \cdot h)\\ &= (u_1 u' S(u_2)) \cdot (u_3 \cdot h)\\ &= (u_1 u' S(u_2) u_3) \cdot h\end{aligned}$$

$$= (u_1 u' \epsilon(u_2)) \cdot h$$
$$= uu' \cdot h.$$

Let us now show that $S(\mathfrak{h})$ is also a $U\mathfrak{g}$-module algebra. Let $x \in \mathfrak{g}$ and $h, h' \in \mathfrak{h}$:

$$\begin{aligned} x \cdot (h \rhd h') &= x \cdot (p(h) \cdot h') \\ &= (xp(h)) \cdot h' \\ &= [x, p(h)] \cdot h' + p(h) \cdot (x \cdot h') \\ &= p(x \cdot h) \cdot h' + h \rhd (x \cdot h') \\ &= (x \cdot h) \rhd h' + h \rhd (x \cdot h'). \end{aligned}$$

This is the $U\mathfrak{g}$-module algebra property in this special case, because x is primitive. Now the general property can be deduced from this using that the action is extended to all of $U\mathfrak{g}$ by successive applications and to all of $S(\mathfrak{g})$ by the derivation property. The properties (a) and (b) are clear from (6.19) and the definition of the rack product. □

Observe that it would have been almost equivalent to show that the augmented Leibniz algebra induces a one-sided crossed module of rack bialgebras $\mathfrak{h} \oplus \mathbb{K}1 \to U\mathfrak{g}$.

In order to generalize this proposition to all crossed modules of Leibniz algebras, let us focus first on the H-module algebra property. For this, we recall Lemma 4.2.2, which is true in a more general framework:

Lemma 6.4.3. *Let $\mathfrak{g}$ be a Lie algebra. A (not necessarily associative) algebra $(A, \rhd)$ is an $U\mathfrak{g}$-module algebra if and only if the vector space A is a $\mathfrak{g}$-module such that elements of $\mathfrak{g}$ act by derivations of the bilinear product $\rhd$ of A.*

Proof. Suppose A is an $U\mathfrak{g}$-module algebra. Then the $U\mathfrak{g}$-module A is a $\mathfrak{g}$-module and we have for all $x \in \mathfrak{g}$ and all $a, b \in A$,

$$x \cdot (a \rhd b) = (x \cdot a) \rhd b + a \rhd (x \cdot b), \tag{6.39}$$

because x is primitive. Therefore, $\mathfrak{g}$ acts by derivations.

Conversely, let A be an algebra which is a $\mathfrak{g}$-module where its elements act by derivations. Then A becomes a $U\mathfrak{g}$-module. We have to show equation (6.39). It holds by construction for the elements of $\mathfrak{g}$. As $U\mathfrak{g}$ is generated as an algebra by the products of elements of $\mathfrak{g}$, it is enough to show for all $x, y \in \mathfrak{g}$ and all $a, b \in A$ that $(xy) \cdot (a \rhd b) = ((xy)_1 \cdot a) \rhd ((xy)_2 \cdot b)$. We compute

$$\begin{aligned} (xy) \cdot (a \rhd b) &= x \cdot (y \cdot (a \rhd b)) \\ &= x \cdot ((y_1 \cdot a) \rhd (y_2 \cdot b)) \\ &= (x_1 \cdot (y_1 \cdot a)) \rhd (x_2 \cdot (y_2 \cdot b)) \\ &= ((x_1 y_1) \cdot a) \rhd ((x_2 y_2) \cdot b) \\ &= ((xy)_1 \cdot a) \rhd ((xy)_2 \cdot b) \end{aligned}$$

where we have used the bialgebra property from the fourth to the last line. □

After these preparations, we can state the theorem.

Theorem 6.4.4. *Let $\mu : \mathfrak{m} \to \mathfrak{n}$ be a crossed module of Leibniz algebras. Then we obtain an induced (two-sided) crossed module of rack bialgebras $\mu : S(\mathfrak{m}) \to S(\mathfrak{n})$.*

Proof. The crossed module of Leibniz algebras $\mu : \mathfrak{m} \to \mathfrak{n}$ induces a crossed module of Lie algebras $\overline{\mu} : \overline{\mathfrak{m}} \to \overline{\mathfrak{n}}$ by passage to the quotient Lie algebras of $\mathfrak{m}$ and $\mathfrak{n}$ with respect to the ideals of squares $Q(\mathfrak{m})$ and $Q(\mathfrak{n})$, respectively. We thus obtain a crossed module of Hopf algebras $\gamma : U\overline{\mathfrak{m}} \to U\overline{\mathfrak{n}}$.

On the other hand, the morphism of Leibniz algebras $\mu : \mathfrak{m} \to \mathfrak{n}$ induces by functoriality a morphism of rack bialgebras $\mu : S(\mathfrak{m}) \to S(\mathfrak{n})$, where the rack products are induced by $U\overline{\mathfrak{m}}$ and $U\overline{\mathfrak{n}}$, respectively, because $p : S(\mathfrak{m}) \to U\overline{\mathfrak{m}}$ and $q : S(\mathfrak{n}) \to U\overline{\mathfrak{n}}$ are augmented rack bialgebras.

Consider the left operation of $\mathfrak{n}$ on $\mathfrak{m}$, and extend it to $\mathbb{K}1 \oplus \mathfrak{n}$ by letting 1 act as the identity. Let us show that this is an action of rack bialgebras. We have for all $n, n' \in \mathfrak{n}$ and all $m \in \mathfrak{m}$,

$$(n_1 \rhd n') \cdot (n_2 \cdot m) = [n, n'] \cdot m + n' \cdot (n \cdot m) = n \cdot (n' \cdot m).$$

Now extend this action to $S(\mathfrak{n})$ by successive application of the elements of $\mathbb{K}1 \oplus \mathfrak{n}$. It remains an action of rack bialgebras. Extend this action to all of $S(\mathfrak{m})$ by the derivation property (with respect to the associative product on the symmetric algebra). This shows that the left operation of $\mathfrak{n}$ on $\mathfrak{m}$ extends to an action of rack bialgebras of $S(\mathfrak{n})$ on $S(\mathfrak{m})$.

Note that the ideal of squares $Q(\mathfrak{n})$ acts trivially by the left operation. Thus the above constructed action of rack bialgebras is in fact an action of the universal enveloping algebra $U\overline{\mathfrak{n}}$. This remark is important, because it permits us to apply the Lemma 6.4.3 in order to have directly the module-algebra property. The module-coalgebra property is clear.

In the same way, the right operation of $\mathfrak{n}$ on $\mathfrak{m}$ is extended to an operation of $S(\mathfrak{n})$ on $S(\mathfrak{m})$. The property $n \cdot (m \cdot n') = (n_1 \cdot m) \cdot (n_2 \rhd n')$, which is true on $\mathfrak{n} \oplus \mathbb{K}1$ is by extension also true for all $n, n' \in S(\mathfrak{n})$. The property $(n \cdot m) \cdot n' = -(m \cdot n) \cdot n'$ leads directly to the corresponding property for the bimodule structure of $S(\mathfrak{n})$ on $S(\mathfrak{m})$. Properties (a) and (b) translate into one another. □

We also have a theorem going in the other direction.

Theorem 6.4.5. *The functor of primitives P sends a two-sided crossed modules of rack bialgebras $\mu : R \to S$ to a crossed module of Leibniz algebras $\mu : P(R) \to P(S)$.*

Proof. The functor of primitives sends a rack bialgebra R to the subspace of primitives $P(R)$, that is, of elements $r \in R$, which satisfy $\Delta(r) = 1 \otimes r + r \otimes 1$. $P(R)$ becomes a Leibniz algebra with respect to the rack product $[r, r'] := r \rhd r'$. Functoriality of this construction implies that the restriction of μ (which we still denote by μ) is a morphism of Leibniz algebras $\mu : P(R) \to P(S)$. The rack bialgebra operations of S on the left and on the right on R can be turned into a two-sided operation of the primitives on the primitives. The properties (1), (2) and (3) translate directly into the properties (LLM), (LML) and (MLL) of a Leibniz bimodule.

The Leibniz algebra $P(S)$ acts on the Leibniz algebra $P(R)$ by derivations, as follows from the property that R is an S-module algebra. For all $s \in P(S)$ and all $r, r' \in P(R)$, we have

$$s \cdot [r, r'] = s \cdot (r \rhd r') = (s_1 \cdot r) \rhd (s_2 \cdot r') = (s \cdot r) \rhd r' + r \rhd (s \cdot r') = [s \cdot r, r'] + [r, s \cdot r'].$$

The remaining two properties to act by derivations follow in the same way. Conditions (a) and (b) also translate directly. □

6.4.3 Crossed modules of racks and rack bialgebras

Here, we consider the links between crossed modules of racks and crossed modules of rack bialgebras, which are induced by the functors of linearization and set-like elements. Recall that we emphasized that for the notions of bimodules we needed involutions. The functor of linearization is supposed to send the fixed involutions on the level of racks to those fixed on the level of rack bialgebras. An instance of this has already been seen in Theorem 6.4.1.

Theorem 6.4.6. *Let $\mu : R \to S$ be a two-sided crossed module of racks. Then the linearized version $\mu : \mathbb{K}[R] \to \mathbb{K}[S]$ (which denote by abuse of notation still by μ) is a two-sided crossed module of rack bialgebras.*

Proof. We have seen earlier that $\mathbb{K}[R]$ and $\mathbb{K}[S]$ are rack bialgebras, and the construction is clearly functorial, thus we dispose of a morphism of rack bialgebras $\mu : \mathbb{K}[R] \to \mathbb{K}[S]$.

The rack S operates on R from the left and from the right by automorphisms

$$s \cdot (r \rhd r') = (s \cdot r) \rhd (s \cdot r').$$

As the elements of S are set-like, this implies directly that $\mathbb{K}[R]$ is a $\mathbb{K}[S]$-module algebra, where by convention for actions of a rack bialgebra on another rack bialgebra, $s \cdot 1_R = \epsilon_S(s)\, 1_R$. As all elements are set-like, it is clear that R is an S-module coalgebra.

The conditions for an $\mathbb{K}[S]$-bimodule structure on $\mathbb{K}[R]$ are seen directly for (1) and (2). For condition (3), we have already observed that the involutions of R and S are sent by linearization to the involutions of $\mathbb{K}[R]$ and $\mathbb{K}[S]$, thus condition $(s \cdot r) \cdot s' = (r(\cdot s)^{-1}) \cdot s'$ implies the condition $(s \cdot r) \cdot s' = -(r \cdot s) \cdot s'$ for a bimodule structure of rack bialgebras.

The properties (a) and (b) follow directly from the corresponding properties for crossed modules of racks. □

In the other direction, we have an analogous theorem.

Theorem 6.4.7. *Let $\mu : R \to S$ be a one-sided crossed module of rack bialgebras. Then the set-like elements* $\mathrm{Slike}(R)$ *and* $\mathrm{Slike}(S)$ *form a one-sided crossed module of racks* $\mu : \mathrm{Slike}(R) \to \mathrm{Slike}(S)$, *where we do not impose that* $\mathrm{Slike}(S)$ *acts on* $\mathrm{Slike}(R)$ *by bijections.*

Proof. The functor Slike sends a rack bialgebra R to its subset $\mathrm{Slike}(R)$ of set-like elements, that is, elements $r \in R$ such that $\Delta(r) = r \otimes r$. By functoriality, $\mathrm{Slike}(R)$ sends the morphism of rack bialgebras μ to a morphism of racks $\mu : \mathrm{Slike}(R) \to \mathrm{Slike}(S)$.

For set-like elements $s, s' \in S$, the rack-bialgebra-action identity reduces to

$$s \cdot (s' \cdot r) = (s \rhd s') \cdot (s \cdot r).$$

Similarly, the fact that R is an S-module algebra implies that

$$s \cdot (r \rhd r') = (s \cdot r) \rhd (s \cdot r'),$$

that is, that S acts by automorphisms on R. The properties (a) and (b) translate directly. □

Remark 6.4.8. We cannot prove that $\mathrm{Slike}(S)$ acts on $\mathrm{Slike}(R)$ by bijections and we have displayed here only a one-sided version for the passage from crossed modules of rack bialgebras to crossed modules of racks. We believe that both points need further investigation in relation to the notion of an *antipode* for rack bialgebras, which should assure that the S-bimodule structure on R induces operations by bijections on the level of set-like elements. We invite the interested reader to think about these matters.

6.5 Categorical racks

6.5.1 From categories to crossed modules

We have seen in Chapter 1 that crossed modules of groups $\mu : M \to N$ are in one-to-one correspondence with category objects in the category of groups Grp, strict 2-groups or categorical groups. We recall the correspondence given in Chapter 1: Given $\mu : M \to N$, we construct a strict 2-group by taking $G_0 := N$ as the group of objects and the semidirect product $G_1 := M \rtimes N$ as the group of morphisms. In the other direction, given a strict 2-group $G_1 \underset{t}{\overset{s}{\rightrightarrows}} G_0$ we take $N := G_0$ and $M := \mathrm{Ker}(s)$, where s is the source map. The map μ is given by the restriction of the target map t to $\mathrm{Ker}(s)$. We will call the passage to $t : \mathrm{Ker}(s) \to G_0$ the *standard construction*. From this well-known construction, we note following lemma.

Lemma 6.5.1. *Let M and N be groups such that N acts on M by automorphisms, and let $\mu : M \to N$ be an equivariant group homomorphism. Then the semidirect product $M \rtimes N$*

carries a unique structure of a category such that the objects are N, the morphisms $M \rtimes N$, the source $(m, n) \mapsto n$, the target $(m, n) \mapsto \mu(m)n$ and the identity $n \mapsto (1_M, n)$.

In fact, we have seen in Chapter 1 some of the three more objects which are equivalent to strict 2-groups; see [64]:

(a) groups G with a subgroup N and two homomorphisms $s, t : G \to N$ with $s|_N = \mathrm{id}_N$, $t|_N = \mathrm{id}_N$ and $[\mathrm{Ker}(s), \mathrm{Ker}(t)] = 1$ (these are called *1-cat groups*);
(b) simplicial groups with Moore complex of length one;
(c) group objects in the category of (small) categories.

We desire to have correspondences of a similar type for crossed modules of racks.

Definition 6.5.1. A *1-cat rack* consists of a pointed rack R, a subrack N and two rack morphisms $s, t : R \to N$ such that $s|_N = \mathrm{id}_N$, $t|_N = \mathrm{id}_N$ and $\mathrm{Ker}(s)$ and $\mathrm{Ker}(t)$ act trivially on each other.

One approach to the construction of a similar correspondence to that described above in the group case is to consider category objects in the category of racks; see [15].

Definition 6.5.2. A *strict 2-rack* or *categorical rack* is a category object in the category of racks. That is, a strict 2-rack consists of two pointed racks R_0 (rack of objects) and R_1 (rack of morphisms) equipped with rack morphisms $s, t : R_1 \to R_0$ (source and target), $i : R_0 \to R_1$ (identity-assigning) and $\circ : R_1 \times_{R_0} R_1 \to R_1$ (composition) such that the usual axioms of a category are satisfied.

One crucial property of the composition in a strict 2-group is the *middle four exchange property* which simply means that the composition $\circ$ is a morphism of groups:

$$(g_1g_2) \circ (f_1f_2) = (g_1 \circ f_1)(g_2 \circ f_2).$$

We observe that this property holds for all morphisms $f_1 : a \mapsto b$, $f_2 : a' \mapsto b'$, $g_1 : b \mapsto c$, $g_2 : b' \mapsto c'$, where the constraints on the domains and codomains reflect the composability.

From this property, one deduces that the kernel of the source map, $\mathrm{Ker}(s)$, and the kernel of the target map, $\mathrm{Ker}(t)$, commute and that the group product on G_1 uniquely determines the composition. Note that as the inclusion of identities the map $G_0 \hookrightarrow G_1$ is a group homomorphism and the identity $1 \in G_1$ is both the identity with respect to the composition and unit with respect to the group product. We see that we have a similar property in the framework of strict 2-racks.

Proposition 6.5.2. *The middle four exchange property for a strict 2-rack implies that* $\mathrm{Ker}(s)$ *and* $\mathrm{Ker}(t)$ *act trivially on each other.*

Proof. For strict 2-racks, the middle four exchange property states

$$(f_1 \rhd f_2) \circ (g_1 \rhd g_2) = (f_1 \circ g_1) \rhd (f_2 \circ g_2)$$

for all $f_1 : a \mapsto b, f_2 : a' \mapsto b', g_1 : b \mapsto c$, and $g_2 : b' \mapsto c'$. By choosing $b' = c' = 1$, $g_2 = 1$ and by using the fact that the identity in $1 \in R_1$ is both unit and identity we deduce that

$$f_1 \rhd f_2 = (f_1 \circ g_1) \rhd f_2$$

for all $f_2 \in \mathrm{Ker}(t)$.

In the special case when $a = b = 1$ with $f_1 = 1$, and thus $g_1 : 1 \mapsto c$, that is, $g_1 \in \mathrm{Ker}(s)$ we obtain from the above that

$$f_2 = g_1 \rhd f_2$$

for all $f_2 \in \mathrm{Ker}(t)$ and all $g_1 \in \mathrm{Ker}(s)$. This means that elements from $\mathrm{Ker}(s)$ act trivially on elements of $\mathrm{Ker}(t)$.

In the same way, choosing $a' = b' = 1$ and $b = c = 1$, $f_2 = 1$ and $g_1 = 1$, and therefore $f_1 \in \mathrm{Ker}(t)$ and $g_2 \in \mathrm{Ker}(s)$, we obtain

$$g_2 = f_1 \rhd g_2,$$

which means that $\mathrm{Ker}(t)$ acts also trivially on $\mathrm{Ker}(s)$. □

Corollary 6.5.3. *A strict 2-rack has an underlying 1-cat rack, that is, a pointed rack R together with a subrack N and two homomorphisms $s, t : R \to N$ such that $s|_N = \mathrm{id}_N$, $t|_N = \mathrm{id}_N$ and* $\mathrm{Ker}(s)$ *and* $\mathrm{Ker}(t)$ *act trivially on each other.*

Proof. Indeed, define $R := R_1$, $N := i(R_0)$, and take source and target maps $s, t : R \to N$. The properties $s|_N = \mathrm{id}_N$ and $t|_N = \mathrm{id}_N$ come from $t \circ i = \mathrm{id}_{R_0}$ and $s \circ i = \mathrm{id}_{R_0}$. By Proposition 6.5.2, we have that $\mathrm{Ker}(s)$ and $\mathrm{Ker}(t)$ act trivially on each other. □

Remark 6.5.3. This correspondence from strict 2-racks to 1-cat racks is actually functorial. We leave it to the interested reader to define the necessary (2-)category structure on both classes of objects in order to make this a mathematical statement. Similar remarks apply to the propositions in this and the next subsection.

Remark 6.5.4. Unfortunately, the four equations following from the middle four exchange property in a strict 2-rack:

(a) $f_1 \rhd f_2 = (f_1 \circ g_1) \rhd f_2$ with the restriction $f_2 \in \mathrm{Ker}(t)$;
(b) $(f_1 \rhd f_2) \circ g_2 = f_1 \rhd (f_2 \circ g_2)$ with the restriction $f_1 \in \mathrm{Ker}(t)$;
(c) $g_1 \rhd g_2 = (f_1 \circ g_1) \rhd g_2$ with the restriction $g_2 \in \mathrm{Ker}(s)$, and
(d) $f_2 \circ (g_1 \rhd g_2) = g_1 \rhd (f_2 \circ g_2)$ with the restriction $g_1 \in \mathrm{Ker}(s)$;

do not seem to enable us to reconstruct the composition starting from the rack product, or vice-versa (as we can in the case for strict 2-groups). We leave this observation as a question for future study.

Proposition 6.5.4. *A strict 2-rack gives rise, via the standard construction, to a crossed module of racks.*

Proof. Recall from Proposition 6.2.1 that a crossed module of racks consists of a rack R, an R-module X and an equivariant map $p : X \to R$. Given a strict 2-rack $(R_0, R_1, s, t, i, \circ)$, we define $R := R_0$. The rack R acts on $X := \mathrm{Ker}(s)$ by

$$r \cdot x := i(r) \rhd x.$$

Indeed, $\mathrm{Ker}(s)$ is preserved by this action since, as s is a morphism,

$$s(r \cdot x) = s(i(r) \rhd x) = s(i(r)) \rhd s(x) = r \rhd 1 = 1.$$

The fact that this is an action follows from the rack identity and the fact that i is a morphism:

$$\begin{aligned} r \cdot (r' \cdot x) &= i(r) \rhd (i(r') \rhd x) \\ &= (i(r) \rhd i(r')) \rhd (i(r') \rhd x) \\ &= i(r \rhd r') \rhd (r' \cdot x) \\ &= (r \rhd r') \cdot (r' \cdot x). \end{aligned}$$

Finally, the map $t|_{\mathrm{Ker}(s)}$ is equivariant, because for all $x \in X$ and all $r \in R$,

$$t(r \cdot x) = t(i(r) \rhd x) = t(i(r)) \rhd t(x) = r \rhd t(x).$$ □

Remark 6.5.5. It is a natural question to ask whether the generalized augmented rack $t|_{\mathrm{Ker}(s)} : \mathrm{Ker}(s) \to R$ constructed in the proof *is* actually a crossed module of racks in the sense of Definition 6.2.2 rather than using Proposition 6.2.1 to *make it* a crossed module of racks, because $\mathrm{Ker}(s)$ already carries a rack structure as a subrack of R_1.

In fact, it is clear that $t|_{\mathrm{Ker}(s)}$ is a morphism of racks, and it is easy to see that the above action of R on X is by automorphisms. The only thing which is not clear is Peiffer's identity.

This means that starting from a strict 2-rack we can always *define* a rack product on $X = \mathrm{Ker}(s)$ such that $t : X \to R$ becomes a crossed module, but there is, a priori, no relation to the *induced* rack product from R_1.

6.5.2 From crossed modules to categories

We will now indicate how one may try to perform the reverse direction, that is, construct strict 2-racks from crossed modules of racks or from 1-cat racks. For this, we use the fact that we are able to pass from crossed modules of racks to augmented racks (or directly to crossed modules of groups) and from augmented racks to crossed modules of groups; see Section 6.2.

Indeed, what we are lacking is an analogue of Lemma 6.5.1 in the pure framework of racks. The most natural approach would be to use the analogue of the semidirect product in the realm of racks, that is, the hemi-semidirect product; see Definition 6.1.12. Actually, this does not work. We are unable to combine a rack R, an R-module X and an equivariant map $p : X \to R$ into a rack structure which gives even a precategory in the category of racks. However, this can be done in some special cases, such as when p has trivial image $\{1\}$, or if the rack product is trivial on R, or if the rack action of $p(X)$ on R is trivial.

We are, nevertheless, able to do the following: Given an augmented rack or a crossed module of racks, one can use the functor As from Section 6.1.1 to associate to it a crossed module of groups; see Section 6.2. To this, one can apply the usual construction to obtain a strict 2-group. Finally, one may use the following proposition from [15].

Proposition 6.5.5 (Carter–Crans–Elhamdadi–Saito [15]). *The functor* Conj : `Grp` → `Racks` *sends strict 2-groups to strict 2-racks.*

As in Section 6.2, one can obtain crossed modules of groups by starting with a crossed module of racks (applying the functor As on both racks) or starting with an augmented rack (applying As only on the G-set). Another way would be to regard the augmented rack as a crossed module of racks—this would result in applying As also on the group G.

Yet another option, this time *without* using the functor As, requires more structure.

Proposition 6.5.6. *Given an augmented rack $p : X \to G$ such that X is a G-module, that is, an abelian group with a linear G-action, and $p : X \to G$ is a homomorphism, then the usual semidirect product construction from Lemma* 6.5.1 *gives rise to a strict* 2*-group.*

Remark 6.5.6. One may ask what happens when we start with a crossed module of racks (or an augmented rack), associate an augmented rack (or a crossed module of groups) to it, perform the 2-group construction, regard it as a strict 2-rack, and then reconstruct a crossed module of racks from it.

Given a crossed module of racks $\mu : R \to S$, we associate to it the crossed module of groups $\mu : \mathrm{As}(R) \to \mathrm{As}(S)$, which can then be regarded as a strict 2-group in the usual way. Then the standard construction of a crossed module of groups from a strict 2-group applies here, and gives $t|_{\mathrm{Ker}(s)} : \mathrm{Ker}(s) \to \mathrm{As}(S)$. It is clear that $\mathrm{Ker}(s) = \mathrm{As}(R)$ (as the map s is the projection from the semidirect product $\mathrm{As}(R) \rtimes \mathrm{As}(S)$ to $\mathrm{As}(S)$, see Lemma 6.5.1). Thus we get back the crossed module of groups between the associated groups. In case we started with a crossed module of racks $\mu : R \to S$ such that the augmented rack $\mu : R \to \mathrm{As}(S)$ satisfies the hypotheses of Proposition 6.5.6, then the crossed module we get from the above construction is $\mu : R \to \mathrm{As}(S)$.

Observe that on the augmented rack $p : X \to G$ from Proposition 6.5.6, there are two rack structures on X. One is the trivial rack structure coming from the conjugation rack with respect to the abelian group structure on X, and the other comes from the $(x, y) \mapsto x \cdot p(y)$-construction.

6.5.3 Crossed modules of racks and trunks

Our main idea to go beyond the constructions from the previous sections is to associate to a crossed module of racks $\mu : R \to S$ not a category, but a *trunk*; see [29].

Definition 6.5.7. A *trunk* is a directed graph Γ together with a collection of oriented squares

$$\begin{array}{ccc} C & \xrightarrow{c} & D \\ {\scriptstyle b}\uparrow & & \uparrow{\scriptstyle d} \\ A & \xrightarrow{a} & B \end{array}$$

called *preferred squares.*

In categorical language, the set of objects consists of the vertices of Γ and the set of morphisms consists of the edges of Γ. There are then source and target maps for an arrow/edge. We notice that we are missing the identity-assigning and composition morphisms. Instead of these, however, we have preferred squares (commutative square diagrams) which, in some sense, replace the composition.

Definition 6.5.8. A *pointed trunk* is a trunk equipped with a chosen edge $e_A : A \to A$ for each vertex A and the following preferred squares for any given edge $a : A \to B$:

$$\begin{array}{ccc} A & \xrightarrow{a} & B \\ {\scriptstyle e_A}\uparrow & & \uparrow{\scriptstyle e_B} \\ A & \xrightarrow{a} & B \end{array}$$

The edges $e_A : A \to A$ are called identities.

We will assume that all our racks are pointed and all our trunks have identities. Now, in order to model racks in terms of trunks, we pass to the so-called *corner trunks*; see [29], page 324:

Definition 6.5.9. A *corner trunk* is a trunk, which satisfies the two corner axioms:

(C1) Given edges $a : A \to B$ and $b : A \to C$, there are unique edges $a \lhd b : C \to D$ and $a \rhd b : B \to D$ such that the following square is preferred:

$$\begin{array}{ccc} C & \xrightarrow{a \lhd b} & D \\ {\scriptstyle b}\uparrow & & \uparrow{\scriptstyle a \rhd b} \\ A & \xrightarrow{a} & B \end{array}$$

(C2) In the following diagram, if the squares $(ABCD)$, $(BDYT)$ and $(CDZT)$ are preferred, then the diagram can be completed, as shown by the dotted lines, such that the squares $(ABXY)$, $(ACXZ)$ and $(XYZT)$ are preferred:

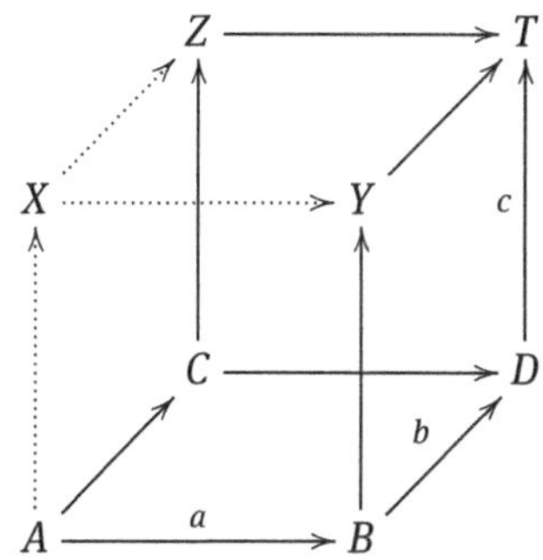

Lemma 6.5.7. *In a corner trunk, the binary operations* $\triangleright$ *and* $\triangleleft$ *satisfy:*
(a) $(a \triangleleft b) \triangleleft (b \triangleright c) = (a \triangleleft c) \triangleleft (b \triangleleft c)$;
(b) $(b \triangleright c) \triangleright (b \triangleright a) = (b \triangleleft c) \triangleright (c \triangleright a)$;
(c) $(b \triangleright a) \triangleleft (b \triangleright c) = (b \triangleleft c) \triangleright (a \triangleleft c)$ *for all arrows* a, b, c.

In fact, a set X with two operations $\triangleright$ and $\triangleleft$ (which are bijective and) satisfy the above three identities, is called a *birack*. Biracks also serve to construct link invariants; see, for example, [19].

The case which is most interesting to us is when one of the two operations $\triangleright$ or $\triangleleft$ is trivial. Then, by Lemma 6.5.7 above, the other operation satisfies a (left or right) rack identity. This means that these types of corner trunks codify racks. The notion of a corner trunk can be pointed in an obvious way.

Example 6.5.10. Any rack $(R, \triangleleft)$ gives rise to a corner trunk $\mathcal{T}(R)$, called the *rack trunk*; see [29], page 327. The trunk $\mathcal{T}(R)$ consists of a single vertex $*$, while the preferred squares are given by the rack operation

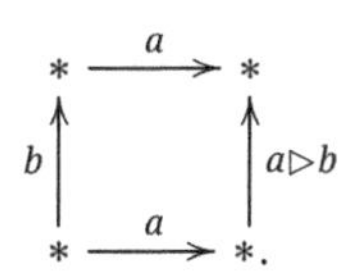

Example 6.5.11. An example of this construction is the *action rack trunk*; cf. [29], page 329. Let X be an R-set where R is a rack. From this data, we construct a trunk $\mathcal{T}_X(R)$. Namely, we take X as the set of vertices and edges of the form $x \xrightarrow{r} r \cdot x$ for $r \in R$ and

$x \in X$. The preferred squares are then of the form:

$$\begin{array}{ccc} r \cdot x & \xrightarrow{r'} & r' \cdot (r \cdot x) \\ {\scriptstyle r}\uparrow & & \uparrow{\scriptstyle r' \triangleright r} \\ x & \xrightarrow{r'} & r' \cdot x \end{array}$$

for all $r, r' \in R$ and $x \in X$. Observe that this fits together in the upper right-hand corner because for our action we have

$$r' \cdot (r \cdot x) = (r' \triangleright r) \cdot (r' \cdot x).$$

As in [29], we see that $\mathcal{T}_X(R)$ is indeed a corner trunk. From the categorical point of view, we will denoted morphisms $x \xrightarrow{r} r \cdot x$ as pairs (x, r), and then we have source and target maps $X \times R \overset{s}{\underset{t}{\rightrightarrows}} X$ given by $s(x, r) = x$ and $t(x, r) = r \cdot x$. As already remarked in *loc. cit.*, the operation expressed by the preferred squares can be expressed as

$$(x, r) \triangleright (x, r') = (r' \cdot x, r' \triangleright r), \quad (x, r') \triangleleft (x, r) = (r \cdot x, r').$$

The first formula resembles the hemi-semidirect product, see Definition 6.1.12, but with a composability condition (the first components have to be equal to x).

In the special case where a rack $(R, \triangleright)$ acts on itself by $y \mapsto x \triangleright y$, the action rack trunk $\mathcal{T}_R(R)$ is called *extended rack trunk* in [29] on page 329.

6.5.4 From crossed modules of racks to trunks

Given a crossed module of racks $\mu : R \to S$, we have, in particular, an S-set R, and we can thus apply the construction from Example 6.5.11 to obtain a corner trunk. This is, in our opinion, the correct "categorical object" associated to a crossed module of racks, as it is a sort of trunk in the category of racks. Thus, we introduce the following.

Definition 6.5.12. A *trunkified rack* consists of a trunk map between an action rack trunk (for R acting on X) and the extended rack trunk of R.

Proposition 6.5.8. *There is a one-to-one correspondence between crossed modules of racks and trunkified racks.*

Proof. We first observe that by Proposition 6.2.1, it is enough to work with generalized augmented racks instead of crossed modules of racks. A generalized augmented rack consists of a rack R, an R-module X and an equivariant map $p : X \to R$. We can associate to these the extended rack trunk $\mathcal{T}_R(R)$, the action trunk $\mathcal{T}_X(R)$ and the induced trunk map $p : \mathcal{T}_X(R) \to \mathcal{T}_R(R)$ (or $p : \mathcal{T}_X(R) \to \mathcal{T}(R)$).

In the other direction, we recover from a trunk map $\mathcal{T}_X(R) \to \mathcal{T}_R(R)$ the rack R and the rack action of R on X. The trunk map gives an equivariant map and, therefore, we recover our crossed module. □

Remark 6.5.13. One can, of course, introduce alternative versions of trunkified racks. For example, one could choose to define a trunkified rack as a trunk in the category of racks.

In order to associate to a crossed module of racks such a trunkified rack, one idea would be to take the image of the above trunk map appearing in the proof of Proposition 6.5.8. Unfortunately, we were unable to show that this gives a trunk in the category of racks, and so this remains a question for further investigation.

We conclude this section with the following scheme which may enable us to eventually associate a strict 2-rack to a crossed module of racks. Let $p : X \to R$ be a crossed module of racks. Denote by cat : `Trunks` $\to$ `Cats` the functor from the category of (small) trunks to the category of (small) categories that associates to a trunk $\mathcal{T}$ the category whose objects consist of the same set of vertices/objects as the trunk, but whose set of morphisms (between two fixed objects) are generated by the set of arrows of $\mathcal{T}$ (between these objects) such that the preferred squares become commutative diagrams.

One should then form the trunk map $p : \mathcal{T}_X(R) \to \mathcal{T}_R(R)$, take the image trunk $\mathrm{im}(p)$ and show that this is a trunk in the category of racks. One should furthermore show that, in general, the functor cat : `Trunks` $\to$ `Cats` sends trunks in the category of racks to categories in the category of racks (i. e., that the functor cat can be enriched in racks). Finally, the image $\mathrm{cat}(\mathrm{im}(p))$ should then be the categorical rack associated to $p : X \to R$. We leave this program to the interested reader.

6.6 Topological applications of crossed modules of racks

We now show that some of these categorical objects that we have associated to crossed modules of racks have topological applications.

6.6.1 The rack space of a crossed module of racks

In the article [29], the authors associate to a rack X the *rack space* BX by taking the geometric realization of the cubical nerve NX of the trunk $\mathcal{T}(X)$ associated to X. They also show that for an action rack trunk $\mathcal{T}_Y(X)$ (with the rack X acting on Y), the canonical map $Y \to \{*\}$ induces a trunk map $\mathcal{T}_Y(X) \to \mathcal{T}(X)$, which gives rise to a covering $B_Y X \to BX$ (see Theorem 3.7 in [29]) where $B_Y X$ is the geometric realization of the cubical nerve of the action rack trunk $\mathcal{T}_Y(X)$.

Now starting with a crossed module of racks $p : X \to R$, we have first of all a trunk map $\mathcal{T}_X(R) \to \mathcal{T}(R)$ inducing the covering $B_X R \to BR$. Then we also have a trunk map

$$p : \mathcal{T}_X(R) \to \mathcal{T}_R(R),$$

where $\mathcal{T}_R(R)$ is the extended rack trunk. This map also induces a map of □-sets between the cubical nerves

$$p : N_X(R) \to N_R(R),$$

and finally a map between the corresponding rack spaces $B_X(R) \to B_R(R)$; see [29], page 331.

Both □-sets $N_X(R)$ and $N_R(R)$ are in fact □-coverings of NR. We will show that p is a covering.

Proposition 6.6.1. *Suppose that $B_R(R)$ is arcwise connected and locally arcwise connected. Then the geometric realization of the natural map $p : N_X(R) \to N_R(R)$ is a covering of topological spaces.*

Proof. The lifting theorem for a continuous map into the base of a covering implies that we only need to show that $\pi_1(B_X R, x) \subset \pi_1(B_R R, p(x))$.

Now recall the fact that the fundamental group $\pi_1(B_Y X, y)$ is just the stabilizer of y, that is,

$$\pi_1(B_Y X, y) = \mathrm{Stab}_y \subset \mathrm{As}(X);$$

see Proposition 4.5 in [29]. In particular, we have $\pi_1(BX, *) = \mathrm{As}(X)$. Thus in order to show the claim, we just need to show that the subgroups $\pi_1(B_X R, x)$ and $\pi_1(B_R R, p(x))$ satisfy

$$\pi_1(B_X R, x) \subset \pi_1(B_R R, p(x))$$

as subgroups of $\mathrm{As}(R)$. This follows from the inclusion of the corresponding stabilizers $\mathrm{Stab}_x \subset \mathrm{Stab}_{p(x)}$. □

In this sense, we can associate to each crossed module of racks a covering of rack spaces. We anticipate that this will serve to enhance link invariants.

Remark 6.6.1. It would be interesting to know what kind of local property these rack spaces have. In order to have a nice theory of covering spaces, one would like to work with topological spaces which are, for example, arcwise connected and locally simply connected.

6.6.2 Crossed modules of racks from link coverings

Here, we are doing in some sense the inverse construction with respect to what we did in the previous subsection. Namely, given a covering (and a link), we associate to it a crossed module of racks.

We start with the construction of a crossed module of racks from a square of augmented racks: Suppose given augmented racks $p_i : X_i \to G_i$ for $i = 0, 1$ and a commutative diagram

$$\begin{array}{ccc} X_1 & \xrightarrow{p_1} & G_1 \\ \downarrow{\scriptstyle\alpha} & & \downarrow{\scriptstyle\beta} \\ X_0 & \xrightarrow{p_0} & G_0 \end{array}$$

where β is a group homomorphism and α is a morphism of group-sets over β, that is, for all $x \in X_1$ and all $g \in G_1$, we have

$$\alpha(g * x) = \beta(g) \cdot \alpha(x).$$

Here, we have written the left actions $g * x$ for the action of G_1 on X_1, and $h \cdot y$ for the action of G_0 on X_0. We then ask: Under which conditions is $\alpha : X_1 \to X_0$ a crossed module of racks?

Proposition 6.6.2. *Suppose that there is a left action of G_0 on X_1, denoted $(g, x) \mapsto g \circ x$, such that:*
(a) *$g * x = \beta(g) \circ x$ for all $g \in G_1$ and all $x \in X_1$;*
(b) *α is equivariant, that is, $\alpha(g \circ x) = g \cdot \alpha(x)$, for all $g \in G_0$ and all $x \in X_1$, and*
(c) *$g \circ (p_1(x) * y) = p_1(g \circ x) * (g \circ y)$ for all $g \in G_0$ and all $x, y \in X_1$.*

Then $\alpha : X_1 \to X_0$ is a crossed module of racks for the augmented racks $p_i : X_i \to G_i$ for $i = 0, 1$.

Proof. The sets X_0 and X_1 become racks via the usual definitions: $x \rhd y := p_1(x) * y$ for $x, y \in X_1$ and $x \rhd y := p_0(x) \cdot y$ for $x, y \in X_0$. The map α is then a morphism of racks since

$$\begin{aligned} \alpha(x \rhd y) &= \alpha(p_1(x) * y) \\ &= \beta(p_1(x)) \cdot \alpha(y) \\ &= p_0(\alpha(x)) \cdot \alpha(y) \\ &= \alpha(x) \rhd \alpha(y). \end{aligned}$$

Using the map p_0, the action $\circ$ of G_0 on X_1 induces an action of X_0 on X_1.

Property (c) clearly translates into the fact that the action $\circ$ is by rack automorphisms:

$$g \circ (x \rhd y) = g \circ (p_1(x) * y) = p_1(g \circ x) * (g \circ y) = (g \circ x) \rhd (g \circ y).$$

Property (b) is the equivariance of map α:

$$\alpha(p_0(x) \circ y) = p_0(x) \cdot \alpha(y) = x \rhd \alpha(y),$$

and Peiffer's identity is satisfied thanks to Property (a)

$$p_0(\alpha(x)) \circ y = \beta(p_1(x)) \circ y = p_1(x) * y = x \rhd y. \qquad \square$$

We can use Proposition 6.6.2 to define a crossed modules of racks in a geometrical/topological context. Before we come to our construction, we recall the fundamental rack of a link from Fenn and Rourke [28], page 358.

A *link* is a codimension two embedding $L : M \subset Q$ of manifolds. We will assume that M is nonempty, that Q is connected (with empty boundary) and that M is transversely oriented in Q. In other words, we assume that each normal disc to M in Q has an orientation, which is locally and globally coherent.

The link is called *framed* if there is a cross-section $\lambda : M \to \partial N(M)$ of the normal disk bundle. Denote by M^+ the image of M under λ. In the following, we will only consider framed links.

Then Fenn and Rourke associate to $L \subset Q$ an augmented rack (called the *fundamental rack* of the link L), which is the space Γ of homotopy classes of paths in $Q_0 := \text{closure}(Q \setminus N(L))$ of L, from a point in M^+ to some base point q_0. During the homotopy, the final point of the path at q_0 is kept fixed and the initial point is allowed to wander at will on M^+.

As Fenn and Rourke observe in [29], the fundamental rack of a link is somewhat stronger than the fundamental group of the link, because in the Wirtinger presentation of the fundamental group of a link, one divides on top of the commutation relations at the intersections by the relations to make it a group, while for the fundamental rack, one divides only by the commutation relations at the intersections.

The set Γ has an action of the fundamental group $\pi_1(Q_0, q_0)$ defined as follows: Let γ be a loop in Q_0 based at q_0 representing an element $g \in \pi_1(Q_0)$. If $a \in \Gamma$ is represented by the path α, define $g \cdot a$ to be the class of the composite path $\gamma \circ \alpha$.

We can use this action to define a rack structure on Γ. Let $p \in M^+$ be a point on the framing image. Then p lies on a unique meridian circle of the normal disc bundle. Let m_p be the loop based at p, which follows the meridian around in a positive direction. Let $a, b \in \Gamma$ be represented by paths α, β, respectively. Let $\partial(b)$ be the element of $\pi_1(Q_0, q_0)$ determined by the homotopy class of the loop $\beta \circ m_{\beta(0)} \circ \beta^{-1}$. The *fundamental rack of the framed link L* is defined to be the set $\Gamma = \Gamma(L)$ with the operation

$$a \rhd b := \partial(a) \cdot b := [\alpha \circ m_{\alpha(0)} \circ \alpha^{-1} \circ \beta].$$

In case the link is evident, but there are different manifolds, we will denote Γ more precisely by Γ_Q.

Fenn and Rourke show in [28], Proposition 3.1, page 359, that Γ is indeed a rack, and go on to show that $\partial : \Gamma \to \pi_1(Q_0, q_0)$ is an augmented rack. We have translated here their right racks into left racks. Furthermore, adding in part of the exact homotopy sequence

$$\pi_2(Q) \to \pi_2(Q, Q_0) \to \pi_1(Q_0) \to \pi_1(Q)$$

they show in Proposition 3.2, page 360, that the associated crossed module of groups (using Proposition 6.2.4) of the augmented rack $\partial : \Gamma \to \pi_1(Q_0, q_0)$ is Whitehead's crossed module of groups

$$\pi_2(Q, Q_0) \to \pi_1(Q_0).$$

We will now extend this theory to coverings on the topological side and crossed modules of augmented racks on the algebraic side.

Consider a covering space $\pi : P \to Q$, and a link $L \subset Q$. Let us furthermore consider a link $L : M \subset Q$ (which we also denote by L_Q) and its inverse image $L_P := \pi^{-1}(M) \subset P$. We will suppose that the link L_p is also framed, and this is in a manner which is compatible with the framing of L_Q.

We have therefore two augmented racks

$$\Gamma_P \to \pi_1(P_0, p_0) \quad \text{and} \quad \Gamma_Q \to \pi_1(Q_0, \pi(p_0)).$$

From now on, we will suppress the base points in the notation.

Theorem 6.6.3. *There is an action of $\pi_1(Q_0)$ on Γ_P such that the conditions of Proposition 6.6.2 are satisfied, that is, the induced map $\pi : \Gamma_P \to \Gamma_Q$ is a crossed module of (augmented) racks.*

Proof. The action is given by the following procedure:

An element c of $\pi_1(Q_0)$ is represented by a based loop γ in Q_0. It lifts to a unique path $\tilde{\gamma}$ in P_0 which ends at the base point $p_0 \in P_0$. Now take an element x in Γ_P, represented by a path ξ. This path ξ projects to a path $\pi(\xi)$ in Q_0, which can be lifted to $\widetilde{\pi(\xi)}$ such that $\widetilde{\pi(\xi)}(1) = \tilde{\gamma}(0)$ meaning that they are composable. The outcome is that ξ (or more precisely some lift of $\pi(\xi)$) can be composed with $\tilde{\Gamma}$, and the composition is then, by definition, the action of the homotopy class $c = [\gamma]$ on $x = [\xi]$.

Observe that the map $\beta : \pi_1(P_0) \to \pi_1(Q_0)$ is induced by $\pi : P \to Q$ and is injective. We identify via β the group $\pi_1(P_0)$ as a subgroup of $\pi_1(Q_0)$, the subgroup of loops in Q_0, which lift to loops in P_0.

We therefore clearly have Property (a), because when lifting the element $\beta(g)$ to a loop in P_0, the action on x becomes $\beta(g) \circ \xi$, which is the action in the augmented crossed module $\Gamma_P \to \pi_1(P_0)$.

The map α (also induced by $\pi : P \to Q$) is clearly equivariant, because π distributes on the factors of the composition of paths.

Finally, Property (c) is illustrated by the following picture:

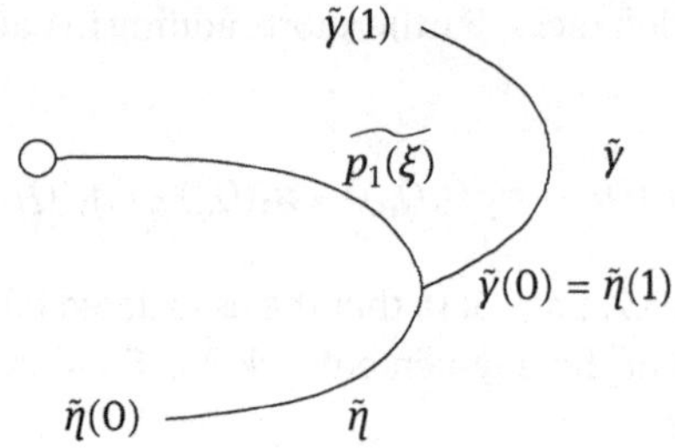

There are two elements x, y of Γ_P and one element g of $\pi_1(Q_0)$ in play here. Therefore, we have two paths ξ, η upstairs, and one loop γ downstairs. The paths ξ and η are pushed down using π and then lifted to $\tilde{\xi}$ and $\tilde{\eta}$.

On the LHS of Property (c), there is an action of $p_1(x)$ applied to the element y, meaning the two paths rejoin each other at the point $p := \tilde{\gamma}(0) \in P$ in the above picture, where one of them is carrying a loop at its left end (which is the $\partial = p_1$ map!). Moreover, the loop γ is lifted to some path $\tilde{\gamma}$ from $\tilde{\gamma}(0)$ to the base point $p_0 = \tilde{\gamma}(1)$.

On the RHS of Property (c), translating into paths in P_0, we have two paths from somewhere to p and then to p_0, which illustrate the action of γ on x and y. But then one takes $p_1 = \partial$ of the path corresponding to the action on x, which means this path gets a little loop on its left end. Then compose the two paths.

One sees that both sides are equal because on the right-hand side, the supplementary round trip along the path corresponding to the lift of γ cancels in the composition. □

Observe that it follows from the constructions in this section that in the above situation, $\pi : P \to Q$ induces simultaneously a map of crossed modules of groups

$$\begin{array}{ccc} \pi_2(P, P_0) & \longrightarrow & \pi_1(P_0) \\ \downarrow{\scriptstyle \alpha} & & \downarrow{\scriptstyle \beta} \\ \pi_2(Q, Q_0) & \longrightarrow & \pi_1(Q_0) \end{array}$$

between the associated crossed modules introduced by Whitehead and a crossed module $\pi_2(P, P_0) \to \pi_2(Q, Q_0)$ as associated to the crossed module of racks $\Gamma_P \to \Gamma_Q$. This follows immediately from the fact that the functor As : `Racks` $\to$ `Groups` sends the fundamental racks associated to the links to the corresponding homotopy groups, see [28], Proposition 3.2, page 360.

Remark 6.6.2. An explicit instance of the theorem is the following. A braid B with n strands in the cylinder $S^1 \times [0, 1]$ may be closed to become a link L in the (plain) torus by identifying the two extremal discs of the cylinder. The fundamental group of this link is then of the form

$$\pi_1(Q \setminus L) = F_n \ltimes F_1 = \langle t, x_1, \dots, x_n \mid \varphi_i(x_1, \dots, x_n) = tx_it^{-1} \rangle,$$

where Q is the torus, F_k is the free group on k generators and φ_i are automorphisms of the free group arising from the closure procedure.

The fundamental group of $Q \setminus L$ of the same braid, but in an infinite cylinder and repeated to infinity is just the free group on n generators F_n. This situation corresponds to the universal covering of the (plain) torus.

The fundamental racks associated to these two situations are free on n generators. Call R_n the free rack on n generators. Now consider the universal covering of the torus. It gives rise to a situation as in the theorem. The corresponding diagram becomes

$$\begin{array}{ccc} R_n & \xrightarrow{i} & F_n = \mathrm{As}(R_n) \\ \downarrow{\scriptstyle \mathrm{id}_{R_n}} & & \downarrow \\ R_n & \xrightarrow{\psi} & F_n \ltimes F_1 \end{array}$$

One obtains more interesting examples by taking an intermediate covering (with a finite number of leaves) instead of the universal covering.

Thus, in some sense we have completed the loop: Crossed modules of groups came from topology and with crossed modules of racks, we come back to topology.

6.7 Exercises

Exercise 6.7.1. Show that any union of conjugacy classes in a group forms a rack with the group-conjugation as operation. Therefore, the set of reflections in the dihedral group D_n (of order $2n$) carries the structure of a rack, called *dihedral rack.*

Exercise 6.7.2. Let $\Lambda := \mathbb{Z}[t, t^{-1}]$ be the ring of Laurent polynomials in the variable t. Show that any Λ-module becomes a rack with the operation $a \rhd b := ta + (1 - t)b$. This is called the *Alexander rack.*

Exercise 6.7.3. Consider the set $R := \{x, y\}$ consisting of two elements x and y with an operation given by

$$x \rhd x = y, \quad x \rhd y = x, \quad y \rhd y = x, \quad y \rhd x = y.$$

Show that R is a rack. The relation $x \rhd y = xyx^{-1}$ in $\mathrm{As}(R)$ implies that $x = y$ in $\mathrm{As}(R)$. Show that $\mathrm{As}(R)$ is isomorphic to $\mathbb{Z}$ an conclude that the canonical map $i : R \to \mathrm{As}(R)$ is thus not necessarily injective.

Exercise 6.7.4. Let A be an associative algebra over a field $\mathbb{K}$ together with a $\mathbb{K}$-linear map $D : A \to A$ satisfying the condition

$$D(a(Db)) = (Da)(Db) = D((Da)b)$$

for all $a, b \in A$.

(a) Show that A becomes a (left) Leibniz algebra for the bracket

$$[x, y] := (Dx)y - y(Dx)$$

for all $x, y \in A$.

(b) Show that the condition on D holds for a derivation $D : A \to A$ of square zero.

Exercise 6.7.5. Investigate whether the functors Aut : $\mathsf{Racks} \to \mathsf{Grp}$ and Bij : $\mathsf{Racks} \to \mathsf{Grp}$ also send crossed modules of racks to crossed modules of groups.

6.8 Bibliographical notes

A general reference for (right) racks is the article of Fenn and Rourke [28]. A general reference for (right) Leibniz algebras is the article of Loday and Pirashvili [66]. A general reference for rack bialgebras is the article of Alexandre–Bordemann–Rivière–Wagemann [2]. This notion grew out of the notion of a rack in coalgebras; see [15]. All notions have been adapted here to their left versions.

Section 6.1 is covered by [11, 2], except the bimodule notions which are new. Proposition 6.1.4 is due to M. Kinyon [60].

The notion of crossed modules of racks is due to Crans–Wagemann [21]. Sections 6.3 and 6.4 are entirely new. Sections 6.5 and 6.6 are again from [21], except the last remark which reflects research together with A. Crans and B. Wiest in February 2015 at Nantes.

A Cohomology of groups

In the following four Appendices, we recall some basic homological algebra and Lie theory, which will be used throughout this book. Most of the time, we will concentrate on *cohomology* as this is the most relevant for crossed modules. A general reference for these matters is [98]. We will only emphasize our point of view and comment on the results without giving full proofs, as all results are well documented in the literature.

A.1 Definitions

Let G be a group and V be a G-module. It will be enough for our purposes to work over a field $\mathbb{K}$, while most of the matters discussed in this Appendix are true over a general commutative ring $\mathbb{K}$. The *pth cohomology group* $H^p(G,V)$ of G with values in V (for an integer $p \geq 0$) is by definition

$$H^p(G,V) := \mathrm{Ext}^p_{\mathbb{K}[G]}(\mathbb{K}, V).$$

Here, $\mathbb{K}[G]$ is the group ring or group algebra of G over $\mathbb{K}$. By definition of the Ext-groups, we have $H^0(G,V) = V^G$, the submodule of G-invariants of V. Moreover, by the Ext-description, the cohomology of G with values in V is the right derived functor of the functor of G-invariants or, equivalently, of the functor $V \mapsto \mathrm{Hom}_G(\mathbb{K}, V)$, where $\mathrm{Hom}_G(V,W)$ denotes the abelian group of G-morphisms between two G-modules V and W.

As a right derived functor, it can be computed using any projective resolution of the base field $\mathbb{K}$. The *bar resolution* is such a free (and thus projective) resolution of $\mathbb{K}$. It has the shape

$$0 \leftarrow \mathbb{K} \leftarrow B_0 \overset{d}{\leftarrow} B_1 \overset{d}{\leftarrow} \dots \overset{d}{\leftarrow} B_n \overset{d}{\leftarrow} \dots$$

In order to simplify matters, we speak here only about the unnormalized bar resolution. Here $B_0 := \mathbb{K}[G]$ and B_n is the free $\mathbb{K}[G]$-module on the set of symbols $[g_1|\dots|g_n]$ with $g_i \in G$. The boundary operator $d : B_n \to B_{n-1}$ in this resolution is defined as $d := \sum_{i=0}^n (-1)^i d_i$ where

$$\begin{aligned} d_0[g_1|\dots|g_n] &= g_1[g_2|\dots|g_n] \\ d_i[g_1|\dots|g_n] &= [g_1|\dots|g_ig_{i+1}|\dots|g_n] \\ d_n[g_1|\dots|g_n] &= [g_1|\dots|g_{n-1}] \end{aligned}$$

for i ranging from 1 to $n-1$. These operators d_i satisfy the simplicial relations and we have therefore $d \circ d = 0$. For the exactness of the complex, we refer to [98], Theorem 6.5.3, page 178. Observe that by applying $\mathrm{Hom}_G(-,V)$ to this resolution, we obtain as cochain spaces $C^p(G,V) := \mathrm{Map}(G^p, V)$. The cohomology space $H^p(G,V)$ is then the

https://doi.org/10.1515/9783110750959-007

cohomology with respect to the coboundary operator $d : C^p(G,V) \to C^{p+1}(G,V)$ given by

$$dc(g_0,\ldots,g_n) := g_0 c(g_1,\ldots,g_n) + \sum_{i=0}^{n-1}(-1)^i c(\ldots,g_i g_{i+1},\ldots) + (-1)^n c(g_0,\ldots,g_{n-1}) \quad \text{(A.1)}$$

for $c \in C^p(G,V)$. A cochain $c \in C^p(G,V)$ is called *normalized* in case for all $1 \leq i \leq p$ and all $g_1,\ldots,g_{i-1},g_{i+1},\ldots,g_p$,

$$c(g_1,\ldots,g_{i-1},1,g_{i+1},\ldots,g_p) = 0.$$

By Theorem 6.5.3 in [98], page 178, the normalized cochains form a subcomplex, which computes the same cohomology. One may therefore always restrict to normalized cochains to compute the same cohomology.

A cochain $c \in C^p(G,V)$ such that $dc = 0$ is called a *p-cocycle*. The space of p-cocycles is denoted $Z^p(G,V)$. Similarly, a cochain $c \in C^p(G,V)$ such that there exists a cochain $c' \in C^{p-1}(G,V)$ with $c = dc'$ is called a *p-coboundary*. The space of p-coboundaries is denoted $B^p(G,V)$. With these notation, we have

$$H^p(G,V) = Z^p(G,V) \,/\, B^p(G,V).$$

A.2 Abelian extensions

Here, we sketch the relation between $H^2(G,V)$ and $\mathrm{Ext}(G,V)$, the set of equivalence classes of extensions of the form

$$0 \to V \xrightarrow{i} E \xrightarrow{\pi} G \to 0,$$

where two such extensions of G by the G-module V are called *equivalent* in case there exists a commutative diagram

$$\begin{array}{ccccccccc} 0 & \longrightarrow & V & \longrightarrow & E & \longrightarrow & G & \longrightarrow & 0 \\ & & \downarrow{\scriptstyle \mathrm{id}_V} & & \downarrow{\scriptstyle \varphi} & & \downarrow{\scriptstyle \mathrm{id}_G} & & \\ 0 & \longrightarrow & V & \longrightarrow & E' & \longrightarrow & G & \longrightarrow & 0 \end{array}$$

Such an extension is also called *abelian extension*, because V (as a G-module) is an abelian group. One may consider more general extensions where this is not the case and these are then simply called *general extension*. A *central extension* is by definition an abelian extension where the subgroup $i(V) \subset E$ is central.

Observe that by the Five Lemma two equivalent extensions have isomorphic groups $E \cong E'$. Note the functoriality of $\mathrm{Ext}(-,-)$ in both variables: A group homomorphism $\psi : G \to G'$ gives rise to a pullback diagram

$$\begin{array}{ccccccccc}
0 & \longrightarrow & V & \longrightarrow & \pi^*E' & \longrightarrow & G & \longrightarrow & 0 \\
 & & \big\downarrow{\scriptstyle \mathrm{id}_V} & & \big\downarrow & & \big\downarrow{\scriptstyle \psi} & & \\
0 & \longrightarrow & V & \longrightarrow & E' & \xrightarrow{\pi} & G' & \longrightarrow & 0,
\end{array}$$

and a G-module homomorphism $\zeta : V \to V'$ gives rise to a pushout diagram

$$\begin{array}{ccccccccc}
0 & \longrightarrow & V & \xrightarrow{i} & E & \longrightarrow & G & \longrightarrow & 0 \\
 & & \big\downarrow{\scriptstyle \zeta} & & \big\downarrow & & \big\downarrow{\scriptstyle \mathrm{id}_G} & & \\
0 & \longrightarrow & V' & \longrightarrow & i_*E & \longrightarrow & G & \longrightarrow & 0.
\end{array}$$

Given two abelian extensions $\xi : 0 \to V \to E \to G \to 0$ and $\xi' : 0 \to V \to E' \to G \to 0$ of the same group G by the same G-module V, the *Baer sum* $\xi + \xi'$ is by definition the extension $(\pi, \pi')^*(i+i')_*(E \times E')$, obtained from the direct product of the two sequences

$$0 \to V \oplus V \xrightarrow{i+i'} E \times E' \xrightarrow{(\pi,\pi')} G \times G \to 0$$

by pushout according to the sum-map $+ : V \oplus V \to V$, $(v, w) \mapsto v + w$, and then by pullback according to the diagonal map $\triangle_G : G \to G \times G$, $g \mapsto (g, g)$. Visually,

$$\begin{array}{ccccccccc}
0 & \longrightarrow & V \oplus V & \xrightarrow{i+i'} & E \times E' & \xrightarrow{(\pi,\pi')} & G \times G & \longrightarrow & 0 \\
 & & \big\downarrow{\scriptstyle +} & & \big\downarrow & & \big\downarrow{\scriptstyle \mathrm{id}_{G\times G}} & & \\
0 & \longrightarrow & V & \longrightarrow & (i+i')_*(E \times E') & \longrightarrow & G \times G' & \longrightarrow & 0 \\
 & & \big\uparrow{\scriptstyle \mathrm{id}_V} & & \big\uparrow & & \big\uparrow{\scriptstyle \triangle_G} & & \\
0 & \longrightarrow & V & \longrightarrow & (\pi, \pi')^*(i+i')_*(E \times E') & \longrightarrow & G & \longrightarrow & 0
\end{array}$$

Using the functoriality in both variables, it can be shown that the Baer sum equips $\mathrm{Ext}(G, V)$ with an abelian group structure; cf. the corresponding statement in the framework R-modules [68], Chapter III.2, page 67.

The link from extensions to cohomology becomes visible by associating a cocycle to an extension. Let $0 \to V \xrightarrow{i} E \xrightarrow{\pi} G \to 0$ be an abelian extension. Observe first that the group G acts on the abelian group V such that V becomes a G-module. Indeed, choose a set-theoretical section $\rho : G \to E$ such that $\rho(1) = 1$ of the quotient map $\pi : E \to G$. Define

$$x \cdot v := i^{-1}(\rho(x) i(v) \rho(x)^{-1})$$

for all $x \in G$ and all $v \in V$. One easily computes $xy \cdot v = x \cdot (y \cdot v)$ using that $\rho(xy)$ and $\rho(x)\rho(y)$ differ at most by an element of $i(V)$, which commutes with $i(v)$ and thus gets canceled in the conjugation. We see that an abelian extension of G by V always comes with the structure of a G-module on V. In case $i(V)$ is central, this action is trivial. In case the extension is not abelian, the thus-defined "action" is only well-defined up to a conjugation in V, that is, we only obtain an outer action.

The failure of $\rho : G \to E$ to be a homomorphism of groups is the quantity

$$\alpha(x,y) := \rho(x)\rho(y)\rho(xy)^{-1}$$

for all $x, y \in G$. We have $\alpha(x,y) \in \mathrm{Ker}(\pi) = \mathrm{Im}(i)$, because π is a homomorphism. Thus there exists $\beta(x,y) \in V$ such that $i(\beta(x,y)) = \alpha(x,y)$. The map $\beta : G \times G \to V$ is a normalized 2-cocycle:

$$\begin{aligned}
i(d\beta(x,y,z)) &= i(x \cdot \beta(y,z)\beta(xy,z)^{-1}\beta(x,yz)\beta(x,y)^{-1}) \\
&= \rho(x)\alpha(y,z)\rho(x)^{-1}\alpha(xy,z)^{-1}\alpha(x,yz)\alpha(x,y)^{-1} \\
&= \rho(x)\rho(y)\rho(z)\rho(yz)^{-1}\rho(x)^{-1}\rho(xyz)\rho(z)^{-1} \\
&\quad \rho(xy)^{-1}\rho(x)\rho(yz)\rho(xyz)^{-1}\rho(xy)\rho(y)^{-1}\rho(x)^{-1} \\
&= \rho(x)\rho(y)\rho(z)\rho(yz)^{-1}\rho(x)^{-1}\rho(x)\rho(yz)\rho(xyz)^{-1} \\
&\quad \rho(xyz)\rho(z)^{-1}\rho(xy)^{-1}\rho(xy)\rho(y)^{-1}\rho(x)^{-1} \\
&= \rho(x)\rho(y)\rho(z)\rho(z)^{-1}\rho(y)^{-1}\rho(x)^{-1} = 1.
\end{aligned}$$

In this computation, the strategy is to switch two elements of the form

$$\rho(a)\rho(b)\rho(ab)^{-1} = \alpha(a,b) = i(\beta(a,b))$$

(or its inverse), because these elements commute as they lie in $i(V)$. From the third to the fourth line, we switched $\rho(xyz)\rho(z)^{-1}\rho(xy)^{-1}$ and $\rho(x)\rho(yz)\rho(xyz)^{-1}$. The rest of the computation are then eliminations of adjacent inverse elements.

The main theorem of this section is the following theorem.

Theorem A.2.1. *The prescription sending an abelian extension to the class of the cocycle induces a map* $\zeta : \mathrm{Ext}(G,V) \to H^2(G,V)$, *which induces an isomorphism of the abelian groups* $H^2(G,V)$ *and* $\mathrm{Ext}(G,V)$.

Proof. This is Theorem 6.6.3 in [98], page 183. □

Remark A.2.1. In the proof of this theorem, one associates an abelian extension to a normalized 2-cocycle $\beta \in Z^2(G,V)$. It is given by the direct product $E = V \times G$ with the following group product:

$$(v,g) \cdot (v',g') := (v + g \cdot v' + \beta(g,g'), gg')$$

for $v, v' \in V$ and $g, g' \in G$. The associativity of this group product is equivalent to the cocycle property of β. We denote this abelian extension by $E := V \times_\beta G$. On the other hand, an arbitrary abelian extension $0 \to V \to E \to G \to 0$ is equivalent to $V \times_\beta G$ for the above constructed β from a section $\rho : G \to E$ of $\pi : E \to G$: The section ρ permits to choose *product coordinates* in E, that is, to describe the group product in E in terms of the set-theoretical decomposition $E \approx V \times G$.

A.3 The Lyndon–Hochschild–Serre spectral sequence

The Lyndon–Hochschild–Serre spectral sequence is associated to a normal subgroup H of a group G and a G-module V. In other words, it is associated to a group G, which is described as a general extension

$$0 \to H \to G \to G/H \to 0.$$

The spectral sequence is a special case of the Grothendieck spectral sequence of the composition of two functors, namely the functors of invariants $(-)^H : G-\mathrm{mod} \to (G/H)-\mathrm{mod}$ and $(-)^{G/H} : (G/H)-\mathrm{mod} \to \mathrm{Ab}$. The spectral sequence can also be described by a filtration on the cochain spaces as given in Section 1.6. We refer the interested reader to Section 6.8 in [98], page 195.

To any spectral sequence in the first quadrant, there exists a 5-term exact sequence of low degree terms. For the Lyndon–Hochschild–Serre spectral sequence, it reads in homology (see 6.8.3, p. 196 in [98]):

$$H_2(G,V) \to H_2(G/H, V_H) \to H_1(H,V)_{G/H} \to H_1(G,V) \to H_1(G/H, V_H) \to 0.$$

As the morphisms in the 5-term exact sequence are given by the coinflation map, corestriction map and the differential, the 5-term exact sequence is natural with respect to group homomorphisms.

A.4 Continuous cohomology

For infinite groups or infinite dimensional Lie groups, one needs to make reference to topology (or measure theory) in order to "tame" the group cohomology spaces. One way to get hold on the cohomology spaces in these cases is to replace the ordinary cochain spaces $C^p(G,V)$ by the spaces of continuous cochains $C^p_{\mathrm{cont}}(G,V)$, that is, spaces of continuous maps from $G^{\times p}$ to V, in case G is a topological group and V a topological G-module. The coboundary operator respects continuous cochains, and the continuous cohomology of G with values in V is by definition the quotient $\mathbb{K}$-vector space,

$$H^p_{\mathrm{cont}}(G,V) := Z^p_{\mathrm{cont}}(G,V) \,/\, B^p_{\mathrm{cont}}(G,V).$$

A basic reference on continuous cohomology is [34]; see also [76], [77] and [79]. In general, continuous group cohomology does not come from a derived functor framework, but note [97].

B Cohomology of Lie algebras

B.1 Definitions

As for groups, the cohomology of Lie algebras is defined in terms of an Ext-functor. Let $\mathfrak{g}$ be a Lie algebra over a field $\mathbb{K}$ and V be a $\mathfrak{g}$-module. The *pth cohomology space* $H^p(\mathfrak{g}, V)$ (for an integer $p \geq 0$) of $\mathfrak{g}$ with values in V is by definition

$$H^p(\mathfrak{g}, V) := \mathrm{Ext}^p_{U\mathfrak{g}}(\mathbb{K}, V),$$

where $U\mathfrak{g}$ is the universal enveloping algebra of $\mathfrak{g}$. As for groups, $H^0(\mathfrak{g}, V) = V^{\mathfrak{g}} := \mathrm{Hom}_{U\mathfrak{g}}(\mathbb{K}, V)$ is the space of invariants, defined as homomorphisms of $U\mathfrak{g}$-modules from the trivial module $\mathbb{K}$ to V, and $H^*(\mathfrak{g}, V)$ can be defined as the right derived functor of the functor of invariants.

As before, there is a standard projective resolution of the trivial $U\mathfrak{g}$-module $\mathbb{K}$. It is the *Koszul complex* of $U\mathfrak{g}$, and has the shape

$$0 \to \mathbb{K} \leftarrow U\mathfrak{g} \leftarrow U\mathfrak{g} \otimes \mathfrak{g} \leftarrow U\mathfrak{g} \otimes \Lambda^2\mathfrak{g} \leftarrow \cdots \leftarrow U\mathfrak{g} \otimes \Lambda^n\mathfrak{g} \leftarrow \cdots.$$

Here, $\Lambda^n\mathfrak{g}$ is the exterior algebra on the vector space $\mathfrak{g}$. Therefore, this resolution is finite for finite dimensional Lie algebras $\mathfrak{g}$. Observe also that the $U\mathfrak{g}$-module $U\mathfrak{g} \otimes \Lambda^n\mathfrak{g}$ (where $U\mathfrak{g}$ acts trivially on the tensor factor $\Lambda^n\mathfrak{g}$) is free, and thus projective.

The map $\mathbb{K} \leftarrow U\mathfrak{g}$ is the augmentation of $U\mathfrak{g}$. All elements, except $\mathbb{K} \subset U\mathfrak{g}$, are sent to zero. The map $U\mathfrak{g} \leftarrow U\mathfrak{g} \otimes \mathfrak{g}$ is the product map $d(u \otimes x) = ux$ for all $u \in U\mathfrak{g}$ and all $x \in \mathfrak{g}$. In the general term, the Koszul differential

$$U\mathfrak{g} \otimes \Lambda^{n-1}\mathfrak{g} \leftarrow U\mathfrak{g} \otimes \Lambda^n\mathfrak{g} : d$$

is given by

$$\begin{aligned} d(u \otimes x_1 \wedge \cdots \wedge x_n) = &\sum_{i=1}^n (-1)^i u x_i \otimes x_1 \wedge \cdots \wedge \widehat{x_i} \wedge \cdots \wedge x_n \\ &+ \sum_{1 \leq i < j \leq n} (-1)^{i+j+1} u \otimes [x_i, x_j] \wedge x_1 \wedge \cdots \wedge \widehat{x_i} \wedge \cdots \wedge \widehat{x_j} \wedge \cdots \wedge x_n. \end{aligned}$$

The derived functor $\mathrm{Ext}^*_{U\mathfrak{g}}(\mathbb{K}, V)$ can therefore be computed via the Koszul resolution as

$$\mathrm{Ext}^p_{U\mathfrak{g}}(\mathbb{K}, V) = \mathrm{Hom}_{U\mathfrak{g}}(U\mathfrak{g} \otimes \Lambda^*\mathfrak{g}, V) = \mathrm{Hom}_{\mathbb{K}}(\Lambda^*\mathfrak{g}, V).$$

The complex (not a resolution in general !) $C^*(\mathfrak{g}, V) := \mathrm{Hom}_{\mathbb{K}}(\Lambda^*\mathfrak{g}, V)$ with the induced differential

$$dc(x_1 \wedge \cdots \wedge x_n) = \sum_{i=1}^n (-1)^i x_i \cdot c(x_1 \wedge \cdots \wedge \widehat{x_i} \wedge \cdots \wedge x_n)$$

https://doi.org/10.1515/9783110750959-008

$$+ \sum_{1\le i<j\le n} (-1)^{i+j+1} c([x_i, x_j] \wedge x_1 \wedge \cdots \wedge \widehat{x_i} \wedge \cdots \wedge \widehat{x_j} \wedge \cdots \wedge x_n) \quad \text{(B.1)}$$

for $c \in C^{n-1}(\mathfrak{g}, V)$ computes thus the Lie algebra cohomology of $\mathfrak{g}$ with values in V. It is called the *Chevalley–Eilenberg* complex. In this sense, Lie algebra cohomology is also called *Chevalley–Eilenberg cohomology*.

B.2 Abelian extensions

The relation of $H^2(\mathfrak{g}, V)$ to the abelian group $\mathrm{Ext}(\mathfrak{g}, V)$ of abelian extensions of $\mathfrak{g}$ by V is very similar to what we have stated in the framework of group cohomology. For example, the abelian group structure on $\mathrm{Ext}(\mathfrak{g}, V)$ is again the Baer sum and defined exactly as in the case of groups.

Let $0 \to V \xrightarrow{i} \mathfrak{e} \xrightarrow{\pi} \mathfrak{g} \to 0$ be an abelian extension. Choosing a linear section $\rho : \mathfrak{g} \to \mathfrak{e}$ of the quotient map $\pi : \mathfrak{e} \to \mathfrak{g}$, one obtains an action of $\mathfrak{g}$ on V by

$$x \cdot v := [\rho(x), i(v)]$$

for all $x \in \mathfrak{g}$ and all $v \in V$. In this sense, an abelian extension has a built-in action of $\mathfrak{g}$ on V (which does not depend on the choice of the section ρ). In a central extension (i. e., in case $i(V)$ is central in $\mathfrak{e}$), this action is trivial. In a general extension, this "action" is only well-defined up to inner derivations, that is, we obtain an outer action.

To an abelian extension $0 \to V \xrightarrow{i} \mathfrak{e} \xrightarrow{\pi} \mathfrak{g} \to 0$, we associate as before a 2-cocycle by considering for $x, y \in \mathfrak{g}$,

$$\alpha(x, y) := [\rho(x), \rho(y)] - \rho([x, y]),$$

the failure of ρ to be a homomorphism of Lie algebras. Then $\beta(x, y) \in V$ such that $i(\beta(x, y)) = \alpha(x, y)$ is a 2-cocycle of $\mathfrak{g}$ with values in V. Indeed,

$$\begin{aligned} -i(d\beta(x, y, z)) &= x \cdot \alpha(y, z) + y \cdot \alpha(z, x) + z \cdot \alpha(x, y) \\ &\quad - \alpha([x, y], z) - \alpha([y, z], x) - \alpha([z, x], y) \\ &= [\rho(x), [\rho(y), \rho(z)]] + \text{c.p.} - [\rho(x), \rho([y, z])] - \text{c.p.} \\ &\quad - [\rho([x, y]), \rho(z)] - \text{c.p.} + \rho([[x, y], z]) + \text{c.p.} \end{aligned}$$

Here, "c.p." means cyclic permutations of $(x, y, z) \in \mathfrak{g}^3$. The first and the last cyclic sums are zero due to the Jacobi identity, while the second and the third cyclic sums cancel each other. This shows $d\beta(x, y, z) = 0$ by injectivity of i.

Once again, the association of a 2-cocycle to an abelian extension does not depend up to a coboundary on the choice of the section ρ and induces a well-defined map

$$\zeta : \mathrm{Ext}(\mathfrak{g}, V) \to H^2(\mathfrak{g}, V).$$

We have as in the case of group cohomology.

Theorem B.2.1. *The above map ζ induces an isomorphism of abelian groups*

$$\mathrm{Ext}(\mathfrak{g}, V) \cong H^2(\mathfrak{g}, V).$$

For the proof, the interested reader may refer to Theorem 7.6.3, page 235, in [98]. We will only comment on the surjectivity of ζ, that is, on the construction of an abelian extension from a 2-cocycle $\beta \in Z^2(\mathfrak{g}, V)$. Given an abelian extension $0 \to V \xrightarrow{i} \mathfrak{e} \xrightarrow{\pi} \mathfrak{g} \to 0$, a section ρ of π permits to describe the Lie algebra $\mathfrak{e}$ in terms of "product coordinates," that is, in terms of the direct sum decomposition $\mathfrak{e} = V \oplus \mathfrak{g}$ as vector spaces. From the fact that i and π are Lie algebra homomorphisms, one sees that the bracket in $\mathfrak{e}$ reads in product coordinates

$$\big[(v,x),(w,y)\big] = \big(x \cdot w - y \cdot v + \beta(x,y), [x,y]\big)$$

for some yet-to-specify term $\beta(x,y)$. Computing the Jacobi identity for this bracket shows that $\beta \in \mathrm{Hom}(\Lambda^2\mathfrak{g}, V)$ must be a 2-cocycle.

For the surjectivity of ζ, we can turn this around: Given a 2-cocycle $\beta \in Z^2(\mathfrak{g}, V)$, the above bracket gives a Lie algebra structure on $\mathfrak{e} = V \oplus \mathfrak{g}$ and, therefore (with the obvious maps $i : V \hookrightarrow \mathfrak{e}$, $v \mapsto (v, 0)$, and $\pi : \mathfrak{e} \to \mathfrak{g}$, $(v,x) \mapsto x$) an abelian extension $\mathfrak{e}$ of $\mathfrak{g}$ by V, which has β as a representative of its associated cohomology class. The corresponding extension is denoted $\mathfrak{e} = V \times_\beta \mathfrak{g}$.

B.3 The Hochschild–Serre spectral sequence

Here, we introduce the Hochschild–Serre spectral sequence. It is also a special case of the Grothendieck spectral sequence, see Section 7.5, page 232, in [98]. We will explain in more detail the approach by filtration of the standard complex and draw for this on [34], Chapter 1, Section 5, pp. 40–44.

Suppose $\mathfrak{n}$ is a Lie algebra and $\mathfrak{l} \subset \mathfrak{n}$ is a subalgebra. Let furthermore be V an $\mathfrak{n}$-module, which is also regarded as an $\mathfrak{l}$-module by restriction. In the usual Chevalley–Eilenberg complex $C^*(\mathfrak{n}, V)$, one introduces the following *filtration*:

$$F^pC^{p+q}(\mathfrak{n}, V) = \{c \in C^{p+q}(\mathfrak{n}, V) \mid c(x_1, \dots, x_{p+q}) = 0 \ \text{ for } \ x_1, \dots, x_{q+1} \in \mathfrak{l}\}. \tag{B.2}$$

We have

$$C^r(\mathfrak{n}, V) = F^0C^r(\mathfrak{n}, V) \supset \cdots \supset F^rC^r(\mathfrak{n}, V) \supset F^{r+1}C^r(\mathfrak{n}, V) = \{0\},$$

and also $dF^pC^{p+q}(\mathfrak{n}, V) \subset F^pC^{p+q+1}(\mathfrak{n}, V)$. Indeed, dc has degree $p + q + 1$ and must be zero as soon as $x_1, \dots, x_{q+2} \in \mathfrak{l}$. To see this, write

$$\begin{aligned} dc(x_1, \dots, x_{p+q+1}) &= \sum \pm c([x_i, x_j], x_1, \dots, \widehat{x_i}, \dots, \widehat{x_j}, \dots, x_{p+q+1}) \\ &\quad + \sum \pm x_i \cdot c(x_1, \dots, \widehat{x_i}, \dots, x_{p+q+1}), \end{aligned}$$

where the hat in $\widehat{x_i}$ means that the element x_i should be omitted from the list. In the first sum, either $x_i, x_j \in \mathfrak{l}$, which implies $[x_i, x_j] \in \mathfrak{l}$ and, therefore, $q+1$ arguments of c are in $\mathfrak{l}$, or one of x_i or x_j is in $\mathfrak{l}$, in which case there are still $q+1$ arguments of c in $\mathfrak{l}$, or neither x_i nor x_j are in $\mathfrak{l}$, in which case there are even $q+2$ arguments of c in $\mathfrak{l}$. In any case, the corresponding term of the sum is zero. The same holds for the second sum.

The $F^pC^{p+q}(\mathfrak{n}, V)$ form thus a filtration of the Chevalley–Eilenberg complex and permit to compute its cohomology by a spectral sequence, the *Hochschild–Serre spectral sequence*; see [46]. The corresponding theorem of Hochschild–Serre reads as follows.

Theorem B.3.1. *The spectral sequence*

$$\{E_r^{p,q}, d_r^{p,q} : E_r^{p,q} \to E_r^{p+r,q-r+1}\}$$

corresponding to the above filtration has the following properties:
(a) $E_1^{p,q} = H^q(\mathfrak{l}, \mathrm{Hom}(\Lambda^p(\mathfrak{n}/\mathfrak{l}), V))$;
(b) $E_2^{p,q} = H^p(\mathfrak{n}, \mathfrak{l}, V)$;
(c) *if* $\mathfrak{l}$ *is an ideal, then* $E_2^{p,q} = H^p(\mathfrak{n}/\mathfrak{l}, H^q(\mathfrak{l}, V))$;
(d) *the spectral sequence converges to* $H^*(\mathfrak{n}, V)$; *the edge homomorphisms*

$$H^q(\mathfrak{n}, V) \to E_\infty^{0,q} \to E_1^{0,q} = H^q(\mathfrak{l}, V),$$

and

$$H^p(\mathfrak{n}, \mathfrak{l}, V) = E_2^{p,0} \to E_\infty^{p,0} \to H^p(\mathfrak{n}, V)$$

may be identified with the natural homomorphisms $H^q(\mathfrak{n}, V) \to H^q(\mathfrak{l}, V)$ *and* $H^p(\mathfrak{n}, \mathfrak{l}, V) \to H^p(\mathfrak{n}, V)$ *respectively;*
(e) *if* $V = \mathbb{K}$ *or if* V *is an associative commutative algebra on which* $\mathfrak{n}$ *acts by derivations, then the spectral sequence is multiplicative; in this case the isomorphisms in (a)–(c) are also multiplicative, as well as the convergence in (d).*

Remark B.3.1. This statement needs some clarifications.

First of all, the space $\mathrm{Hom}(\Lambda^p(\mathfrak{n}/\mathfrak{l}), V)$ is an $\mathfrak{l}$-module by setting for all $l \in \mathfrak{l}$ and all $\bar{x}_i \in \mathfrak{n}/\mathfrak{l}$ with $i = 1, \dots, p$:

$$(l \cdot f)(\bar{x}_1, \dots, \bar{x}_p) = l \cdot f(\bar{x}_1, \dots, \bar{x}_p) - \sum_{i=1}^{p} f(\bar{x}_1, \dots, [l, \bar{x}_i], \dots, \bar{x}_p).$$

Observe here that $\mathfrak{n}/\mathfrak{l}$ is the quotient $\mathfrak{l}$-module of the $\mathfrak{l}$-module $\mathfrak{n}$ by the adjoint $\mathfrak{l}$-module $\mathfrak{l}$. This induces then a natural action on $\mathrm{Hom}(\Lambda^p(\mathfrak{n}/\mathfrak{l}), V)$, using furthermore the action of $\mathfrak{l}$ on V.

The relative Lie algebra cohomology in (b) is the cohomology of the complex

$$C^*(\mathfrak{n},\mathfrak{l},V) = \{c \in C^*(\mathfrak{n},V) \mid c(x_1,\dots,x_q) = 0 \text{ and } dc(x_1,\dots,x_{q+1}) = 0 \text{ if some } x_i \in \mathfrak{l}\}.$$

This is not the same as the relative cohomology in Section 2.9.

If $\mathfrak{l}$ is an ideal, then the space $H^*(\mathfrak{l},V)$ is an $\mathfrak{n}/\mathfrak{l}$-module: Indeed, it is an $\mathfrak{n}$-module as well as $\mathfrak{l}$, and it is trivial as an $\mathfrak{l}$-module, because each Lie algebra acts trivially on its cohomology.

In order to get acquainted with the differentials in the spectral sequence, let us compute its first terms: Every cochain $c \in C^{p+q}(\mathfrak{n},V)$ determines a map $\Lambda^q\mathfrak{l} \to \mathrm{Hom}(\Lambda^p\mathfrak{n},V)$ by sending $l_1\wedge\cdots\wedge l_q$ to the homomorphism

$$h_{c;l_1\wedge\cdots\wedge l_q} : n_1\wedge\cdots\wedge n_p \mapsto c(l_1,\dots,l_q,n_1,\dots,n_p).$$

The inclusion $c \in F^pC^{p+q}(\mathfrak{n},V)$ is equivalent to the assertion that for all $l_1\wedge\cdots\wedge l_q \in \Lambda^q\mathfrak{l}$, the image of $h_{c;l_1\wedge\cdots\wedge l_q}$ is contained in

$$\mathrm{Hom}(\Lambda^p(\mathfrak{n}/\mathfrak{l}),V) \subset \mathrm{Hom}(\Lambda^p\mathfrak{n},V),$$

because, q elements of $\mathfrak{l}$ already being arguments of c, $h_{c;l_1\wedge\cdots\wedge l_q}$ is zero as soon as one argument lies in $\mathfrak{l}$, and this means exactly that it factorizes to a map in $\mathrm{Hom}(\Lambda^p(\mathfrak{n}/\mathfrak{l}),V)$.

We obtain a map

$$\varphi : F^pC^{p+q}(\mathfrak{n},V) \to C^q(\mathfrak{l},\mathrm{Hom}(\Lambda^p(\mathfrak{n}/\mathfrak{l}),V)),$$

defined by restricting the first q arguments of a given cochain to $\mathfrak{l}$. The map φ is an epimorphism, as is easily seen by choosing a linear supplement of $\mathfrak{l}$ in $\mathfrak{n}$ to extend an element of $C^q(\mathfrak{l},\mathrm{Hom}(\Lambda^p(\mathfrak{n}/\mathfrak{l}),V))$ to a cochain on all of $\mathfrak{n}$.

The kernel of φ is then $F^{p+1}C^{p+q}(\mathfrak{n},V)$, because a cochain is mapped to zero if and only if it vanishes already on q elements of $\mathfrak{l}$. We get an isomorphism

$$F^pC^{p+q}(\mathfrak{n},V)/F^{p+1}C^{p+q}(\mathfrak{n},V) \to C^q(\mathfrak{l},\mathrm{Hom}(\Lambda^p(\mathfrak{n}/\mathfrak{l}),V)).$$

This isomorphism commutes with the differentials; see Chapter 1, Section 5, pp. 40–44 in [34].

In their article [45], Hochschild and Serre use the spectral sequence and some basic results about reductive Lie algebras to obtain in their Theorem 13 a general formula for the cohomology of a finite dimensional complex Lie algebra $\mathfrak{g}$, starting from its Levi decomposition $\mathfrak{g} = \mathfrak{r} \rtimes \mathfrak{s}$, which describes $\mathfrak{g}$ as the semi-direct product of a semi-simple subalgebra $\mathfrak{s}$ with its solvable radical $\mathfrak{r}$. The formula reads

$$H^n(\mathfrak{g},V) = \bigoplus_{p+q=n} H^p(\mathfrak{s},\mathbb{C}) \otimes H^q(\mathfrak{r},V)^{\mathfrak{s}}.$$

While the cohomology spaces $H^p(\mathfrak{s},\mathbb{C})$ of a semi-simple Lie algebra are in principle well known, the spaces $H^q(\mathfrak{r},V)$ and their invariants $H^q(\mathfrak{r},V)^{\mathfrak{s}}$ are much less known.

B.4 Continuous cohomology

For an infinite dimensional Lie algebra, one needs again topology in order to "tame" the cohomology. The infinite dimensional Lie algebras occurring in this book are the Lie algebras $\mathrm{Vect}(M)$ of (smooth) vector fields on some differentiable manifold M. The Lie algebra $\mathrm{Vect}(M)$ is a *topological Lie algebra*, that is, the underlying vector space is a topological vector space and the Lie bracket is continuous. The smooth vector fields $\mathrm{Vect}(M)$ carry the C^∞-topology, that is, the topology of uniform convergence of the representative function and all of its derivatives of the vector field on compact subsets of M. This topology renders $\mathrm{Vect}(M)$ a Fréchet–Lie algebra.

For such a topological Lie algebra $\mathfrak{g}$ and a continuous $\mathfrak{g}$-module V, one may consider *continuous cohomology*, that is, the cohomology spaces

$$H^p_{\mathrm{cont}}(\mathfrak{g}, V) := Z^p_{\mathrm{cont}}(\mathfrak{g}, V) \,/\, B^p_{\mathrm{cont}}(\mathfrak{g}, V),$$

that is, the vector space quotient of the continuous cocycles by the continuous coboundaries. In order to be precise, the space $\Lambda^p \mathfrak{g}$ must be completed to be a Fréchet space, and the continuous cochains are all continuous maps from this completion $\widehat{\Lambda^p \mathfrak{g}}$ to V. The Chevalley–Eilenberg coboundary operator sends continuous cochains to continuous cochains and makes $H^p_{\mathrm{cont}}(\mathfrak{g}, V)$ well-defined. These cohomology spaces are defined in an ad hoc manner and do not come from a derived functor framework. The correct framework is more one of relative homological algebra; see [68], Chapter IX.

The subject of the continuous cohomology of the Lie algebra of vector fields is classical and has among its main protagonists I. M. Gelfand and D. B. Fuchs. The interested reader is invited to learn more about this subject in the book [34].

C Cohomology of associative algebras

C.1 Definitions

Let us first of all recall *the algebra of multiplications* $M(L)$ of an associative algebra L; see [41]. By definition,

$$M(L) := \{(u,v),\ u,v \in \mathrm{End}(L),\ \forall l_1,l_2 \in L : l_1 u(l_2) = v(l_1)l_2,\\ u(l_1 l_2) = u(l_1)l_2,\ v(l_1 l_2) = l_1 v(l_2)\}.$$

The set $M(L)$ becomes a vector space as subspace of $\mathrm{End}(L) \oplus \mathrm{End}(L)$. It becomes an associative algebra as a subalgebra of $\mathrm{End}(L) \times \mathrm{End}(L)^{\mathrm{op}}$.

A special subalgebra (in fact, an ideal) of $M(L)$ is the subspace $\mathrm{Inn}(L) \subset M(L)$ of inner multiplications, that is, of pairs of endomorphisms of the form (l^l, l^r) where $l \in L$ and $l^l : L \to L$ denotes the endomorphism of left multiplication by l: $l^l(l') := ll'$, while $l^r : L \to L$ denotes the endomorphism of right multiplication by l: $l^r(l') := l'l$ for all $l' \in L$. There is a natural morphism of algebras $\varphi : L \to M(L)$, $\varphi(l) := (l^l, l^r)$ whose image is exactly $\mathrm{Inn}(L)$. $\mathrm{Inn}(L)$ is an ideal: Let $(u,v) \in M(L)$ and $(l^l, l^r) \in \mathrm{Inn}(L)$. Then $(u,v)(l^l,l^r) = (u(l)^l, v(l)^r)$, because $u(l^l(l')) = u(ll') = u(l)l'$ and $v(l^r l') = v(l'l) = l'v(l)$ for all $l' \in L$.

The corresponding quotient algebra is the *algebra of outer multiplications*

$$\mathrm{Out}(L) := M(L) / \mathrm{Inn}(L).$$

It plays the same role for associative algebras as does the group of outer automorphisms for groups and the Lie algebra of outer derivations for Lie algebras. We have an exact sequence

$$0 \to Z \to L \stackrel{\varphi}{\to} M(L) \to \mathrm{Out}(L) \to 0,$$

where $Z := Z(L)$ is the *nucleus* of L, that is, the subspace of L of elements $l \in L$ such that for all $l' \in L$, $ll' = l'l = 0$. Despite the notation, Z is *not* the center of L.

C.2 Hochschild cohomology

The cohomology of associative algebras follows the general scheme of the cohomologies of algebraic structures. General standard references for Hochschild cohomology include [61] and [98].

Let A be an associative algebra over the field $\mathbb{K}$ (we will not discuss the case of a general commutative ring $\mathbb{K}$). The cohomology of A is defined as a certain Ext functor, but in the category of A-bimodules or (left) $A \times A^{\mathrm{op}}$-modules. Let V be an A-bimodule.

https://doi.org/10.1515/9783110750959-009

Then the *pth Hochschild cohomology space* $HH^p(A, V)$ for an integer $p \geq 0$ is by definition

$$HH^p(A, V) := \mathrm{Ext}^p_{A\times A^{\mathrm{op}}}(A, V).$$

As for the cohomology of groups and Lie algebras, we also have a standard resolution of A as an A-bimodule. The *bar resolution* of the associative algebra A is a resolution of the type

$$0 \leftarrow A \overset{\mu}{\leftarrow} A \otimes A \overset{d}{\leftarrow} A \otimes A \otimes A \overset{d}{\leftarrow} \cdots \overset{d}{\leftarrow} A \otimes A^{\otimes n} \otimes A \overset{d}{\leftarrow} \cdots.$$

Here, the first map $\mu : A \otimes A \to A$ is the multiplication map: $\mu(ab) = ab$ for all $a, b \in A$. Any term $A \otimes A^{\otimes n} \otimes A$ is an $A \times A^{\mathrm{op}}$-module by left multiplication by $a \in A$ on the first and right multiplication on the last factor. They are thus free A-bimodules (for $n \geq 0$). The differential $d : A \otimes A^{\otimes n} \otimes A \to A \otimes A^{\otimes n-1} \otimes A$ is given by $d = \sum_{i=0}^n d_i$ with

$$\begin{aligned} d_0(a_0 \otimes \cdots \otimes a_{n+1}) &= a_0 a_1 \otimes a_2 \otimes \cdots \otimes a_{n+1}, \\ d_i(a_0 \otimes \cdots \otimes a_{n+1}) &= a_0 \otimes \cdots \otimes a_i a_{i+1} \otimes \cdots \otimes a_{n+1}, \\ d_n(a_0 \otimes \cdots \otimes a_{n+1}) &= a_0 \otimes \cdots \otimes a_n a_{n+1} \end{aligned}$$

for $i = 1, \ldots, n-1$. Note the similarity with the bar resolution for groups. Once again, the operators d_i satisfy the simplicial relations and verify therefore $d \circ d = 0$ for $d = \sum_{i=0}^n (-1)^i d_i$. In fact, the above complex is exact for a unital associative algebra A; see Proposition 1.1.12, page 12, in [61]. Therefore, Hochschild cohomology of an associative unital algebra A with values in an A-bimodule V can be computed by

$$HH^p(A, V) = \mathrm{Hom}_{A\times A^{\mathrm{op}}}(A \otimes A^{\otimes *} \otimes A, V) = \mathrm{Hom}_{\mathbb{K}}(A^{\otimes *}, V)$$

with the induced differential. The complex $(CH^p(A, V) := \mathrm{Hom}_{\mathbb{K}}(A^{\otimes p}, V), d)_{p\geq 0}$ is called the *Hochschild complex*, and the induced differential reads for $c \in HC^n(A, V)$,

$$\begin{aligned} dc(a_1, \ldots, a_{n+1}) = {} & a_1 \cdot c(a_2, \ldots, a_{n+1}) - c(a_1 a_2, a_3, \ldots, a_{n+1}) \pm \cdots + (-1)^n c(a_1, \ldots, a_n a_{n+1}) \\ & + (-1)^{n+1} c(a_1, \ldots, a_n) \cdot a_{n+1}. \end{aligned} \tag{C.1}$$

From the description of the bar resolution, it is clear that a 3-cocycle on A with values in V can be viewed as a map of A-bimodules $c : A^{\otimes 5}/\mathrm{Im}(d) \to V$.

C.3 Square zero extensions

As before, there is an isomorphism of abelian groups between the second cohomology space $HH^2(A, V)$ and the abelian group (with respect to the Baer sum) of "abelian" extensions of A by V. In the context of associative algebras, "abelian" extension are here *square zero extensions*, that is, the multiplication on V is trivial: $vw = 0$ for all $v, w \in V$. This is the only multiplication we have on the A-bimodule V. Denote the abelian group of square zero extensions of an associative algebra A by an A-bimodule V by $\mathrm{Ext}(A, V)$.

Again, in a square zero extension

$$0 \to V \xrightarrow{i} E \xrightarrow{\pi} A \to 0,$$

we obtain an A-bimodule structure on V by choosing a section $\rho : A \to E$ of $\pi : E \to A$. Indeed, one defines

$$a \cdot v := \rho(a)i(v), \quad v \cdot a := i(v)\rho(a)$$

for all $v \in V$ and $a \in A$. As before, one can use the section ρ to identify the underlying vector space of E with the direct sum $E \cong V \oplus A$ and translate the associative product on E into these "product coordinates." The product in the square-zero extension E takes then the following form:

$$(v, a)(v', a') = (a \cdot v' + v \cdot a' + \omega(a, a'), aa'),$$

because the product in V is zero. One verifies easily that the associativity of this product is equivalent (modulo the bimodule property and associativity in A) to the 2-cocycle condition of $\omega : A \otimes A \to V$.

One associates a 2-cocycle to an arbitrary square zero extension by taking a section $\rho : A \to E$ of the quotient map $\pi : E \to A$ and computing the failure of ρ to be a homomorphism of associative algebras. In this way, one obtains a well-defined map

$$\xi : \mathrm{Ext}(A, V) \to HH^2(A, V).$$

Theorem C.3.1. *The map ξ is an isomorphism of abelian groups*

$$\mathrm{Ext}(A, V) \cong HH^2(A, V).$$

Proof. For a proof, the interested reader may consult Theorem 9.3.1, page 312, in [98]. □

D Lie groups and their cohomology

D.1 Lie groups

All our manifolds are smooth, that is, C^∞, finite dimensional, separated, paracompact, second countable and connected (if of dimension ≥ 1). All our manifolds are real or complex. A general reference is [94]; see also [78] concerning the infinite dimensional Lie groups appearing in our book.

Definition D.1.1. A *Lie group* G is a group G, which possesses the structure of a smooth manifold such that the multiplication map $G \times G \to G$ and the inversion map $G \to G$ are smooth.

To a Lie group G, one associates its *Lie algebra* $\mathfrak{g} := T_1 G$, the tangent space of G at the unit element $1 \in G$. The smooth conjugation map $\mathrm{conj}_h : G \to G, g \mapsto hgh^{-1}$, can be derived in both variables at $1 \in G$ in order to define a bracket $[,] : \mathfrak{g} \times \mathfrak{g} \to \mathfrak{g}$, which turns out to be a Lie bracket. Another way to define the bracket is as the bracket of invariant vector fields, using that $T_1 G$ identifies with the space of translation-invariant vector fields on G. For these matters, compare Section 2.3, page 51, in [94].

There is a close relationship between Lie groups and Lie algebras, expressed by Lie's theorems. Lie's first theorem (Theorem 2.5.2, p. 58, in [94]) is the 1–1 correspondence between Lie subalgebras of the Lie algebra $\mathfrak{g}$ of a Lie group G and Lie subgroups of G. Here, a (not necessarily closed) Lie subgroup $H \subset G$ is a subgroup H of G, which is a Lie group and such that the inclusion map $i : H \hookrightarrow G$ is embedding of analytic manifolds.

Lie's first theorem is shown using Frobenius theorem in order to integrate the distribution of subspaces in the tangent bundle of G given by a Lie subalgebra $\mathfrak{h} \subset \mathfrak{g}$ into a submanifold. The submanifolds which one obtains in this way are in general not regularly embedded, but *quasiregularly embedded*; see Theorem 1.3.6, page 30, in [94]. The property of being quasiregularly embedded (see [94] p. 17) is exactly the property of being an initial submanifold; see Definition 5.1.4.

Lie's second theorem (see Section 2.7 in [94]) asserts the 1–1 correspondence between homomorphisms of Lie groups $f : G \to H$ such that G is simply connected and homomorphisms of the corresponding Lie algebras $T_1 f : \mathfrak{g} \to \mathfrak{h}$. Lie's third theorem (Theorem 3.15.1, p. 230, in [94]) asserts the possibility of integrating a finite dimensional real or complex Lie algebra into a Lie group whose Lie algebra is the one we started with. This integration becomes unique by imposing that the Lie group should be simply connected. This can always be achieved by passing to the universal covering group. Therefore, when one talks about *the* Lie group associated to a Lie algebra by Lie's third theorem, it is always the simply connected one which is meant.

The functoriality of the integration procedure is best understood in the procedure proposed in [23]; see Theorem 1.14.3, page 79. The Lie group is constructed as a set of paths in the Lie algebra with the product imposed by viewing paths in the Lie algebra

https://doi.org/10.1515/9783110750959-010

as logarithmic derivatives of paths in the Lie group (without assuming the existence of the Lie group); see Proposition 1.14.1, page 73. The (finite-dimensional) Lie group arises then as the quotient of this infinite-dimensional Banach Lie group by the normal subgroup given by the paths of zero integral.

D.2 Defining a Lie group structure

In Section 5, we need a theorem which constructs a Lie group structure on a group. The theorem we use is the following.

Theorem D.2.1. *Let G be a group and $K = K^{-1}$ be a symmetric subset of G. Assume furthermore that K is a smooth manifold such that:*

(a) *there exists an open 1-neighborhood $V \subset K$ with $V^2 = V \cdot V \subset K$ such that the group product $\mu_V : V \times V \to K$ is smooth;*
(b) *the inversion map $\eta_K : K \to K, k \mapsto k^{-1}$ is smooth, and*
(c) *for each $g \in G$, there exists an open 1-neighborhood $K_g \subset K$ with $\operatorname{conj}_g(K_g) \subset K$ and such that the conjugation map*

$$\operatorname{conj}_g : K_g \to K, \quad x \mapsto gxg^{-1}$$

is smooth.

Then there exists a unique Lie group structure on G for which there exists an open 1-neighborhood $K_1 \subset K$ such that the inclusion map $K_1 \hookrightarrow G$ induces a diffeomorphism onto an open subset of G.

If G is generated (as a group) by each 1-neighborhood $U \subset K$, then condition (c) may be omitted.

Proof. This follows from Theorem A.4 and Remark A.5 in [76]. □

D.3 Lie group cohomology

As a Lie group is an object which mixes group theoretical and manifold properties, one would like to see these two kinds of features in a meaningful cohomology theory for Lie groups. There are many ways to do this. In [97], the interested reader will find the definitions of some of these ways and the circumstances in which they coincide.

One important way to take into account the Lie group structure in a cohomology theory is to consider *locally smooth cochains*. Let G be a Lie group and V be a *smooth G-module*, that is, a G-set with a manifold structure such that the action map is smooth. The set V here is in general not a $\mathbb{K}$-vector space. For example, $V = S^1$ may be handy.

To G and V, we associate the complex of locally smooth cochains. A group cochain $c \in C^p(G, V) = \mathrm{Map}(G^p, V)$ is called *locally smooth* in case there exists a unit neighborhood $U \subset G^p$ such that $c|_U$ is smooth. The standard group cohomology coboundary operator d from Appendix A sends locally smooth cochains to locally smooth cochains, thus we have a complex of locally smooth cochains. The associated cohomology

$$H_s^p(G, V) := Z_s^p(G, V) \,/\, B_s^p(G, V)$$

is the quotient space of the locally smooth cocycles by the locally smooth coboundaries. The spaces $H^p(G, V)$ do not fit in general together to be the derived functor of some functor of invariants.

Bibliography

[1] H. Abbaspour, F. Wagemann, *On 2-holonomy*. J. Aust. Math. Soc. 110 (2021), no. 2, pp. 145–174.

[2] C. Alexandre, M. Bordemann, S. Rivière, F. Wagemann, *Structure theory of rack-bialgebras*. J. Gen. Lie Theory Appl. 10 (2016), no. 1, Art. ID 1000244, 20 pp.

[3] J. Baez, A. Lauda, *Higher-dimensional algebra. V. 2-groups*. Theory Appl. Categ. 12 (2004), pp. 423–491.

[4] J. Baez and A. Crans, *Higher-dimensional algebra. VI. Lie 2-algebras*. Theory Appl. Categ. 12 (2004), pp. 492–538.

[5] J. Baez, D. Stevenson, A. Crans, U. Schreiber, *From loop groups to 2-groups*. Homology, Homotopy Appl. 9 (2007), no. 2, pp. 101–135.

[6] H.-J. Baues, E. G. Minian, *Crossed extensions of algebras and Hochschild cohomology*. Homology, Homotopy Appl. 4 (2002), no. 2, pp. 63–82.

[7] H.-J. Baues and T. Pirashvili, *Comparison of MacLane, Shukla and Hochschild cohomologies*. J. Reine Angewandte Math. 598 (2006), pp. 25–69.

[8] Yu. N. Bespalov, *Crossed modules and quantum groups in braided categories*. Appl. Categ. Structures 5 (1997), no. 2, pp. 155–204.

[9] C. Blohmann, A. Weinstein, *Group-like objects in Poisson geometry and algebra*. Contemp. Math. 450 (2008), pp. 25–39.

[10] A. M. Blokh, *On a generalization of the concept of a Lie algebra*. Dokl. Akad. Nauk. USSR 165 (1965), pp. 471–473.

[11] M. Bordemann, F. Wagemann, *Global integration of Leibniz algebras*. J. Lie Theory 27 (2017), no. 2, pp. 555–567.

[12] R. Brooks, P. Trauber, *The van Est theorem for groups of diffeomorphisms*. Hadronic J. 1 (1978), no. 3, pp. 916–934.

[13] N. Bourbaki, *Éléments de mathématique. Algèbre. Chapitres 1 à 3*. Hermann, Paris (1970), xiii+635 pp.

[14] N. Bourbaki, *Éléments de mathématique. Fasc. XXXVII. Groupes et algèbres de Lie. Chapitre II: Algèbres de Lie libres*. Actualitś Scientifiques et Industrielles, 1349. Hermann, Paris (1972), 320 pp.

[15] J. Carter, A. Crans, M. Elhamdadi, M. Saito, *Cohomology of the Categorical Self-Distributivity*. J. Homotopy Relat. Struct. 3 (2008), no. 1, pp. 13–63.

[16] J. M. Casas, M. Ladra, T. Pirashvili, *Crossed modules for Lie-Rinehart algebras*. J. Algebra 274 (2004), no. 1, pp. 192–201.

[17] J. M. Casas, R. F. Casado, E. Khmaladze, M. Ladra, *Universal enveloping crossed module of a Lie crossed module*. Homology Homotopy Appl. 16 (2014), no. 2, pp. 143–158.

[18] J. M. Casas, N. Inassaridze, E. Khamaladze, M. Ladra, *Adjunction between crossed modules of groups and algebras*. J. Homotopy Relat. Struct. 9 (2014), no. 1, pp. 223–237.

[19] J. Ceniceros, M. Elhamdadi, M. Green, S. Nelson, *Augmented biracks and their homology*. Internat. J. Math. 25 (2014), no. 9, 1450087, 19 pp.

[20] A. Connes, *Noncommutative Geometry*. Academic Press, Inc., San Diego, CA (1994).

[21] A. S. Crans, F. Wagemann, *Crossed modules of racks*. Homology Homotopy Appl. 16 (2014), no. 2, pp. 85–106.

[22] J. Dixmier, *Algèbres Enveloppantes*. Cahiers Scientifiques, Gauthier Villars (1974).

[23] J. J. Duistermaat, J. A. C. Kolk, *Lie Groups*. Springer Universitext. Springer, Berlin, Tokyo (1999).

[24] S. Eilenberg, S. MacLane, *Relations between homology and homotopy groups of spaces*. Ann. of Math. (2) 46 (1945), pp. 480–509.

[25] S. Eilenberg, S. MacLane, *Cohomology theory in abstract groups. II. Group extensions with a non-Abelian kernel*. Ann. of Math. (2) 48 (1947), pp. 326–341.

https://doi.org/10.1515/9783110750959-011

[26] J. Faria Martins, *Crossed modules of Hopf algebras and of associative algebras and two-dimensional holonomy*. J. Geom. Phys. 99 (2016), pp. 68–110.

[27] R. Fenn, C. Rourke, *Racks and links in codimension two*. J. Knot Theory Ramifications 1 (1992), no. 4, pp. 343–406.

[28] R. Fenn, C. Rourke, B. Sanderson, *Trunks and classifying spaces*. Appl. Categ. Structures 3 (1995), no. 4, pp. 321–356.

[29] J. M. Fernández Vilaboa, M. P. López López, E. Villanueva Novoa, *Cat^1-Hopf algebras and crossed modules*. Commun. Algebra 35 (2007), pp. 181–191.

[30] Y. Frégier, F. Wagemann, *On Hopf 2-algebras*. Int. Math. Res. Not. IMRN 2011 (2011), no. 15, pp. 3471–3501.

[31] E. Frenkel, X. Zhu, *Gerbal representations of double loop groups*. Int. Math. Res. Not. IMRN 2012 (2012), no. 17, pp. 3929–4013.

[32] B. Fresse, *Homotopy of Operads and Grothendieck–Teichmüller Groups. Part 1. The Algebraic Theory and Its Topological Background*. Mathematical Surveys and Monographs, 217. American Mathematical Society, Providence, RI (2017), xlvi+532 pp.

[33] P. J. Freyd, D. N. Yetter, *Braided compact closed categories with applications to low-dimensional topology*. Adv. Math. 77 (1989), no. 2, pp. 156–182.

[34] D. B. Fuchs, *Cohomology of Infinite Dimensional Lie Algebras*. Consultant Bureau, New York (1987).

[35] M. Gerstenhaber, *A uniform cohomology theory for algebras*. Proc. Nat. Acad. Sci. 51 (1964), pp. 626–629.

[36] M. Gerstenhaber, *On the deformation of rings and algebras II*. Ann. of Math. 84 (1966), pp. 1–19.

[37] S. I. Goldberg, *Extensions of Lie algebras and the third cohomology group*. Canadian J. Math. 5 (1953), pp. 470–476.

[38] L. Guieu, C. Roger, *L'algèbre et le groupe de Virasoro*. With an appendix by Vlad Sergiescu. Les Publications CRM, Montreal, QC (2007).

[39] J. Helmstetter, *Série de Hausdorff d'une algèbre de Lie et projections canoniques dans l'algèbre enveloppante*. J. Algebra 120 (1989), no. 1, pp. 170–199.

[40] G. Hochschild, *On the cohomology theory for associative algebras*. Ann. of Math. (2) 47 (1946), pp. 568–579.

[41] G. Hochschild, *Cohomology and representations of associative algebras*. Duke Math. J. 14 (1947), pp. 921–948.

[42] G. Hochschild, *The automorphism group of a Lie group*. Trans. AMS 75 (1952), pp. 209–216.

[43] G. Hochschild, *Lie algebra kernels and cohomology*. Amer. J. Math. 76 (1954), pp. 698–716.

[44] G. Hochschild, *Cohomology classes of finite type and finite dimensional kernels for Lie algebras*. Amer. J. Math. 76 (1954), pp. 763–778.

[45] G. Hochschild, J.-P. Serre, *Cohomology of group extensions*. Trans. Amer. Math. Soc. 74 (1953), pp. 110–134.

[46] G. Hochschild, J.-P. Serre, *Cohomology of Lie algebras*. Ann. Math. 57 (1953), pp. 591–603.

[47] G. Hochschild, *Algebraic groups and Hopf algebras*. Illinois J. Math. 14 (1970), pp. 52–65.

[48] H. Hopf, *Fundamentalgruppe und zweite Bettische Gruppe*. Comment. Math. Helv. 14 (1942), pp. 257–309.

[49] S.-T. Hu, *Cohomology theory in topological groups*. Michigan Math. J. 1 (1952), pp. 11–59.

[50] J. Huebschmann, *Automorphisms of group extensions and differentials in the Lyndon–Hochschild–Serre spectral sequence*. J. Algebra 72 (1981), no. 2, pp. 296–334.

[51] J. Huebschmann, *Poisson cohomology and quantization*. J. Reine Angew. Math. 408 (1990), pp. 57–113.

[52] M. Kapranov, *Membranes and higher groupoids*. arXiv:1502.06166.

[53] C. Kassel and J.-L. Loday, *Extensions centrales d'algèbres de Lie*. Ann. Inst. Fourier Grenoble 32 (1982), no. 4, pp. 119–142.

[54] C. Kassel, *Quantum Groups*. Graduate Texts in Mathematics, 155. Springer (1995).

[55] M. Kinyon, A. Weinstein, *Leibniz algebras, Courant algebroids, and multiplications on reductive homogeneous spaces*. Amer. J. Math. 123 (2001), no. 3, pp. 525–550.

[56] M. K. Kinyon, *Leibniz algebras, Lie racks, and digroups*. J. Lie Theory 17 (2007), no. 1, pp. 99–114.

[57] A. Klimyk and K. Schmüdgen, *Quantum Groups and Their Representations*. Texts and Monographs in Physics. Springer-Verlag, Berlin, 1997.

[58] U. Krähmer, F. Wagemann, *Racks, Leibniz algebras and Yetter–Drinfel'd modules*. Georgian Math. J. 22 (2015), no. 4, pp. 529–542.

[59] J.-L. Loday, *Cohomologie et groupe de Steinberg relatifs*. J. Algebra 54 (1978), no. 1, pp. 178–202.

[60] J.-L. Loday, *Spaces with finitely many non-trivial homotopy groups*. J. Pure Applied Algebra 24 (1982), pp. 179–202.

[61] J.-L. Loday, *Cyclic Homology*. Springer, Berlin (1992).

[62] J.-L. Loday and T. Pirashvili, *Universal enveloping algebras of Leibniz algebras and (co)homology*. Math. Ann. 296 (1993), no. 1, pp. 139–158.

[63] J.-L. Loday, *Série de Hausdorff, idempotents eulériens et algèbres de Hopf*. Exposition. Math. 12 (1994), no. 2, pp. 165–178.

[64] R. C. Lyndon, *The cohomology theory of group extensions*. Duke Math. J. 15 (1948), pp. 271–292.

[65] S. MacLane, *Cohomology theory in abstract groups III*. Annals Math. 50 (1949), no. 3, pp. 736–761.

[66] S. MacLane, J. H. C. Whitehead, *On the 3-type of a complex*. Proc. Nat. Acad. Sci. U. S. A. 36 (1950), pp. 41–48.

[67] S. MacLane, *Historical note*. J. Algebra 60 (1979), pp. 319–320.

[68] S. MacLane, *Homology*. Classics in Mathematics. Springer (1995).

[69] S. MacLane, *Categories for the Working Mathematician*, Second Edition. Graduate Texts in Mathematics, 5. Springer (1997).

[70] S. Majid, *Braided matrix structure of the Sklyanin algebra and of the quantum Lorentz group*. Comm. Math. Phys. 156 (1993), no. 3, pp. 607–638.

[71] S. Majid, *Strict quantum 2-groups*. arXiv:1208.6265.

[72] A. I. Mal'cev, *Nilpotent torsion-free groups*. Izvestiya Akad. Nauk. SSSR. Ser. Mat. 13 (1949), pp. 201–212.

[73] S. Montgomery, *Hopf Algebras and Their Actions on Rings*. CBMS Regional Conference Series in Mathematics, 82. Published for the Conference Board of the Mathematical Sciences, Washington, DC; by the American Mathematical Society, Providence, RI (1993).

[74] M. Murray, *Another construction of the central extension of the loop group*. Comm. Math. Phys. 116 (1988), no. 1, pp. 73–80.

[75] M. Murray, D. Stevenson, *Higgs fields, bundle gerbes and string structures*. Comm. Math. Phys. 243 (2003), no. 3, pp. 541–555.

[76] K.-H. Neeb, *Abelian extensions of infinite-dimensional Lie groups*. In: *Travaux mathématiques. Fasc. XV*, pp. 69–194, Trav. Math., XV. Univ. Luxemb., Luxembourg (2004).

[77] K.-H. Neeb, *Non-abelian extensions of topological Lie algebras*. Comm. Algebra 34 (2006), no. 3, pp. 991–1041.

[78] K.-H. Neeb, *Towards a Lie theory of locally convex groups*. Japan. J. Math. 1 (2006), pp. 291–468.

[79] K.-H. Neeb, *Non-abelian extensions of infinite-dimensional Lie groups*. Ann. Inst. Fourier Grenoble 57 (2007), no. 1, pp. 209–271.

[80] T. Nikolaus, C. Sachse, C. Wockel, *A smooth model for the string group*. Int. Math. Res. Not. IMRN 2013 (2013), no. 16, pp. 3678–3721.
[81] B. Noohi, *Integrating morphisms of Lie 2-algebras*. Compos. Math. 149 (2013), no. 2, pp. 264–294.
[82] V. Ovsienko and C. Roger, *Generalizations of Virasoro group and Virasoro algebra through extensions by modules of tensor-densities on S^1*. Indag. Mathem. N. S. 9 (1998), 2, pp. 277–288.
[83] S. Porst, *Strict 2-groups are crossed modules*. arXiv:0812.1464.
[84] A. Pressley and G. Segal, *Loop Groups*. Oxford Mathematical Monographs. Oxford Science Publications. The Clarendon Press, Oxford University Press, New York (1986).
[85] D. E. Radford, *The structure of Hopf algebras with a projection*. J. Algebra 92 (1985), no. 2, pp. 322–347.
[86] J. G. Ratcliffe, *Free and projective crossed modules*. J. London Math. Soc. (2) 22 (1980), no. 1, pp. 66–74.
[87] J. G. Ratcliffe, *On the second transgression of the Lyndon–Hochschild–Serre spectral sequence*. J. Algebra 61 (1979), no. 2, pp. 593–598.
[88] S. Rivière *Sur l'isomorphisme entre cohomologies de Chevalley–Eilenberg et de Hochschild*. Available at: http://www.theses.fr/2012NANT2092.
[89] S. Rivière, F. Wagemann, *On the string Lie algebra*. Algebr. Represent. Theory 18 (2015), no. 4, pp. 1071–1099.
[90] C. Schommer-Pries, *Central extensions of smooth 2-groups and a finite-dimensional string 2-group*. Geom. Top. 15 (2011), pp. 609–676.
[91] J.-P. Serre, *Lie Algebras and Lie Groups. 1964 Lectures Given at Harvard University*. Lecture Notes in Mathematics, 1500. Springer-Verlag, Berlin (2006).
[92] S. Stolz, P. Teichner, *What is an elliptic object?* In: *Topology, Geometry and Quantum Field Theory*, pp. 247–343. London Math. Soc. Lecture Note Ser., 308. Cambridge Univ. Press, Cambridge (2004).
[93] M. E. Sweedler, *Hopf Algebras*. Benjamin, New York (1969).
[94] V. S. Varadarajan, *Lie Groups, Lie Algebras, and Their Representations*. Reprint of the 1974 edition. Graduate Texts in Mathematics, 102. Springer-Verlag, New York (1984).
[95] F. Wagemann, *A crossed module representing the Godbillon–Vey cocycle*. Lett. Math. Phys. 51 (2000), pp. 293–299.
[96] F. Wagemann, *On Lie algebra crossed modules*. Comm. Algebra 34 (2006), no. 5, pp. 1699–1722.
[97] F. Wagemann, C. Wockel, *A cocycle model for topological and Lie group cohomology*. Trans. Amer. Math. Soc. 367 (2015), no. 3, pp. 1871–1909.
[98] C. Weibel, *An Introduction to Homological Algebra*. Cambridge Studies in Advanced Mathematics, 38. Cambridge University Press (1994).
[99] J. H. C. Whitehead, *Combinatorial homotopy. II*. Bull. Amer. Math. Soc. 55 (1949), pp. 453–496.
[100] G. W. Whitehead, *Elements of Homotopy Theory*. Graduate Texts in Mathematics, 61. Springer-Verlag, New York-Berlin (1978).
[101] D. Wigner, *An identity in the free Lie algebra*. Proc. Amer. Math. Soc. 106 (1989), no. 3, pp. 639–640.

Index

1-cat rack 352
2-group 63
2-vector space 159
$\mathfrak{g}$-kernel 117

abelian extension of Lie algebras 374
abelian group extension 368
abelian representative 193
abstract Bianchi identity 99, 106, 166
action rack trunk 357
action-cocommutative crossed module 259
admissible pair 249
Alexander rack 365
algebra of multiplications 194, 379
algebra of outer multiplications 171, 194, 379
associated group 321
associative algebra kernel 194
augmented Leibniz algebra 326
augmented rack 272, 321
augmented rack bialgebra 329

bar resolution 367, 380
BCSS model of the string group 52

Cartan cocycle 50, 128, 133, 142, 143, 165
cat^1-associative algebra 206
cat^1-group 67
cat^1-Hopf algebra 255
cat^1-Lie algebra 158
categorification 63, 157
central extension of Lie algebras 374
central group extension 368
characteristic class 279
Chevalley–Eilenberg cohomology 374
connecting homomorphism 31, 113, 225
continuous group cohomology 371
continuous Lie algebra cohomology 378
corner trunk 356
crossed bimodule 170
crossed comodule of Hopf algebras 239
crossed module of associative algebras 169
crossed module of groups 1
crossed module of Hopf algebras 229
crossed module of Leibniz algebras 345
crossed module of Lie algebras 85
crossed module of Lie groups 289
crossed module of Lie–Rinehart algebras 313
crossed module of rack bialgebras 342
crossed module of racks 336, 340

deviation class 36
differential graded Lie algebra 310
dihedral rack 365

effaceable 124
elementary equivalence 3, 87, 171
embedded quantum groupoid 261
equivalence of crossed modules 3, 87, 171
equivalence of extensions 17, 101
equivalent Lie group extensions 280
Eulerian idempotent 135
extended rack trunk 358
extendible kernel 117
extension of Lie groups 279

factor system 18, 102
fiber integral 149
finite type cohomology class 126
formal de Rham algebra 122
free crossed module 71
free presentation of a group 34
fundamental rack 362

$\mathfrak{g}$-kernel 105
G-kernels 20
general extension of algebras 189, 198
general extension of Lie algebras 101, 374
general group extension 16, 30, 368
generalized augmented rack 337
Gerstenhaber's theorem for algebras 173
Gerstenhaber's theorem for groups 5
Gerstenhaber's theorem for Lie algebras 89, 155, 167
Godbillon–Vey cocycle 52, 145, 146
Goldberg's theorem 116
group cohomology 367
group of outer automorphisms 3
group-like element 232

H-comodule coalgebra 228, 271
H-module algebra 228, 271
H-module coalgebra 229
hemi-semidirect product 323
hindrance 195

Hochschild cohomology 380
Hochschild–Serre formula 125, 377
Hochschild–Serre spectral sequence 128, 375
Hopfish algebra 238

ideal of squares 325
initial subgroup 276
initial submanifold 383
inner automorphisms 3
irreducible coalgebra 236

Kac–Moody cocycle 166
Kac–Moody group 50
Koszul complex 373

L_∞-algebra 163
λ-densities 146
left center 325
Leibniz algebra 324
Leibniz bimodule 345
Lie algebra cohomology 373
Lie algebra kernel 117
Lie group 383
Lie rack 323
Lie–Rinehart algebra 312
Lie's third theorem 383
locally smooth 1-cocycle 273
locally smooth cohomology 248, 384, 385
loop algebra 166, 246
Lyndon–Hochschild–Serre spectral sequence 39, 371

Magnus Lie algebra 304
Malcev theory 310
middle four exchange law 64, 160, 208, 224, 352
multiplication algebra 171

n-homotopy type 79
normalized section 16
nucleus 117, 171, 195, 379

obstruction class 24, 107, 295
obstruction crossed module 26, 108, 297
outer action 2, 16, 85, 101, 105
outer bimodule action 170, 189
outer derivations 86
outer multiplication algebra 224

Peiffer identity 1, 85, 169, 230
pointed rack 320
pointed trunk 356
pre-cat^1-Hopf algebra 253
primitive element 232, 327, 330
principal construction 30, 112
product coordinates 13, 19, 371, 381

R-bimodule 341
rack 319
rack bialgebra 328
rack bimodule 322
rack space 359
rack trunk 357
Radford's isomorphism theorem 250
Radford's recognition theorem 249
relative cohomology 152
relative crossed module of groups 54
relative equivalence 55, 151
relative group cohomology 56

semi-strict Lie 2-algebra 162
semidirect product of Hopf algebras 249
set-like element 331
skeletal 163
smash coproduct 249
smash product 248
smooth factor system 281
smooth G-kernel 274
smooth G-module 384
smooth outer action 274
special kernel 198
split extension 17
split Lie group extension 280
split Lie subgroup 279
square zero extension 381
standard extension 104
Stolz–Teichner model of the string group 49
strict 2-group 63
strict 2-rack 352
strict associative 2-algebra 208
strict Lie 2-algebra 159
strict quantum 2-group 262
string group 46, 128, 133
string Lie algebra 128, 133, 165
sum of $\mathfrak{g}$-kernels 117
sum of crossed modules 4, 88, 172

transgression 41, 130
transgressive element 41, 130
trunk 356
trunkified rack 358

Virasoro algebra 147

Yetter–Drinfeld module 251, 272
Yoneda product 30, 112, 142, 192

De Gruyter Studies in Mathematics

Volume 81
Ľubica Holá, Dušan Holý, Warren Moors
USCO and Quasicontinuous Mappings, 2021
ISBN 978-3-11-075015-7, e-ISBN 978-3-11-075018-8,
e-ISBN (ePUB) 978-3-11-075022-5

Volume 46
Steven Lord, Fedor Sukochev, Dmitriy Zanin
Singular Traces. Volume 1: Theory, 2021
ISBN 978-3-11-037779-8, e-ISBN 978-3-11-037805-4,
e-ISBN (ePUB) 978-3-11-039231-9

Volume 80
Xiao-Xiong Gan
Formal Analysis. An Introduction, 2021
ISBN 978-3-11-059785-1, e-ISBN 978-3-11-059945-9,
e-ISBN (ePUB) 978-3-11-059860-5

Volume 79
Rajendra V. Gurjar, Kayo Masuda, Masayoshi Miyanishi
Affine Space Fibrations, 2021
ISBN 978-3-11-057736-5, e-ISBN 978-3-11-057756-3,
e-ISBN (ePUB) 978-3-11-057742-6

Volume 78
Kai Liu, Ilpo Laine, Lianzhong Yang
Complex Delay-Differential Equations, 2021
ISBN 978-3-11-056016-9, e-ISBN 978-3-11-056056-5,
e-ISBN (ePUB) 978-3-11-056040-4

Volume 77
Laura Costa, Rosa Maria Miró-Roig, Joan Pons-Llopis
Ulrich Bundles. From Commutative Algebra to Algebraic Geometry, 2021
ISBN 978-3-11-064540-8, e-ISBN 978-3-11-064768-6,
e-ISBN (ePUB) 978-3-11-064580-4